ANIMAL GROWTH
and NUTRITION

ANIMAL GROWTH
and NUTRITION

Edited by

E. S. E. HAFEZ
Washington State University
Pullman, Washington

and

I. A. DYER
Washington State University
Pullman, Washington

113 Illustrations and 26 Plates

1969

Lea & Febiger · *Philadelphia*

SBN   8121–0109–X

Library of Congress Catalog Card Number 70:78538

PRINTED IN THE UNITED STATES OF AMERICA

Preface

GROWTH, an essential and peculiar characteristic of all living matter, has interested investigators in the biological sciences. For three decades a wealth of data has accumulated concerning the physiological and nutritional aspects of animal growth. However, progress in understanding mechanisms of growth at the cellular and tissue levels has been slow and tortuous.

In the last ten years because of advances in instrumentation, microanatomy, physiology, genetics, endocrinology, biochemistry, and biophysics, progress has been particularly rapid. Nonetheless, studies are scattered in such a wide spectrum of journals that students of growth can hardly keep abreast of all the advances. *Animal Growth and Nutrition* attempts to combine the pertinent physiochemical concepts of growth and the nutritional requirements controlling it into one volume for students of general biology, nutrition, physiology, and veterinary medicine.

The book is divided into four parts. The first part describes differentiation, prenatal growth, and fetal nutrition. The second part presents the fundamentals of postnatal growth and development. The third part is concerned with body composition, particularly with the structure and ultrastructure of muscle, bone, fat, wool, and hair. The fourth part deals with the nutritional requirements of farm mammals and poultry. Several chapters are based primarily on studies of laboratory animals. We attempted to include recent techniques for measuring physiological and nutritional aspects of growth. We did not, however, attempt to provide a comprehensive bibliography, although selected classical papers and review articles are listed at the end of each chapter.

It has been our good fortune to have had the assistance of interested contributors, who prepared their chapters meticulously. Each contributor was asked to submit an outline of his chapter. These outlines were edited carefully to avoid unnecessary duplication and to make the textbook as comprehensive as possible. Each chapter was reviewed by at least three scientists. We hope that we have taken their suggestions and criticisms into account, but in no way should they be held responsible for the present contents. Sincere thanks are due to Dr. James Carlson, Dr. Luigi Giacometti, Dr. Leo S. Jensen, Dr. Anne McLaren, Dr. Allen T. Ralston, Dr. Imogen E. Russe, Dr. T. M. Sutherland, Dr. H. D. Wallace, and Dr. Paul C. Harrison.

The unpublished data in Chapter 1 (E. S. E. Hafez) were obtained during investigations supported by Washington State University College of Agriculture Research Center; U. S. Public Health Service research grant HD 00585 from the National Institute of Child Health and Human Development; and the Banta Research Fund. Chapter 9 (H. J. Curtis) contains results from research carried out at Brookhaven National Laboratory under the auspices of the U. S. Atomic Energy Commission. For Chapter 15 (W. P. Flatt & P. W. Moe) thanks are due to Dr. R. L. Baldwin for assistance in the preparation of the section on metabolism, and to Mr. J. L. Flatt, Blue Mountain College, Blue Mountain, Mississippi, for preparation of illustrations. Sincere thanks are also extended to Messrs. H. L. Hudson, J. F. Spahr, T. J. Colaiezzi, and to the staff of Lea & Febiger for their help in preparing and producing *Animal Growth and Nutrition*.

Pullman, Washington   E. S. E. HAFEZ
                                                                   I. A. DYER

Contributors

ALBRIGHT, J. L., Department of Animal Sciences, Purdue University, Lafayette, Indiana 47907.

BAUMGARDT, B. R., Department of Animal Science, The Pennsylvania State University, University Park, Pennsylvania 16802.

BERRY, R. K., University of Kentucky-Maysville Community College, Maysville, Kentucky 41056.

BINNS, W., Poisonous Plant Research Laboratory, Animal Disease and Parasite Research Division, Agricultural Research Service, U.S.D.A., 1150 East 14th North, Logan, Utah 84321.

BRISKEY, E. J., Department of Meat and Animal Science, The University of Wisconsin, Madison, Wisconsin 53706.

CARLSON, J. R., Department of Animal Sciences, Washington State University, Pullman, Washington 99163.

CURTIS, H. J., Department of Biology, Brookhaven National Laboratory, Associated Universities, Inc., Upton, Long Island, New York 11973.

DOLNICK, E. H., Animal Fiber Development Investigations, Agricultural Research Service, Animal Husbandry Research Division, U.S.D.A., Beltsville, Maryland 20705.

DYER, I. A., Department of Animal Sciences, Washington State University, Pullman, Washington 99163.

EMERY, R. S., Department of Dairy, Michigan State University, East Lansing, Michigan 48823.

FLATT, W. P., Agricultural Research Service, Animal Husbandry Research Division, U.S.D.A., Beltsville, Maryland 20705 (Present address, Department of Animal Sciences, University of Georgia, Athens, Georgia 30601).

FULLER, M. F., The Rowett Research Institute, Bucksburn, Aberdeen, Scotland AB2 9SB.

GALL, G. A. E., Department of Animal Science, University of California, Davis, California 95616.

HAFEZ, E. S. E., Department of Animal Sciences, Washington State University, Pullman, Washington 99163.

HANSARD, S. L., Department of Animal Husbandry-Veterinary Science, The University of Tennessee, Knoxville, Tennessee 37901.

JAMES, L. F., Agricultural Research Service, Animal Disease and Parasite Research Division, U.S.D.A., 1150 East 14th North, Logan, Utah 84321.

JENSEN, L. S., Department of Animal Sciences, Washington State University, Pullman, Washington 99163.

KEELER, R. F., Agricultural Research Service, Animal Disease and Parasite Research Division, U.S.D.A., 1150 East 14th North, Logan, Utah 84321.

MOE, P. W., Agricultural Research Service, Animal Husbandry Research Division, U.S.D.A., Beltsville, Maryland 20705.

PFANDER, W. H., Department of Animal Husbandry, University of Missouri, Columbia, Missouri 65201.

ROUBICEK, C. B., Department of Animal Science, The University of Arizona, Tucson, Arizona 85721.

RUSSELL, T. S., Department of Statistical Services, Washington State University, Pullman, Washington 99163.

WHITEHAIR, C. K., Department of Pathology, Michigan State University, East Lansing, Michigan 48823.

VAN KAMPEN, K. R., Agricultural Research Service, Animal Disease and Parasite Research Division, U.S.D.A., 1150 East 14th North, Logan, Utah 84321.

ZOBRISKY, S. E., Cooperative State Research Service, U.S.D.A., Washington, D.C. 20250.

Contents

Contents

Introduction to Animal Growth

By E. S. E. Hafez

THE word "growth" has been used to describe many biological phenomena. Population growth involves the reproduction of animals; body growth includes the multiplication of cells (hyperplasia) or an increase in cell size (hypertrophy); cell growth involves the replication of molecules; and the replication of molecules entails the mobilization of precursors. The growth of muscle and adipose tissue differs from the growth of other organs in the body. Muscles enlarge with age by hypertrophy. Adipose tissues grow partly by the addition of new cells, but largely by the continued intracellular accumulation of lipids (Fig. 1). Consequently, the dimensions of adipose cells vary with age, plane of nutrition, and size of the animal.

Hypertrophy and hyperplasia are consequences of DNA replication and protein synthesis. Animal growth, as it pertains to this volume, refers to an increase in linear size, weight, accumulation of adipose tissue, and retention of nitrogen and water.

This introduction centers around the biodynamics of cell populations, cytogenetics, the regulation of growth, patterns of growth rate, differential growth, body composition, and the effects of nutrition on growth.

I. THE BIODYNAMICS OF CELL POPULATIONS

During animal growth, cells increase in number and size, and extracellular substances accumulate. Hyperplasia and hypertrophy do not always occur simultaneously. Cell division can occur without any increase in protoplasm: the result is a larger number of smaller cells. On the other hand, protoplasm can be synthesized in the absence of cell division, in which case the cells are larger.

Growth is intimately related to the anabolic synthesis of a wide variety of cell constituents: nuclei, nucleoli, chromosomes, centrioles, mitochondria, cytoplasmic organelles, enzymes, and cell membranes. The synthesis of macromolecules is closely associated with energy-yielding reactions.

During prenatal life the embryo produces a galaxy of specialized cells including those of muscle, cartilage, skin, and nerve. This phenomenon is called *differentiation* and involves complex changes in cell structure and function. The rates and patterns of cellular differentiation are apparently carefully regulated so that the numbers of each cell type are optimal for tissue function.

1

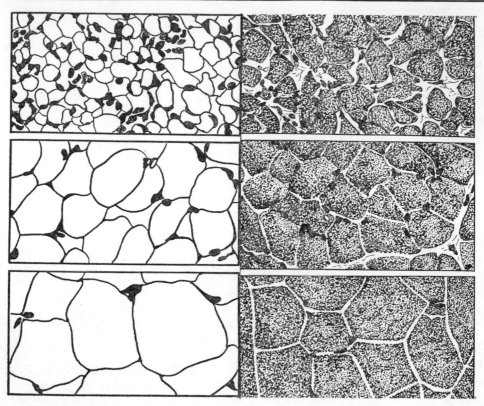

FIG. 1.   The effects of age on the histological characteristics of tissues in the rat.   (*Left*) Adipose tissue of 17-, 34-, and 95-day-old rats (from top to bottom, respectively).   Growth occurs largely by cellular hypertrophy (317 ×).   (*Right*) Transverse sections of skeletal muscle from 17-, 34-, and 95-day-old rats (from top to bottom, respectively).   Growth is achieved by hypertrophy of the muscle fibers (337 ×).   (*From Enesco & Leblond, 1962. J. Embryo. Exp. Morphol. 10:530.*)

Animal cells vary in diameter from the 4-micron (μ) leukocyte to the 180-μ ovum. Cells also vary in shape and in the way they pack. Some form such flat sheets (epithelia) as skin and the linings of blood and lymph vessels. Muscle cells are slender, elongated, and frequently spindle-shaped. Red blood cells are disc-shaped. Nerve cells are irregular in shape and have long axons (Fig. 2). The size and shape of cells are related to their functions. For example, the round erythrocyte has a large surface area for gas exchange, but is small enough to penetrate the finest capillaries. Other types of cells (*e.g.*, keratinized epidermal cells) are most useful to the animal when they are dead.

Embryonic cells multiply exponentially prior to implantation of the embryo. However, the rate of cell multiplication varies during prenatal development. During the last stages of gestation in the rat the number of new cells produced each day may increase by 60 percent over the number existing the previous day. This rate falls to 20 percent just before birth, and to only 2 percent at puberty. The rate of mitosis is affected by various factors, *e.g.*, heredity, sex, nutrition, and ambient temperature. Mitosis occurs in 14 to 100 minutes, depending on the tissue (Table 1). In general, mitotic activity is inversely proportional to the degree of cellular specialization. Most mitotic divisions occur during sleep, when sympathetic activity is minimal (Bullough & Laurence, 1964).

Mitosis is minimal in most adult tissues, but continues at a steady low rate to keep the size of the existing cell population con-

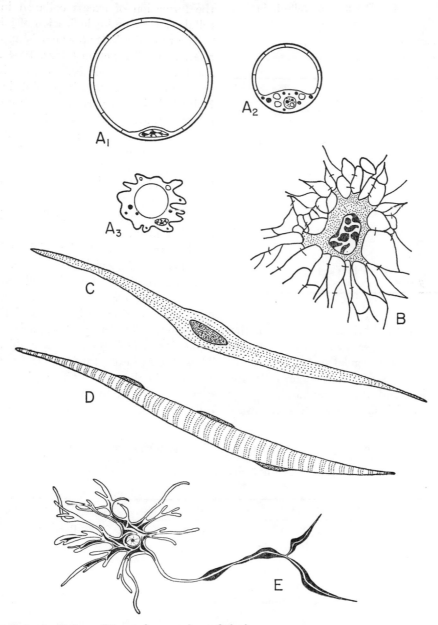

Fig. 2. Animal cells have different shapes and morphologies:
A. White adipose tissue: (A_1) A developing cell. The nucleus is round and the cytoplasm contains several fat droplets. (A_2) A mature cell. The nucleus is compressed between the cell membrane and the large fat droplet. (A_3) A fat-depleted cell.
B. Bone cell within the Haversian system of a membrane bone.
C. Smooth muscle cell
D. Striated muscle cell
E. Nerve cell with axon (the long branch) and dendrites (the short branches).
(*A, B, and C adapted from Swanson, 1965, The Cell. New York, Prentice-Hall, Inc.*)

Table 1.   The Duration of Mitosis in Some Animals

(*Adapted from Growth. Chapter 3; Cells & Tissues, 1962. Altman & Dittmer, Eds., Washington, D. C., Fed. Amer. Soc. Exp. Biol.*)

Species of Animal	Tissue	Duration of Mitosis (min)
Mouse (*Mus musculus*)	Spleen	59
	Adrenal gland	14
	Epidermis	30
Rat (*Rattus*)	Connective tissue	25–45
	Intestinal epithelium	68
Rabbit (*Oryctolagus cuniculus*)	Connective tissue	35–65
Fowl (*Gallus*)	Fibroblasts and myoblasts	55–100

stant.   This equilibrium depends on the activity of hormones, sympathetic nerves, and specific growth-regulating factors. *Negative growth* occurs during old age: cells are gradually lost and are not replaced.   Aging is associated with such metabolic changes as the formation of excess collagen in extracellular spaces and calcification of joints.

The rate of *cellular turnover* is apparently related to cell function (Goss, 1964).   Constant replacement of cells occurs in some parts of the body, but is impossible in others. Epithelial tissues, particularly those in the respiratory system and gastrointestinal tract, are constantly renewed: it is estimated that the intestinal mucosa of the rat is replaced every 38 hours.   The life span of some red blood cells is about 120 days.   All nerve and muscle cells of the body have been formed at birth and will theoretically continue to function until death.

The turnover rates of nucleic acids and proteins are of special significance.   In the rat, for example, fetal weight and deoxyribonucleic acid (DNA) synthesis increase proportionally during early prenatal life, but near the time of birth protein increases more rapidly than DNA (Winick & Noble, 1965). Thus, the ratio of protein to DNA changes as the animal grows.   The rapid postnatal growth of organs involves the production of cells, with little or no change in tissue weight

Table 2.   Postnatal Growth in Various Mammalian Tissues

(*Adapted from Growth.   Chapter 3:   Cells & Tissues, 1962.   Altman & Dittmer, Eds., Washington, D. C. Fed. Amer. Soc. Exp. Biol.*)

Phenomenon	Muscle	Liver	Brain and Spinal Cord
Cell division	Scant; confined to nuclei	Rate varies	Rare
Mode of growth	Enlargement, and possibly some splitting of fibers; hypertrophy of sarcoplasm in pre-existing cells	In early life, considerable contribution from cell division; later, from increase in cell size	Growth of axons; myelination of fiber tracts
Life span	Coextensive with normal function	Unknown	Coextensive with normal function
Replacement	Normally none	Dividing cells persist in small numbers in the adult, presumably to replace cell losses	Confined to neuroglia
Regenerative capacity	Protoplasmic outgrowth from pre-existing fibers, in which nuclei may divide by amitosis; outgrowth begins 3rd day after injury; no new formation from indifferent cells	In rat, extirpated two-thirds of liver regenerated in 21 days by cell division in parenchymal cells and, to a smaller extent, in duct cells; division also occurs in Kupffer cells and connective tissue cells	Very large, but confined to neuroglia; some axon formation; regeneration of motor axons occurs; several factors affect rate

per unit of DNA (Enesco & Leblond, 1962). The emphasis later shifts to increases in cell size.

II. CYTOGENETICS

The nucleus is the functional center of the cell, controlling the cell's growth, differentiation, and biochemical processes. The growth characteristics of a species are contained in *genes*. The replication of chromosomes, and of the genes they contain, is related to the structure of the DNA molecule. This molecule has a high molecular weight and is made up of many smaller molecules linked together in a precise way. These molecules include the sugar, deoxyribose, phosphoric acid, and four nitrogenous bases, two of which are pyrimidines (thymine and cytosine) and two purines (adenine and guanine) (Fig. 3). DNA directs cell growth: a "messenger RNA" is made in the nucleus and travels to the cytoplasmic ribosomes with genetic directions for the sequence in which amino acids will be assembled into the proteins which are responsible for most of the cell's functional capability. Nucleic acids (DNA and RNA) are essential for protein synthesis and are thus indirectly responsible for all cell functions. Protein synthesis occurs in the cytoplasm and the nucleus of all cells of the body. All the cells in a tissue contain similar nucleic acids and proteins. Nucleic acids make it possible for cells to multiply geometrically and to mutate occasionally, whereas proteins enable the cell to perform a variety of functions.

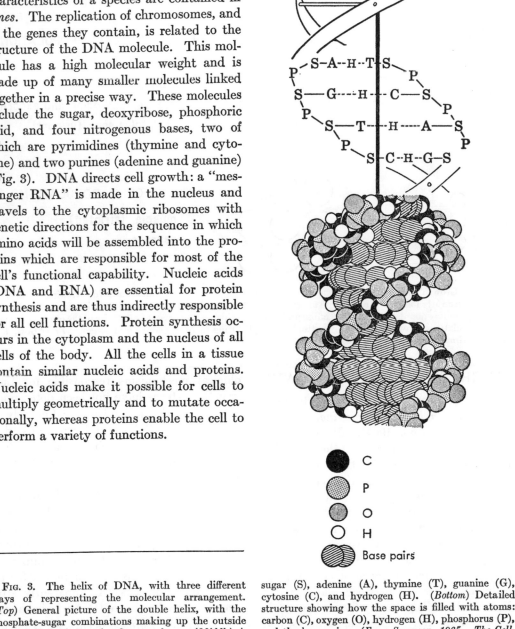

C

P

O

H

Base pairs

FIG. 3. The helix of DNA, with three different ways of representing the molecular arrangement. (*Top*) General picture of the double helix, with the phosphate-sugar combinations making up the outside spirals and the base pairs the cross-bars. (*Middle*) A somewhat more detailed representation: phosphate (P), sugar (S), adenine (A), thymine (T), guanine (G), cytosine (C), and hydrogen (H). (*Bottom*) Detailed structure showing how the space is filled with atoms: carbon (C), oxygen (O), hydrogen (H), phosphorus (P), and the base pairs. (*From Swanson, 1965. The Cell, New York, Prentice-Hall, Inc.*)

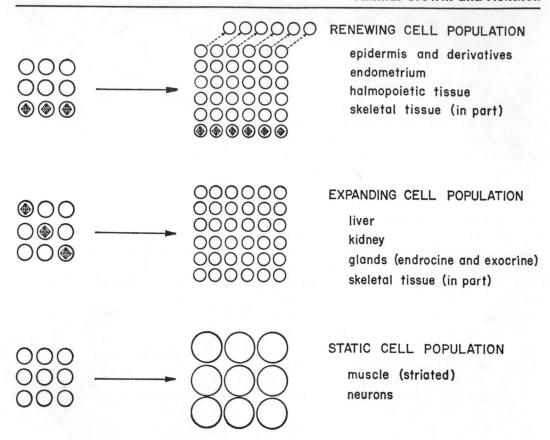

RENEWING CELL POPULATION

epidermis and derivatives
endometrium
halmopoietic tissue
skeletal tissue (in part)

EXPANDING CELL POPULATION

liver
kidney
glands (endrocine and exocrine)
skeletal tissue (in part)

STATIC CELL POPULATION

muscle (striated)
neurons

Fig. 4.   Growth characteristics in renewing, expanding, and static cell populations.   In reviewing populations, cells are constantly lost or destroyed, and are replenished by proliferation of stem cells restricted to a "generative" zone.   Expanding populations grow by diffuse proliferation, which ceases once the adult size of the organ is attained. Static populations grow by cellular hypertrophy, except during early developmental stages.   (*From Goss, 1963. Adaptive Growth.   New York, Academic Press.*)

III. THE REGULATION OF GROWTH

Body cells can be divided, in terms of growth, into *renewing, expanding,* and *static* populations (Goss, 1964).   Renewing and expanding tissues differ in (a) the time when cells proliferate and (b) the tissues from which the cell population arises (Fig. 4).   In renewing tissues, cells are continually lost and replaced.   In expanding systems, cell division continues until the adult size of an organ is reached.   Static tissues are expanding systems in which cell division is restricted to the early stages of development, although cellular hypertrophy may occur later.   Mitotically static cells generally live as long as the animal does and include muscle fibers, neurons, and retinal rods and cones.   The physi-

ological mechanisms responsible for these growth patterns have not been determined.

Homeostatic mechanisms exist in body organs so that cellular activity is adjusted to systemic demands.   Such feedback mechanisms limit organ size and restore this size after tissue has been lost.   These mechanisms may involve a direct biochemical equilibrium on which functional and hormonal effects are merely superimposed.   Normal growth and cell division require hormones such as somatotropin and insulin.   Insulin is important for protein synthesis and, therefore, for the hypertrophy of cells.   It apparently stimulates the production of RNA within the cell.

Compensatory growth occurs when part of an organ stops growing or when part of it is experimentally removed. When Flint removed 60 percent of the small intestine from dogs, the rest of it enlarged, its functional parts increased in diameter (but not in length), and individual villi grew larger although they did not increase in number (Fig. 5).

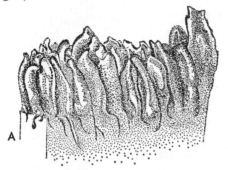

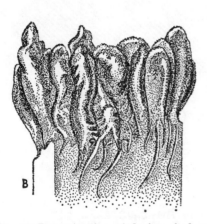

FIG. 5. Reconstructions of the intestinal mucosa of dogs. The control animal (A) has normal villi, whereas villi of a dog subjected to partial excision of the intestine (B) are hypertrophic. (*From Flint, 1910. Trans. Conn. State Med. Soc. p. 283.*)

Several kinds of growth are stimulated by physiological demands for increased functional activity. Erythropoiesis, for example, increases when the body requires more oxygen; kidney hyperplasia occurs when the body fluids contain excess sodium; lymphatic organs enlarge during infections; and muscles hypertrophy if repeatedly exercised (Goss, 1964). These responses are partially mediated by the pituitary trophic hormones, which promote the growth or secretory activity of target organs. If an organ is

chronically subjected to abnormally high functional demands, it may enlarge excessively. Iodine deficiency, for example, causes goiter.

Some organs are capable of compensatory hypertrophy following partial ablation, *e.g.*, the liver, kidney, and endocrine (adrenals, ovaries, thyroid) and exocrine (pancreas, salivary glands) glands. Others cannot hypertrophy, *e.g.*, skeletal muscle, bone, cartilage, skin, orbital glands, limbs, teeth, sense organs, and most nervous tissues.

Compensatory growth increases the functional capacity of organs and tissues. Since organs are composed of many functional units, it is these that are most responsive to the demands for compensation.

IV. GROWTH RATE

The growth rate of the suckling offspring depends upon its own physical and social environment and that of its mother. The season of birth affects both maternal and neonatal environments. For example, grazing animals born near spring are heavier at weaning than animals born in other seasons.

In lambs, the egg transfer technique has been used to determine how much size and growth are affected by maternal and genetic factors. Maternal effects, reinforced as they are by differences in the milk supply after birth, last until the fat lamb is sold. Figure 6 shows the variation which is due to the heredity of the sire and the dam and to maternal influences at various stages during the early life of single lambs.

Postweaning growth rate is affected primarily by hereditary factors, ambient temperature, the animal's ability to adapt to its environment, social stress, and availability of feed. Growth is most rapid at an optimal ambient temperature: 20 °C to 25 °C for pigs; 27 °C for chickens, rats, and mice. Differences in strain occur, however, at least in mice, since some grow better at 32 °C, while others grow better at 21 °C. High temperatures also depress the growth of Shorthorns more than that of acclimated Brahmans (Fig. 7).

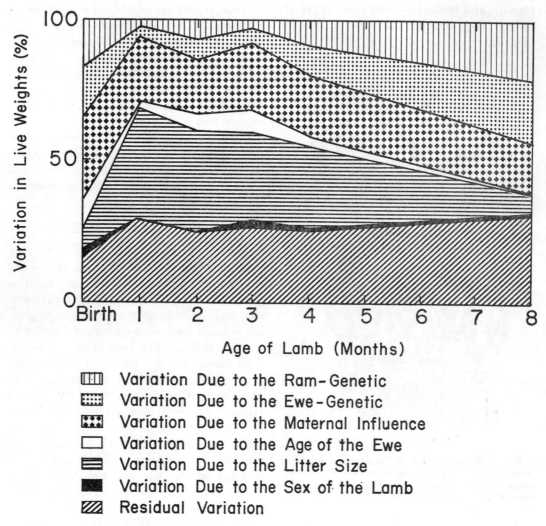

- ▭ Variation Due to the Ram-Genetic
- ▦ Variation Due to the Ewe-Genetic
- ▨ Variation Due to the Maternal Influence
- ▢ Variation Due to the Age of the Ewe
- ▤ Variation Due to the Litter Size
- ■ Variation Due to the Sex of the Lamb
- ▨ Residual Variation

Fig. 6. Variations in the live weight of lambs from birth to 8 months, due to genetics, maternal influence, age of the ewe, litter size, and the lamb's sex. (*From Hunter, 1956. J. Agric. Sci. 48:36.*)

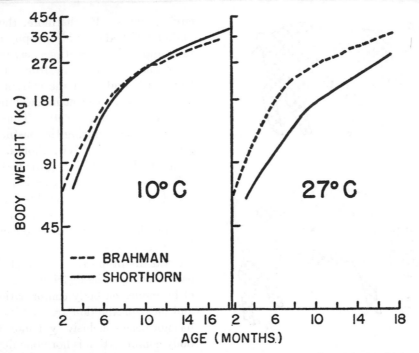

Fig. 7. The influence of environmental temperature on the growth rate of cattle. Note that high temperature depress growth in Shorthorns more than in Brahmans. (*From Johnston et al., 1957. Mo. Agric. Expt. Sta. Res Bull. No. 646.*)

V. DIFFERENTIAL GROWTH

The proportional dimensions of body organs seldom remain the same throughout prenatal and postnatal life (Fig. 8). The different growth centers of the body become active at different times and exhibit different rates of activity. These factors are coordinated and produce the predetermined form characteristic of the species. Huxley defines the size relationship (*allometry* or *heterogony*) between the whole and its parts mathematically as $y = bx^k$, where y is the size of an organ, x is body size, multiplied by the coefficient b (the fraction of the body size that the organ represents) and raised to the power k. When $k = 1$, the growth rates of the organ and the body are the same, *i.e.*, isometric. Allometric (heterogonic) growth may be positive or negative, depending on whether the rate of organ growth is greater ($k > 1$) or less ($k < 1$) than that of the body (Goss, 1964).

The growth of one part of an animal is usually controlled by the activities of other parts. All parts of the animal do not stop growing simultaneously. In fact, some tissues continue to grow throughout the life of the animal. A satisfactory explanation of the self-limiting character of overall growth has yet to be found.

Body organs and tissues grow at different characteristic rates to the size determined by the genetic makeup of the animal. Some grow rapidly early in life, while others begin to grow later and do not achieve their maximum proportions until late in life. The growth rate of each organ and tissue increases to a maximum and then declines. These maximum rates of growth occur in a definite sequence. For example, the central nervous system reaches its maximum growth rate first, bone follows, and muscle and adipose tissue reach their maximum last.

When animals are reared in artificial environments and are deprived of specific needs, certain appropriate organs atrophy. For example, the lymphatic system develops poorly if animals are raised in germ-free

FIG. 8. Changes in the proportions of the pig as it matures (from top to bottom). In order to compare changes in proportions separately from changes in size, all the animals have been reduced to the same shoulder height. (*From Hammond, 1932. J. Roy. Agric. Soc.* *93:131.*)

environments. Furthermore, the tissues of undernourished animals atrophy at different rates and body proportions may thus change.

Growth gradients run from the cranium backwards and from the extremities of the limbs upward to meet in the lumbar vertebrae (Hammond, 1960). Thus, the loin grows the most during the postnatal period, followed by the pelvis, thorax, and neck, while the head and legs grow least.

The size and conformation of modern breeds of livestock have been changed by artificial selection. The modern beef animal is generally smaller and more compact than its predecessors, but this trend has not increased the proportion of muscle in the carcass or changed its distribution. The effects of hormones on body conformation, growth, and meat production are well-known, but it has not been conclusively demonstrated that body conformation is hormone-dependent.

VI. BODY COMPOSITION

Body composition varies with the species, breed, age, sex, plane of nutrition, and physical environment. For example, the percentage of adipose tissue in the body increases significantly after maturity (Fig. 9). With age, changes in body composition are brought about by differential growth gradients which exist in various parts of the body. Such changes can be accelerated with high protein feeding (Weil & Wallace, 1963).

The male has less adipose tissue than the female, and is consequently able to convert feed to body weight more efficiently since the production of 1 kg of adipose requires more feed than the production of 1 kg of muscle or bone. Castrated male cattle and sheep also have more muscle and less adipose tissue than their normal female counterparts.

Bone. Bone develops either directly from connective tissue or is preformed as cartilage which undergoes gradual ossification. The pattern and rate of ossification vary in different bones. Skull bones, and consequently the volume of the cranial cavity, grow rapidly. Long bones grow in length at the

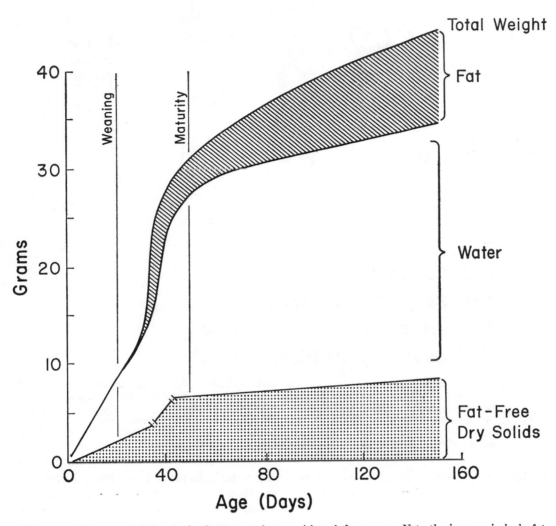

Fig. 9. Developmental changes in the postnatal composition of the mouse. Note the increase in body fat content during growth. There are two linear relationships between the fat-free dry solid content of the animal and its chronological age. These are separated by a growth spurt which occurs prior to maturity. (*Adapted from Cheek & Holt, 1963. Amer. J. Physiol. 205:913.*)

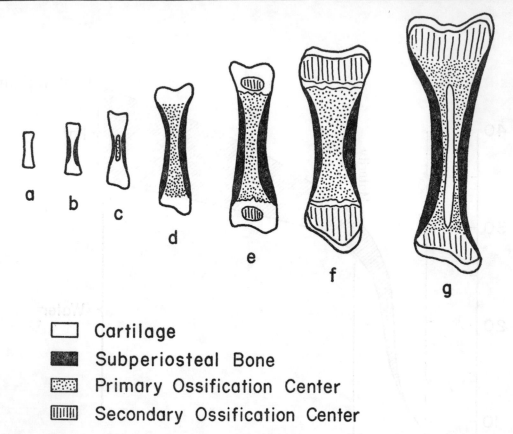

☐ Cartilage

▨ Subperiosteal Bone

▧ Primary Ossification Center

▥ Secondary Ossification Center

FIG. 10. The growth patterns of limb bones. a. Original cartilaginous model. b. Appearance of subperiosteaˡ bone. c. Initial bone function in the "primary" ossification center. d. Extension of the latter towards the epiphysis. e. Appearance of secondary (epiphysial) ossification centers. f. Ossification complete except at the epiphyseal plates. g. Disappearance of the latter. The only remaining cartilage occurs on the articular surfaces. Appearance of the medullary cavity. Note the onset of subperiosteal ossification in (b) to (g). (*Redrawn from B. Mathias Duval in Carleton & Short, 1961. Schafer's Essentials of Histology, London, Longmans.*)

"epiphyseal plates" near either end. Long bones grow in thickness when the periosteum, which surrounds the diaphysis of long bones, lays down bone around the outside of the shaft. The process begins at a "primary ossification center" in the middle of the shaft and extends progressively toward the two ends, the epiphyses (Fig. 10). At later stages, "secondary ossification centers" develop along the shaft.

The bones of the male are thicker than those of the female, particularly in parts of the body which develop early (head, neck, and shanks). Castration, or removal of androgen secretion by the testis, reduces the thickness of the bone to that which occurs in the female. The bones of modern breeds of livestock, improved for meat production, are shorter and thicker than those of un-improved livestock (Fig. 11). This change in the bone shortens the leg and increases the depth of the flesh which surrounds it.

Muscle. Skeletal muscle is composed of many cells or *fibers* and is generally attached to bones (Fig. 12). The muscle fiber is one of the largest cells in the body. Skeletal muscle is composed of red and white fibers, which vary morphologically, biochemically, and physiologically (Barany *et al.*, 1965; Sreter *et al.*, 1966). Red muscle fibers are physiologically slow (tonic activity) and have a high oxidative enzyme activity compared to the white muscle cells which are

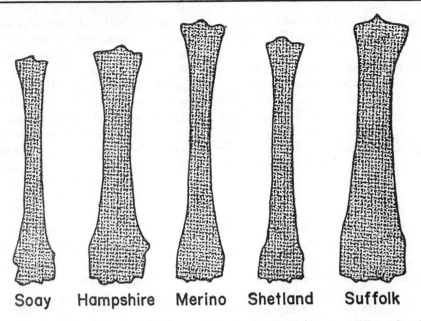

Soay   Hampshire   Merino   Shetland   Suffolk

FIG. 11.   Differences in the shape of the hind cannon bone and femur in rams from different breeds of sheep. Unimproved breeds have slender bones which resemble those in the young of improved breeds.   (*From Hammond, 1960. Farm Animals. London, Edward Arnold.*)

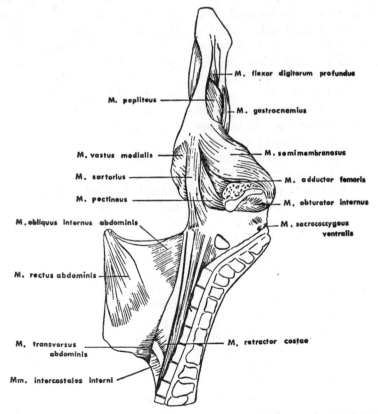

FIG. 12.   A medial view of the muscles in the hindquarters of the ox.   The *gracilus* muscle has been removed. (*From Butterfield and May, 1966.   Muscles of the Ox, Brisbane.   Queensland, University of Queensland Press.*)

physiologically fast (phasic activity) and have a phosphorylase activity (Dubowitz & Pearse, 1960).

The different muscles vary in fat and major fatty acid content. For example, the *longissimus dorsi* muscle contains more G16 and G18 fatty acids than do the *triceps brachii* and *semimembranosus* muscles. The *longissimus dorsi* muscle contains less oleic acid than does the *semimembranosus* (Marchello *et al.*, 1968).

Adipose Tissue. Adipose cells contain enormous quantities of fat in the form of droplets invested by a thin film of cytoplasm. Fat cells undergo cyclical changes in which fat accumulates in small droplets, which coalesce to form a single large droplet; the latter subsequently decreases in size. The length of this cycle varies with the species, location in the body, and plane of nutrition. Adipose tissue, as its cyclical changes suggest, is a dynamic part of the body which is influenced by genetic, nutritional, metabolic, endocrine, and climatic factors. Its distribution within the body has physiological significance. It is abundant in the abdominal wall, the cavities of long bones, and around the kidneys, but rarely accumulates within the cranial cavity, the scrotum, or the eyelids. The physical and chemical characteristics of adipose tissue in meat-producing animals are intimately related to the edibility and nutritive value of meat products. The physical characteristics of adipose vary with age. For example, the adipose tissue of old pigs is firmer than that of young pigs, especially if the growth rate of the young pig is fast and the adipose tissue is synthesized from carbohydrates.

The adipose tissue of ruminants is relatively stable in composition and less subject to the influence of nutrition and physical environment than that of many monogastric species. Much variability exists in the amount of various lipid fractions and fatty acid composition at different sites in a single animal as well as between species.

The degree of unsaturation, the amount of natural antioxidants present, and the color of adipose tissue are influenced by breed, feed, and managerial factors. Its content of natural antioxidants also influences the stability of adipose tissue. The color of bovine adipose tissue is also influenced by breed, and, to some extent, by the ration. In sheep, the triglycerides of adipose tissue are highly saturated intramuscularly. The addition of roughage to the diet increases the palmitic, heptadecanoic, and stearic acid concentrations and decreases the branched-chair pentadecanoic or myristoleic, palmitoleic, oleic, and linoleic acid concentrations of depot triglycerides (Ziegler *et al.*, 1967), *i.e.*, roughage enhances the saturated fatty acid content of depot triglycerides.

VII. GROWTH RETARDATION

Growth retardation is associated with the failure of DNA to replicate or reduce cellular protein synthesis. It is caused by hormone imbalances, chromosome anomalies, radiation, drugs, toxins, hypoxia, sympathetic overactivity, or infection. Reduced protein synthesis may be due to low protein intake, protein loss (*e.g.*, in nephrosis), impaired amino acid transport into cells (*e.g.*, in renal diseases and electrolyte imbalances), hormone deficiency, or abnormal ratios of amino acids (*e.g.*, hyperglycemia) [Cheek, 1968(b)].

Since the sympathetic nervous system controls body functions in part, it is not surprising that it also influences the rate of body growth. The increased sympathetic activity associated with stress retards the growth of fattening animals. Epinephrine released during stress may alter the secretion of growth hormone. If the ventromedial nuclei in the hypothalamus are destroyed, the growth hormone level in the blood also declines.

The free amino acids from which proteins are synthesized occur intracellularly and extracelluarly (*i.e.*, within the body fluids). Retarded growth may be associated with subnormal concentrations of amino acids in either or both of these "pools" as well as with protein deficiency. It may also be associated with increased concentrations in both pools when the conversion of amino acids to proteins is blocked. The relative concentrations of amino acids are also important, since the rate of cellular protein synthesis may decrease if the concentration of one amino acid is unusually high or low.

Growth retardation accompanies metabolic diseases in which the body is unable to use or synthesize protein, those in which abnormal ratios of amino acids occur in the extracellular fluids, or those in which transport of amino acids from the gastrointestinal tract to the plasma or into body cells is impaired. Most human patients with retarded growth, hypopituitarism, Turner's syndrome, and congenital heart disease have subnormal concentrations of amino acid in plasma or muscle [Cheek, 1968 (a)].

VIII. THE EFFECTS OF NUTRITION ON GROWTH

Nutrition refers to the acquisition, degradation, absorption, and metabolism of the foodstuffs necessary for survival, growth, production, and reproduction. The bioenergetics of an animal are influenced by such factors as appetite and satiety; the integrative activity of the central nervous system; basal metabolism; specific dynamic action; the activity of voluntary muscles; the use of carbohydrates, proteins, and lipids in body growth, maintenance, and reproduction; and the storage of surplus energy in adipose tissue.

All energy used in the biological processes of the animal ultimately comes from solar radiation. An animal loses a large percentage of ingested energy through respiration and in its feces and urine. Most of the energy from consumed feed is used to maintain the cell's cytoplasm (*e.g.*, used in the construction of peptide bonds). Cell division can be temporarily suspended by starvation, to return to normal only with restoration of adequate nutrition.

Caloric intake and nitrogen retention (as approximated by N intake-fecal and urinary N) increases steadily in postnatal life until

sexual maturity is reached. They decrease thereafter. Restriction of caloric intake greatly curtails intracellular DNA synthesis, although cells may continue to increase in size. On the other hand, if dietary protein is restricted but caloric intake is normal, cell size is reduced but DNA accumulation continues (Mendes & Waterlow, 1958). Thus, both protein and total caloric intake are important.

Vitamins and minerals are essential for normal growth and morphogenesis. Severe deficiencies or excesses may cause striking changes in the metabolism and development of an animal. For example, iodide deficiency not only produces hypothyroidism *in utero*, but also inhibits normal brain development. Nutritional requirements vary with the species, breed, age, reproductive stage, social stress, disease, parasites, and physical environment. Various physiological mechanisms are involved.

Estimates of nutritional requirements have been made for several species at different ages, levels of production, and using different experimental techniques (U.S. National Research Council). The nutritional aspects of growth in domestic animals are particularly significant to biologists, not only for reasons of comparative biochemistry and physiology but also because these animals are a primary source of food.

IX. RESEARCH TECHNIQUES

Several methods have been used to study developmental changes in body weight, conformation, and composition in living and slaughtered farm animals (reviewed by Hankins *et al.*, 1960; Setchell *et al.*, 1965).

Body Weight. Body weight is often used to evaluate the response of farm animals to a wide variety of diets, environments, and management practices. However, several factors obscure the accuracy of live weight used in this manner: variation in the content of the gastrointestinal tract and bladder, adaptation of the animals to the experimental conditions, fleece, frequency and time of weighing, and the accuracy of the scales.

The fasting *metabolic body weight* (metabolic body size) is useful for investigations of comparative growth. It permits determination of the metabolic rate of a fasting animal via its body weight. This measurement is of value in comparing animals of different age or genotype. In fact, the minimal amount of energy required for body maintenance is often expressed as a function of metabolic body weight (Kleiber, 1961; Blaxter, 1962). Metabolic rate (R in kcal/day) is directly related to body weight (W in kg) according to the formula: $R = 70W^{0.75}$.

Body Conformation. Studies of body conformation are generally based either on the development of carcass proportions, as influenced by nutrition and overall rate and pattern of growth, or on subjectively defined standards of desirability. Such linear body measurements as depth of chest, length of the hind cannon, length of the pelvis, and width of the hocks have been used. These criteria, however, are of little value in studies of the biodynamics of growth.

Body Composition of Living Animals. Techniques used to evaluate body composition of live animals are useful for selecting breeding stock, evaluating meat-producing animals before slaughter, determining the nutritional adequacy of various diets, and ascertaining the composition of gains in body weight.

Body water, fat, and protein of living animals can be measured by determining (a) body water and empty body weight; (b) the body's specific gravity; (c) the analysis of tissue biopsies; or (d) ^{40}K-counting to estimate fat or muscle content. Body water content is measured by injecting known amounts of tritiated water (TOH), deuteriated water (D_2O), or antipyrine into the body and allowing them to equilibrate with the body water. Air displacement techniques are used to estimate the specific gravity of both animals and man. A helium

dilution procedure has also been used to determine specific gravity (Hix *et al.*, 1967). The thickness of subcutaneous fat and the area of the *longissimus dorsi* muscle, which are both indicators of the fatness of the carcass, can be estimated in live animals with ultrasonics (Watkins *et al.*, 1967). Although potassium is distributed throughout the fat-free body and is the most abundant intracellular ion, little is associated with the accumulating lipids of growing animals. Large amounts of potassium occur in muscle tissue, as compared to other tissues, so that whole-body potassium is also a quantitative index of muscle (Lohman & Norton, 1968).

Several coring devices have been used to obtain tissue biopsy samples. In cattle, 30-gm samples are routinely taken from subcutaneous fat and *semitendinosus* muscle and analyzed for nitrogen, moisture, fat, and ash. This provides useful information about meat quality and other factors related to animal production.

Carcass Composition. In *slaughter experiments* for studies on development, sample animals are slaughtered at predetermined times during the experiment. Small animals are minced directly for chemical analysis. The carcasses of large animals are broken up into individual organs, tissues, and anatomical or "butcher" units, and relevant measurements are made on each part. An important source of error arises by using half carcasses and extrapolating to the whole animal. Errors in splitting the carcass, for example, will affect bone more than muscle and adipose tissue and will thus give erroneous muscle:bone ratios. The method for analyzing the body or carcass depends upon the species and expected differences among experimental treatments. Whole carcasses can be minced for complete analysis or they can be dissected into bone, muscle, and adipose tissue.

Organoleptic or sensory methods are used to evaluate palatability in meat research. Great variation exists in the methods of selecting judging panels, preparing samples, setting up rating scales, and analyzing data statistically.

REFERENCES

Barany, M., Barany, K., Reckard, T., & Volpe, A. (1965). Myosin of fast and slow muscles of the rabbit. Arch. Biochem. Biophys. *109*:185–191.

Blaxter, K. L. (1962). The fasting metabolism of adult wether sheep. Brit. J. Nutr. *16*:615–626.

Bullough, W. S., & Laurence, E. B. (1964). Mitotic control by internal secretions: the role of the chalone-adrenalin complex. Exp. Cell Res. *33*:176–194.

Cheek, D. B. (1968). (a) *Human Growth*. Philadelphia, Lea & Febiger.

Cheek, D. B. (1968). (b) *Ibid.* Chap. 43, p. 616.

Dubowitz, Y., & Pearse, A. G. E. (1960). Reciprocal relationship of phosphorylase and oxidative enzymes in skeletal muscle. Nature (Lond.) *185*: 701–702.

Enesco, M., & Leblond, C. (1962). Increase in cell number, as a factor in the growth of the organs and tissues of the young male rats. J. Embryol. Exp. Morph. *10*:530–562.

Goss, R. J. (1964). *Adaptive Growth*. New York, Academic Press.

Hammond, J. (1960). *Farm Animals. Their Growth, Breeding, and Inheritance*. 3rd ed. London, Edward Arnold.

Hankins, O. G., Gaddis, A. M., & Sulzbacher, W. L. (1960). Meat research techniques pertinent to animal production research. pp. 194–228. In: *Techniques and Procedures in Animal Production Research*. Amer. Soc. Anim. Prod., Boyd Printing Co., Albany, N.Y.

Hix, V. M., Pearson, A. M., Reineke, E. P., Gillett, T. A., & Giacoletto, L. J. (1967). Determination

of specific gravity of live hogs by suppressed zero techniques. J. Anim. Sci. *26*:50–57.

Kleiber, M. (1961). *The Fire of Life*. Chap. 10, p. 177, New York, John Wiley & Sons.

Lohman, T. G., & Norton, H. W. (1968). Distribution of potassium in steers by K measurement. J. Anim. Sci. *27*:1266–1272.

Marchello, J. A., Dryden, F. D., & Ray, D. E. (1968). Variation in the lipid content and fatty acid composition of three bovine muscles as affected by different methods of extraction. J. Anim. Sci. *27*:1233–1238.

Mendes, C. B., & Waterlow, J. C. (1958). The effect of a low protein diet and of refeeding on the composition of the liver and muscle in the weanling rat. Brit. J. Nutr. *12*:74–88.

Setchell, B. P., Bassett, J. M., Briggs, P. K., & Panaretto, B. A. (1965). *Techniques for Investigating Body Growth in Field Investigations with Sheep, a Manual of Techniques*. G. R. Moule, D.V.Sc. (Ed.), Melbourne, Australia, C.S.I.R.O.

Sreter, F. A., Seidel, J. C., & Gercely, J. (1966). Studies on myosin from red and white skeletal muscles of the rabbit. I. Adenosine triphosphatase activity. J. Biol. Chem., *241*:5772–5776.

Watkins, J. L., Sherritt, G. W., & Ziegler, J. H. (1967). Predicting body tissue characteristics using ultrasonic techniques. J. Anim. Sci. *26*:470–473.

Weil, W. B., & Wallace, W. M. (1963). The effect of variable food intakes on growth and body composition. Ann. N. Y. Acad. Sci. *110*:358–373.

Winick, M., & Noble, A. (1965). Quantitative changes in DNA, RNA, and protein during prenatal and postnatal growth in the rat. Develop. Biol. *12*: 451–466.

Ziegler, J. H., Miller, R. C., Stanislaw, C. M., & Sink, J. D. (1967). Effect of roughage on the composition of ovine depot fats. J. Anim. Sci. *26*:59–63.

I. Differentiation and Prenatal Aspects

1. Differentiation and Prenatal Aspects

Chapter 1

Prenatal Growth

By E. S. E. Hafez

PRENATAL growth results from a series of orderly differential processes that transform a single-celled zygote into an animal. The following discussion deals with the dynamics of differentiation, organogenesis, differential growth, hormonal regulation of prenatal growth, and genetic and physiological factors affecting birth weight.

I. PATTERNS AND MECHANISMS OF PRENATAL GROWTH

A. Cleavage and Differentiation

The mammalian egg is enormous compared with the somatic cells of the body, especially since its cytoplasm contains reserve nutrients or deutoplasm (yolk). After fertilization, the zygote divides and redivides several times without increasing the volume of cytoplasm. The process of cellular division without an increase in volume is called *cleavage* and the cleaving cells are called *blastomeres*. Cleavage by its very nature is characterized by a progressive decrease in cell size (Fig. 1–1). The early cleavages are synchronous and occur at specific times after fertilization. During later cleavages, cell division is asynchronous and different parts of the embryo may cleave at different rates. Although there is no size increase during this period, large amounts of deoxyribonucleic acid (DNA) and other nuclear constituents are synthesized.

When the dividing cells reach a certain number that is characteristic of the species, they are arranged in a one-cell thick layer around a *blastocoelic* (segmentation) *cavity*. The embryo, then called a *blastocyst*, is a spherical or flattened hollow ball. The blastomeres vary in size, yolk content, and cytoplasmic organization, but no tissues and certainly no organs exist yet. The blastocyst expands rapidly in some species as it absorbs fluids from its oviductal and uterine surroundings.

Differentiation involves the progressive specialization of cells both structurally and functionally. The first apparent differentiation occurs in the blastocyst. During the first week of gestation the blastocyst differentiates into two parts: one covers its surface and is known as the *trophoblast;* the other is a small aggregation of cells which lies beneath the trophoblast and is known as the *inner cell mass* (embryonic disc). The trophoblast is largely concerned with the establish-

21

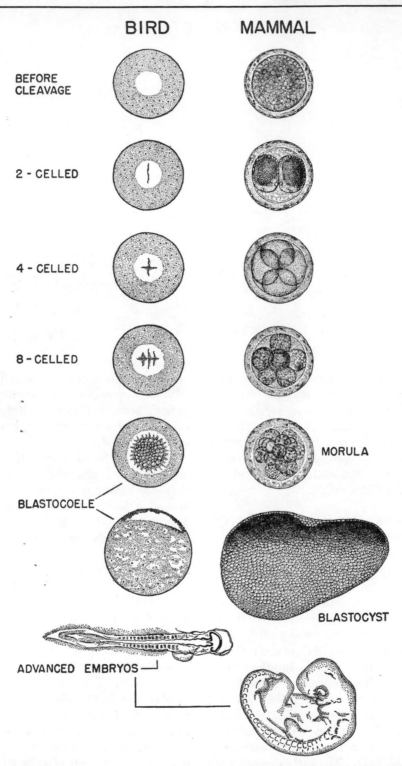

BIRD   MAMMAL

BEFORE
CLEAVAGE

2 - CELLED

4 - CELLED

8 - CELLED

MORULA

BLASTOCOELE

BLASTOCYST

ADVANCED EMBRYOS

FIG. 1–1.   Patterns of cleavage and early prenatal development in a mammal and a bird.   Note the progressive decrease in cell size.

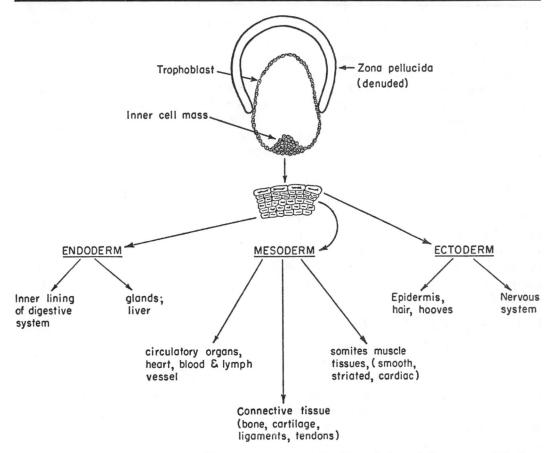

FIG. 1–2. The derivation of various body organs by progressive differentiation and divergent specialization. The origin of all fetal organs can be traced back to the primary germ layers which originate from the inner cell mass.

ment of a placenta whereas the inner cell mass will develop into the embryo proper. The blastocyst then becomes a *gastrula*, which is successively two-layered and three-layered. *Ectoderm* (primary outer) and *endoderm* (inner) layers differentiate during the two-layered stage. A third layer, the *mesoderm*, then develops between these from the primitive streak (ectoderm). By various modifications, these three primary germ layers give rise to the organs of the embryo (Fig. 1–2). Further developmental processes convert the almost undifferentiated embryonic cells into the histological types peculiar to each organ.

Cell divisions during embryonic development lead to a marked increase in the number of cells. Nuclear multiplication involves the synthesis of considerable amounts of DNA, which is one of the main constituents of chromosomes and chromatin. Disruption of the mechanisms which regulate cell division may result in various abnormalities. Under certain conditions animal cells reproduce rapidly and haphazardly, producing neoplasms; or they respond to abnormal developmental cues, creating congenital abnormalities.

Cells can develop into more than one type of tissue before gastrulation. Thereafter, however, their fate is determined: at the end of gastrulation, a piece of presumptive nervous system, for example, will differentiate only into nervous tissue, although earlier it could have developed into several other tissues.

The environment of embryonic cells determines the tissue into which they will differentiate. The neural plate, for example, develops only where ectoderm contacts the roof of the archenteron. Epidermis will not develop into skin epithelium, but will degenerate unless connective tissue underlies it. However, cells cannot be influenced or *induced* by surrounding tissues (the *inductor*) to develop into one tissue or another unless they are at an appropriate stage of development, *i.e.*, cells must be *competent* to develop into the tissue in question. Cells are competent for only a brief time: if presumptive ectoderm, for example, is not stimulated to differentiate into nervous tissue while it is competent to do so, it will differentiate into epidermis.

B. Organogenesis (Organ Formation)

Organogenesis refers to the establishment of organ rudiments through the interaction of germ layers. It often begins when the cells are still undifferentiated. Organs are first blocked out in their general form and then molded and shaped in detail by such processes as invagination (infolding), evagination (outfolding), budding, or hollowing out.

Cells from each of the three germ layers aggregate into *primordial cell masses* from which special organs will ultimately be formed. Subtle changes precede or accompany this regrouping of cells (Patten, 1964). The innermost layer (endoderm) forms the inner lining of the gut, its glands, and the bladder. The outermost layer (ectoderm) forms a middorsal ridge along the anterior-posterior axis of the blastodisc quite early in development. This elongated ridge of *neural ectoderm* subsequently gives rise to the brain, spinal cord, and other derivatives of the nervous system. Ectodermal cells lateral to the neural ectoderm develop into the skin and its derivatives, such as the mammary glands, hair, and hoofs. The mesodermal layer gives rise to muscles, bones, connective tissues, and vascular systems.

A series of *somites* (body segments) devel-

ops along each side of the spinal cord from the outer (somatic) layer of the mesoderm. The most cephalic ones of the series form earliest. In cattle, differentiation of somites starts on the 19th day of pregnancy; the number of somites gradually increases to 25 on the 23rd day, 40 on the 26th day, and 55 on the 32nd day. The striking resemblance of embryos from different vertebrate classes suggests that the processes involved in their development are basically similar (Fig. 1–3).

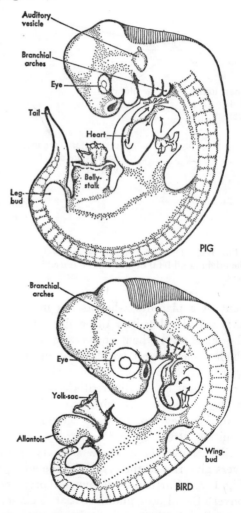

FIG. 1–3. Embryos of a pig and a bird at corresponding developmental stages. The striking resemblance of the embryos to one another suggests that the processes involved in their development are fundamentally similar. (*After William Patten in B. M. Patten. Foundations of Embryology.* McGraw Hill Book Company, New York.)

Most organs are not functional when they first appear in the embryo. The embryonic structures commence to function only after further differentiation. The phases of cellular differentiation, in which the fetal tissues develop their definitive characteristics, are

Table 1–1. Differentiation and Organogenesis in the Bovine Embryo

(*From Greenstein & Foley, 1958. Int. J. Fertil. 3:67.*)

Day of Gestation	Embryological Characteristics
16	Blastocyst bilaminar and unattached in the uterine horn; endodermal and ectodermal layers definitely separate; trophoblast elongates rapidly; vesicae 1.5 mm in diameter; germinal disc 0.4 mm in diameter
late 16	Mesodermal elements form splanchnopleure, somatopleure, and extraembryonic coelom
17–18	Trophoblast differentiates rapidly; primitive groove, and primitive node appear
18	Amnion complete; embryo elongates cephalocaudally; longitudinal axis of embryo established
19	Somite formation and neurulation begin; occipital somites, neural plate, and early neural folds appear; mesoderm thickens on both sides of emerging notochord
20	Neural tube partially closed; flexures develop in main body axis; number of somite pairs variable (0–12); blood islands and heart primordia form
late 20	Allantois appears
21–22	Neural tube continues to close; optic cups appear; forebrain evident; first visceral groove appears; heart begins to beat; cervical somites appear; primitive endoderm differentiates into fore-, mid-, and hindgut; liver primordium present below heart.
late 22	14 somite neurala
23	25 somite neurala; neural tube closed; allantois well developed; 2 distinct visceral arches present
24	Anterior limb buds present at level of heart; mesonephric tubules differentiate
25	Third branchial arch added
26	Initial transition to tailbud stage; cephalic and caudal flexures make embryo C-shaped; stomach, bile duct, and gallbladder distinct; allantois several centimeters long.

known as *histogenesis*. For example, cells which become muscle develop contractile fibrils; cells which form the skeleton lay down calcium salts as the matrix of young bone.

In cattle, the earliest formation of most organs and body parts occurs during the 2nd through 6th week of gestation (Table 1–1). The digestive tract, lungs, liver, and pancreas develop from the primitive gut during this period (Steen & Montagu, 1964). The differentiation and development of embryonic and fetal structures are called *developmental horizons* and are useful criteria for determining the age of embryos and fetuses (Table 1–2). The ages when muscular, skeletal, nervous, and urogenital systems begin to develop, for example, are well known.

After the initiation of organogenesis, there is a period of rapid increase in organ size. This increase in dimensions, which results from cell proliferation and increase in cell size, is the growth process.

C. Measurement of Prenatal Growth

The size of the fetus is judged by total fetal length, crown-rump length, curved crown-rump length (Fig. 1–4), or fetal weight. The patterns of prenatal growth are best

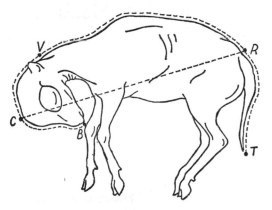

FIG. 1–4. Measurements used for estimating the age and growth rate of mammalian fetuses. Measurements include total length (*BCVRT*), crown-rump length (*C-R*), curved crown-rump length (*CVR*), vertebral column length (*VR*), and vertebral column and tail length (*VRT*). (*From Harvey, 1959. In: Reproduction in Domestic Animals: H. H. Cole & P. T. Cupps (Eds.), New York, Academic Press, Inc.*)

studied either as mathematical formulas or as growth curves. Algebraic equations facilitate analysis and yield certain biological information. During prenatal development, several growth curves can be employed for the fetus as a whole, or for different organs: (a) absolute growth, (b) relative growth, or (c) specific growth (Fig. 1–5).

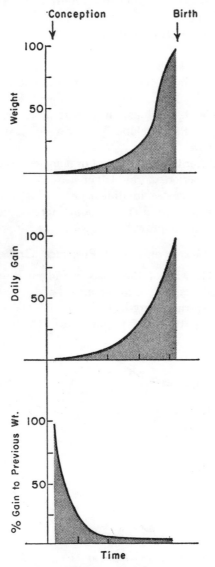

FIG. 1–5. The prenatal growth curves of farm animals. (*Top*) Absolute live weight change; adult weight is considered as 100. (*Middle*) Weight gain per unit time; maximum gain per unit time is considered as 100. (*Bottom*) Weight change expressed as a percentage of the previous weight.

Absolute growth is the change in volume, crown-rump length, or weight of the fetus per unit of time, whereas relative growth is the absolute growth per initial dimension (Fig. 1–6). In the curve of specific growth, the logarithm of size is plotted against age. The absolute growth of the fetus is not linear, but increases exponentially with age until birth. In cattle, more than half of the increase in fetal weight occurs during the last 2 months of gestation (Fig. 1–7). The relative growth of the fetus increases during early pregnancy, but then declines about midway through the gestation period.

Prenatal growth curves vary in detail from species to species, but they resemble each other mathematically. Changes in the rate of prenatal growth during gestation seem to indicate that fetal size is itself a regulator of growth, *i.e.*, that prenatal growth is in some measure self-regulating. The level of the sustained prenatal growth and the period over which it occurs result in the variety of sizes among living organisms.

The growth of one fetal organ is often expressed as a mathematical function of the growth of the whole fetus or of other fetal organs. Such mathematical relationships among the differential growth rates of organs clearly influence the shape and form of the fetus and should concern students of prenatal growth.

D. Differential Growth

The birth weight of the fetus is approximately 60 percent of the total weight of the conceptus (Table 1–3; Fig. 1–8). The fetal membranes (amnion, chorioallantois, and chorionic villi) and placental fluids (amniotic and allantoic) make up the remaining 40 percent. Different parts of the conceptus grow at different rates during gestation. For example, the fetal membranes grow rapidly during the early stages of gestation and their absolute weight reaches a maximum during midpregnancy. The increase in the volume of placental fluids corresponds, more or less, to that of the fetus. Toward the end of

Table 1-2. Some Outstanding Horizons in the Prenatal Development of Domestic Animals

(Adapted from Salisbury & VanDemark, 1961. Physiology of Reproduction and Artificial Insemination of Cattle. San Francisco, Freeman, for the cow; Cloette, 1939. Onderstep. J. Vet. Sci. Anim. Ind. 13:417, for the sheep; and Patten, 1948. Embryology of the Pig. Philadelphia, The Blakiston Co., for the sow.)

Developmental Horizons	Cow (Days)	Ewe (Days)	Sow (Days)
Morula	4–7	3–4	3.5
Blastula	7–12	4–10	4.75
Differentiation of germ layers	14	10–14	7–8
Elongation of chorionic vesicle	16	13–14	9
Primitive streak formation	18	14	9–12
Open neural tube	20	15–21	13
Somite differentiation (1st)	20	17 (9 somites)	14 (3–4 somites)
Fusion of chorioamniotic folds	18	17	16
Chorion enters nonpregnant horn	20	14	—
Heartbeat apparent	21–22	20	16
Closed neural tube	22–23	21–28	16 (11 somites)
Allantois prominent (anchor shaped)	23	21–28	16–17
Forelimb bud visible	25	28–35	17–18
Hindlimb bud visible	27–28	28–35	17–19
Differentiation of digits	30–45	35–42	28 or more
Nose and eyes differentiated	30–45	42–49	21–28
Cotyledons first appear on chorion	30	—	—
Allantois replaces exocoelom within pregnant horn	32	21–28	—
First attachment (implantation)	33 or more	21–30	24 or more
Allantois replaces all of exocoelom	36–37	—	25–28
Eyelids close	60	49–56	—
Hair follicles 1st appear	90	42–49	28
Horn pits apparent	100	77–84	—
Teeth erupt	110	98–105	(160 mm pig)
Hair around eyes and muzzle	150	98–105	—
Hair covers body	230	119–126	—
Birth	280	147–155	112

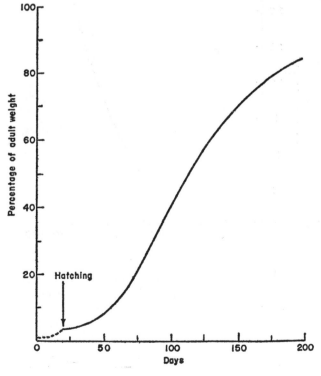

FIG. 1-6. Growth of the chick during embryonic and posthatching periods. (*After Weiss & Kavanau, in Sussman, 1964. Growth and Development. New York, N. Y., Prentice-Hall, Inc.*)

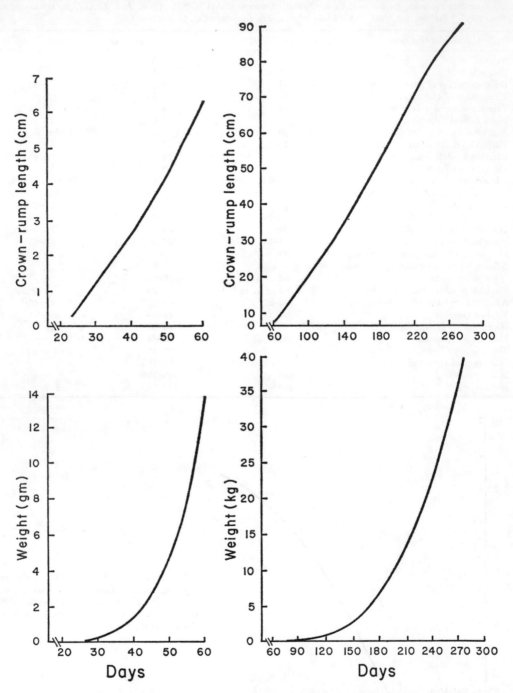

Fɪɢ. 1–7.   Changes in the crown-rump length and weight of bovine fetuses during the first two months of life (*left*) and during mid- and late gestation (*right*). The change in length is nearly constant and linear with time. (*Data from literature summarized by Salisbury & VanDemark, 1961 Physiology of Reproduction & Artificial Insemination of Cattle. San Francisco, Freeman.*)

Table 1–3. Maximum Volumes of Placental Fluids and Weights of Placental Membranes in Domestic Animals

(*From the Literature.*)

Volume	Cattle	Sheep	Swine
Amniotic fluid (maximum volume, ml)*	1252	698	50–280
Allantoic fluid (maximum volume, ml)*	318	762	10–240
Weight of placenta at parturition (kg) . . .	3.9	0.2	0.2
Birth weight (kg) . . .	40	5	1.8
Ratio of placenta weight to birth weight. . . .	1:10	1:25	1:9

* Fluid volumes vary widely among individulas, as indicated by the ranges shown for swine. The amniotic fluid volume of sheep reaches a maximum near mid-pregnancy, then decreases to about 200 ml at term; allantoic fluid is maximal at term.

gestation, however, the placental fluids tend to decrease in volume, although the fetus is still growing.

Fetal organs grow at different and non-constant rates. Moreover the same kind of tissue (muscle, bone, or fat) grows at different rates in different parts of the body and at different stages of gestation. In the early stages of embryonic development, the cephalic region grows rapidly, so that the head of the fetus is consequently disproportionately large. During the later stages of fetal life, when growth in the cephalic region is

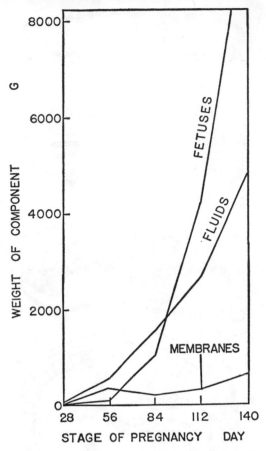

Fig. 1–8. Weight changes of fetuses, fetal fluids, and membranes during pregnancy in ditokous ewes. (*Adapted from Robinson, 1957. Chapter 18. In: Progress in Physiology of Farm Animals. J. Hammond, (Ed.), London, Butterworth's Scientific Publications.*)

Table 1–4. Prenatal Development of the Muscular and Skeletal Systems of the Pig

(*From Growth, Including Reproduction and Morphological Development. 1962. P. A. Altman & D. S. Dittmer (Eds.), Washington, D. C.: Fed. Amer. Soc. Exp. Biol.*)

Age (weeks)	Crown-rump Length (cm)	Muscular System	Skeletal System
3	1–2	Skeletal muscle myoblasts elongate; smooth muscle myofibrillae appear in myoblasts of the esophagus; cross striations appear in skeletal muscle cells	Sternum appears
4	3	Heart myoblasts become spindle-shaped; smooth muscle nuclei elongate in longitudinal muscle layer of esophagus and stomach	Presternum appears; mandible ossifies
5	4	Skeletal muscle bundles present; muscularis mucosae (smooth muscle) present	Humerus and femur ossify ribs, and the centra and arches of vertebrae start to ossify
9	14	———	Spinous processes of vertebrae ossify

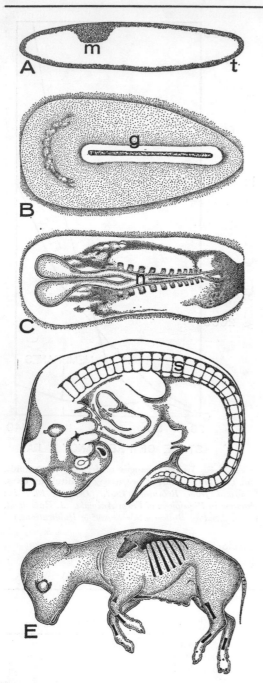

less rapid, the proportions of the neonate are established (Fig. 1–9). Since the various organs of the fetus grow at vastly different rates, the conformation of the fetus is continually changing.

The growth gradients of the fetus follow a definite order. For example, at birth the head, limbs, and forequarters are relatively more developed than the muscles (Table 1–4; Fig. 1–10). Differential growth may be related to functional needs. For example, the very rapid growth of the mesodermal elements which form the embryo's heart is correlated with the necessity for circulating blood and nutrients to all parts of the growing fetus.

Muscle is the major body tissue of the fetus. In the early stages of prenatal growth, muscle cells increase in number (hyperplasia) (Joubert, 1956). During the later stages of prenatal and also in postnatal life, the increase in the weight of body muscles is due to an increase in the size of muscle cells (hypertrophy) (Fig. 1–11). This means that the maximum adult size of the animal has been fixed by the beginning of the fetal stage, since breed differences in size are due to differences in muscle cell numbers and not

Table 1–5.   Number of Blastomeres and Blastocyst Diameter in Two Breeds of Rabbits

(From Castle & Gregory, 1929.   J. Morphol. 48:81–104.)

Hours after Copulation	Small Breed	Large Breed
Average Number of Blastomeres		
32	4.1	4.4
40	8.3	9.9
41	8.6	11.6
48	14.0	21.8
Blastocyst Diameter		
144	40.5 μ	47.8 μ

FIG. 1–9. Conformational changes in prenatal cattle.
A. Ten-day blastocyst stage. The inner cell mass (m) develops into the embryo and the trophoblast (t) develops into the placenta.
B. Fifteen-day-old embryo with primitive groove (g) and the first differentiation of ectoderm, mesoderm, and endoderm.
C. Twenty-day-old embryo showing the two lobes of the brain at the anterior end of the neural groove (n); note the somites.

D. Thirty-day-old embryo with well developed somites (s).
E. Ninety-day-old fetus showing ossification centers in the skeletal system.

PLATE 1

The ultrastructure of the human *quadriceps* muscle (striated muscle) before (A) and after (B) exhaustive bicycle exercise for 101 min. Note that the glycogen particles in the sarcoplasm disappear during exercise. The length of the sarcomere (←⟶) is about 1.5 μ. (*Photos by P. D. Gollnick, C. D. Ianuzzo, D. Williams, & T. R. Hill.*)

PLATE 2

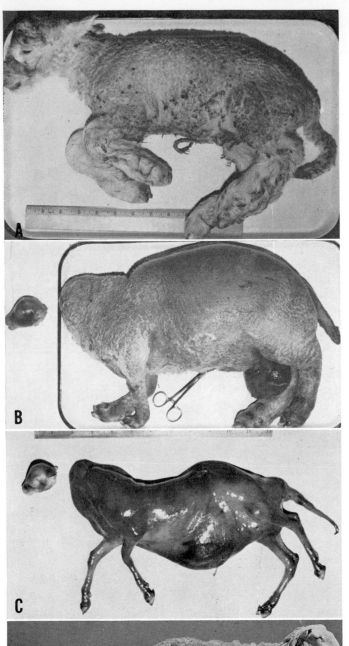

A, B, & C. Effect of fetal decapitation on prenatal growth.

A. Normal term lamb: gestation time, 147 days; birth weight, 3 kg; length (spine over midshoulder to tail base), 22.4 cm.

B. Lamb decapitated *in utero* on gestation day 85; head preserved in formalin at left; mother went into labor on day 168 (24 days past term) and delivered by section on day 169; birth weight (without head), 6 kg (lamb was grossly edematous); length, 27.9 cm.

C. Lamb decapitated *in utero* on gestation day 74; head preserved in formalin at left; born on day 147 (normal term); birth weight (without head), 0.8 kg; length, 19.8 cm. (*Lanman & Schaffer, 1968. Fertil. & Steril. 19:598.*)

D. The effect of maternal nutrition on the size of the lamb at birth. At the left is a purebred Welsh lamb with its mother. At the right is a purebred Welsh lamb of the same age which was transplanted as a fertilized egg into the Border Leicester ewe standing beside it. Despite the same heredity, the lamb which developed in the uterus of the larger mother is bigger than the normally developed lamb. (*Courtesy of Esso Petroleum Co. & Dr. George Hunter.*)

PLATE 3

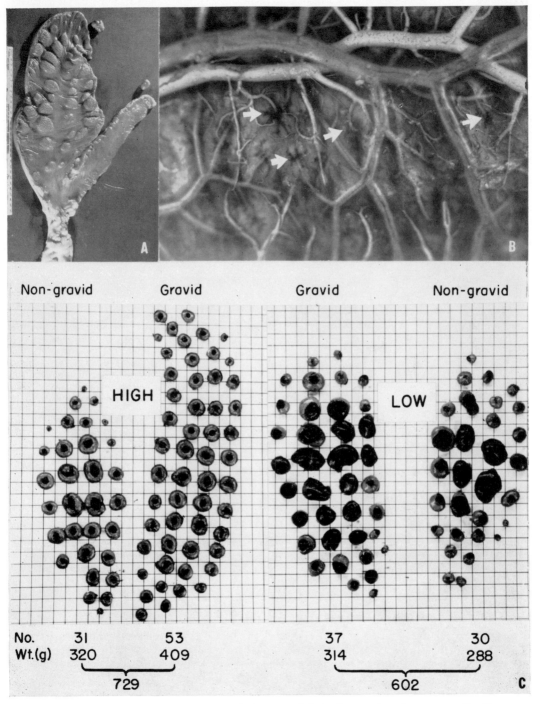

A. Bovine uterus at 105 days of pregnancy (twin fetuses and their placentas have been removed). Pregnancy developed in one uterine horn. Note the compensatory growth of maternal caruncles in the gravid horn due to the sterility of the opposite uterine horn.

B. The distribution of fetal blood vessels on the surface of the cow's placentome. Many small asterisk-shaped holes (*arrows*) occur on the surface of the placentome. Small fetal arteries and veins are embedded in these holes (latex injected specimen, × 9.2). (*Tustsumi & Hafez, 1966. Cornell Vet. 56:527.*)

C. Sheep caruncles at 90 days of gestation. Note the effect of maternal nutrition on the number and weight of eroded caruncles. (*From Everitt, 1965. Ph.D. thesis, University of Adelaide, S. Australia.*)

PLATE 4

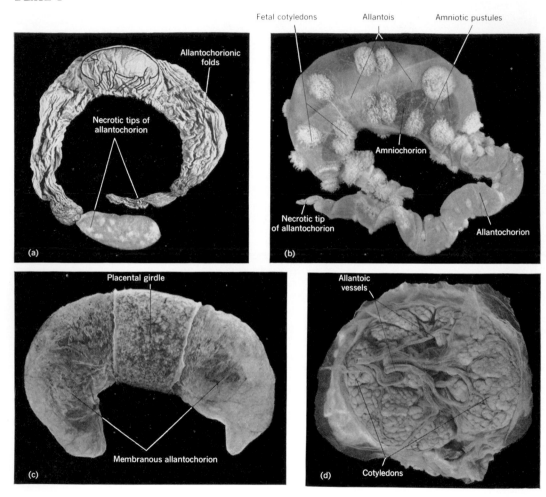

Types of mammalian placentas: a, epitheliochorial-diffuse from sow; b, syndesmochorial-cotyledonary from cow; c, hemochorial-zonary from cat; d, hemoendothelial-speroidal from rabbit. (*From Amoroso, 1952. Marshall's Physiology of Reproduction. A. S. Parkes (Ed.), London, Longmans.*)

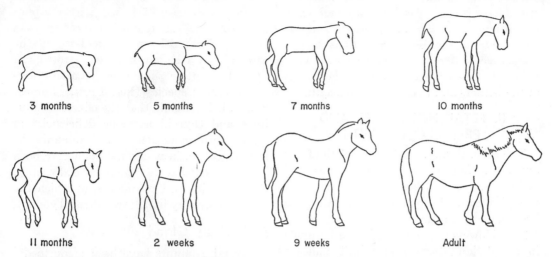

3 months   5 months   7 months   10 months

11 months   2 weeks   9 weeks   Adult

FIG. 1–10.   The changes which occur in the proportions of the Welsh pony during prenatal life.   In order to show relative changes, all the drawings have been reproduced at the same cranium size (eye-to-ear length).   (*From Hammond, 1935.   Empire J. Exp. Agric. 3:1*).

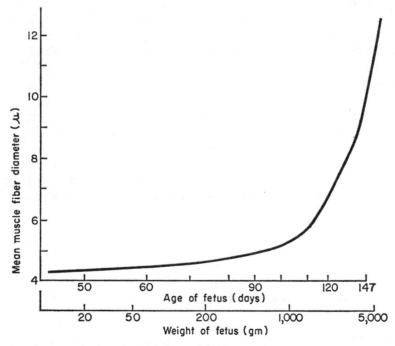

FIG. 1–11.   Growth of muscle tissue in fetal sheep.   Muscles grow by cell division up to the age of 90 to 100 days, that is, the number of cells increase.   After 100 days, there is a marked increase in the size of the cells. (*From Joubert, 1956.   J. Agric. Sci. 47:59.*)

cell size (Hammond, 1961). Skeletal growth of the fetus is fairly constant, although some dimensions increase more rapidly than others, with the result that body proportions change. Some breed differences, particularly between large and small breeds, occur in the height at the withers and in crown-rump length.

II. FETAL HORMONES AND PRENATAL GROWTH

The cells, tissues, and organs of the fetus apparently function in concert by means of hormones. Several hormones, *e.g.*, those of the pituitary, thyroid, and adrenal, have been detected in the endocrine glands of the fetus. However, the presence of a hormone in a fetal gland does not necessarily indicate that it is actually secreted into the blood and plays a physiological role.

Several techniques, including surgical removal of a fetal gland and destruction of the gland by x-irradiation, have been employed to study the effects of fetal hormones on prenatal growth. Other investigations involve the effects of chemically purified hormones, either separately or in combination, on the normal fetus, as well as on fetuses from which one or more endocrine glands have been removed. Most of these studies have been conducted on laboratory animals. In view of the pronounced species differences in the endocrinology of prenatal growth (Fig. 1–12), caution should be used when applying such data to farm animals and man.

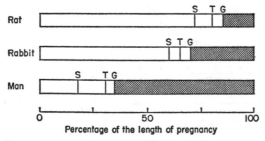

Fig. 1–12. Species differences in some endocrine and metabolic parameters of the mammalian fetus. Note that the dates when somatic sexual differentiation (S), colloid storage in the thyroid (T), and glycogen storage in the liver (G) begin are plotted as a percentage of the gestation period. (*From Jost, 1966. In: The Pituitary Gland. W. Harris & B. T. Donovan (Eds.), Vol. II., Berkeley, U. of California Press.*)

Experimental attempts should be made to correlate the onset of growth with the temporal course of differentiation of cell types in the anterior pituitary, particularly in the case of growth hormone (somatotropin). Special attention should be given to differences among species (Fig. 1–12). For example, it is possible that the effects of pituitary and thyroid hormone deficiencies in fetal life are more marked in those animals, such as cattle, in which the young are born in a relatively mature state (precocial young), than in animals, like rodents, in which the neonates are very immature (altricial young).

A. Pituitary Gland

Several attempts have been made to determine the role of the fetal pituitary gland in prenatal growth. Two experimental approaches have been used: removal of the fetal hypophysis, and observations of anomalous fetuses lacking pituitary glands. Pituitary growth hormone is not apparently necessary for the differentiation, morphogenesis, or *in utero* growth of some mammals. Cyclocephalic[1] or triocephalic[2] young pigs and rabbits have normal birth weight, but lack the pituitary gland. Anencephalic[3] fetuses have normal birth weight, if adjustment for the missing head is made. Experimentally decapitated sheep fetuses grow to normal size (Plate 2). Somatotropin-deficient children born to mothers with the same deficiency have normal birth length and weight. The whole body dry weight and the liver weight are about the same in decapitated and control rabbits (Jost, 1966), although decapitated fetuses do not store glycogen in their livers, whereas normal fetuses do.

In rats, however, growth hormone from the developing hypophysis governs, in part, the prenatal growth of the fetus. Decapita-

[1] Severe defects of the facial skeleton and brain characterized by the absence of a nose and connected rudimentary eyes situation in a single orbit.

[2] Absence of eyes, nose, and mouth: the head is merely a small spheroidal mass with no brain.

[3] Absence of cerebrum and cerebellum with absence of the flat bone of the skull.

tion of the rat fetus *in utero* causes retardation of prenatal growth. This effect can be prevented by giving the headless fetus subcutaneous injections of bovine somatotropin (*cf.* Wells, 1965). It is estimated that somatotropin from the fetal hypophysis of the rat is normally responsible for about 20 percent of the fetal growth that occurs shortly before birth.

In contrast to the mammalian fetus, the chick embryo, which is hypophysectomized (removal of all of the forebrain region) at 2 days of incubation, develops somewhat normally in body proportions and shape, but its size is considerably smaller than that of normal controls of the same age. There is also evidence that injection of the exogenous growth hormone may stimulate growth of the avian embryo.

It would appear from these experiments that fetal growth does not require growth hormone in all species. Indeed, pituitary somatotropin may be unable to cross the placenta, and a substitute, which influences fetal development, may be produced in its place. At any rate, differentiation, morphogenesis, and intrauterine growth are probably independent of such a substitute, as well as the somatotropin of the mother and her fetus. However, growth hormone may play a physiological role in the fetus: it may, for example, regulate metabolic processes. In this respect, Houssay found that the pituitary glands of 3-month-old bovine fetuses were diabetogenic (caused diabetes).

The functioning of the hypophysis-adrenal axis and the hypophysis-thyroid axis before birth has been demonstrated by experiments *in vivo*. If a fetal rat is subjected to left adrenalectomy, the right adrenal undergoes a significant, compensatory hypertrophy (Eguchi *et al.*, 1964). Such compensatory hypertrophy of the fetal adrenal can be prevented by cortisone treatment. In a normal fetus, an implanted pellet of cortisone not only retards the growth of the adrenal, but also retards the growth of the hypophysis. Hypophysioprivus by subtotal decapitation of the fetus *in utero* also retards the growth of the fetal thyroid. This effect may be prevented by giving the headless fetus subcutaneous injections of thyrotropin (*cf.* Wells, 1965).

B. Thyroid Gland and Gonads

In contrast to pituitary somatotropin, other hormones such as thyroxine and androgens are required, at least for morphogenesis. Thyroxine deficiency, for example, prevents a tadpole from metamorphosing into a frog. In man and rat, the brain fails to develop normally when thyroxine is deficient. The morphogenesis of the external genitalia (*e.g.*, formation of a scrotum and penis) is androgen-dependent, although linear growth of the body is not.

The thyroid of the fetus is able to synthesize and secrete thyroid hormone. This hormone is normally available to the fetus during the later stages of pregnancy. However, whether it plays any essential role in the development or metabolism of the fetus is not known. Observation of cretinous infants, however, suggests that the human fetus requires thyroid hormone to complete its normal development. Possibly, this hormone is only necessary for the final maturation stages of certain fetal tissues. Even massive amounts of thyroid hormone do not completely correct all the defects of thyroid hormone deficiency during prenatal development (Carr *et al.*, 1959).

The deficiencies in growth and development produced by a lack of the thyroid hormone are not identical to those produced by hypophysectomy. Normal growth and development thus appear to depend partly upon the direct action of thyroid hormone and partly upon the action of other hormones, particularly growth hormone, acting in concert with thyroid hormone (Shellabarger, 1964). The physiological mechanisms by which thyroid hormone and other hormones promote prenatal growth and development are unknown.

III. FACTORS AFFECTING PRENATAL GROWTH

The birth weight of litter-bearing species is generally smaller than that of single-bearing animals. In the latter, birth weight is related more to adult weight than to the length of the gestation period. For example, in cattle and man the gestation period is 280 days, but the birth weights are 35 and 3.4 kg, respectively. Species with a high birth weight and longer gestation period tend to be born at a relatively higher stage of physiological maturity. Prenatal growth is influenced by several factors, namely, heredity; size, parity, and nutrition of the mother; litter size; placenta size; and ambient temperature.

A. Heredity

The ultimate size of the fetus is determined by its own genotype, and that of its mother and litter mates. The maternal contribution to variability in fetal size is greater than the paternal contribution. Species, breed, and strain differences in fetal size are well known. Breed differences in the rate of prenatal growth may be traced to early embryonic stages, even when there are no differences in egg size. For example, differences in the prenatal growth rates of small- and large-size breeds of rabbits can be detected in the first 2 days of pregnancy (Table 1–5).

B. Size and Age of Dam

Faster prenatal growth is directly related to large maternal size. This phenomenon has been demonstrated in horses, cattle, sheep, and rabbits. Walton and Hammond (1938) made reciprocal crosses between large Shire and small Shetland breeds of horses and compared the foals. At birth, the crossbred foal from the Shire mare was three times as large as the crossbred foal from the Shetland mare (Fig. 1–13). The difference between the reciprocal crosses was still well-marked

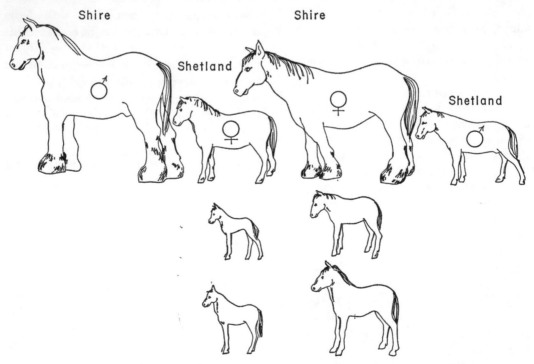

Fig. 1–13. Maternal size influences the size of the foal in reciprocal crosses between the large Shire horse and the small Shetland pony: all to the same scale. *Top line*: parents: Shire stallion X Shetland mare, and Shire mare X Shetland stallion. *Middle line*: their respective foals at birth. *Bottom line*: the foals at 1 month of age. (*Walton & Hammond, 1938. Proc. Roy. Soc. Biol. 125:311.*)

after 4 years, at which time the foal from the Shire mare was one and one-half times heavier than the foal from the Shetland mare. Furthermore, the differences in size persisted in adult life. The same size differences occur in crosses between the horse and donkey: the mule with the large horse as a dam is much larger than the hinny with the small donkey as a dam. Similar results were obtained from reciprocal crosses between South Devon (large) and Dexter (small) cattle. When a small but normally developed cow is bred to a large bull, the smaller maternal environment should restrict fetal size so that parturition is possible. On the other hand, when the mother is the larger parent, she will provide a beneficial influence on the offspring, at least so far as birth size is concerned.

The genetics of the sire determine the upper limit of size at birth when the mother is large but the size of the placenta limits the size in the small mother. That this is due to maternal nutrition and not to cytoplasmic inheritance is shown by embryo transfer experiments (Plate 2). Hunter (1956) made reciprocal crosses between large Border Leicester and small Welsh breeds of sheep. There was a slight maternal effect in the reciprocal crosses: the crossbred lambs from the large mother were somewhat heavier than those from the small mother. A comparison of the crossbred and the purebred lambs from the large and the small mothers, respectively, showed that a small ram had less influence on the size of the lamb from the large ewe than a large ram had on the size of the lamb from the small ewe.

The length of time that these maternal influences on size persist is determined by the stage of development at which the young are born (Hammond, 1961). In the horse, in whom the legs reach full length before birth, differences in birth size persist into adult life (Walton & Hammond, 1938). However, in sheep, whose cannon bone is not fully developed at birth, maternal effects on size diminish with age although weight differences persist for many months (Hunter,

1956). In species with a relatively long gestation period, the maternal tissue competes with fetal growth for a longer period and consequently has more effect on the size of the fetus. Thus, the maternal effect on fetal size is more apparent in horses and cattle than in sheep and laboratory mammals. In litter-bearing species, the maternal influence may be obscured by variations in litter size.

The effect of the mother's age on fetal size has not been fully investigated. Young dams which have not reached adult size continue to grow during their first pregnancy and thus compete with the fetus for available nutrients. The maternal environment also apparently changes with parity and possibly the degree of development and vascularity of the uterus. In addition, the young born from aged animals with excessive internal fat (which prevents full expansion of the pregnant uterus) are often small.

C. Maternal Nutrition

The fetus is privileged from a nutritional point of view. Even with severe maternal undernourishment, it continues to grow and will achieve birth weight which is close to normal. Under certain circumstances, however, the weight of the fetus is proportional to the caloric intake of the mother. In sheep, for example, the level of nutrition during the later stages of pregnancy has marked effect on birth weight. If the ewe is undernourished during the last third of pregnancy, she produces stunted lambs even though she was well-fed earlier in pregnancy. Conversely, a high level of nutrition in late pregnancy results in normal-sized lambs (Wallace, 1948). Babies of ill-fed mothers are on the whole somewhat lighter at birth than babies of well-fed mothers; however, the former group gives birth to many more premature babies, which of course are lighter than they would be had pregnancy run its full course.

In the last third of pregnancy, variations in fetal weight reflects differences in genetic factors, litter size, nutritional status, and health of the dam (Plate 3). When the mother is adequately fed, birth weight is likely to be at its upper genetic limit. Nonetheless, the birth weight of single lambs from mothers fed on a very high plane of nutrition during the second half of pregnancy is not greater than the birth weight of singles from ewes fed on a moderate plane. Male lambs grow larger before birth than females. The difference in birth weight is greater between male-female twins than between like-sexed twins or single-born male and female lambs.

Poor maternal nutrition during late gestation causes a reduction in the glycogen content of fetal muscles and, particularly, of the fetal liver. Fetal glycogen stores normally build up during late gestation and serve as a source of energy immediately after birth. Thus, poor maternal nutrition may increase neonatal mortality.

The rate of cell division within organs varies with the stage of development. An organ is apparently most vulnerable to nutritional stress when its rate of cellular division is highest. Early maturing organs such as the brain and central nervous system would, for example, be most vulnerable to earlier nutritional stresses.

Restricted maternal nutrition affects the different organs of the fetus in different degrees. The growth of ovine fetuses is retarded after malnutrition of the ewe during the second half of gestation, but the growth of the nervous system, heart, and skeleton is less affected than general body growth; kidneys, lungs, and muscle are affected to the same degree as general body growth; and skin (presumably including subcutaneous fat), spleen, thymus, and liver to a greater extent than body growth (Barcroft, 1947). Similar trends exist in the newborn of malnourished rodents. When the vessels of one uterine horn in the rat are ligated, prenatal growth declines; the brain of the fetus is affected less and the liver more than is body weight (Wigglesworth, 1964).

The effects of maternal undernutrition on postnatal growth can be identified if the neonates from underfed mothers are fostered to normally fed mothers after birth. Post-natal growth is reduced because of retarded prenatal growth, but this reduction may not become apparent until relatively late in the growth period. The weight at weaning, for example, may be normal, but anomalies and delays in behavior and locomotor performance appear at early stages of development before the growth deficit becomes apparent. Furthermore, animals with retarded prenatal growth have very poor feed efficiency during postnatal growth.

D. Litter Size

In polytocous species, increased litter size reduces the rate of prenatal growth because of variations in the functions of the placenta and the duration of pregnancy. The retardation of fetal growth in animals belonging to large litters becomes more marked as pregnancy proceeds. For some unknown reason, the effect of litter size on fetal weight is more pronounced in laboratory rodents than in the pig.

In monotocous species, multiple births usually retard prenatal growth; the combined weights of multiple-born young, however, exceed the weight of singletons (Fig. 1–14). Ovine singletons may be 120 percent and triplets 90 percent of the weight of twins

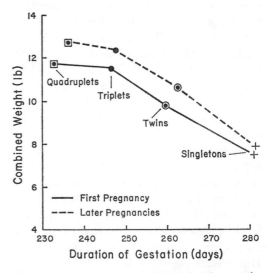

Fig. 1–14. Average fetal weights during the gestation period of the first and later pregnancies in man. (*Redrawn from Berbos et al.; after McKeown & Record, 1952. J. Endocrinol. 8:386.*)

Table 1–6. Birth Weight, Body Weight at One Year of Age, and Milk Production of Twin-Born Holstein Heifers Compared to Their Dams and Maternal Half-Sisters

(Data of A. B. Schultze.)

Animals	Birth Weight (kg)		Body Weight at 1 Year of Age (kg)		Milk Production (kg)*	
	Range	Mean	Range	Mean	Range	Mean
Dams	36–43	39	268–346	311	4,944–9,109	7,596
Twin-born heifers	22–38	31	244–329	299	4,811–8,214	6,593
Maternal half-sisters of twins	37–42	40	304–325	314	4,775–7,375	6,430

* Corrected milk production: total skimmed milk, containing 3.5 percent fat, for a lactation period of 305 days.

at birth. The birth weight of twin calves is also less than that of singles. Although such twin calves still weigh less than single-born calves at the age of 1 year, their milk production is not affected (Table 1–6).

Why prenatal growth is slower in large litters is by no means clear. This slowness is not due to the mechanical effect of many fetuses in one uterine horn, but may be due to a limited supply of some blood nutrient from the mother. However, increasing the caloric intake of the mother does not cause significant increases in fetal weight beyond the normal limits imposed by the fetal genotype.

The influence of litter size on fetal size is manifested through a "local effect" and a "general effect" (Fig. 1–15). The "local

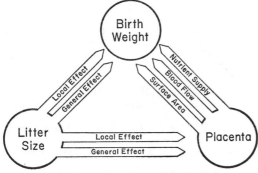

FIG. 1–15. Factors affecting prenatal development and birth weight. The "local effect" denotes the influence of other conceptuses in the same uterine horn, while the "general effect" denotes the influence of other conceptuses in both uterine horns.

effect" refers to the influence which other fetuses in the same horn have on prenatal growth. The "general effect" refers to the influence of other fetuses in the uterus independent of their distribution between the horns. Thus, an increase in the number of fetuses occupying one uterine horn could have both local and general effects on a fetus, whereas an increase in the number in the opposite horn could have only a general effect. The relative importance of the local and general effects varies between species.

In polytocous species, overcrowding within one area of the uterus is prevented by intrauterine transport and spacing of the embryos along the uterine horn. Transuterine migration of embryos occurs in ungulates and apparently equalizes differences in the number of eggs ovulated from each ovary (Hafez, 1964).

Variation in the growth rate of human twins is usually greater in monochorionic than in dichorionic pairs (Fogel *et al.*, 1965). This phenomenon is not necessarily associated with zygosity; *i.e.*, there is no difference between like- and unlike-sexed dichorionic twins, although some of the former must be monozygotic. Almost all monochorionic twin placentas have vascular anastomoses between the twin circulations, but their effectiveness in equalizing nutritional supplies is variable (Gruenwald, 1967). Placental anastomoses sometimes alter blood flow, leading to the so-called "intrauterine transfusion syndrome," in which the donor twin is small and anemic and the recipient is large and hypervolemic (blood volume greater than normal) (Benirschke, 1961; Rausen *et al.*, 1965). The significance of this syndrome in the incidence of the "runt" in farm animals is not clear.

E. Placental Size

The placenta is often regarded as a fetal organ. Since all nutrients which reach the fetus are transported by the placenta, it is often assumed that prenatal growth is determined by the placenta and that a small placenta retards prenatal growth. However, it is possible that a fetus which is small because of an inherently low growth potential or a poor supply of food from the mother has a small placenta for the same reason that its other organs are small (*cf.* Gruenwald, 1967).

The period of maximal fetal growth is preceded by a period when placental growth is highest. In man the placental weight (P) has been described in terms of fetal weight (W) by an allometric equation $P = aW^b$ (Aherne, 1966). More specifically, the exponent b was 0.67. Placental size can be limited by various maternal processes, and such placental restrictions may subsequently retard fetal growth. This indirect maternal effect, mediated through the placenta, may account for the close relationship between placental size and fetal weight in certain species. The stunted growth of neonate piglets, often seen in large litters, is related to the small size of the placenta.

Anatomically, the placenta acts somewhat like a blood reservoir in the maternal circulation. The blood flow in the uterine artery may, therefore, be lower if there are several placentas in one uterine horn than if there is only one. Decreased blood flow could reduce the oxygen supply to at least the more distantly placed placentas. Such fetal vascular abnormalities as "intrauterine transfusion syndrome" markedly affect prenatal growth.

F. Ambient Temperature

High ambient temperature during pregnancy affects fetal size in some species. Exposure of pregnant ewes to heat stress reduces fetal growth, and the degree of reduction is proportional to the length of the exposure. Dwarfing of lamb fetuses is caused specifically by temperature and not by the reduced feed intake which occurs during pregnancy: examination of the bones, organs, and endocrine glands of stunted lambs shows that the dwarfing effect is distinct from that of malnutrition (Yeates, 1953). The heat-induced dwarfs are well-proportioned miniatures compared to the long-legged, thin lambs from underfed ewes.

Animals which are undersized at birth are physiologically premature and are more subject to neonatal mortality owing to their poor thermoregulatory mechanisms and their inability to withstand stress in a new environment. The exact implications of physiological prematurity have not been clearly defined.

REFERENCES

Aherne, W. (1966). A weight relationship between the human foetus and placenta. Biol. Neonat. *10*: 113–118.

Barcroft, J. (1947). *Researches on Prenatal Life.* Springfield, Ill., Charles C Thomas.

Benirschke, K. (1961). Twin placenta in prenatal mortality. New York State J. Med. *61*:1499–1508.

Carr, E. A. Jr., Beierwaltes, W. H., Raman, G., Dodson, V. N., Tanton, J., Betts, J. S., & Stambaugh, R. A. (1959). The effect of maternal thyroid function on fetal thyroid function and development. J. Clin. Endocrin. *19*:1–18.

Eguchi, Y., Eguchi, K., & Wells, L. J. (1964). Compensatory hypertrophy of right adrenal after left adrenalectomy: observations in fetal, newborn and week-old rats. Proc. Soc. Exp. Biol. Med. *116*:89–92.

Fogel, B. J., Nitowsky, H. M., & Gruenwald, P. (1965). Discordant abnormalities in monozygotic twins. J. Pediat. *66*:64–72.

Gruenwald, P. (1967). Growth of the human foetus. In: *Advances in Reproductive Physiology.* A. McLaren (Ed.), Vol. II, New York, Academic Press.

Hafez, E. S. E. (1964). Transuterine migration and spacing of bovine embryos during gonadotropin-induced multiple pregnancy. Anat. Rec. *148*: 203–208.

Hammond, J. (1961). Growth in size and body proportions in farm animals. In: *Growth in Living Systems*. New York, Basic Books, Inc.

Hunter, G. L. (1956). The maternal influence on size in sheep. J. Agric. Sci. *48*:36–60.

Jost, A. (1966). Anterior pituitary function in foetal life. Chap. 9. In: *The Pituitary Gland*. G. W. Harris &. B. T. Donovan (Eds.), Vol. II, Berkeley, University of California Press.

Joubert, D. M. (1956). An analysis of factors influencing post-natal growth and development in the muscle fibre. J. Agric. Sci. *47*:59–102.

Patten, B. M. (1964). *Foundations of Embryology*. 2nd ed. New York, McGraw-Hill Book Co.

Rausen, A. R., Seki, M., & Strauss, L. (1965). Twin transfusion syndrome; a review of 19 cases studied at one institution. J. Pediat. *66*:613–628.

Shellabarger, C. J. (1964). The effects of thyroid hormones on growth and differentiation. Chap. 9. In: *The Thyroid Gland*. Vol. I, R. Pitt-Rivers & W. R. Trotter (Eds.), London, Butterworth.

Steen, E. B., & Montagu, A. (1964). Urinary, respiratory, and nervous systems, sensations and sense organs, endocrine and reproductive systems. *Anatomy and Physiology*. Vol. II, New York, Barnes & Noble, Inc.

Wallace, L. R. (1948). The growth of lambs before and after birth in relation to the level of nutrition. J. Agric. Sci. *38*:243–302, 367–401.

Walton, A., & Hammond, J. (1938). The maternal effect on growth and conformation in Shire horse-Shetland pony crosses. Proc. Roy. Soc. (Biol.) 125–311.

Wells, L. J. (1965). Fetal hormones and their role in organogenesis. pp. 673–680. In: *Organogenesis*. R. L. DeHaan & H. Ursprung (Eds.), New York, Holt, Rinehart & Winston.

Wigglesworth, J. S. (1964). Experimental growth retardation in the fetal rat. J. Path. Bacteriol. *88*: 1–13.

Yeates, N. T. M. (1953). The effect of high air temperature on reproduction in the ewe. J. Agric. Sci. *43*:199–203.

Chapter 2

Fetal Nutrition

By S. L. Hansard and R. K. Berry

THE significance of early nutritional and environmental influence upon subsequent behavior, developmental patterns, and productive performance in animals is of universal interest. Continued advances in methodology and instrumentation have provided new approaches to the intricate maternal-fetal physiological mechanisms involved during pregnancy and are permitting quantitation of periodic nutritional needs during gestation in terms of maternal dietary requirements. Although food energy is commonly the critical nutrient involved, the various micro- and macronutrients essential to normal tissue differentiation and growth become of greater concern as trends toward animal confinement increase. The following discussion is relevant to the maternal-fetal nutritional relationships as the embryo develops and grows as a functional part of the maternal products of conception.

I. MATERNAL ADJUSTMENTS TO PREGNANCY

Although the nutritional burden of reproduction is related primarily to changes that occur in the uterus subsequent to fertilization, the presence of a growing fetus pro-gressively adds an extra physiological load to the dam. Much of the maternal response is due to this increased burden. Pregnancy stimulates feed intake, increases nutrient absorption and utilization efficiency, and in young females, maternal tissue growth, nutrient accretion, and weight gain may exceed that accredited to the products of conception. This partition of absorbed nutrients by the gravid female has been explained by Robinson (1957) on the basis of tissue metabolic rate. Therefore, a suggested maternal nutrient distribution would probably be in the following order: (a) fetal growth, (b) uterus and placenta, (c) blood volume and hemoglobin (iron and protein), (d) mammary development, (e) fetal storage, and (f) maternal storage. Perhaps (e) and (f) should be reversed, but the relative demands of the fetus would probably cause time differences for these functions.

A. Pregnancy Anabolism

These specific maternal effects during normal pregnancy are regarded as *pregnancy anabolism*. In an animal of constant weight, pregnancy anabolism is commonly calculated by subtracting the weight at conception from that immediately following parturition. In young females, therefore, nutrient retention due to growth may mask actual weight increase due to pregnancy and would necessarily be combined in weight gains during pregnancy. Two hormones apparently govern this phenomenon of pregnancy: initially, progesterone, followed by prolactin, which is secreted by the placenta. Size of the placenta is determined by fetal number or size. However, litter size and maternal weight gain in swine appear to be independent of each other. It is probable that energy intake is the limiting factor in pregnancy anabolism, but true specific anabolism of pregnancy must be eventually quantitated to include net maternal accretion of the individual nutrients.

During gestation, basal metabolism, total body water, and circulating blood volume increase, especially during the last trimester, while total body lipids decrease, and as much as 80 percent of the absorbed nitrogen and minerals are deposited in the body of the dam (Mitchell, 1962). Although pregnancy anabolism increases in absolute value throughout gestation, it tends to decrease in relative proportions (Fig. 2–1). The weight gain that is maternal stores decreased from 50 at trimester 1 to 33 percent at term. In the sow, after 100 days of gestation two thirds of the nitrogen and body weight increase and 95 percent of the gross energy store increase were in the carcass and offal, with the remainder being deposited in the products of

conception. Special consideration, therefore, must be given to specific nutrient requirements of the dam during pregnancy, not only for body maintenance but also for constant replenishment and possible storage for use during late gestation and subsequent milk production. Age and nutritional status at the time of conception, additional needs for development and growth of the products of conception, including the fetus, mammary development, colostrum formation, and ready reserves for lactation are of special concern. The many factors affecting nutrient anabolism, and the reciprocal effects of this accretion on litter size, vigor of the fetus, milk production and postnatal growth rate and behavior, need further investigation before it will be possible to finally quantitate maternal-fetal nutrient requirements during the complete reproductive cycle. Maternal requirements for energy are known to change little throughout pregnancy; however, mineral and protein requirements exceed usual maintenance levels, increasing progressively with gestation age and fetal development.

B. Endocrine Adjustment to Pregnancy

Hormones in the correct proportions are essential for maintenance of pregnancy and for embryo survival, development, and growth. Maternal adjustments are facilitated and reflected by changes in hormone levels in body fluids. Anterior pituitary concentrations of FSH, LH, and lactogenic hormones increase with gestation. The thyroid is stimulated and basal metabolism increases. The placenta secretes large quantities of chorionic gonadotropin, estrogens, and progesterone, all of which facilitate normal pregnancy and nutrient storage. Progesterone secretion is increased, especially near the end of pregnancy, and it may increase as much as tenfold during gestation. This hormone is essential for maintenance of pregnancy and causes decidual cells to develop in the endometrium. These cells are thought to play an important role in the nutrition of the embryo. Ovariectomized

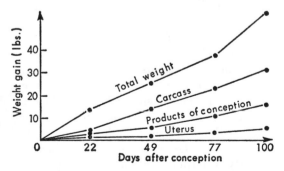

Fig. 2–1. Pregnancy anabolism partition in gravid swine. (*After Salmon-Legagneur & Rerat, 1962. Nutrition of Pigs and Poultry. Morgan & Lewis, Eds., London, Butterworth.*)

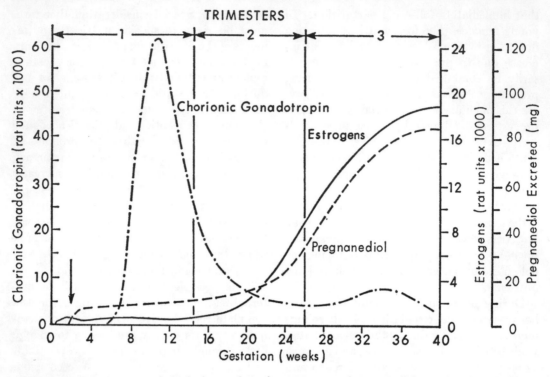

FIG. 2-2. Secretion rates of chorionic gonadotropin, estrogens, and pregnanediol as a function of stage of gestation in man (*Adapted from Guyton, 1966. Textbook of Medical Physiology. Philadelphia, W. B. Saunders.*)

cows require 50 to 75 mg of progesterone daily to maintain pregnancy, whereas sheep need only 5 to 10 mg. Rabbits require 1 mg for implantation and 2 to 3 mg daily for continued gestation (Hafez, 1968). The placenta is probably permeable to all hormones, including the steroids, insulin, and gonadotropins. These relationships are shown graphically in Figure 2-2 as a function of gestation time in man. As placenta and endocrine functions develop, an interplay of hormone action takes place between the dam, the fetus, and the placenta, and this balance, maintained by production and excretion, results in normal gestation.

Until parturition, the fetus then dictates the maternal metabolic processes, and nutrient utilization and deposition proceed in definite phases in accordance with the needs for the biochemical and physiological reactions of gestation.

C. Enzyme Activity during Gestation

Embryonic tissue metabolism is characterized mainly by the activity of the enzymes concerned with the synthesis of proteins and lipoproteins making up the organic matter of the tissues. The proteins are mainly derived from glucose and amino acids from the blood. This activity requires energy which is initially limited, but as development proceeds there is evidence of a shift in energy sources in the chick embryo from carbohydrates to proteins and from proteins to fat. Enzyme systems involved during early growth appear to be transient in that they occur, and then activity is diminished as the various phases of development are completed. Some enzymes that have a low rate of activity in the embryo initially increase greatly when fetal organs become functionally mature and then remain active throughout life. Others are active early and disappear with the cessation of growth. Glucuronyl transferase is a late-developing enzyme that affects detoxification of a number of toxic compounds, including bilirubin. It is absent in the fetal liver and present only in traces in the newborn, but the amount increases rapidly in the neonate. Changes in activity during development of the many different types of enzymes are discussed by Richter (1961) and are shown diagrammatically in Figure 2–3.

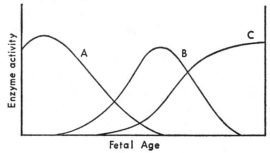

FIG. 2–3. Diagrammatic illustration of enzyme activity during gestation concerned with growth (A), function (B), and maturation (C) of the conception products. (*After Richter, 1961. Brit. Med. Bull. 17:118.*)

Enzymes primarily concerned with growth are most active in the early embryo and give activity curves of type A; enzymes concerned with function develop later and give curves of type B. A number of enzymes which show sharp peaks of activity at the intermediate stages before tissues are functionally mature may be concerned in effecting the morphological and other changes associated with the processes of maturation. As development proceeds, and as different organs become functionally mature, the metabolism of the growth which predominates in the fetus is gradually replaced by functional metabolism of more highly specialized enzyme activity curves of type C. These ascending and descending curves reflect the gradual change from one metabolic pattern to another during fetal maturation.

II. MATERNAL-FETAL RELATIONSHIPS

A favorable fetal environment is essential for nourishment, normal development, and growth during the gestation period. The mammalian ovum is small and has insufficient yolk to survive beyond the very early stages of cleavage unless additional food material is provided. Prenatal life, therefore, is initiated with segmentation of the fertilized ovum and its subsequent movement to the superficial tissues of the uterus. It is retained here, and, by a modification of the maternal tissue and a part of the ovum itself, a complex which functions as an "organ of physiological exchange" between the maternal tissue and the fetus is formed (Amoroso, 1961). The nourishment of the fetus is accomplished by the close juxtaposition of the maternal-fetal capillaries which are separated by various placental membranes differing in number and type with the species.

A. Mechanisms for Fetal Nutrition

1. TRANSITIONAL PHASE

Initially the zygote is free in the uterus, and until implantation occurs, it absorbs foodstuffs for maintenance directly from surrounding fluids. In the rat the blastocyst remains small and becomes lodged in the uterine lumen, where it implants itself by passing through the epithelium and becomes cut off completely from the uterine lumen (Nalbandov, 1964). This period from free-floating blastocyst to definite implantation varies with the species. However, in the cow, ewe, and mare, the transition is not very clear-cut. In the sow, blastocysts may be free in the uterus from 10 to 18 days before implantation. This time permits the ova to migrate and equalize in the horns; however, mechanisms of control have not yet been elucidated.

As the blastocyst increases in size it can no longer absorb sufficient "uterine milk" to supply nutritive material adequately by diffusion. Implantation is thus a step toward the effectual formation of embryonic membranes to give the embryo access to nutrient supply via the maternal circulation system. During this transition from the *histotrophic* to the *hemotrophic* stage, the placenta is formed by fusion of the fetal organ to the maternal tissues (Nalbandov, 1964). Shortly after attachment the organization of placental membranes begins and the chorion, amnion, and allantois are formed. Nutrition after this time depends upon maternal absorption and subsequent placental transfer, the major route for prenatal nurture in farm animals.

2. PLACENTAL FUNCTIONS

During the early stages of pregnancy and until the fetal liver becomes functional, the placenta serves essentially the same purpose for the developing embryo that the maternal liver serves for the adult animal. Mitchell (1962) has described the placenta as representing the fetal organ for respiration, nutrition, and excretion. It serves not only to *protect* the fetus by *controlling its environment* and *nutrition* through selective permeability, but acts as a ready *reservoir* for nutrients, *synthesizing* many and subsequently *transferring* them to the fetus as needed. Throughout pregnancy nutrients must pass from maternal circulation through the placental barriers into the fetal circulation, yet the blood of the fetus and dam never come in contact with each other. They are close enough at the junction of the chorion and endometrium, however, for nutrients and oxygen to pass from maternal to fetal blood and for waste products to pass in the opposite direction. Blood moves from the fetus to the placenta by way of the umbilical artery, carrying urea, carbon dioxide, and other fetal metabolites. These metabolites are removed by the placenta and subsequently transported to the maternal circulation. The umbilical vein gleans the oxygen and nutrient material that have passed the placental barriers and transports them to the fetus.

The *placental barrier* refers to the degree of maternal-fetal placental blood intimacy, and *placental permeability* is usually measured as rate and quantity of materials passing from the dam to the fetus. A most characteristic fact about the placenta, however, is that it progressively undergoes change with gestation age. Circulating placental blood reservoirs decrease, walls become thinner, permeability increases, and absolute functional activity is reduced. For convenience of discussion, therefore, and for comparison of change at different gestation ages in different species, the period is arbitrarily divided into the three trimesters of pregnancy. These are herein characterized as periods of (a) formation and differentiation, (b) fetal development, and (c) fetal growth. Triggering mechanisms for rate changes and sequence scheduling in the various species are unknown, but they are probably genetically programmed through hormone regulators and RNA function in embryogenesis.

3. NUTRIENT TRANSPORT

Maternal control of fetal environment and nurture has prevented a direct approach to many of the interesting and fascinating problems of fetal, placental, and maternal physiology. However, progress on basic metabolic procedures for rate and quantitation of nutrient transfer has been continuous. Investigations have been hampered by anatomical inaccessibility and the complex problem of studying fetal and maternal circulations simultaneously. Indirect procedures for estimation of body compartment size and the measurement of composition changes present many unique advantages in animal studies (Hansard, 1963). Several procedures have been developed for measuring the rate and total quantity of nutrients transferred from dam to the fetus (Barcroft, 1946; Robinson, 1957; Park, 1961; Reynolds, 1965; and Hansard, 1965). Methods range from whole body analyses of the dam and fetus to the use of indicators and radioactive-labeled substances for measuring physical and chemical changes in the body, and for quantitating specific mechanisms, membrane permeability, and transport rates to provide needed information on pregnancy anabolism, fetal accretion, and utilization rates.

Gases, including oxygen and carbon dioxide, freely traverse the placenta by simple diffusion and/or exchange. The principal conditions of gaseous exchange involved (Widdas, 1961; Barcroft, 1946; Metcalfe, 1964) are illustrated in Figure 2–4. However, these effects of oxygen tension, pressure gradients, and fetal metabolism do not fully explain all of the mechanisms involved. During late gestation, for example, maternal blood, largely depleted of oxygen, will readily equilibrate with the countercurrent fetal blood carrying a much higher content of oxygen.

A simple working hypothesis in such relationships is that the substance being transferred combines chemically with a component of the cell membranes, which possibly acts as a carrier complex for transport. This *facilitated diffusion* has been evidenced with those substances insoluble in the lipid membrane of cells such as sugars, amino acids, and salts.

Water moves rapidly from dam to fetus against osmotic gradients and serves to maintain body temperature and electrolyte

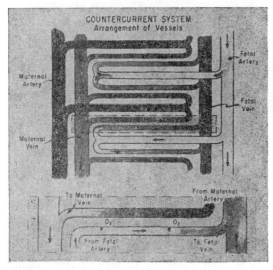

FIG. 2–4. Schematic presentation of the anatomical pattern of maternal and fetal blood flow, changes in oxygen tension (*intensity of shading*) between maternal aterial and venous blood, and the capillary countercurrent blood flow in the rabbit placenta. Location of the functional units (enlarged below) is indicated above by the stippled area. (*From Metcalfe et al., 1964. Fed. Proc. 23:774.*)

balance and to transport water-soluble nutrients. The process of transport for diffusible substances from maternal to fetal blood usually involves a large turnover in excess of current fetal needs. This has been demonstrated as especially true with water and certain electrolytes and less evident with larger molecules and cations. The ratio of the quantity of any substance supplied to the fetus from maternal plasma to the amount retained by the fetus has been called the "safety factor" for that substance. This "safety factor" for water varies from 700 at 14 weeks to 3800 at 30 weeks in a human fetus. This means that of 3800 parts of water transferred to the fetal circulation at 30 weeks only one part is retained, and the remainder is returned to maternal blood.

It has been apparent from radionuclide studies that nutrient "feed-back" from fetus to dam is much greater during early gestation and for highly exchangeable molecules, while "turnover coefficients" between fetus and dam decrease markedly as pregnancy advances, especially for those essential building components readily sequestered by fetal tissue during the phase of rapid growth. For most physiologically important molecules, however, active transport has been postulated as being involved in transfer across the placental barriers. This theory is evidenced in instances in which concentration gradients permit the embryo or fetus to accumulate a higher nutrient concentration than the one in maternal blood. This transport, therefore, is an energy-requiring process which must be linked with metabolism. These processes are few and only those for sodium, potassium, and possibly calcium have been authenticated in animal tissues. Transport is facilitated by many contributing factors including (a) catalyzing by placental enzymes, (b) chelating organic molecules, (c) exchange processes, (d) possibly membrane pulsations, and (e) metabolic processes in the placenta and the developing fetus itself. Specific contributions of these factors to nutrient accretion by the fetus are evidenced in other sections of this book.

Fortunately, most of the essential *inorganic nutrients* readily traverse the placenta, and utilization and deposition patterns by the fetus essentially parallel those of the dam. The macrominerals absorbed by the dam are transferred to the fetus at an accelerated rate throughout gestation and closely follow estimated needs for increased growth. Fetal blood levels may exceed maternal values, and, as noted for calcium and phosphorus, passage is often made against a concentration gradient. During early gestation, fetal iron is supplied in the sow and ruminant animals from uterine secretions, while for the hemochorial placental types, maternal blood hemoglobin provides the major source for iron. Other microminerals, including selenium and fluorine, appear to traverse the placenta and are either stored or utilized by the fetus (Reynolds, 1965).

Proteins are enzymatically hydrolyzed, and the nonprotein nitrogenous products diffuse across the placenta as proteoses and amino acids for resyntheses at the cellular level on the fetal side. Passage appears to be against a concentration gradient, since level of nonprotein nitrogen is usually higher in fetal than maternal blood. Many proteinases have been demonstrated in the placenta of all species examined and are probably involved in protein degradation after they have passed across the membranes. Antibodies and unhydrolyzed protein may traverse the placenta in trace quantities in man, rabbit, and guinea pig, but ungulates including the cow, sheep, goat, pig, and horse depend upon colostrum and postnatal synthesis for passive immunity (Hemmings, 1961). The intracellular mechanism by which protein selection is achieved has not been clarified, but the living membranes through which the protein is selectively transported from dam to fetus provide a biological means of separation that can be used for experimental purposes.

Energy is stored in the placenta as glycogen and provides a ready source for the developing fetus (Shelly, 1961). Fetal blood glucose levels usually exceed that of the dam, but in man, rat, and guinea pig, the level approximates 70 to 80 percent of that of the dam. In ungulates, fructose is essential for fetal survival, but unlike glucose it does not cross the placenta. It apparently originates in the placenta and represents an irreversible reserve of carbohydrates in the fetus.

Hydrolyzed *fats* traverse placental membranes as free fatty acids and glycerol for resynthesis on the fetal side. However, there is no evidence that unaltered glycerides pass the placenta as chylomicron, the emulsified fat droplets present in maternal blood. The levels of lipid in fetal blood exceed the levels in maternal blood in the guinea pig and rabbit, but the reverse has been observed in man and sheep. Fetal fat is richer in palmitic acid and contains less oleic and stearic acid than does maternal blood. Fat depots are formed only during the final stages of pregnancy (McCance & Widdowson, 1961).

Vitamins traverse the placental membranes and are essential for fetal growth and development. The relationship of maternal vitamin A to normal *in utero* development was observed early in swine, but, possibly because of lipid retention, placental transfer is dubious. However, membranes appear to be more permeable to the ester than to vitamin A alcohol. There has been little evidence that vitamin D traverses the placenta; however, maternal levels apparently enhance nutrient availability, and effects of vitamin A and D upon the developing fetus indicate that small quantities do reach the fetus. Vitamin E is necessary for development of the early ovum, and for prevention of muscular alteration in lamb fetuses. If vitamin K is transmitted, the rate of transfer is slow. However, localized postnatal hemorrhages have been observed in fetuses from dams low in vitamin K.

Water-soluble vitamins, including dehydro-ascorbic acid, are apparently readily transferred to the normal fetus at rates and in quantities essential for development and growth. These appear to exceed the requirements for maternal maintenance (Giroud, 1968).

B. Factors Affecting Placental Nutrient Transfer

There are various factors, both maternal and fetal, contributing to the rate and quantity of nutrients traversing the placental membranes for deposition in the fetus. Those receiving most emphasis include: (a) maternal nutrient availability for fetal transfer, (b) placental type and subsequent membrane permeability, (c) stage of gestation and physiological age or fetal needs, and (d) size, weight, and charge of particles being transferred. However, additional factors that may affect, directly or indirectly, placental transport rate and quantity include maternal age and nutritional history, time intervals between pregnancies, number of fetuses, length of gestation, certain absorbed toxins, and ambient temperature.

1. MATERNAL NUTRIENTS AVAILABLE FOR TRANSFER

The fetus is dependent directly upon maternal sources for all nutrients. These supplies are either endogenous or available from current feed intake. Thus fetal supply is dependent upon readily available maternal source. Dams deprived of essential nutrients over long periods of time and/or limited in current dietary supply could provoke fetal deficiencies and subsequent structural or functional abnormalities. Factors depressing maternal absorption and retention or utilization would likewise limit availability. This result is clearly shown in studies involving ingestion or injection of radioactive minerals for subsequent quantitation of placental transfer. After oral administration of Ca^{45} to a pregnant heifer, 70 percent is excreted and 30 percent is retained for tissue distribution in the dam and her fetus. However, if the same dose is given intravenously, 15 percent is excreted and 85 percent is retained for maternal-fetal distribution. In-

terpretation of radiochemical data is facilitated, therefore, by calculating transfer on the basis of that quantity which is actually retained by the dam, rather than as a percent of the administered dose, thus avoiding the obvious errors due to availability differences. These procedures emphasize the need for evaluation of the quantitative utilization in terms of what the dam does with that portion available for placental transfer. A normal animal differs in this respect from one on a deficient ration, especially if body stores have been depleted of a specific element or nutrient. Although it is generally accepted that the fetus has priority over the dam for the nutrients, the avidity of depleted maternal bones and organs may drastically reduce the quantity available for transfer. Likewise, other factors affect absorption and subsequent availability. For example, the quantity and percentage of dietary calcium, phosphorus, potassium, sodium, and glucose actually absorbed are greater than those of cobalt, iron, zinc, iodine, or vitamins. Since both needs and supply vary widely, consideration is necessary in dietary formulation to avoid nutrient combinations, antagonists, chelates, and the like, that could restrict maternal availability for subsequent fetal utilization.

2. PLACENTAL TYPES

The various families of mammals develop different types of placentas, depending upon the number of fetuses, length of gestation, the internal structure of the uterus, and the degree of fusion between maternal and fetal tissue (Hafez, 1968). These have been classified for comparative purposes in many different ways. One of these is summarized in Table 2–1 and illustrated in Plate 4 and Figure 2–5 (Amoroso, 1961). This schematization on the basis of the number of tissue layers separating the maternal and fetal blood supplies has been regarded as oversimplified by many histologists in that it ignores both the cytological organization and the regional differences in the placental bar-

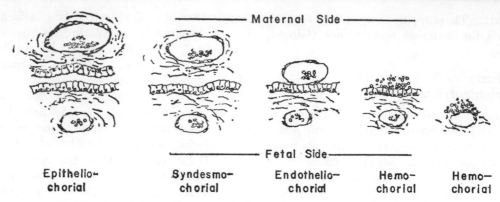

Fig. 2–5. Classification of placental types on the bases of opposing maternal and fetal tissues, illustrating anatomical arrangements. (*Adapted from Flexner & Gelhorn, 1942. Amer. J. Obstet. Gynec. 43:965.*)

Table 2–1. Classification of Mammalian Placental Types on Bases of Number of Tissue Layers Separating Maternal and Fetal Circulations

(*Adapted from Grosser, 1927. Fruhentwicklung, etc., Munchen.*)

Placental Types	Number of Tissue Layers		Species Examples
	Maternal Placenta	Fetal Placenta	
Epithelio-chorial	3	3	Swine, horse
Syndesmo-chorial	2	3	Cattle, sheep, goat
Endothelio-chorial	1	3	Cat, dog, mink
Hemochorial	–	3	Man, rat, rabbit, guinea pig

riers that could best be correlated with biochemical observations. However, simplification affords the descriptive flexibility necessary for discussion of nutrient availability for fetal growth and development. Specific details of the functional differences of these placental types is well documented in the literature (Grosser, 1927; Huggett, 1941; Amoroso, 1952, 1961; Hafez, 1968).

3. STAGE OF GESTATION

The limited demands for nutrients during early gestation would be primarily for developmental purposes, and the supply must be balanced and readily available. Observed variations between individuals during this period are, therefore, small. With increased age and size, however, competition for maternal supplies increases, and subsequent differences due to litter size and maternal dietaries become greater.

As gestation age and fetal size increase, the demand for nutrients required in the growth process of the fetus increases proportionately. Total increase in nitrogen (protein), water, and ash (minerals) is, for example, greater than that for body fat. The levels of bone building minerals (calcium, phosphorus) exceed levels of minor elements as reflected by growth changes of component body parts in animals of all placental types. Sodium freely traverses the placental barrier; Figure 2–6 graphically illustrates gestation effects on both the rate and quantity transferred to the products of conception in swine, the goat, rat, and rabbit. Calculated on the bases of milligrams of sodium per gram of placenta per hour, 300 times more sodium traverses a unit weight of placenta in the rat than in swine, and between swine and goats the difference was approximately 16 times, with quantity increasing with fetal age and size. Values for the rabbit and guinea pig were only slightly less than those for the rat, but were 10 times those for the cat. Sodium transfer rate parallels fetal growth rate, but only when compared with number of tissue layers intervening between maternal-fetal circulation does placental type actually become a factor.

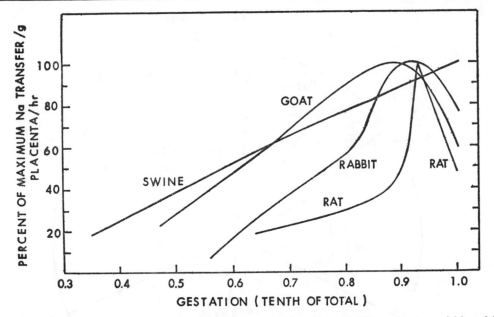

Fig. 2–6. Effects of gestation age and placental type upon placental permeability to sodium. (*Adapted from Flexner & Gelhorn, 1942. Amer. J. Obstet. Gynec. 43:965.*)

Sodium transfer decreases as the number of intervening tissue layers increases. In swine, as the need for growth increases, both quantity and rate increase progressively up to parturition, but quantity transferred in by other species during the last few days of gestation decreases.

4. Size, Weight and Charge of Nutrients

There is evidence that placental permeability is indirectly associated with particle size. Large molecules such as phospholipids which contain phosphorus do not traverse the rat placenta, while the inorganic ions of sodium, phosphorus, sulfur, calcium, and strontium freely penetrate the membranes and are subsequently deposited in fetal tissues. Transfer rate of alkaline earth in the hemochorial placenta during 18 to 21 days gestation have been shown to be directly related to molecular weight of the injected radioactive metal, and occurred in the following order: $Ca^{45} > Ba^{140} > Pu^{239} > Ra^{226}$. Data from the literature indicate that the placenta is more permeable to cations (calcium, sodium, iron, magnesium, zinc) than to anions (iodine, phosphate, sulfate),

and considerably more calcium than sulfur or iodine traverses the placenta of all species. However, the difficulty of completely separating the contribution of other influencing factors such as animal needs, number and size of fetuses, maternal nutritional history, and related factors complicates interpretation of true relationships. Comparison on the basis of unit weight, too, is less desirable than on the basis of per unit of absorbing surface, but the latter is more difficult to measure.

These facts further point out that placental permeability is a complex phenomenon that is difficult to ascribe to any specific chemical or physical mechanism, and that rate and quantity of a substance traversing the membranes at any period is dependent on a combination of the several maternal and fetal factors, physical, chemical, and biological, contributing to the net transport to the developing fetus.

C. Partition of Maternal Influence on Fetal Growth

Systematic investigations of the intricate relationships between dam and fetus have

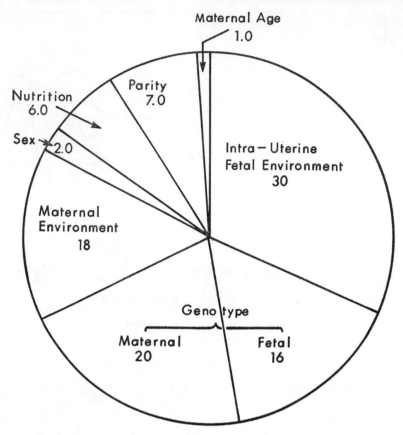

Fig. 2–7. Partition of causes of variations in animal birth weights; numbers indicate percent; the relative contribution of maternal environment to the developing fetus is apparent. (*After Penrose, 1955. Sci. Prog. Lond. 43:12.*)

suggested that many possible factors influence the environment of the fetus. The problems of separating nutritive effects from the numerous other contributing factors governing normal or abnormal fetal development, and quantitating the minimum requirements and specific interrelationships, however, are recognized. In farm animals, criteria for measurement of deviations from normal have been based for the most part on fetal survival and weight at birth. In recognition of these difficulties for evaluating major contributions of the dam to fetal growth, Penrose (1955) attempted to quantitate some of the causes for variability in birth weights as is illustrated in Figure 2–7. Numbers indicate percent, and contributions

of maternal and fetal environment are demonstrated in relation to genotypes and specific items within these individual groups. Evidence indicates that nutritional protection for the fetus begins even before conception, and must continue throughout the reproductive life of the animal. The achievement and maintenance of maternal-fetal nutrient equilibrium, therefore, are equally as important as maternal nutrient balance. Teratologic research concerned with extreme malnutrition during gestation has resulted in considerable information about the effects of specific vitamin, mineral, and amino acid deficiencies. The more generalized studies dealing with protein and energy levels have also contributed information.

III. PRENATAL DEVELOPMENT, GROWTH, AND NUTRIENT ACCRETION

During pregnancy, uterine growth appears to be resolved into two parts to accommodate the organ's contents. One is dependent on circulating internal secretion and the other on pressure effects of the fetus and its membranes. The latter effects are evidenced by changes in relative mass and volume of the components of the various compartments.

A. Products of Conception

1. CHANGES IN MASS

The fetus, fluids, and membranes show little changes in weight during the first few weeks, but become progressively larger, and nutrient accretion increases with advancing pregnancy. The effects of gestation age upon weight changes in swine, man, cattle, and sheep are graphically illustrated by trimesters for comparative purposes in Figure 2–8. In sheep and swine, mature weight for placental membranes is apparently reached during the second trimester and then decreases in late pregnancy. In man and cattle, however, growth continues well into the third trimester. In man and swine, membrane weights exceed those of fluids to parturition, but in sheep and cattle, weight ratios do not vary far from one throughout gestation.

The developing fetus gains little weight until the placenta is firmly established, but early in the second trimester development is augmented, and, in cattle, fetal length increases 7 times and weight increases 30 times during this period to equal that of the placenta and membranes. Thereafter, growth continues at an increasing rate, and fetal weight again doubles during the last fifth of the gestation period. The total quantity of fetal fluids rises rapidly early in pregnancy, and the ratio of the allantoic to amniotic fluids increases. Fetal mass changes follow an exponential law, reaching the maximum during late gestation (see Chapter 1), and

nutritional requirements diminish per unit weight with age. With time, therefore, the absolute requirements change because the mass of the fetus is progressively increasing to the time of parturition. The proportionally greater increase in fetal size and combined fluid volume over membrane weight during the final phase of gestation would imply a stretching and expansion of the placenta, thus creating a thinning of the laminae of the epithelial cytoplasm separating fetal and maternal vascular plexuses, and suggest an anatomical explanation for the marked increase observed in nutrient exchange between dam and fetus (Amoroso, 1961).

2. CHANGES IN COMPOSITION

Body composition changes during reproduction reflect nutrient needs and consist of deposition of absorbed nutrients in the new tissue formed, plus that expended or retained and stored by the dam. The tissues formed include the fetus, its fluids and membranes, enlargement of the uterus, and mammary development. Periodic composition values have provided basic data for estimating the nutritional demands of pregnancy. Maternal deposition in terms of energy, crude protein, and ash have been computed from data of Mitchell *et al.* (1931) for 100 kg swine and plotted in Figure 2–9 to illustrate progressive nutrient accretion as a function of gestation age. Little energy was stored during early gestation, but 270 kcal were deposited during the last week of pregnancy. Mammary energy storage does not usually exceed 10 percent of that in the total conception products. The added energy requirement for reproduction in swine does not exceed 6 percent until the last part of gestation, when it is increased to 15 percent. Trends for protein and mineral needs more or less parallel energy values and are reflected in the total weight changes (Fig. 2–1). Protein accretion computed for pregnant (200 kg) swine is tabulated in Table 2–2; it suggests increased needs for maternal adjust-

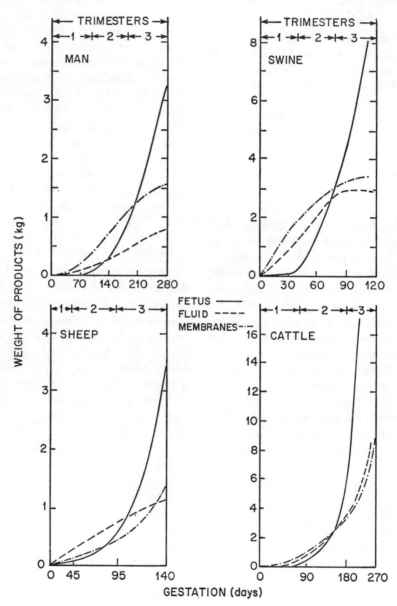

Fig. 2–8.   Weight changes during gestation reflecting total nutrient anabolism rates and quantities in products of conception standardized to 227, 45, 150 and 70 kg body weight for the cow, sheep, swine and man, respectively (*Data form Hansard et al., 1966.   J. Nutr. 89:335 and unpublished data.   Wallace, 1948.   J. Agric. Sci. 38:93, 243 & from the literature.*)

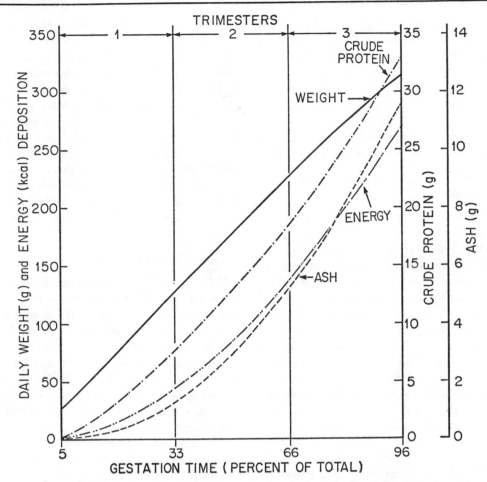

FIG. 2–9. Gestation age effects upon daily weight, energy, protein, and mineral deposition in total products of conception of 100 kg gravid gilts. (*Data from Hansard et al., unpublished, and Mitchell et al., 1931. Ill. Agr. Exp. Sta. Bull. 375, & 465.*)

Table 2–2. Protein Requirements of the Pregnant Sow (200 kg)

(*Corrected by the formula of Mitchell et al., 1931 Ill. Agr. Exp. Sta. Bull. 375; and from Lenkeit, 1963 Proc. 6th Int. Nutr. Congr., Edinburgh p. 397.*)

Gestation (Weeks)	Daily Weight Increase Protein (gm) Uterus and Contents	Mammary Gland[1]	For Weight Increase[2]	Maintenance (gm)	Total Protein Requirement (gm)
6–7	9	—	—	92	106
7–8	11	—	—	92	108
8–9	13	—	—	92	112
9–10	15	—	—	92	115
10–11	18	20	57	92	149
11–12	20	20	61	92	153
12–13	23	20	66	92	158
13–14	26	20	71	92	163
14–15	28	20	75	92	167
15–16	30	20	77	92	169

[1] Final weight 6 kg, 16 percent protein.
[2] Assuming 65 percent coefficient.

ments for the late period of gestation and for lactation. These data reflect partition of the total products of conception, mammary gland, weight increase, and gestation maintenance requirements during pregnancy. This procedure for calculating quantitative requirements, furthermore, does not consider those nutrients in the products of conception mobilized from possible maternal reserves.

B. Fetal Development and Nutrient Accretion

Nutrient accretion by the fetus occurs at an increasing rate as pregnancy advances. This progressive composition change is made up of weight increments of the various tissues, organs, and body parts, all of which grow at rates differing from one another, and varying with the species and stage of gestation (Robinson, 1957). The lamb and calf make little actual weight gain during the first half of uterine life, and the pig is born anatomically younger than either. Partition of the blood-transported nutrients among fetal tissues appears to be variable, but depends primarily upon the relative metabolic activity as reflected by blood supply and by oxygen uptake. These demands are easily met by placental exchange of endogenous nutrients during early gestation, as evidenced by the small fetal weight variations in sheep, cattle, and swine during this period of orientation and differentiation. This period appears to be the most critical stage for survival, however, since approximately two thirds of the total prenatal deaths occur prior to the twenty-fifth day of gestation.

Fetal needs have been considered to include the requirements for maintenance, for differentiation, and for subsequent development and growth. Although the requirements for maintenance may be readily determined, they may also be masked by the larger requirements for differentiation and growth.

1. ORIENTATION AND ORGAN DIFFERENTIATION

Although the fetus is well buffered against nutritional deficiencies, crucial periods dur-

ing early development include those when organs are being formed and nutrient needs for balance and quality surpass nutrient quantity (see Chapter 1). In the mammalian embryo, needs for growth appear to be greater than those for differentiation. Deficiency of pantothenic acid, for example, retards growth, but differentiation may progress normally. Factors other than nutrition may likewise exert effects on these biological timetables (Fig. 2–7), and may delay or prevent normality. Fetal-maternal relationships are therefore most critical during this early period.

2. GROWTH AND DEVELOPMENT

Following differentiation, these fetal structures are grossly similar to those of the newborn. However, cellular development is usually incomplete and requires most of the remaining period of gestation for growth. By midpregnancy, kidney and intestinal functions are in evidence. By the final trimester, the fetus is ingesting and absorbing large quantities of amniotic fluid, and intestinal and kidney functions approach those of the newborn (Guyton, 1966). However, control of extracellular fluids, electrolyte balance, and acid-base balance does not reach full development until early postpartum. As the fetus enters the stage of rapid growth and as the placenta approaches the period of decreased growth, mechanisms contributing to variable fetal growth rates do not appear to operate as uniformly, and fetal and/or maternal factors may thus limit both size and weight of the fetus (Reynolds, 1965).

Progressively growth changes in terms of body chemical constituents have been measured by intermittent determinations of body balance for the individual elements. This method, however, does not permit the estimation of the instantaneous rate of deposition, and is inadequate for the growth phase in which velocity changes are occurring. These have been best described mathematically by use of appropriate equations for the experimental observations (Mitchell, 1962; Moustgaard, 1962).

A generalized equation for prenatal growth

measurements, based upon the concepts of Brody in his description of postnatal growth of mammals, has been developed by Weinbach (1941). In this evaluation it is assumed that fetal growth is proportional to the weight already attained, plus the weight equivalent of "the impulse to grow"; the sum of the two is the "effective weight" of the embryo or fetus. Expressed mathematically:

$$dw/dt = k(A + w)$$

where dw/dt is the instantaneous rate of growth in weight per unit of time; $(A + w)$ is the "effective weight" for growth; w is the weight at time t; k is a proportionality constant; and A is the weight equivalent of the "impulse to grow."

Integration and subsequent simplification result in the final form of the formula for prenatal growth:

$$w = Ae^{k(t - t')} - A$$

in which e is the base of the natural logarithms, and t' is an age parameter, indicating the point at which the curve crosses the age axis; when $t = t'$, $w = o$. These constants for prenatal growth have been calculated and are compared for several species in Table 2–3.

Table 2–3. Prenatal Growth Constants of Various Species of Mammals

(After Weinbach, 1941. Growth 5:215.)

Species	k^1 (% per day)	A^2 (gm)	t'^3 (days)
Cattle	1.9	1775.0	98.1
Chick	18.0	3.2	6.4
Guinea pig	6.1	10.7	28.3
Man	1.2	471.5	94.0
Mouse	37.6	0.1	9.7
Rabbit	16.5	14.1	20.0
Rat	45.4	0.1	12.9
Swine	1.5	521.3	48.0

[1] k is a proportionality constant; assuming fetal growth (w) is proportional to weight already attained, plus A, the "weight equivalent" of the impulse to grow.

[2] t' is an age parameter in days and e is the base of the natural logarithm.

[3] Upon integration (see text) the formula for fetal or prenatal growth (w) becomes:

$$w = Ae^{k(t - t')} - A$$

3. Fetal Nutrient Accretion

Accretion of the various nutrients by the developing fetus is reflected in the fetal size and weight changes during gestation (Fig. 2–9), and changes in chemical composition with advancing pregnancy (Fig. 2–10). This chemical growth involves the anabolic rather than the catabolic processes of the animal body, and is measured in terms of nutrient storage in the tissues. Initially, fetuses are high in water content, but as development occurs, they are progressively dehydrated or diluted by sequestered nutrients. The human fetus weighing less than 2 gm has been found, on analysis, to contain 93 to 95 percent water, or slightly more than adult plasma. At 200 gm, however, the proportion of water decreases to 85 percent of the fat-free body weight, and continues to decrease to term, when total water content accounts for less than 80 percent of the total body mass (Brozek, 1965).

Fetal requirements for most vitamins are equal to or exceed those of the dam, and apparently the functions of the vitamins are similar in both fetus and dam. The B vitamins, especially B_{12} and folic acid, are essential for red blood cell formation and for fetal development. Vitamin C is necessary for the formation of bone matrix and the fibers of connective tissues. The fat-soluble vitamins are apparently essential although there is little evidence for direct placental transfer. Vitamin A does not accumulate in the placenta, but vitamins E and D do (Mitchell, 1962). Vitamin E is necessary for normal development of the early ovum, and vitamin A is essential for viable young and storage of maternal colostrum. Vitamin K is required for the formation of factor VII and prothrombin to prevent postnatal hemorrhages.

Although nitrogen content increases throughout gestation, its concentration at birth is only 75 percent of that in the adult. Muscle tissue grows slowly and at birth represents about 25 percent of the body weight, or one-half that of the adult. In contrast, fat accumulation is species-de-

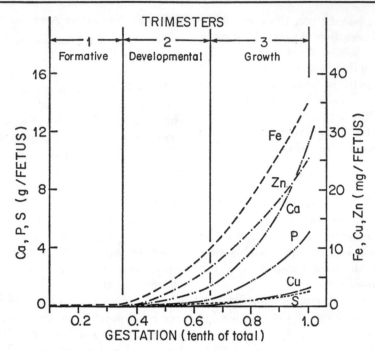

F<small>IG</small>. 2–10. Composition changes in the developing swine fetus for selected macro- and microminerals with gestation age. (*Data from the literature.*)

pendent and little increase is observed until late pregnancy.

The accumulation of selected minerals in the total swine fetus as a function of gestation age is shown graphically in Figure 2–10. Skeletal ossification is initiated late in trimester 1, and up to this time body calcium and phosphorus are low. Early in the second trimester, the amounts of fetal calcium and phosphorus are about equal, but as the fetus develops, the rate of skeletal growth exceeds the rate of soft tissue growth and as calcification occurs the proportion of calcium exceeds that of phosphorus. By late gestation, skeletal weight equals 15 to 20 percent of the total body weight although both calcium and phosphorus are well below the concentrations they will reach in adult animals. In the pig, calcium and phosphates accumulate slowly until the last trimester, which corresponds to the period of accelerated bone ossification and weight increase. Magnesium, like phosphorus, is distributed between the hard and soft tissue, neither of which has reached chemical maturity at

birth. Potassium increases progressively to term, but its fetal values are lower than the adult values (Widdowson & Dickerson, 1964).

Iron generally accumulates in the developing fetus more rapidly than calcium or phosphorus. Most of this iron is in the form of hemoglobin, which begins to be formed early in the first trimester. Approximately one third of the iron in the fetal liver is stored and available for hemoglobin formation after birth. Total iron in the bovine fetus increases from 5 mg/kg at 180 days gestation to 28 mg/kg at 270 days pregnancy. In swine, sheep, and cattle, the amounts of iron, copper, zinc, and sulfur in the fetal liver are higher throughout the gestation period than in the maternal liver. Available evidence indicates that adaptive changes to minimize nutrient requirements occur in the gravid female. More information concerning the range of maternal intake is needed, however, before the normality for pregnancy can be reestablished.

The comparative whole body analyses of

Table 2–4. Body Composition of Newborn Mammals

(*Adapted from Widdowson, 1950. Nature 166:626.*)

Species	Average Weight (gm)	Gm per 100 gm Fresh Weight			
		Water	Protein	Fat	Ash
Calf	16,600	78	16	2.6	4.0
Cat	118	81	15	1.8	2.6
Guinea pig	80.1	73	15	10.0	2.1
Lamb	4,340	81	13	2.4	3.2
Man	3,564	76	12	10.7	2.8
Mouse	1.6	83	13	2.1	2.1
Rabbit	54	85	11	2.0	2.3
Rat	5.9	86	11	1.1	2.1
Swine	1,460	84	11	1.1	3.5

newly born mammals of nine species are tabulated in Table 2–4; these illustrate total accretion and uniformity in fetal body composition. Fat content is the major variable, and, except for the higher levels in man and the guinea pig, averages 2.6 percent or less.

The ash component is broken down further into selected minerals present in the fetus, placenta, and fluids. These values in swine are presented in Table 2–5, and further emphasize the nutritional importance of both macro- and microminerals in the developing fetus.

Normal pregnancy is a period of maternal *anabolism*. The changes it causes must be duly considered in any true evaluation of

Table 2–5. Mineral Composition of Products of Conception in Swine at 110 Days Gestation

(*From Hansard, 1965 in Swine in Biomedical Research. Frayn, Seattle.*)

Mineral[1]	Products of Conception		
	Placenta	Placental Fluids[2]	Average Fetus
Calcium (gm)	0.59	0.62	9.9
Chlorine (mEq)	88	87	50
Copper (mg)	1.60	0.30	2.5
Iron (mg)	30.50	0.80	34.8
Magnesium (gm)	0.09	0.02	0.3
Phosphorus (gm)	0.62	0.15	5.3
Potassium (mEq)	20	15	50
Sodium (mEq)	97	108	92
Sulfur, inorg. (gm)	0.03	0.04	1.0
Zinc (mg)	55.80	1.40	26.8

[1] Mineral calculated per kg wet weight.
[2] All placental fluids were combined for sampling after removal of fetuses.

gestation. Comparative data from both pregnant and nonpregnant animals, however, must be evaluated before representative mathematical equations can be formulated to quantitate net nutrient requirements for the developing fetus.

IV. PRENATAL INFLUENCE UPON POSTNATAL PERFORMANCE

There has been increasing evidence that both pre- and neonatal environments affect physical development and exert a profound influence upon subsequent animal performance (Widdowson & Dickerson, 1964). There is direct evidence that balanced and enriched nutrition may favorably influence genetic potentials in all species. Early studies of the effects of prolonged maternal undernutrition on fetal size have suggested that they are most severe during the last stages of gestation, and that they are directly related to the stage of maturity (McCance, 1963). Increased awareness of variations within species, and the economic implications of the feedlot response to rations have stimulated considerable interest in maternal diet manipulation and the quantitation of subsequent effects on the fetus.

A. Dietary Manipulation

Several approaches have been employed in an effort to distinguish and then quantitate the influences of fetal nutrition on the subsequent weight gain and performance of the neonate. Fetal requirements are known to exceed maternal needs on a body weight basis, and extended periods of minimum nutrition are known to exert definite effects on the fetus (Wallace, 1946). However, body weight cannot always be considered dependable for evaluating functional development. Individual nutrients have specific functions during growth and development. Nutrient intake appears to have a relatively narrow and critical range. When it falls below this range, development is arrested and anomalies result; above this minimal level the fetus is normal. The maternal capacity to buffer reactions within these ranges and prevent the triggering of mechanisms detrimental to normal pregnancy is apparently dependant upon an abundance of maternal stores and adjustments to pregnancy. Demands of growth, however, rapidly reveal almost any defect in nutrition, and, when produced, cause damage to the developing embryo. The nature and severity of the damage depend upon the nature of the deficiency, the time it is induced, and its duration.

A most common procedure for inducing stress has been to deprive the dam of specific dietary nutrients, either by limiting total diet or by feeding antimetabolites, chelates, or inhibitors, during all or particular parts of gestation, and subsequently to evaluate gross and microscopical malformations, behavioral abnormalities, or biochemical anomalies in the embryo or fetus. Either de-

ficiencies or excesses deleterious to embryo survival at critical periods of development may induce physical or mental abnormalities that alter function, decrease nutrient utilization, and, postnatally, retard growth rate, organ development, and performance. Many of these malformations, the most spectacular repercussion of deficiencies, have been characterized for specific nutrients in several of the animal species and have been well documented in the literature (Palladan, 1966; Giroud, 1968).

B. Specific Nutrient Effects

Growth of the total fetus is the result of weight increments of the various organs and tissues, all of which grow at different rates (Wallace, 1946), and are age-dependent. Limited nutrition during late pregnancy exerts a marked effect upon the birth weight of lambs, and correlations with postnatal response to rations have been high. The effects of feed restriction on organ weights in the fetal sheep are shown in Table 2–6, and reflect retarded development and the differential effects on organ maturity. Carbohydrates are the principal source of energy and are usually the first limiting nutrients for the fetus.

Development differs for individual organs, and nutrient limitations may affect composition and structure without significantly affecting weight. Cell division may be limited

Table 2–6. Relative Weight (%) of Fetal Sheep Organs, and Relative Loss in Late Pregnancy
due to Maternal Nutritional Restriction

(After Robinson, 1957, in Hammond's Physiology of Farm Animals. London, Butterworth.)

Organ or Tissue	Weight, as % of Normal Weight at Birth, at Fetal Age in Days					Weight Loss in % Due to Undernutrition (3rd Trimester)
	72	*97*	*120*	*135*	*140*	
Blood	4	13	34	80	100	57
Brain	9	40	120	94	100	16
Heart	6	24	62	68	100	43
Intestines	3	12	45	83	100	44
Kidney	5	24	38	76	100	49
Liver	11	57	116	137	100	92
Muscle	12	30	63	90	100	57
Skeleton	5	29	61	92	100	45
TOTAL	5	22	62	91	100	53

in the fetus by specific dietary deficiencies, yet the growth of cells may continue normally. Vulnerability to nutritional stress appears to be greatest at periods when the rate of cell division is at a maximum. Structural malformations may well be irreversible. They can interfere with potential organ function and thus drastically affect postnatal performance. Restricted gestation diets for rats produce progeny that exhibit impaired responses to feeding, to hypothermal stress, to maintenance of energy stores, and to antigenic stimulation (Chow & Lee, 1964). Observed decreases in utilization efficiency of dietary protein and total feed by rats suggest variations in biochemical and physiological characteristics related to functional development. Results with swine have been less conclusive. however, When gravid gilts were fed 5 percent protein or protein-free diets (0.5 percent) on day 24 to 28 of gestation, neither body weight nor total serum protein of the progeny was lowered (Ripple et al., 1965). Subsequent milk production was adversely affected by low protein, however, and suggested a possible hormone effect.

As more direct information about the significance of maternal anabolism of specific nutrients is made available and as maternal-fetal relationships are elucidated, newly developed procedures may permit separate quantification of maternal dietary composition for optimal fetal nutrition.

REFERENCES

Amoroso, E. C. (1952). *Marshall's Physiology of Reproduction*. A. S. Parkes (Ed.), London, Longmans.

Amoroso, E. C. (1961). Histology of the placenta. Brit. Med. Bull. *17*:81–91.

Barcroft, J. (1946). *Researchers on Prenatal Life*. Oxford, Blackwell Scientific Publication.

Brozek, J. (Ed.) (1965). *Human Body Composition. Approaches & Applications*. Vol. VII, New York, Symp. Publ., Pergamon Press.

Chow, B. F., & Lee, C. J. (1964). Effect of dietary restriction of pregnant rats on body weight gains of the offsprings. J. Nutr. *82*:10–18.

Giroud, A. (1968). Nutrition of the embryo. Fed. Proc. *27*:163–184.

Hafez, E. S. E. (1968). Gestation, parturition and prenatal development. Chap. 10. In: *Reproduction in Farm Animals*. E. S. E. Hafez (Ed.), 2nd ed. Philadelphia, Lea & Febiger.

Hansard, S. L. (1963). Radiochemical procedures for estimating body composition in animals. Ann. N. Y. Acad. Sci., *110*:229–245.

Hemmings, W. A. (1961). Protein transfer across foetal membranes. Brit. Med. Bull. *17*:96–101.

Huggett, A. St. G. (1941). The nutrition of the fetus. Physiol. Rev. *21*:438–462.

McCance, R. A. (1963). The bearing of early nutrition on later development. Proc. 6th Intern. Nutrition Congr. 74–81, Edinburgh.

McCance, R. A., & Widdowson, E. M. (1961). Mineral metabolism of the foetus and newborn. Brit. Med. Bull. *17*:132–143.

Mitchell, H. H. (1962). The nutritive requirements for mammalian reproduction. Chap. 9. In: *Comparative Nutrition of Man and Domestic Animals*. Vol. I, New York, Academic Press.

Moustgaard, J. (1962). Fetal nutrition in the pig. Chap. 13. In: *Nutrition of Pigs and Poultry*. J. T. Morgan & D. Lewis (Eds.), London, Butterworth.

Nalbandov, A. V. (1964). Chap. 9. In: *Reproductive Physiology*. San Francisco, Freeman.

Palladan, B. (1966). Swine in teratological studies. In: *Swine in Biomedical Research*. pp. 51–78, Seattle, Frayn.

Park, C. R. (1961). *Membrane Transport and Metabolism*. A. Kleinzeller & A. Kotyk (Eds.), New York, Academic Press.

Reynolds, S. M. R. (1965). Maternal-fetal relationship and placental exchange. Chap. 29. In: *Physiology of the Uterus*. New York, Hafner.

Ripple, R. H., Harmon, B. G., Jensen, A. H., Norton, H. W., & Becker, D. E. (1965). Response of the gilt to levels of protein as determined by nitrogen balance. J. Anim. Sci. *24*:209–215.

Shelly, H. J. (1961). Glycogen reserves and their changes at birth. Brit. Med. Bull. *17*:137–143.

Wallace, L. R. (1946). The effect of diet on fetal development. J. Physiol. *104*:34–42.

Widdas, W. F. (1961). Transport mechanisms in the foetus. Brit. Med. Bull. *17*:107–111.

Widdowson, E. M., & Dickerson, J. W. T. (1964). Chemical composition of the body. Chap. 17. In: *Mineral Metabolism*. C. L. Comar & F. Bronner (Eds.), New York, Academic Press.

II. Postnatal Growth and Development

Chapter 3

Genetics of Growth

By G. A. E. Gall

I. DEFINITION OF GROWTH

A. Biological Definition

EVERY living organism represents a hierarchy of organized systems. The problem of animal growth must be examined from different viewpoints and at various levels. Consequently, what is termed "growth" from a certain viewpoint need not be termed so from another. It is not even a scientific term with defined and constant meaning, but a popular label that varies with the accidental traditions and purposes of the individual using it. It has come to connote any and all of the following: reproduction, increase in dimension, linear increase, gain in weight, gain in organic mass, cell multiplication, mitosis, cell migration, protein synthesis, and perhaps others. For example, if the growth of a tumor is investigated, this naturally cannot be considered as normal growth of the organism as a whole, but must be examined as the multiplication of a specific cell population. The adipose tissue of an aging man grows but the individual may not increase in total mass.

Growth in the biological sense is intimately connected with metabolism and the organization of living systems in self-reproducing units. If the molecule of a chemical compound increases in size, it changes into another chemical compound. An organism, however, represents a system consisting of units which may increase in number but remain of the same kind within their respective species. Only this fact allows for growth as increase in size of a system which remains specifically identical. The more important aspects of the subject of growth can be enumerated as follows:

1. Growth as the synthesis of substances. First of all, growth signifies the synthesis of high molecular weight organic compounds, especially proteins. This does not mean that the proteins are merely synthesized but that they are produced in such a way that their species, tissue, cellular, and even individual specificities are maintained.

2. Growth as identical reproduction. Among the components of the cell there are certain systems that show the capacity of self-reproduction. These may be termed the "elementary biological units." Genes and viruses are the most important representatives of these units. These systems further show the characteristic of directing cellular

processes, and in so far as genes are concerned, of directing particularly protein synthesis.

3. Growth as a cellular process. The growth of an organism takes place as a cellular process by cell multiplication (mitosis), increase in cell size, and the formation of intercellular substances. The excess of anabolism over catabolism leads to an increase in cell size. In spite of its continuous ability for synthesis, however, a cell cannot exceed a certain size. After this size has been reached, growth of the cell either ceases or cell division takes place.

4. Growth of the organism as a whole. Animal development leads to the establishment of certain shapes or forms. In early developmental stages, morphogenetic changes take place essentially by segregation and differentiation of tissues. In later stages they are mainly caused by relative growth, that is, different growth rates exhibited by various components of the body. In this sense the problem of organic form is a problem of growth.

B. Quantitative Definition

To arrive at a rational theory of growth, several principles must be considered: (1) Every living organism or living system in general is an open system maintaining itself by continuously importing and exporting, building up and breaking down component material. This is evident from the facts that (a) metabolism is a basic characteristic of living systems and (b) the turnover of building material takes place in the animal body. (2) Growth, that is, the increase in body mass in time, is not unlimited. As a general rule there is first a rapid increase which gradually slows down until the organism reaches a steady state or is "adult." What is measured as growth is the outcome of processes of considerable complexity whether envisioned from the biochemical, physiological, or morphological viewpoint. Furthermore, the growing organism undergoes changes in many respects, such as chemical composition, content of water, proteins,

and other compounds, the ratio between protein and fat, and changes in shape of the body. (3) Animal growth can be considered a result of a counteraction of the processes of anabolism and catabolism. There will be growth as long as anabolism, the building up of materials, prevails over catabolism, the breaking down of materials. The organism reaches a steady state if and when both processes are equal.

Any attempt to define a quantitative theory of growth must first emphasize an essential fact; no general growth curve exists which could be expressed in a universal formula. Rather, the course of growth is different in different organisms, in different dimensions, and in the organism as a whole and its parts. Such variations are to be expected because the physiological basis of growth is different in each case. Since growth depends ultimately on metabolism it should be within the scope of a theory of growth to derive these differences from comparative physiology and metabolism.

From the latter idea and a small number of empirical facts, a family of growth equations can be derived by deductive mathematical reasoning (Bertalanffy, 1960). The rate of synthesis and degradation of body material is related to body mass by a mathematical function. Experimental evidence shows that as a rule an approximation of the rate of physiological processes can be expressed as a function of body mass, as in the simple expression:

$$dw/dt = aw^m - bw^n$$

This expression simply states that a change in body weight for a given change in time is equal to the rate of anabolism, a, times weight to the power m minus the rate of catabolism, b, times weight to the power n. Thus the rate of change of body weight can be expressed as a power function of the body mass present. In other words the expression states: the change in body weight, w, is given by the difference between the processes of building up materials and breaking down materials; a and b are constants of anabolism and catabolism, respectively, and the exponents m and n indicate that the latter are proportional to some power of body weight. All that remains is to derive values for the exponents m and n.

1. CATABOLISM

The degradative processes expressed by the constant b are represented by the continuous loss of building material as it takes place in every living cell. Physiologically, this process means the continuous loss of cells and cell parts; biochemically, it means the continuous degradation of building materials as measured by isotope techniques, nitrogen excretion, or protein loss during starvation. In the adult state this degradation is compensated by regeneration so that the catabolic rate equals the turnover rate. The catabolic constant, b, of the growth equation can be equated to the turnover rate of total protein; however, turnover is not limited to protein. The constant in the equation refers to the resultant of all growth-limiting factors including manifest protein losses, changes in water content, changes in fat deposition, and factors of aging.

Consideration of the facts that a loss of weight during starvation is proportional to body weight and that the nitrogen or protein content remains nearly constant in starving animals, yields the physiologically plausible explanation that the rate of catabolism can be assumed to be directly proportional to body weight. The constant, n, can then be equated to 1 (one), without much loss of generality, thus reducing the number of unknown parameters in the expression.

2. ANABOLISM

The relationship between metabolic rate and body size belongs to the classical topics of physiology. According to the surface rule, weight-specific metabolic rate, that is, the intensity of metabolism as measured by oxygen consumption per unit weight, decreases with increasing body weight. If, however, metabolism is calculated per unit of body surface, approximately constant values occur for all body weights. The surface rule can be explained in terms of homeothermy. Since all warm-blooded animals heat their bodies to a temperature of approximately $37°$ C and since heat output takes place through the body surfaces, the same number of calories must be produced per unit surface in order to maintain the body temperature constant.

It is difficult to measure exactly the outer surfaces of animals and it is questionable whether these are responsible for some observed reductions in metabolic rate which occur with increases in body size. If two bodies are geometrically similar, however, any surface can be expressed as the two-thirds power of weight multiplied by a constant because the cube root of volume or weight yields a linear dimension and therefore its square has the dimension of a surface. Hence, a surface area of geometrically similar bodies can be obtained by multiplying the two-thirds power of weight by a suitable constant. This can be expressed in a formula:

$$s = bw^{2/3}$$

Accordingly, the surface rule of metabolism states that metabolic rate is proportional to the two-thirds power of weight. This rule was first stated for mammals in general; however, more recently the relationship between body size and metabolic rate has been shown to be more complex. In intraspecific comparisons, that is, comparisons within a species, a crude overall relation of respiration to the two-thirds power of weight can be found. However, in interspecific comparisons, that is, comparisons of mammals of different species, the metabolic rate varies with the three-fourths power of body weight rather than the two-thirds power as would correspond to the surface rule.

3. THE GROWTH EQUATION

With the foregoing discussion, the basic growth equation can be simplified to include the derived values for the exponents m and n:

$$dw/dt = aw^{2/3} - bw$$

The solution of this equation gives an equation for increase in weight and can be written in the following form:

$$w = [\sqrt[3]{w^*} - (\sqrt[3]{w^*} - \sqrt[3]{w_0})e^{-kt}]^3$$

In this equation w_0 equals weight at time, $t = 0$, w^* equals final weight and k equals $b/3$.

It is possible to derive an equation for growth in a linear dimension, that is, growth in length, from the equation for growth in weight. To make this transformation it is necessary to assume that growth is proportional, that is, that an increase in volume as represented by weight results from a proportional increase in all linear dimensions such that the volume or weight can be obtained by taking the cube of any one linear dimension. Such an equation would be:

$$l = l^* - (l^* - l_0) \, e^{-kt}$$

The symbols in this equation are similar to those in the weight equation, that is, l_0 equals the length at time, t_0, l^* equals final length and k is equal to $b/3$. It should be noted that these equations contain only empirical parameters, namely, the final values which can be determined from growth curves, and the catabolic constant which can be determined by physiological experiment.

The equation resulting from the insertion of m equal to two-thirds, that is, a metabolic rate proportional to two-thirds power of body weight, has the following main characteristics: (1) Growth rates increase and growth eventually attains a steady state. Thus, although there is a continuous increase in body weight with time until a steady state is reached there is a general decrease in the rate of growth. (2) The curves for weight growth and linear growth are characteristically different. The resultant curve for weight growth is sigmoid and has a point of inflection at about two thirds of the final weight. The curve for linear growth is a decaying exponential curve without a turning point, that is, there is no inflection point. (3) The inflection point in the weight curve

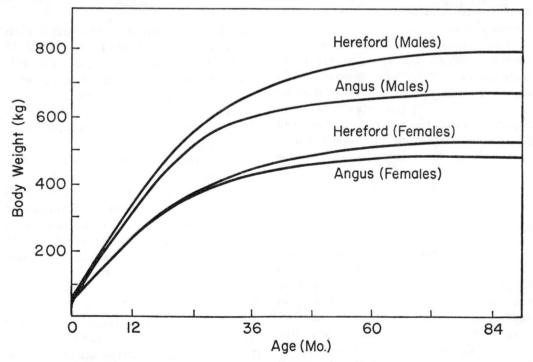

Fig. 3–1. Effect of sex on growth curves of Hereford and Angus cattle. Note the increased growth rate of the males. (*Adapted from Brown et al., 1956. Arkansas Agric. Exp. Sta. Bull. No. 570 & 571.*)

implies that the rate of growth in weight is initially increasing and beyond the inflection point the rate is decreasing (Fig. 3–1). This characteristic course of growth in body weight is easily understood. If an animal increases in size with little change in shape, its surfaces will increase approximately in proportion to the second power of length, but its volume and mass will increase approximately in proportion to the third power of length. Hence, the ratio between surface and weight is continually shifting in favor of weight and the ratio of surface to weight is continually decreasing. Consequently, as long as the animal is small, surface-proportional anabolism (anabolism proportional to weight to the two-thirds power) prevails over weight-proportional catabolism (catabolism based on weight to the first power) and the animal grows. The larger it grows the narrower the difference between anabolism and catabolism becomes and eventually a steady state will be reached wherein anabolism and catabolism balance each other.

C. Biological Implications

The discussion above demonstrates the complexity of the phenomenon referred to as growth. Great care must be taken in defining any particular growth characteristic that may be considered in a genetic discussion, or, for that matter, in any biological discourse involving growth. One must be aware not only of the chronological time period under consideration but also of the environmental conditions prevailing, the genetic or nutritional modifications which have occurred, and their effects on the growth process. It is obvious that manipulation of the growth pattern of an individual, whether it be by genetic or some other means, will result in a modification in the relative rate of anabolism and catabolism. In other words, the effect observed will be the result of a change in basic physiological and biochemical properties in the individual.

Some general comments are in order relative to the rate of growth expected at any given time, t. These can be visualized from an examination of the equation given above. Firstly, the rate of growth is dependent not only on the initial weight but also on the final weight that the organism will attain. Secondly, the rate of increase in weight is dependent on the amount of growth remaining after time, t. In other words, it is dependent on the difference between the weight at the time, t, and final weight.

An important consequence of the growth laws discussed is the equifinality encountered in animal growth. It is well known that the same species-characteristic final size may be reached after a temporary suspension of growth has occurred or when initial sizes at the beginning of the growth period differ. The first applies to animals whose diet was temporarily insufficient in quantity or in quality of nutrients such as carbohydrate, protein, or vitamins. Naturally, this equifinality of growth prevails only if no lasting damage remains after the period of malnutrition. The second holds true for animals from litters of varying size. The weight of an animal born in a large litter is generally lower at birth than that of an animal born in a small litter, but both animals attain the same weight eventually. These observations lend support to the previous statement that growth rate is not a function of time, that is, it does not automatically decrease as age increases, but rather, it is a function of the body size reached. Hence, after an inhibition has been lifted, growth resumes at a rate which corresponds to the body size at the time that the inhibition is lifted and not to the age at that time.

The principle of equifinality is also applicable to the balance of organs and of chemical constituents in the animal organism. A certain ratio between the size of the individual organs and the size of the body must prevail if the organism is to function properly. However, different species are born in widely different stages of development. In order to reach a functional balance of organs and constituents eventually, relative growth of the organs has to be different in different species. This implies that the same final result, a balance of organs and chemical constituents, will be reached in different ways by different organs in different species. Many such phenomena are not amenable to quantitative treatment; however, they can be envisioned as different expressions of the same basic principles.

II. HERITABILITY

A. Variation

When we consider that probably no two calves, no two piglets, no two blades of grass, nor any two men have ever been exactly alike, we get a glimpse of the resourcefulness of nature. It is fortunate that this is so because without this perpetuation of variability progress would be impossible. Variation is the raw material which the breeder has available for herd or flock improvement. Variation among animals in size, rate of growth, efficiency of feed utilization, carcass characteristics, wool quality, and color has been observed and recorded numerous times. Of two steers, one may gain 0.86 kg/day during a 120-day feeding period while another gains 1.16 kg/day during the same period; of two cows in a milking herd, one may produce 4080 kg of milk in a year, the other may produce 6800 kg; of two litters from two sows, one litter gains 55 kg from 225 kg of feed, the other litter requires only 160 kg of the same feed for a 55-kg gain.

Both heredity and environment are important in producing differences among individual animals. By studying variation alone we are not able to determine what proportion of the variation is certainly due to environment and what proportion is due to heredity. Nevertheless, the relative importance of hereditary and environmental influences on the variation in individual traits can be determined. Environmental differences include the nongenetic variation resulting from managerial, nutritional, and climatic differences. Some animals may have been born in large litters and others in small ones; some may have had better care than others; and some may have been born where the

temperature was extremely hot and others where it was extremely cold.

1. DISCONTINUOUS AND CONTINUOUS VARIATION

Traits generally can be grouped into two kinds: those which show qualitative differences, that is, differences which fall into a few clearly defined classes, and those which show quantitative differences, that is, those which show slight gradations from small to large with essentially no perceptible divisions existing between classes. The polled or horned condition in cattle is an example of a qualitative trait. This type of trait is said to exhibit discontinuous variation since it can be divided into discrete classes. Milk production, body size, and rate of gain in the feedlot are examples of quantitative traits, and these are said to show continuous variation. The variation is continuous in the sense that only the measuring device used in characterizing the value of the trait for a particular individual determines any discontinuities that we may observe. In fact, quantitative or continuous variation suggests many small gradations which integrate almost imperceptibly into each other.

The inheritance of qualitative traits is usually due to one or a few pairs of genes whose final expression is not greatly influenced by external environmental factors. Quantitative traits, which include most of the economically important traits in higher animals, are under the control of a very large number of genes, probably thousands. The presence of this large number of genes makes possible countless combinations, only one of which can be expected in any given individual animal. In addition, the hereditary variation results from the action of the genes and gene combinations in response to the environmental conditions provided to the individual during development. Thus the environmental conditions tend to have a blending effect on the phenotypic expression directed by the genes. The collective effect of the large number of genes and of environmental influences tends to shade differences between animals with slightly different genotypes causing the population of animals to appear continuous for a particular trait.

The nature of growth traits demands that they be measured quantitatively, for example, in pounds or kilograms. A single measurement can describe the weight of a pig. A group of pigs will vary in their weight; a complete description of the weight of the entire group of pigs requires an enumeration of the weight of each pig. Such a mass of data for a large herd is unmanageable and a statistical description of the population is desired. This description in its simplest form consists of a measure of central tendency, the mean, and a measure of the variability of the population, the variance. Such descriptions of the entire population of measurements are called parameters. Ordinarily we do not have measurements of the entire population. By studying a sample of observations from the population we can obtain estimates of the population parameters called statistics. The size of the sample studied will affect the reliability of a statistical description of the population.

The mean is simply the arithmetic average of all the values included in the sample. The degree of variability exhibited by a population can be estimated by calculating the average squared deviation of the individual measurements from the sample mean. Since the deviations are squared, variance is a positive value with zero as a lower limit. Formulas for the mean and variance can be given as:

$$\bar{x} = (X_1 + X_2 + X_3 + \ldots + X_n)/n$$

$$s^2 = [(X_1 - \bar{x})^2 + (X_2 + \bar{x})^{-2} + \ldots + (X_n - \bar{x})^{-2}]/(n-1)$$

where n represents the number of observations included in the sample.

B. Hereditary Variance

We have observed that both heredity and environment influence the observed variation in records of performance. The hereditary variation results from the action of the genes and gene combinations in response to the environmental conditions provided the individual members of the population. These environmental conditions may be minor with respect to the gross environment at the time; however, each individual is exposed to what might be called microenvironments which are acting at a particular time. Only the variation caused by heredity can be transmitted from parent to offspring or selected for inbreeding programs. Most important growth traits are usually influenced by an unknown number of genes. Estimates of the number of genes influencing such traits as body size vary from 10 to 200. Moreover, their mode of action, whether additive, multiplicative, interactive, etc., is also rarely known either chemically or descriptively. Fortunately, this is not an insurmountable handicap to animal breeders since it is possible to partition the hereditary variance into components representing the additive effects, multiplicative effects, and interactive effects of all genes affecting the character.

To provide a description of the subdivision of hereditary variance we can think of the phenotype of an individual for some particular trait, say body size, as being made up of a component due to the genotype of the individual and a component due to the environment to which that individual was exposed. If P equals the phenotype of the individual, G equals the genotype of the individual, and E equals the environmental effects, we can write the formula:

$$P = G + E$$

From this equation we can derive three components of variance,

$$\sigma_P^2 = \sigma_G^2 + \sigma_E^2$$

where σ_P^2 is the phenotypic variance observed in the population of individuals, σ_G^2 represents the heredity variance, that is, the total genetic variance causing the phenotype and σ_E^2 represents the variance in the phenotype due to environmental influences.

The factor G in the above formulations represents the total effects of all genes concerned with the particular trait in question.

It is possible to further subdivide the components of variance by subdividing the total genetic variance, σ_G^2, into components representing particular modes of gene action. Thus we can formulate

$$\sigma_G^2 = \sigma_A^2 + \sigma_D^2 + \sigma_I^2$$

in which σ_A^2 represents the genetic variance due to additive gene effects, σ_D^2 represents the genetic variation due to multiplicative effects (dominance deviations), and σ_I^2 represents the interaction effects (epistasis). We now can write a formula for the phenotype of an individual which includes the subdivision of modes of gene action. Thus,

$$P = A + D + I + E$$

with variance

$$\sigma_P^2 = \sigma_A^2 + \sigma_D^2 + \sigma_I^2 + \sigma_E^2$$

It is a common practice to combine the variance due to dominance effects and epistatic effects with the variance due to environmental influences such that

$$\sigma_P^2 = \sigma_A^2 + \sigma_E'^2$$

This formulation is desirable since selection programs are effective only when they are based on the additive portion of genetic variability.

C. Heritability Estimates

Heritability indicates the percentage of response or improvement expected by a breeder from exerting a given amount of selection pressure on a trait in his herd or flock. Mathematically, heritability has two definitions: (1) heritability in the broad sense is the ratio of the total genetic variance to the total phenotypic variance, that is, $h_b^2 = \sigma_G^2/\sigma_P^2$, and (2) heritability in the narrow sense is the ratio of the additive genetic variance to the total phenotypic variance, that is, $h_n^2 = \sigma_A^2/\sigma_P^2$. It is obvious that these two estimates of the proportion of total variance which is due to heredity represent a maximum and a minimum. Most heritability estimates calculated by animal breeders fall somewhere between

these two defined expectations due to the errors involved in attempting to calculate the preferred estimate, heritability in the narrow sense.

An example of the use of heritability is in order. If in a breeder's unselected beef herd the average weight at birth is 215 kg and bulls averaging 250 kg are selected from the herd to mate with selected heifers averaging 230 kg, the average selection differential between the selected group and the general average is $[(250 + 230)/2] - 215 = 25$ kg. If the heritability of birth weight was 100 percent, the progeny of the selected group of parents would average 25 kg above the mean of progeny from an unselected herd. However, the average of the offspring is usually nearer to the mean of the herd than it is to that of their parents. In other words, heredity has less than 100 percent effect. A realis-

Table 3–1. Some Average Heritability Estimates of Farm Animals

(Summarized from many published sources.)

Low Heritability 10–20%	Medium Heritability 30–40%	High Heritability 50–60%
Beef Cattle		
	Birth weight	Weight at 18 mos. on pasture
	Weaning weight	Mature cow weight
	Pasture gain at 18 mos.	Feed lot gain
	Tenderness	Efficiency of feed lot gain
	Carcass grade	Dressing percent Rib eye area
Swine		
Weaning weight of litter	Pig weight at 5 to 6 mos.	Length of body
	Post weaning daily gain	Back-fat thickness
	Efficiency of gain	Yield of lean cuts
	Dressing percent	Rib eye area
Sheep		
Birth weight	Weaning weight	Mature staple length
	Yearling weight	
	Av. daily gain	
	Grease-fleece weight	
	Fiber diameter	

tic estimate of the heritability of birth weight in cattle would be 30 percent. Then we would expect the progeny of the selected group to average $25 \times 0.30 = 7.5$ kg above the mean of the progeny of the unselected herd. Some average heritability estimates for growth traits of farm animals are given in Table 3–1.

III. MATING SYSTEMS

The study of mating systems has to do with an aspect of genetics and animal improvement that is quite apart from the selection process. There are many ways in which breeding animals may be paired for matings. They may be paired simply by picking each member of the pair at random. This system is often used for control populations in genetic experiments. They may be paired by relationships, for example, brother to sister, sire to daughter, son to dam; these matings produce progeny which are called inbred. They may be paired to be different genetically, as by mating a Shorthorn bull to a Brahman cow; these matings produce species hybrids or outbreds. They may be paired in order to be dissimilar in appearance, such as mating a lean, short boar to a fat long gilt; this pairing tends to produce intermediate types of offspring and is sometimes called corrective mating. Phenotypically similar animals may be paired such as would occur by mating selected individuals; such pairs cause more extreme types in the total population. Different mating systems are used to accomplish different ends. In most programs, selection is used in conjunction with a particular mating system.

Inbreeding and crossbreeding or hybridizing are the best known mating systems. They are the opposite of one another in pairing procedure and in their genetic effects.

A. Inbreeding

Inbreeding increases the proportion of gene pairs that are homozygous and decreases the proportion that are heterozygous. Inbreeding itself does not change gene frequency, but in small populations the gene frequency fluctuates rather extremely, and, by chance, certain genes become lost and others fixed; all animals in the population become homozygous for that allele. In large populations the fluctuations of gene frequency are less extreme and few of the alleles are lost. As inbreeding progresses without selection, a large population becomes a series of subpopulations or lines, each likely to be homozygous for a number of different alleles.

Just what effect selection has on the degree of homozygosity in a specific situation is problematic. A number of situations have been reported in which heterozygosity in inbred lines is apparently greater than would be expected. These cases point to the possibility that, in some instances, selection during inbreeding favors the heterozygote over either homozygote. How frequently heterozygote superiority is manifest is not known; it is likely to be more frequent for traits influencing reproductive capacity than growth characteristics. Inbreeding on the average tends to depress performance. Since there may be heterozygote superiority, it is probable that selection for performance traits during inbreeding favors animals more heterozygous than average. Selection continued over a period of several generations might result in substantially more heterozygosity in an inbred line than would be expected from a calculated inbreeding coefficient if (1) gene pairs with major effects exhibited heterozygote superiority and (2) the inbreeding system was a mild one permitting selection for traits other than reproductive fitness.

The coefficient of inbreeding expected from the mating of some relatives is shown in Table 3–2. The most rapid system of inbreeding is by mating full brother to full sister; the relationship between the mating pairs is one half, resulting in an inbreeding coefficient of one quarter in the progeny. The value of one quarter indicates that on the average, 25 percent of the loci will be homozygous for one allele or another. The effects of continued inbreeding over subsequent generations are cumulative. In the

Table 3–2. Coefficient of Inbreeding from
Mating of Relatives

Mating	Relationship between Mates	Inbreeding Coefficient of Progeny
Uncle-niece	.125	.062
Half brother-sister	.250	.125
Sire-daughter	.500	.250
Full brother-sister	.500	.250
Full brother-sister second generation	.600	.375
Full brother-sister third generation	.727	.500

third generation of full brother-sister mating the relationship between the mating pairs has increased to .727, with a resulting inbreeding coefficient of the progeny of one half.

During the past 20 to 25 years, considerable experimental work has been reported on inbreeding effects in swine, cattle, and sheep. Evidence from these species is reinforced by a larger amount of experimental results involving laboratory animals such as mice. Considering the combined effects of the inbreeding of the dam and the litter, a reduction in birth weight of about 0.015 kg (1.2 percent) per pig can be expected for each 10 percent increase in the inbreeding coefficient. A reduction of approximately 1.1 kg (1.8 percent) per 10 percent increase in inbreeding can be expected for average weight at 154 days of age (Dickerson et al., 1954; Bereskin et al., 1968). Similar results have been found for beef cattle; a reduction of 4 percent in weaning weight and 3 percent in yearling weight can be anticipated for each 10 percent increase in inbreeding (Dinkel et al., 1968).

With the exception of a few cases in laboratory animals and a few inbred lines of swine, the results of experiments unanimously indicate a reduction in reproductive efficiency with inbreeding. Generally, an extremely high percentage of the inbred lines started in such experiments have been discarded because of low reproductive performance.

Efforts to find the physiological basis for reduced reproductive efficiency have suggested that inbreeding delays puberty in both sexes, increases early embryonic death rates, reduces the number of ova shed by the females, and slows testicular developments. In spite of the rather drastic average effects of inbreeding on reproduction, certain inbred lines have performed reasonably well.

Although vigor is difficult to define quantitatively, observations indicate that inbreeding lowers vitality. Some inbred lines show major increases in death rate with increased inbreeding; death losses in other lines have been little affected. The additional observation has been made that inbred farm animals are much more susceptible to critical environmental conditions than are noninbred animals.

Although growth rates of most species are apparently not severely affected by inbreeding, the general conclusion from all experimental results is that the effect of inbreeding is to reduce the net performance of the population. It is the effect on net performance that a breeder must consider if he undertakes a particular inbreeding system. The reductions are definitely large enough to discourage the routine use of inbreeding in commercial herds and flocks.

B. Crossbreeding

In spite of occasional good results obtained from inbreeding, it is becoming increasingly apparent that the development of highly productive inbreds in farm animals is not likely. Rather, inbreeding must find its usefulness as an aid in producing seed stock which can be used as parents for the production of crossbred or outbred commercial animals. Seed stocks must be developed with sufficient attention paid to the performance of the inbreds so that the results of crossing will be predictable. The use of inbreds for commercial production is being exploited more widely with poultry than with other farm animals.

In livestock, reproductive rate and weaning weight have been indicated by some to be the most important traits economically and these are traits that show a considerable amount of heterosis. The general rule is that traits of high heritability do not show as high a degree of heterosis as do traits of low heritability. Similarly, traits that show little heterosis also exhibit a low inbreeding depression. We may then accept the general proposition that inbreeding and heterosis are opposite and exclusive of one another in their effects. They may be considered as a gradation of the effects from a high degree of homozygosity to a high degree of heterozygosity. For traits that show heterosis, the more distant the relationship between the animals in the cross, the higher is the degree of hybrid vigor realized. Crosses between breeds exhibit greater hybrid vigor than crosses between families within a breed. Crosses between inbred lines from different breeds may give even greater heterosis. However, if crosses are made between extremely unrelated animals, as between different species or different genera, there may be a high rate of embryonic loss and other incapabilities that limit the cross.

Outcrossing is the term applied to mating unrelated animals within the same pure breed. It is the most common mating system in use by American breeders of seed stock herds. The practice is combined with selection in the sense that the breeder chooses the best available but unrelated sires from another herd for use on females of his herd in an outcrossing system. Although unspectacular and in many ways less effective than some other breeding systems it has nevertheless been responsible for a high percentage of the changes in livestock since breeds evolved. Many past accomplishments can be used to illustrate the power of this breeding system.

Crossbreeding involves the mating of animals from different established breeds. Technically the term applies only to the first crosses of purebreds but it is generally applied to most widely used systems. These include the crisscrossing of two breeds or the rotational crossing of three or more breeds,

the crossing of purebred sires of one breed to high grade females of another, and to the crossing of crossbred sires to crossbred females.

Crossbreeding for commercial production may be practiced for either or both of two reasons. The first is to take advantage of hybrid vigor, which may make the crossbreds more productive than either of the parental breeds even though the latter are of similar type. That is what a commercial hog producer is doing when he uses two or more American breeds of hogs in a cross and what the commercial cattle producer is doing when he uses two or more European breeds of cattle. The second reason for crossbreeding is to take advantage of the good qualities of two or more breeds of distinctly different types. This is what the Western sheepman is doing when he crosses the vigorous, hardy, Rambouillet ewes to Hampshire or Suffolk rams. He hopes to get improved carcass quality and increased growth rate in the lambs from their sires while continuing to take advantage of the good range qualities and the mothering ability of the Rambouillet ewes. Similar reasoning is used by the commercial beef producer when he crosses a beef breed such as the Hereford with a dairy breed such as the Holstein-Friesian. He is attempting to improve the growth rate and milk production of the offspring by using the Holstein and still retain the good carcass qualities and ranging ability of the Hereford.

In considering crossbreeding, it must be emphasized that the highest productivity in a commercial livestock enterprise usually depends upon maximizing both heterosis and the frequency of occurrence of desirable genes with additive effects. This suggests that selection must be practiced within the breeds intended for use in a cross. Except for reproductive characters, most traits of economic importance are influenced more by additive gene action than by heterosis. As an example, growth rate in cattle is highly heritable in the narrow sense. Some breeds have average growth rates of 25 to 50 percent greater than those of other breeds. As a contrast to these wide differences between breeds, increases in growth rate through heterosis in most cross-breeding experiments have been only 5 percent or less. As a more extreme example, many carcass traits such as back-fat thickness in swine are highly heritable but exhibit little or no heterosis. For highly heritable characters such as this, breed differences are transmitted to the crossbreds as well as the purebred progeny. Since little or no heterosis can be expected from crossing two breeds which are widely different in growth rate, we cannot expect the progeny of such a cross to exceed the rate of the better parent breed. The gene action will combine in an additive manner such that the progeny of the cross will be near the average of the two parental breeds. Such a definition of heterosis of course implies that heterosis is of economic value only if the productivity of the progeny of the cross exceeds that of the better parental breed.

One of the characteristics of crossbreds is that although they exceed the average of their purebred parents only slightly for any one trait, by virtue of the fact that small advantages in each trait are cumulative, they greatly exceed parental averages in total net merit. This is particularly true with reference to growth traits when the dam of the progeny is herself a crossbred animal. Since most reproductive characters such as litter size and milk production respond to crossing, the progeny of crossbred dams tend to show increased body weight at weaning and increased growth rate due to the improved mothering ability of the dams. This has been observed particularly in swine, for example (O'Ferrall et al., 1968).

Table 3–3 presents a summary of results obtained from crossbreeding swine with regard to size and weight of litters at three stages of development. It may be concluded that crossbred mothers nurse their litters more effectively than do purebred mothers. Growth rate of the young is about equal for first generation crossbreds and backcrosses or three-breed crosses as soon as the litter is independent of the mother for its feed supply. When these advantages are combined with the heterotic effect achieved for litter size, the net value of a crossbred herd is much superior to that realized from pure breeding. It must be emphasized that certain breed crosses show greater heterosis than others, probably because of differences in genetic diversity and degree of heterozygosity.

IV. SELECTION

The objective of selection for any performance trait is to increase the frequency of desirable genes affecting that trait. This is accomplished by selecting animals that are above herd average in genetic merit. Differential reproduction is the basis for change in gene frequency and genetic improvement

Table 3–3. Crossbreeding in Swine. Performance of Single Cross, Backcross, and Three-Breed Cross Progeny Compared with Parental Purebreds

(From the Literature.)

		Advantage of Crosses (Percent of Purebreds)		
	Purebread Average	Single Crosses	First Backcrosses	Three-Breed Crosses
Litter weight (kg)				
At birth	10.3	107	115	120
At weaning	91.2	112	133	142
At 20 weeks	266.1	111	116	115
Litter size				
At birth	8.7	105	109	114
At weaning	6.2	108	122	124
At 20 weeks	6.4	109	109	109

since superior animals have a higher percentage of desirable genes. The increase of desirable genes in one generation is added to those of the previous generation; for this reason genetic improvement tends to be permanent.

A. Factors Affecting Rate of Improvement

The factors affecting the rate of improvement in growth characteristics from selection are heritability, selection differential, genetic association among traits, and generation interval. The rate of improvement is also affected by the method of selection used and the mating system employed; however, the effect of these factors is indirect in the sense that the method of selection and mating system used will affect the selection differential and generation interval.

1. Heritability

Heritability is the proportion of the differences between animals that are transmitted to the offspring; thus it is the proportion of the total variation that is due to additive gene effects. The higher the heritability for any trait, the greater the rate of genetic improvement or the more effective selection will be for that trait. For traits of equal economic value, those with high heritability should receive more attention in selection than those with low heritability. Every attempt should be made to subject all animals from which selections are made to as nearly the same environment as possible. This practice will increase the effectiveness of selection.

The heritability of any trait can be expected to vary slightly in different herds, depending on the genetic variability present and the uniformity of the environment. The heritability estimates in Table 3–1 were obtained under controlled environmental conditions and represent average estimates. However, estimates from different research centers have been reasonably consistent. These heritabilities represent average expectations for many herds provided the general

environment is similar for all cattle within the herd. The estimates indicate that selection should be reasonably effective for most performance traits.

2. Selection Differential

Selection differential is a difference between the selected individuals and the average of all animals from which they were selected. If the average weaning weight of a herd of swine is 20 kg, and the individuals retained for breeding average 23 kg, the selection differential is 3 kg. Selection differential is determined by the proportion of progeny needed for replacements, the number of traits considered in selection, and the differences that exist among the animals in the herd.

Reproductive rate greatly affects the selection differential since up to 40 percent of the females must be kept for replacements in order to maintain the size of the herd. An even higher proportion of females must be retained if the herd is to be expanded. The larger the number of replacements that must be retained, the lower will be the selection differential for each trait. Consequently, the greatest opportunity for selection exists among males because a smaller percentage must be saved for replacement purposes. It must be remembered, however, that delaying the replacement of an animal extends the generation interval and thus reduces the improvement that can be made per year.

3. Genetic Association Among Traits

A genetic correlation among traits can exist as the result of genes favorable for the expression of one trait tending to be either favorable (positive correlation) or unfavorable (negative correlation) for the expression of another trait. For example, a positive correlation exists between rate of gain in beef cattle and the efficiency of gain and a negative correlation exists between the area of the loin-eye and length in swine. If the association is favorable among the traits on which selection is based, the rate of improvement in total merit is increased. Conversely, if genetic antagonism exists among the traits, the rate of improvement from selection is reduced. Often it is possible to eliminate a trait from consideration in a selection program if it is highly correlated genetically with a second trait. In the first example given, selection for rate of gain in beef cattle will result in a strongly correlated improvement in efficiency of gain; this association has an advantage for two reasons: it reduces the number of traits which must be included in the selection program and it eliminates a trait which is rather difficult to measure (Koch *et al.*, 1963). However, if two characters are negatively associated genetically and both are of economic importance, both traits must be included in the selection program to insure that the improvement in one trait does not result in a decline in performance for the other trait.

Information concerning the existence of genetic correlations must be considered carefully in designing a selection program. The existence of negative genetic correlations is not always detrimental. For example, a negative association between rate of gain in beef cattle or swine and the length of time required to reach market weight would be a favorable negative genetic correlation. Available information indicates that favorable associations exist between most of the economically important traits in farm animals. The traits studied include those that contribute to both productive efficiency and carcass merit.

4. Generation Interval

The fourth major factor that influences rate of improvement from selection is the generation interval—the average age of all parents when their progeny are born. The generation interval averages approximately 18 to 24 months in most swine herds and $4\frac{1}{2}$ to 6 years in most beef herds.

The progress made per generation in any trait is equal to the superiority of the selected individuals over the average of the population from which they came (selection differ-

ential) multiplied by the heritability of the trait. The improvement per year can be obtained by dividing that estimate by the average length of a generation. For example:

$$\text{the annual progress for a trait} = \frac{\text{heritability} \times \text{selection differential}}{\text{generation interval}}$$

If heritability of weaning weight in sheep is 40 percent, the selected sheep (males and females) average 3.0 kg more than the average of all animals, and the generation interval is four years, the rate of improvement per year in weaning weight would be $(0.40 \times 3.0)/4$, or 0.3 kg. It is desirable, therefore, to keep the generation interval as short as possible consistent with obtaining near maximum selection differentials.

B. Methods of Selection

Selection may be based on individual performance information, pedigree information, progeny test or family performance information, or on a combination of all three.

Selection on an individual's own performance will result in the most rapid improvement when heritability is high. An example of such a trait is growth rate. The advantage of selecting on individual performance is that it permits a rapid turnover of generations and thus shortens the generation interval.

Pedigree information is most useful in selecting among young animals before their own performance or their progeny's performance is known. Pedigree information may also be used in selecting for characters that are measured late in life such as longevity and resistance to disease or in selecting for traits expressed only in one sex such as mothering ability—for example, selecting bulls from cows that have produced calves with a high average weaning weight. In using pedigree information, only the closest relatives should receive consideration since the more distant relatives can influence the heredity of the individual only through the close relatives such as the sire and dam.

Information about the poorly performing ancestors should be considered along with the well performing ones because, on the average, the influence of all grandparents on the herdity of the individual is the same (Lush, 1947). Pedigree information should be given less attention after information about an individual's own performance or its progeny's performance is available.

Use of progeny test information results in the most accurate selection if the test is adequate. Progeny tests are most needed in selecting: for carcass traits if good indicators are not available in the live animal; for sex-limited traits such as mothering ability since individual performance information is not available on males; and for traits with low heritability. The disadvantages of progeny testing include the less intense culling possible because of the small proportion of animals that can be adequately progeny tested, the longer generation interval required to obtain progeny test information, and the decreased accuracy as compared to individual performance if not enough progeny are tested or if they are improperly evaluated.

In summary, all three types of information —individual performance, pedigree, and progeny test—should be used in selecting in most economic species. Pedigree information is useful if selections have to be made early in life or in selecting for traits expressed only in one sex. Individual performance information should be used for traits with high heritability that can be measured in the individual. After it becomes available, progeny test information should be used for all traits. It is most needed for traits of low heritability, for certain carcass traits, and for sex-limited traits. A good policy is to make initial selections on the basis of pedigree and individual performance and let the extent of an individual's use in a herd in later years be determined by progeny test performance. The latter procedure is most useful in evaluating sires.

Three types of selection can be employed when more than one trait is involved in the selection program—tandem selection, selection based on independent culling levels, and selection based on an index of net merit.

Tandem selection is selection for one trait at a time. When a desired level of performance is reached for this trait, a second trait is given primary emphasis. When the desired level of performance is reached for the second trait, a third trait is given primary emphasis, and so on. Tandem selection is the least effective of the three types and is not recommended. Its major disadvantage is that, by selecting for only one trait at a time, some animals extremely poor in other traits will be retained as replacements. In addition, relaxing selection for one trait while selection is practiced for a second may result in some decline in the performance of the initial trait.

Independent culling levels require that specific levels of performance be attained for each trait before an animal is kept as a replacement. This is the second most effective type of selection; however, it has a major disadvantage. In requiring specific levels of performance in all traits, it does not allow for slightly substandard performance in one trait to be offset by superior performance in another.

Selection based on an index of net merit gives weight to the various traits in proportion to their relative economic importance and their heritability, and takes into account any genetic association existing among the various traits. This is the most effective type of selection because it allows slightly substandard performance in one trait to be offset by outstanding performance in another. Also, by giving additional weight to traits of higher heritability or of greater economic importance, greater improvement in net merit can be attained. The use of the index or some modification of it is the preferred type in most situations.

Although the type of mating system followed is not considered part of the selection method, it is an integral part of herd management during any selection program. The principal consideration given to the mating system is the control of inbreeding. Many breeders fear the consequences of inbreeding because it intensifies what is already present in the herd, including poor traits as well as good. However, inbreeding is not the cause of poor traits, because the genes responsible for any undesirable effects were already present. Inbreeding normally has some adverse effect on most performance traits and results in some reduction in general vigor; however, herds of reasonable size in which several sires are used can be maintained closed to outside breeding for relatively long periods without any appreciable increase in inbreeding or decline in performance associated with inbreeding.

C. Results of Selection

Selection should be made directly for those traits or characteristics that will result in either (1) improved production efficiency, that is, a greater output of meat per unit of feed consumed, or (2) improved qualtiy of product. Every industry must produce a product of the quality desired by consumers at a price they can afford. To date, the stock industry has done well in these respects. There is reason to believe that the adoption of modern selection procedures and the abandonment of some practices that have hindered progress in the past will result in further improvement of the genetic potential of our farm species.

The basic problem of selection is one of estimating as accurately as possible the genotypes of the available potential breeding animals. The animals with probable superior transmitting ability for those characters considered important must then be intermated in combinations calculated to result in maximum improvement in the next generation. In commercial herds, interest centers on production of both superior market animals and replacement females that will improve performance in the next generation. In purebred or seed stock herds the sole ob-

jective is to produce animals of improved breeding value.

Reportedly, the characters included in standards of excellence by breed associations have in some cases been associated with higher production, but others are seemingly unrelated to productive traits. For example, there are indications that open-faced ewes are more productive than woolly-faced ewes. However, significant relationships between coat color and weaning traits have not been found in sheep and the number of skin folds is seemingly not associated with production (Terrill, 1949; Terrill & Kyle, 1952). It has been shown that registration limitations can seriously restrict the effectiveness of selection in small flocks of sheep. Selection for weaning weight and for fleece weight has been relatively ineffective in changing production or creating differences between selected lines and unselected lines since little improvement can be achieved in these traits (Botkin & Stratton, 1967). The main reason for lack of effectiveness of selection in one study was that only 50 percent of all the offspring were eligible for registration in the breed association. This seriously limited the selection pressure that could be obtained. It is interesting to note that in this study it was found that lambs disqualified for registration averaged about the same in weaning weight as lambs eligible for registration.

Surveys have almost always shown that American consumers want meat with a high percentage of lean, a minimum of waste fat, sufficient marbling, and a high degree of tenderness. Since fat may be associated with some degree of tenderness and flavor, all these requirements are difficult to meet. Within the carcass, some cuts—the rib, loin and round—are preferred by consumers. It would be desirable to produce livestock with as high a percentage of these cuts as possible. Only in very recent years has serious attention been given to the possibility of using scientific approaches in attempts to breed for improved carcass quality in livestock species. Inevitably, research has exposed fallacies in the traditional standards without developing evidence that will present clear-cut recommendations to improve procedures. Carcass evaluation is, however, a fast-developing field and the alert student will give careful consideration to new developments.

A significant development in this respect was the acceptance of the back-fat probe technique in live hogs which gave a practical method by which swine breeders could obtain objective estimates of the back-fat thickness of potential breeding stock (Hazel & Kline, 1952). This technique can replace, to a large extent, the estimation of carcass merit in swine from characteristics of body conformation and type, which were the only important criteria available to breeders before this time. Although selection based on these latter criteria was fairly successful, progressive breeders recognized that more accurate measures were needed to continue improvement. Selection experiments have

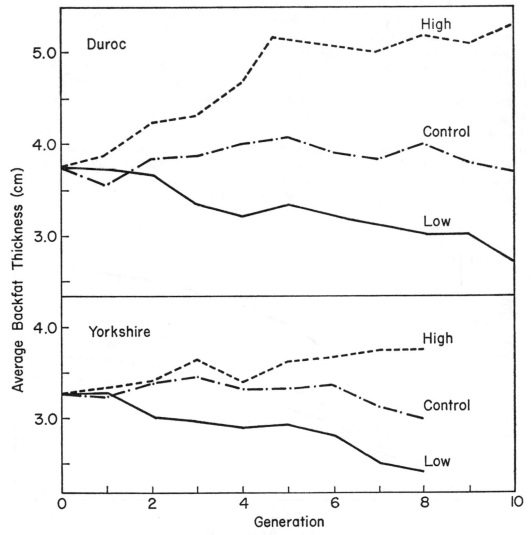

Fig. 3–2. Diagram showing the results of selection for high and low fatness in Duroc and Yorkshire swine. Selection for high fatness (*top line*). Control. No selection for fatness (*middle line*). Selection for low fatness (*bottom line*). (*Data from Hetzer & Harvey, 1967. J. Anim. Sci. 21:1244.*)

been carried out using the back-fat probe technique. The results have been very successful and indicate that selection practiced for decreased back-fat thickness in swine is highly effective when this technique is used (Fig. 3–2). One study involving five generations of selection with Poland China pigs resulted in approximately a 20 percent decrease in the back-fat thickness as measured by the probe technique (Gray *et al.*, 1968). High genetic correlations exist between the various locations on the shoulder, the loin, and the ham, where a back-fat probe may be taken, and indicate that many of the same genes affect back-fat at these sites.

Selection for increased body weight results in a modification of the growth pattern of the individuals. Figure 3–3 compares the growth curves of mice selected for increased body weight with those of unselected random-bred and inbred mice. The general pattern of growth is similar for both the random-bred and inbred mice; the inbred mice are slightly smaller due to inbreeding depression. The selected mice, however,

grow much more rapidly and consequently reach a greater mature size. The duration of the growth period has not been altered by selection. All the mice approach maturity, that is, a steady state, at approximately the same chronological age (Fig. 3–4) (Gall & Kyle, 1968).

Under closed mating systems, experiments involving selection for heritable quantitative growth traits in laboratory animals including mice, rats, the fruit fly (*Drosophila melanogaster*), and the flour beetle (*Tribolium castaneum*) have often shown responses for 20 to 30 generations. The level at which failure to respond occurs is often referred to as a plateau and such populations are said to have "plateaued." Whether any economically important growth traits of farm livestock have reached a point where selection is no longer effective is not certain. The evidence available would indicate that they have not, and that carefully designed and executed selection programs will be most effective.

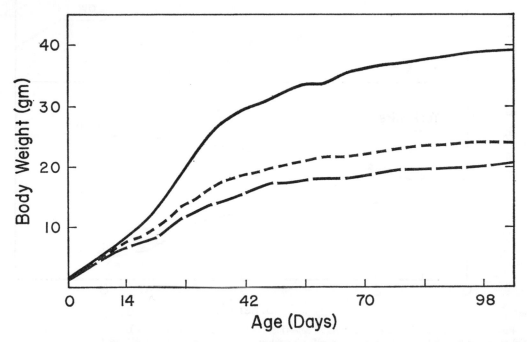

Fɪɢ. 3–3. Diagram illustrating the growth curves of selected and unselected mice. Average body weight of males selected for high body weight at 60 days of age (*top line*). Body weight of unselected random-bred mice (*middle line*). Body weight of unselected inbred mice (*bottom line*).

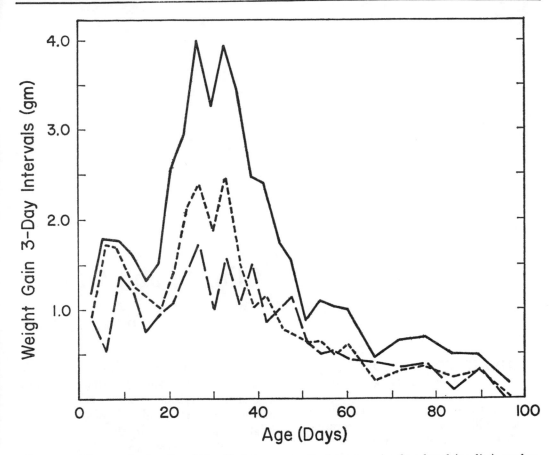

Fig. 3–4. Diagram showing the relationship between rate of gain and age for the selected (*top line*), random-bred (*middle line*), and inbred mice (*bottom line*) shown in Figure 3–3. The rate but not the duration of growth was influenced by selection.

REFERENCES

Bereskin, B., Shelby, C. E., Rowe, K. E., Urban, W. E. Jr., Blunn, C. T., Chapman, A. B., Garwood, V. A., Hazel, L. N., Lasley, J. F., Magee, W. T., Mc-Carty, J. W., & Whatley, J. A., Jr. (1968). Inbreeding and swine productivity traits. J. Anim. Sci. 27:339–350.

Bertalanffy, Von L. (1960). Principles and Theory of Growth. In: *Fundamental Aspects of Normal and Malignant Growth*. W. W. Nowinski (Ed.), New York, Elsevier Publishing Co.

Botkin, M. T., & Stratton, P. O. (1967). Factors limiting selection effectiveness in small flocks of Columbias and Corriedales. J. Anim. Sci. 26: 971–975.

Dickerson, G. E., Blunn, C. T., Chapman, A. B., Kottman, R. M., Krider, J. L., Warwick, E. J., & Whatley, J. A., Jr. (1954). Evaluation of selection in developing inbred lines of swine. Mo. Agr. Exp. Sta. Res. Bull. No. 551.

Dinkel, C. A., Busch, D. A., Minyard, J. H., & Trevillyan, W. R. (1968). Effects of inbreeding on growth and conformation of beef cattle. J. Anim. Sci. 27:313–322.

Gall, G. A. E., & Kyle, W. H. (1968). Growth of the laboratory mouse. Theor. Appl. Genetics 38: 304–308.

Gray, R. C., Tribble, L. F., Day, B. N., & Lasley, J. F. (1968). Results of five generations of selection for low back fat thickness in swine. J. Anim. Sci. 27: 331–335.

Hazel, L. N., & Kline, E. A. (1952). Mechanical measurement of fatness and carcass value on live hogs. J. Anim. Sci. 11:313–318.

Koch, R. M., Swiger, L. A., Chambers, D., & Gregory, K. E. (1963). Efficiency of feed use in beef cattle. J. Anim. Sci. 22:486–494.

Lush, J. L. (1947). Family merit and individual merit as bases for selection. Amer. Natur. 81:241–261, 362–379.

O'Ferrall, G. J. M., Hetzer, H. O., & Gaines, J. A. (1968). Heterosis in preweaning traits of swine. J. Anim. Sci. 27:17–21.

Terrill, C. E. (1949). The relation of face covering to lamb and wool production in range Rambouillet ewes. J. Anim. Sci. 8:353–361.

Terrill, C. E., & Kyle, W. H. (1952). The effect of selection on the incidence and heritability of neck folds in Rambouillet weaning lambs. J. Anim. Sci. 11: 744 (Abstract).

Chapter 4

Climate and Growth

By M. F. Fuller

I. CLIMATE, PLANT GROWTH, AND ANIMAL LIFE

IT is little wonder that in the dawn of human civilization sun worship became established as an early form of religion, for even the most primitive human intelligence could associate the rising of the sun with the daily flood of warmth and light that is the *sine qua non* of life. The sun is the ultimate source of all the energy in our food, which, in the process of growth, we convert into the substance of our bodies. Our climate, too, owes its nature and vagaries to the sun, which, as it heats the land and the seas, creates the variations of temperature and pressure which cause winds to blow, clouds to form, and rain to fall. The complexity of these patterns throughout the earth is immense; nevertheless, they do exist, and, to some extent, their consequences can be predicted.

A. Weather and Climate

The distinction between weather and climate is primarily one of time scale. Weather can be defined as a specific, temporary combination of certain meteorological factors. These include air temperature, humidity, air movement (wind speed), radiation (both visible and invisible), barometric pressure and precipitation (rain, snow). Weather is also characterized by the rate at which each of these factors varies; living organisms respond not only to changing conditions but also to rates of change. Weather changes continuously, from hour to hour, from day to day, and from week to week. In the longer term these variations become smaller and smaller, so that if we compare one year with the next we find that the average air temperature or rainfall is quite similar. This long-term pattern is climate. Even in respect to climate, the variability factor is important; two regions with the same average meteorological conditions can be very different as places to live in if one is subject to rapid changes and the other to relatively steady conditions.

These regional and seasonal variations of climate produce effects that are more far-reaching than is apparent from a simple meteorological record. The age-long action of weather on the native landscape has built up a characteristic ecology, in which climate and the original geological structure of the land have interacted to give a particular type of soil. The soil interacting with the climate has allowed characteristic forms of plant life to flourish. Upon this plant life certain kinds of animals have come to depend for their food. This chain reaction in the evolution of a natural habitat, which connects climate successively with soils, plants, and animals, may be extended to include man, who has latterly demonstrated his particular abilities not only to exploit, but also to control, at least partially, his environment. These general relationships are illustrated by Figure 4–1.

By its control of the food supply, the climate of a region has a profound influence on the animals living there. Animal life is geared to the seasonal rhythm of plant growth and decline. The birth of the young in spring, for example, coincides with the onset of rapid plant growth, and the shortage of plant food in winter or in the dry season is the most important determinant of animal populations.

In addition to its effect on food supplies, climate affects living creatures directly through its control of their heat exchanges,

and thence, through nervous, hormonal, and behavioral mediation, all productive processes. It is this aspect which is the principal subject of this chapter.

It is worth noting here that direct and indirect effects of climate often go hand in hand. Here in Scotland, for example, it is a sunny day of early summer, the temperature is around 20° C, and the grass is growing rapidly. Animals grazing outside have ample food of the highest quality and are subject to no climatic stress. Four months ago, however, the land was covered with snow, the temperature barely rose above the freezing point, and only the remains of last year's herbage were to be found beneath the snow. For wild animals the consequence of these fluctuations can be extremely severe. For domestic animals, with which this book is mainly concerned, the situation is more favorable. In his efforts to increase their productivity, man affords his stock a measure of protection from the vicissitudes of the weather and from the seasonal pattern of alternating glut and dearth of food. In spite of man's intervention, however, in many regions of the world large numbers of domes-

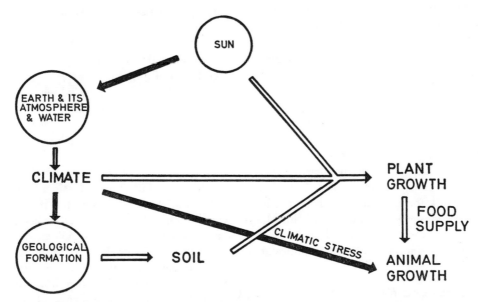

Fig. 4–1. Simplified scheme of the chain of cause and effect by which climate affects animal growth. Black arrows = 'act upon,' and white arrows = 'produce,'

tic animals still live with little protection from either the direct or indirect stresses of their climatic environment.

B. Components of Climate: The Thermal Environment

Within the broad definition of climate it is convenient to isolate particular factors or groups of factors which are of special importance to the animal. Light, for example, and the seasonal changes of day length have an important role in the control of the breeding cycle of sheep, of coat shedding in cattle and sheep, and of egg production and growth of poultry. Climatic factors are also known to be involved in the spread of certain kinds of infection and in animals' susceptibility to disease. The most important grouping of climatic factors, however, is that which defines the animal's thermal environment. It is comprised of air temperature, humidity, air movement, and infrared radiation, both from the sun and from the environment at large. The thermal environment also includes the thermal characteristics of those objects in the environment with which the animal comes into contact.

These factors create what Brody (1945) called the "physiologically-effective temperature" of the environment in contrast to the physical temperature measured with an ordinary thermometer. If we think in terms of using the animal as our thermometer, then by recording its reactions to different combinations of these climatic factors, we arrive at a scale of "physiologically-effective" temperature. On such a scale, two environments could be equivalent even though their physical (*i.e.*, air) temperature differed widely. Scales of this kind have been worked out for human beings by methods which are described at the end of this chapter; so far, for reasons which will become clear, no satisfactory scale has been designed for animals.

C. Macro- and Microclimate

A further complication arises in the definition of the thermal environment actually experienced by the animal. Measurements made with the usual physical instruments define the climatic environment at large, which we call the macroclimate; much more difficult to define is the microclimate, the private climate or, rather, weather, which surrounds the animal in a precise place at a precise time. The microclimate is particularly important in natural environments in which animals may be able to shelter themselves comfortably from sun, wind, rain, or snow at times when measurements with meteorological instruments would suggest that the environment was intolerable.

Even in the case of animals kept indoors the same problems of definition occur, and later we shall consider in detail how the animal modifies the general thermal environment to produce a microclimate with a "physiologically-effective" temperature better suited to its requirements. For convenience in subsequent discussion, the term "environmental temperature" is used to describe the "physiologically-effective temperature" actually experienced by the animal.

II. CLIMATE AND THE WARM-BLOODED ANIMAL

A. Homeotherms and Poikilotherms

The animal kingdom may be divided into two kinds of animals: the *poikilotherms*, or cold-blooded animals, in which body temperature fluctuates according to the environmental temperature, and the *homeotherms*, or warm-blooded animals, which maintain their body temperature almost constant in the face of wide fluctuations of environmental conditions. Like most classifications in the natural world, this division has its borderline cases, some of which, such as hibernating animals, are of great physiological interest. The domestic animals with which this book is concerned are all homeotherms; indeed, most of the successful large animals of the world today are homeotherms. This has not always been the case. The giant dinosaurian reptiles of prehistory which once dominated the earth were poikilotherms, like the lizards of today that resemble them. These great animals, however swift and strong they were in the heat of the sun, must have become dull and slow in the cold of winter. The newly evolved homeotherm, as able in the cold as in the warm to fight, flee, eat, and reproduce, must have been possessed of overwhelming advantage. It is reasonable to state that the homeotherms' emancipation from the fetters of their thermal environment is the explanation for their dominance today.

B. Heat Exchange and the Regulation of Body Temperature

1. HEAT PRODUCTION AND HEAT LOSS

Living animals produce heat. The biochemical processes by which they do so are discussed more fully in Chapter 15. Here it is sufficient to state that animal heat is in essence a waste product just as heat is a waste product of a car engine. The business of living and growing involves the transformation of food molecules into molecules suitable for the synthesis of new body substances or into forms which are suitable for doing the various kinds of mechanical, electrical, and chemical work necessary to the maintenance of life. Since these transformations are less than 100 percent efficient, an appreciable part (up to 80 percent) of the energy of the food molecule is lost as heat. Even the mechanical, electrical, and chemical energy is largely degraded to heat in the course of doing its work. In order that its body temperature shall not rise, the animal must lose this heat at the same rate as it is produced. Because the homeotherm normally lives in an environment which is at a temperature lower than that of its body, it continuously loses heat to the environment according to the physical laws of conduction, convection, and radiation. The heat lost by these three routes together is called the *sensible heat loss*. The rate of sensible heat loss (H_s) is proportional to the animal's surface area (A) and to the difference in temperature between the animal and the environment:

$$H_s = cA \ (T_r - T_e) \qquad \text{Equation 4:1}$$

where T_r is the deep body (rectal) temperature and T_e is the environmental temperature. The proportionality constant (c) is the rate at which heat is lost from a unit area of the animal's surface for each degree of temperature difference between animal and environment. It therefore has the dimensions of a conductance, and its reciprocal, $1/c = i$, is the average insulation of a unit area of the animal's surface. This equation is often described as a statement of Newton's law of cooling. In point of fact, as Kleiber (1961, p. 140) points out, it does not describe cooling but steady state heat loss.

In addition to the sensible heat loss, animals also lose heat by the evaporation of water from the moist surfaces of the skin and from the respiratory tract. This heat (H_e), although of course it flows to the animal's skin surface as sensible heat, is there converted to latent heat. About 580 kcal of latent heat is required to evaporate one liter of water. In hot environments, evaporation becomes the most effective, and ultimately, the only means of heat dissipation. In the cold, evaporation falls to a certain irreducible minimum; the total heat loss, therefore, increases with falling temperature linearly and at the same rate as sensible heat loss. These ideas are put together in the simplified scheme shown in Figure 4–2, which also shows the basal heat production, that is, the minimum metabolism of the fasting animal, and the amount of heat produced when the animal is fed. It is clear that there is a certain range of temperature within which the environment has no effect on heat production. This is the *thermoneutral* or *comfort zone*. The lower limit of this range is the *critical temperature*, $C_1 \ldots .C_4$. At this temperature the animal's thermoneutral heat production equals exactly the minimal heat loss by evaporation and the environmentally-induced sensible heat loss. When the tem-

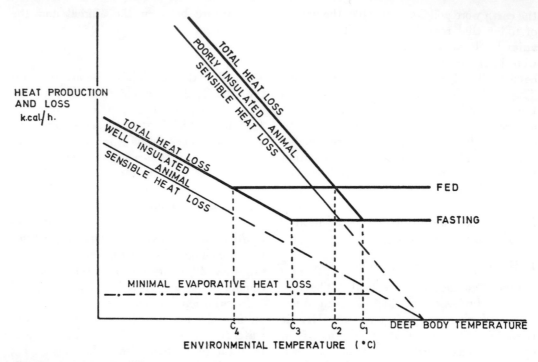

Fig. 4–2. Schematic representation of the relation between environmental temperature and the energy exchanges of homeotherms in cold environments. The heavy lines show the total heat loss of animals well and poorly insulated, fasting, and fed. The horizontal part of this line defines the zone of thermal neutrality, in which range heat loss is unaffected by environmental temperature. The lower limit of the thermoneutral zone is the critical temperature, C_1, C_2, C_3, or C_4, below which more heat must be produced to maintain body temperature.

perature falls below this point, heat loss tends to exceed basal heat production and the animal must produce more heat in order to prevent its body temperature from falling.

2. Physical Regulation

Above the critical temperature, basal heat production exceeds the environmentally-induced heat loss, and, in order to prevent a rise of body temperature, the animal must continue to dissipate this heat. This it does largely by increasing the rate of vaporization. It might be thought from the foregoing and from Equation 4:1 that the rate of heat loss is determined solely by the environment. In fact, two of the terms we considered to be constants in Equation 4:1 actually are not, but are regulated by the animal in response to changing conditions.

First, the animal's surface area (A) is not the total skin surface area but the effective

area for heat loss. If the animal lies with its legs under its body these areas are not presented to the environment, and do not form part of the effective surface area. This is one aspect of *behavioral* temperature regulation. Second, the term "c," the conductance of the tissues, can be increased by varying the rate of blood flow to the skin. This partially short-circuits the *internal insulation* of the subcutaneous tissues. Also, by raising the hairs of the coat the thickness of the layer of air entrapped by the coat can be increased, thereby increasing the *external insulation*. By far the most powerful mechanism of heat dissipation in species which can sweat is the increase of the evaporative heat loss. Provided the air is dry, animals can prevent their body temperature from rising even in environments in which the air temperature is much above body temperature. Indeed, it is clear from Equation 4:1

that however the terms "A" and "c" are altered, when environmental temperature reaches body temperature

$$(T_r - T_e) = 0$$

and sensible heat loss falls to zero, evaporation becomes the only remaining avenue of heat dissipation. These mechanisms—for varying effective area, internal and external insulations, and evaporative loss—are collectively called "physical temperature regulation," a term coined by the famous physiologist, Rubner. By means of physical temperature regulation, the animal is able to match heat loss to resting heat production within the range of the thermoneutral zone.

3. CHEMICAL REGULATION

As environmental temperature falls through the thermoneutral zone towards the critical temperature, physical temperature regulation is employed increasingly so that when the critical temperature is reached no further heat conservation is possible.

In environments below the critical temperature the animal produces more heat to satisfy the environmental demand. This *thermoregulatory heat production* is a second aspect of body temperature regulation; Rubner called it "chemical regulation." At this point evaporative loss is at its minimum, and heat loss increases linearly with falling temperature (increasing by c kcals per degree centigrade) until eventually the animal is producing heat at the maximum rate of which it is capable. This is the *summit metabolism.* If the temperature falls further still, body temperature falls and death ensues.

At environmental temperatures above body temperature, the animal gains sensible heat from its environment which, in addition to its own heat production, it must dissipate by evaporation. Once its maximum rate of evaporative heat loss is reached, any further increase of environmental temperature causes body temperature to rise; metabolic rate then increases further by the van't Hoff effect, thus setting up a vicious circle with a

lethal outcome. A rise of more than a few degrees in brain temperature causes irreversible damage.

4. THE EFFECT OF FOOD INTAKE

Any muscular activity, including eating, involves an increase of heat production. The heat increment of feeding, measured as the increase in heat production per unit increment of food ingested, is the heat arising as a by-product of the inefficient transformation of food molecules into the end products of their utilization.

When the thermoneutral heat production is increased as a consequence of the ingestion of food, it intersects the total heat loss line at a lower environmental temperature, which is therefore the critical temperature in those circumstances.

Figure 4–3 shows some results of experiments with sheep by Graham and his colleagues (1959) at the Hannah Institute. These experiments bring out clearly certain points which have emerged in the foregoing discussion.

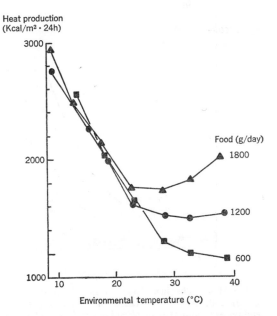

FIG. 4–3. The heat production of a closely clipped sheep at three feeding levels in relation to environmental temperature. (*From Graham et al., 1959. J. Agric. Sci. Camb. 52:13.*)

First, as the food intake is increased the thermoneutral heat production rises and the critical temperature becomes progressively lower. Second, below the critical temperature, heat loss per unit of surface area is independent of feeding level. The heat increment of the food "replaces" part of the thermoregulatory heat production. Third, an increase of feeding level lowers not only the critical temperature, but the zone of thermal neutrality as a whole. Notice how the rate of heat production of the sheep on the highest feeding level increases again at around 30° C, while on the lowest level of feeding it is still falling. Tolerance of hot environments is clearly impaired by high levels of feeding.

5. PARTITION OF HEAT LOSS

In Figure 4–4 the heat losses of the same sheep are divided into their sensible and

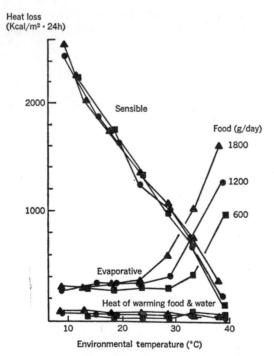

FIG. 4–4. The heat losses of the sheep shown in Figure 4–3, divided into the sensible and evaporative components and that of warming the food and water ingested. (*From Blaxter et al., 1959. J. Agric. Sci. Camb. 52:25.*)

evaporative components. An additional component is the heat employed to warm food and water to body temperature. The reason for including this component rather than that of the heat lost when urine and feces are excreted is perhaps not immediately obvious. Whether we regard the ingesta as increments to the body mass or as outside the body proper, the fact remains that it is in the process of ingestion rather than excretion that the average temperature of the body tends to fall; anything which tends to decrease body temperature we regard as a component of heat loss.

So far we have, by implication, considered animals in a steady state, that is, those in which heat loss and heat production are equal, and in which there is no change in the heat content of the body. In practice, absolutely steady states are rarely achieved, although they are approached in the long term. Although the animal maintains the temperature of the deep tissues of its body very constant in the long run, short-term fluctuations do occur, as in fever or severe climatic stress. Furthermore, the outermost tissues of the body have a temperature very different from that of the central or "core" tissues. Indeed, it is only by reducing cutaneous blood flow and allowing skin temperature to fall that animals are able to utilize the potential thermal insulation of their subcutaneous tissues. Thus, although the deep body temperature may remain constant irrespective of outside conditions, the mean body temperature certainly does not. The rate of heat storage, which can be positive or negative, must be taken into account when relating heat loss to heat production.

As the animal grows, both its resting heat production and its insulation increase. As a result the thermoneutral zone widens during growth, and, at the same time, moves lower down the temperature scale. This can be seen in Figure 4–5, which shows the results of experiments with chickens.

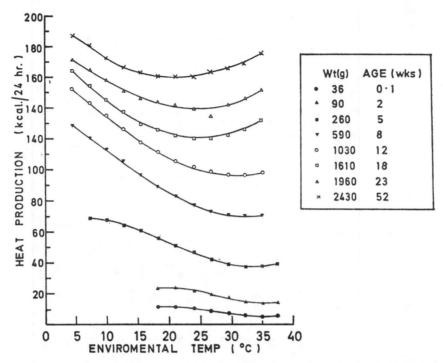

	Wt(g)	AGE (wks)
•	36	0·1
▲	90	2
■	260	5
▼	590	8
○	1030	12
□	1610	18
▲	1960	23
×	2430	52

FIG. 4–5.   The heat production of chicks growing from 1 day to 1 year of age, showing the shift of the thermo-neutral zone to a lower temperature range as the birds grow.   (*Data from Barott & Pringle, 1946. J. Nutr. 31:35.*)

6. THE COMPONENTS OF THERMAL INSULATION

The insulation (i) of a unit area of an animal's surface is the sum of three component insulations in series. (Insulations in series are added just as electrical resistances in series are added; if they are in parallel the total insulation is the reciprocal of the sum of the reciprocals.)

in which T_r, T_s, T_i and T_a are the temperatures of the rectum, the skin, the coat surface and the air, and H_t, H_s and H_{ct} are the total, sensible, and convective heat losses respectively. Notice that the denominator of each equation is different; this is because the coat does not impede the transfer of latent heat, nor does the boundary layer of air form a barrier to radiation. For convenience, the

The components are:

(i) the insulation of the tissues (i_t):

$$i_t = \frac{A\,(T_r - T_s)}{H_t}$$

(ii) the insulation of the coat or fleece (i_f):

$$i_f = \frac{A\,(T_s - T_i)}{H_s}$$

and (iii) the insulation of the boundary layer of air surrounding the animal (i_a):

$$i_a = \frac{A\,(T_i - T_a)}{H_{ct}}$$

Table 4–1. The Overall Thermal Conductance, Components of Thermal Insulation, Minimum Evaporative Heat Loss and Thermoneutral Heat Production of Various Domestic Animals

(From the Literature.)

Species and Breed	Wt (kg)	Surface Area (m²)	Overall Conductance (kcal/m².24 hr)	Coat Type	Insulation (°C.m².24 hr/Mcal)			Heat Production (Mcal/m².24 hr)		Minimum Evaporative Heat Loss (Mcal/m².24 hr)
					Tissue	Coat	Air	At Maintenance	Fully Fed	
Pig										
Large white	1.5	0.13	133	Normal	2.1	–	8.3	1.6	2.8	0.16
Large white	4	0.23	100	Normal	–	–	–	1.6	2.8	0.09
Large white	8	0.36	100	Normal	–	–	–	1.6	2.8	0.10
Large white	10	0.42	100	Normal	–	–	–	1.6	2.8	0.11
Landrace	23	0.71	98	Normal	4.6	–	7.2	1.6	2.8	0.11
Large white	150	2.31	36	Normal	–	–	–	1.4	2.5	—
Sheep										
Merino	5	0.26	58	Fine	8.2	–	8.5	1.2	1.6	0.48
Merino	5	0.26	43	Coarse	7.5	–	14.7	1.2	1.4	0.48
Down	45	1.14	81	Clipped	3.5	nil	7.7	1.3	2.3	0.31
Down	45	1.14	31	5 cm	3.5	27.9	–	1.3	2.3	0.31
Blackface	45	1.14	68	Clipped	5.7	nil	7.7	1.3	2.3	0.24
Blackface	45	1.14	29	5 cm	5.7	27.9	–		2.3	0.24
Cattle										
Ayrshire	42	1.08	53	Normal	4.4	7.7	5.7	1.9	3.0	0.34
Jersey	390	4.78	62	Normal	–	–	–	2.6	3.7	0.28
Brahman	470	5.42	62	Normal	–	–	–			0.24
Brown Swiss	550	6.02	52	Normal	–	–	–			0.21
Galloway	500	5.67	41	Normal	6.7	–	16.2	2.0	3.2	0.35

air insulation is usually treated as if it were a barrier to radiation, so that

$$i_a = \frac{A\,(T_i - T_a)}{H_s}.$$

It is this term which will be employed here.

Because of the somewhat arbitrary position of the coat surface, particularly in sparsely coated animals, it is sometimes preferable to make a single estimate of the total external insulation, that is:

$$\frac{A\,(T_s - T_a)}{H_s}.$$

Figure 4–6 illustrates these separate components of insulation and the temperature gradients which they create. Table 4–1 gives estimates of the tissue and external insulations of various animals at successive stages of growth.

In an environment below thermal neutrality, overall conductance and evaporative heat loss are minimal and approximately constant. Estimates of their value are given in Table 4–1, together with approximate values of thermoneutral heat production for animals at maintenance and at full feed. From the three sets of values, the critical temperature can be calculated as follows:

Since:

$$H_t = c\,(T_r - T_e) + H_{e\ (min)},$$

at the critical temperature H_t is the thermo-neutral heat production and $T_{cr} = T_e$. Then:

$$H_t - H_{e(min)} = c\,(T_r - T_e)$$

and

$$T_e = T_r - \frac{(H_t - H_{e(min)})}{c}.$$

As an example, for the young, fully-fed calf of 42 kg, $c = 53$, $H_t = 3.0$, and $H_{e(min)} = 0.34$. Then $T_{cr} = -11°$ C.

7. BEHAVIORAL THERMOREGULATION

Animals rarely experience the conditions of the environment at large. In almost every situation some aspect of their behavior makes their thermal situation different from that which would be expected from physical measurements of the environment. The search for shelter by animals living outdoors was mentioned earlier; in some cases, it is only by sheltering that survival is possible at all. Certain small Arctic mammals such as weasels and lemmings survive the Alaskan winters with temperatures of $-50°$ C even though their own critical temperatures are around $15°$ C. Without shelter they would have to maintain a three- to fourfold increase of metabolism; instead, they burrow beneath the snow into the earth which retains the heat they produce and creates a comfortable

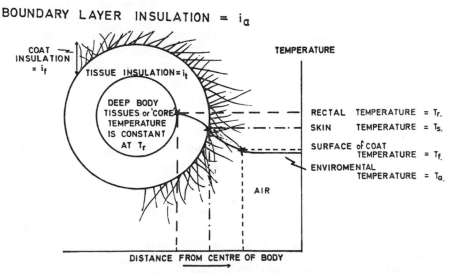

BOUNDARY LAYER INSULATION = i_a

Fig. 4–6. Diagram to show the components of thermal insulation of an animal and the gradients of temperature which exist between the animal's central tissues and its environment.

microclimate around them. This type of regulation is at its most sophisticated in that interesting tropical mammal, man, whose critical temperature is around 28° C, but who lives comfortably in every region of the world in shelters of his own making. He also lacks a coat of any insulation value, but overcomes this deficiency by wearing clothes so that his skin remains comfortably warm.

Young animals are particularly vulnerable to cold. This is largely due to their high ratio of surface area (for heat loss) to mass (for heat production). Their practice of huddling together, seen best in newborn animals which normally exist in groups such as nests of birds or litters of pigs or puppies, represents another form of behavioral thermoregulation. Those areas of one member of the group which are in contact with another member lose practically no heat. Figure 4–7 shows how, when four to six pigs are allowed to huddle together in a group, their heat production (measured by their oxygen consumption) does not increase in

the cold as much as that of a single pig. As the environment warms up, the pigs spread out, and their heat production becomes the same as that of the single animal. The value of this conservation of heat in terms of growth is demonstrated by Sorensen's finding that pigs kept at 3° C in groups grew at a rate of 0.63 kg/day, but when they were kept singly, their daily gain was only 0.45 kg.

8. The Physics of Heat Exchange

The total heat loss to the environment is the sum of the losses by conduction (H_{cd}), convection (H_{cv}), radiation (H_r) and evaporation (H_e), the total of the first three being the sensible heat loss.

Conduction. Heat is lost by conduction to all those objects with which the animal is in contact which have a temperature lower than the animal's surface. Such objects are the earth or floor and the food and water the animal ingests.

The fundamental equation for steady state conduction is:

$$H_{cd} = A_c(T_1 - T_2) \cdot \frac{k}{d} \qquad \text{Equation 4:2}$$

where A_c is the area of contact, T_1 and T_2 are the temperatures of the bodies in contact, k is the thermal conductivity, and d the thickness of the intervening material.

The proviso of steady state conditions assumes that there is no change in the temperature of the body or of the object which is conducting heat away from it. In practice, the ground beneath the animal warms up when the animal is lying on it, so that the conductive loss diminishes the longer the animal remains lying.

Convection. Heat loss by convection involves the transfer *en masse* of air (or liquid) away from the body and its replacement by a cooler mass.

We distinguish between *natural convection*, in which the air rises away from the body by virtue of its increased warmth and therefore

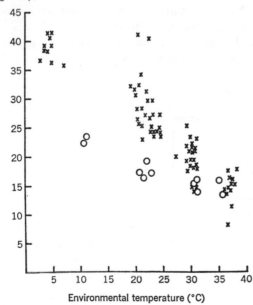

Oxygen consumption
(ml./Kg · min)

Environmental temperature (°C)

Fig. 4–7. The effect of huddling on the oxygen consumption of young pigs at various environmental temperatures. x = single pigs, 0 = pigs in groups of four to six. (*From Mount, 1960. J. Agric. Sci. Camb. 55:101.*)

lower density, and *forced convection*, such as is produced by wind or by a fan. Heat is first transferred by conduction from the body to the air surrounding it; for this reason, conduction and convection are often treated together. In this context, however, it seems preferable to confine the term conduction to the transfer of heat not involving convection, for example, to the floor and to the food ingested.

Heat loss by forced convection from bodies with constant surface temperature (T_i) is proportional to the temperature difference between body and air, and to a power of the air velocity which is close to 0.5:

$$H_{cv} = kA\,(T_i - T_e)\,.\,\sqrt{V} \qquad \text{Equation 4:3}$$

Animals do not, however, maintain a constant surface temperature as wind speed increases; this means that heat loss is proportional to a power of wind speed which is less than 0.5. In consequence, even a slight air movement has a considerable effect on heat loss. Furthermore, the cooling of the animal's surface by wind means that radiative loss is simultaneously diminished.

Radiation. The emission of infrared radiation is given by the equation:

$$H_r = A\,\sigma\,eT^4 \qquad \text{Equation 4:4}$$

in which A is the animal's effective or *profile* surface area, σ is the *Stefan-Boltzmann constant* $= 1.37 \times 10^{-12}$ cal/sec cm^2, e is the *emissivity* of the surface, and T^4 is the fourth power of the surface temperature in degrees absolute ($^\circ$ C $+ 273^\circ$).

Similarly, the net radiation exchange between the animal and its environment is given by:

$$H_r = A\,\sigma\,(e_1 T_1{}^4 - e_2 T_2{}^4) \qquad \text{Equation 4:5}$$

Here, e_1 and T_1 are the emissivity and absolute temperature of the animal surface, e_2 and T_2 the emissivity and mean radiant temperature of the environment. The emissivity of a surface is the fraction of incident radiation at a given wavelength which it absorbs: if it absorbs all radiation its emissivity

is 1.0 and the surface is called a *black body*; a *perfect reflector* has an emissivity of 0.0. As an example, at a wavelength of 9μ, highly polished silver has an emissivity of 0.02; the value for most natural surfaces is nearly 1.0; for human and animal skin it is 0.99.

The wavelength of maximum radiation is given by the surface temperature according to Wien's displacement law. For the sun (with a temperature of 6000° K) it is around 0.5μ (visible), and for the animal with a temperature of about 300° K it is 9 to 10μ (near infrared).

By measuring the parameters of equation 4:5 the agreement of the theoretical model with experimental findings can be assessed. The measurement of the average surface temperature involves making readings of skin temperature at many different points and weighting these according to the fraction of the total area which they represent.

Evaporation. Evaporative heat loss may be estimated by the equation

$$H_e = A_w\,(E_s - E_a)\,.\,\sqrt{V}$$

in which E_s and E_a are the vapor pressures of water at the skin and of the surrounding air, and V is the air velocity. A_w is the wetted area of the body, which is under physiological control. (Notice that this is the only equation in which the humidity of the air appears as a variable. Thus, it is only on evaporative heat loss that humidity has any effect. Since below thermal neutrality evaporative heat loss is minimized by the restriction of cutaneous moisture secretion, it is only in neutral and hot environments that humidity is of importance to thermoregulation. In such environments, however, it imposes an absolute limit on thermoregulation, for in a saturated atmosphere evaporation is impossible.)

The value of the equations describing each of the four channels of heat loss is principally in showing how heat losses vary with the major climatic factors, with the first power of temperature for conduction and convection, with the fourth power of radiant temperature, and with the square root of air velocity. The use of such equations for predictive purposes is as yet a science in its infancy. In many cases, constants can only be estimated empirically by measurement of the dependent variable—heat loss—itself. A further complication is that some of the parameters used such as skin temperature, skin water vapor pressure, and others are themselves determined by the conditions of heat loss.

C. Appetite and Food Intake

From the nutritional point of view, the most important way in which animals react to a change in their climatic environment is by adjusting their voluntary intake of food. Since heat production increases directly with the amount of food consumed, it is clear that this behavior can greatly modify the relation between the animal's environment and its energy metabolism and growth. Section IIB4, "The Effect of Food Intake," described how both the position and the width of the thermoneutral zone varied with the amount of food consumed; this is clearly seen in Figure 4–8. In addition, an increased food intake compensates, in part at least, for the increased energy expenditure incurred in a cold environment and thereby mitigates to some extent the adverse effects of low temperatures upon growth. Growth rate is therefore less impaired by cold weather when animals have food *ad libitum* than when they are restricted to the same amounts of food in all circumstances.

Figure 4–8 shows the results of experiments with growing animals of several species whose food intakes have been measured at various temperatures. The general pattern is probably the same in all domestic species. It is characterized by an inflexion which occurs near the critical temperature. Below this point food consumption increases initially at a rather constant rate with falling temperature, although in extreme cold it approaches a limit set by the animal's inherent capacity to ingest and digest food. Above the thermoneutral zone, food intake

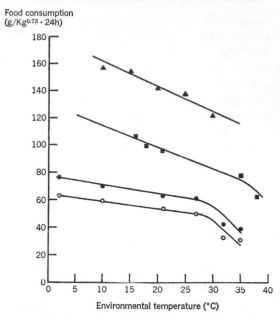

Food consumption
(g/Kg^{0.73} · 24h)

FIG. 4–8. The effect of environmental temperature on food intake. ▲—▲: pigs aged 2–10 weeks; ■—■ = chicks aged 5–12 days; ●—● = Holstein calves; and ○—○ = Jersey calves. (Data from Fuller, 1965. Brit. J. Nutr. 19:531; Winchester & Kleiber, 1938. J. Agric. Res. 57:529; & Johnson & Yeck, 1964. Mo. Agric. Exp. Sta. Bull. No. 865.)

declines steeply with rising temperature and may fall to zero in animals subjected to thermal stress.

The Extent of Compensation to Cold by Increasing Food Intake. It is of interest to inquire to what extent the augmentation of energy intake in the cold can compensate for the concomitant increase in heat production. The evidence which is available suggests that slight cooling of the environment may elicit an overcompensation for the extra heat produced; at lower temperatures, however, the rate of increase of caloric intake with falling temperature seems always to be less than that of heat production. An example of these relationships is seen in Figure 4–10. Here, food intake increased more rapidly between 38° and 35° than at lower temperatures; heat production, in contrast, increased uniformly throughout the temperature range. As a result, energy retention reached its maximum value at about 31° C, which was below the thermoneutral zone of these birds. It is unfortunate that information on this issue is so scarce, for it would be of considerable value in animal husbandry. As Bianca and Blaxter (1961) remark in their review, the relationship between food intake, heat production, and environmental temperature may be of great importance in deciding upon the optimum temperatures at which to keep livestock.

D. Energy Utilization

The energy an animal receives in its food can follow one of three courses: it can be lost in the excreta, emitted as heat, or retained in the form of body substance. As we have seen, the environment affects both food intake and heat production, and, therefore, energy retention. Its effects on energy losses in the excreta are, by comparison, slight. A fall of digestibility of about 0.1 percentage units per ° C fall of temperature has commonly been observed, although in some experiments no effect has been found. Energy losses in urine may be slightly increased in the cold, particularly in circumstances where more urea is excreted. No effect on methane production has been found. The sum of these changes is therefore a slight but consistent tendency towards lower values of metabolizable energy in the cold.

1. THE SOURCE OF THERMOREGULATORY HEAT PRODUCTION

Although the increase in heat production outside the thermoneutral zone may be derived from the oxidation of carbohydrate, fat, or protein, it is very important in studying growing animals to know in what proportion these different nutrients contribute to heat production and what factors determine their respective contributions. The special role of protein in this connection will be considered more fully in the next section.

In general, whether fasting or fed, animals produce heat for thermoregulation by combustion of body fat. This is understandable in the fasting animal, but it is by no means

clear why an animal receiving large amounts of carbohydrate food should not use this directly for heat production.

2. ENERGY RETENTION

In animals kept at different temperatures but given the same amount of food, the amounts of energy retained vary inversely with their heat production. Figure 4–9

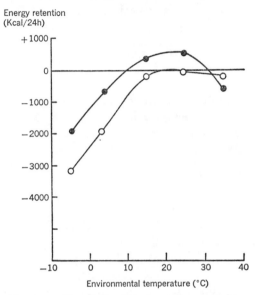

FIG. 4–9. The relation between environmental temperature and the energy retention of two steers given a daily ration of 4.5 kg of hay and 1.8 kg of oats. ●—● = Steer Amos; ○—○ = Steer Andy. (*From Blaxter & Wainman, 1961. J. Agric. Sci. Camb. 56:81.*)

shows results obtained with steers by Blaxter and Wainman at the Hannah Institute. The two animals used showed the same general pattern, but Amos was clearly less tolerant of heat than Andy, as witnessed by the steep decline in his energy retention above 25° C.

When food is given *ad libitum*, the extra food which is (usually) consumed in the cold means that the decline in energy retention with falling temperature is less severe: in moderate cold, food intake may even increase more than heat production, in which case more energy may be retained than in the thermoneutral zone. An example of this is shown in Figure 4–10, which illustrates the effect of temperature on each phase of energy

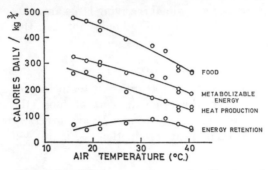

FIG. 4–10. The consumption and utilization of dietary energy by chicks at various environmental temperatures. (*After Winchester & Kleiber, 1938. J. Agric. Res. 57:529.*)

utilization by growing chicks. Here, energy retention reaches its highest value at a temperature lower than that at which heat production is minimal.

3. NET ENERGY

The net energy of food is defined (see Chapter 15) as the increase in energy retention corresponding to a unit increment of food energy supplied.

The experiment with sheep shown in Figure 4–3 demonstrates that, at the high food intake, heat production begins to increase with rising temperature while it is still falling at the low intake. The determination of the heat increment must be made at a temperature at which there is no environmental stress. For the sheep shown in Figure 4–3, for example, there is no thermoneutral temperature common to all feeding levels. In this case, estimation of the "true" net energy of the diets would be made by keeping the well-fed sheep at a lower temperature than the poorly fed ones. As temperature falls, heat production becomes independent of feed intake; there is no heat increment and net energy is equal to metabolizable energy. At high temperatures, heat production increases most rapidly at the high feeding level and gives a reduced net energy.

E. Protein Utilization

The fact that protein, in addition to its

key role in growth, can be used as an energy source gives it special importance in connection with environmental stress. In one sense, growth is virtually synonymous with protein synthesis; if protein is diverted from synthetic processes to provide heat for thermoregulation, growth will be correspondingly impaired.

Much of the work on the participation of protein in chemical regulation has been carried out with adult animals and has led to the conclusion that thermoregulatory heat can be produced without increasing protein catabolism. In adult animals, synthesis of tissue proteins proceeds slowly, at a pace just sufficient to meet the inevitable losses from skin and hair, the endogenous losses in the digestive tract, and other such necessary replacement. These essential syntheses are little affected by variation in energy expenditure so that environmental cold has correspondingly little effect on protein utilization.

Growing animals, on the other hand, are characterized by high rates of synthesis of tissue proteins, the form in which a large fraction of their dietary protein is retained.

It is well known that the utilization of dietary protein is improved when extra dietary energy is supplied (see Chapter 16). This is the so-called "protein sparing" action of dietary energy. The mechanism underlying this effect is not yet fully understood, but it is recognized that there are at least three factors at work. The first is implicit in the term "protein sparing," namely, the more nonprotein calories are supplied, the smaller is the contribution of protein to the total energy supply. The pathways of carbohydrate, fat, and protein dissimilation converge in the tricarboxylic acid cycle. Since turnover rates are to some extent determined by substrate concentration, the rate of protein catabolism depends, at least partly, on the proportion of the total energy which it supplies. Second, a certain amount of energy, in the form of ATP, is required to activate the amino acids and to form the peptide bonds which link them together in the protein molecule. Energy is also required to synthesize nonessential amino acids from simpler molecules. Although these requirements are not great—they have been estimated at 1 to 2 kcal per gram of protein formed—they may give rise to some dependence of protein synthesis on energy supply. Third, recognition of the superiority of carbohydrate to fat in promoting nitrogen retention has led to the suggestion that the carbohydrate effect is mediated by insulin, which is known to stimulate protein synthesis. Exposure to a cold environment, by increasing energy expenditure, reduces the amount of energy which is available to provide the protein sparing effect. It may therefore be that the effect of environmental cold on protein metabolism can be explained entirely in these terms. Whatever the mechanism, it has been found that in many cases a low temperature reduces the nitrogen retention of growing animals. How much it is reduced in a given environment depends on the amount of food being given. At one extreme, when a maintenance ration is given, all the energy supplied, including its protein, will be catabolized. (This is not quite true; animals kept at energy equilibrium gain a little protein and lose a little fat.) At the other extreme, when animals are given so much food that the demands of thermoregulation account for only a small proportion of their metabolizable energy, then any effect of cold on protein utilization will be minimal or absent. Part of the diminution of the effect of temperature at high food intakes is, of course, due to the greater heat increment of feeding, which "replaces" part of the thermoregulatory heat production found at lower feeding levels.

The second factor affecting the utilization of dietary protein in a cold environment is the protein content of the diet. In general, the higher the protein content of the diet, the more nitrogen retention depends on energy supply and the more severely it is impaired in an adverse environment.

III. EFFECT OF CLIMATE ON GROWTH

Growth is a phenomenon of change—in size, weight, shape, composition, and structure. In the face of such a variety of concomitant events, it is convenient to distinguish between two general aspects of growth: one, the accretion of body substance; the other, the alterations of form and function which proceed largely independently of changes in the size and weight of the body as a whole. These two aspects are often distinguished as growth and development.

A. Growth: Increase in Body Weight

In previous sections the effects of the environment on food intake, energy utilization, and protein metabolism were described. The summation of these is reflected in the animal's weight gain. Since the major influence of the environment is on energy exchange, it might be expected that the highest weight gain would be attained at the temperature at which energy retention is at a maximum. For animals given the same amount of food at different temperatures, this is true. When food is available *ad libitum*, however, it is usual that more of it is consumed in the cold and at least some of the extra protein it contains may be used for tissue growth. While 1 kcal retained as fat gives a weight gain of 0.11 gm, as protein it results in the production of some 0.8 gm of tissue because of the water retained with it. This differential of about 7:1 in the caloric density of body gains gives overwhelming importance to protein. How much of the extra protein consumed in the cold is used for anabolic purposes depends on dietary circumstances and on the magnitude of the increase of food intake in relation to the higher energy requirement. In the example given in Table 4–2, drawn from work with chicks, energy retention at 21°C was only 58 percent of that at 32°C; however, as a result of increased food intake, protein deposition and also growth rate reached maximum values at 21°C.

Table 4–2.   Gains of Protein, Energy and Body Weight by Chicks 6 to 15 Days of Age in Relation to Environmental Temperature

(Data from Kleiber & Dougherty, 1933.
J. Gen. Physiol. 17:701.)

Temperature (°C)	21	27	32	38	40
Protein deposition (g/24 hr)	1.10	1.08	0.97	0.79	0.68
Energy retention (kcal/24 hr)	6.9	10.4	11.8	8.7	6.7
Body weight gain (g/24 hr)	4.88	4.64	4.39	2.97	2.91

the results of 5 experiments in which pigs have been kept at a series of different temperatures. In each trial there was a clearly defined range of temperature within which they grew most rapidly. For the younger pigs, which were given feed *ad libitum*, this range was between 20° C and 25° C; in the case of the older animals, the optimum temperature for growth was between 20° C and 12° C. A further point of interest in these data is that the temperature giving maximum growth is clearly lower, the higher is the overall growth rate of the animal. This effect plainly echoes that noted previously, of the downward shift of the thermoneutral zone with increasing food intake.

At temperatures above the optimum a decrease in the growth rate of the young pigs is evident; with the older ones high temperatures were not investigated, but other work has shown that with them the effect is even more pronounced.

With meat-producing animals, the relation between environmental temperature and growth rate is of considerable economic importance. Buildings for intensively housed species such as chickens and pigs are designed to maintain temperatures which allow the most rapid growth consistent with acceptable carcass quality. Figure 4–11 shows

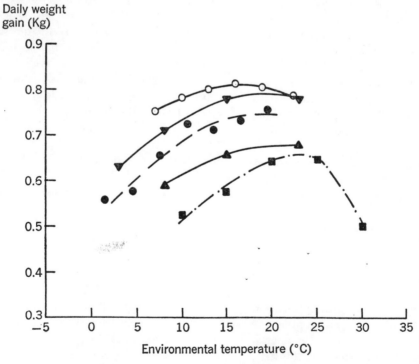

FIG. 4–11.   The daily weight gain of pigs as a function of environmental temperature. O—O = pigs from 30 to 90 kg; ▼—▼ and ▲—▲ = pigs from 30 to 90 kg on two feeding regimes; ●— —● = pigs from 30 to 110 kg; ■– · – · ■ = pigs from 2 to 10 weeks of age (4 to 40 kg). *(Data from Comberg, 1959. Züchtungskunde 31:462; Sorensen, 1960. Tagungsber. Dtsch. Akad. Landw. 23:79; Siegl, 1960. Arch. f. Tierz. 3:188; & Fuller, 1965. Brit. J. Nutr. 19:531.)*

Table 4–3. Growth of Brahman, Shorthorn and Santa Gertrudis Calves
at Temperatures of 10° C and 27° C

(*Data from Ragsdale et al., 1957. Mo. Agric. Exp. Sta. Res. Bull. 642.*)

Breed	Brahman		Santa Gertrudis		Shorthorn	
Temperature (° C)	10	27	10	27	10	27
Growth rate between 3 and 14 months (kg/day)	0.70	0.74	0.89	0.80	0.81	0.56

Other species have seldom been investigated in this respect. Comparisons have been made between different designs and degrees of shelter for cattle; these are reviewed by Bianca and Blaxter (1961). Without meteorological records, such comparisons are difficult to translate into general terms because the various structures provide, in addition to different temperatures, differing degrees of protection from wind, rain, snow, and sun. There exist also marked differences between breeds in the temperature range over which the most rapid growth is achieved. The comparisons made in Missouri between Shorthorn, Brahman, and Santa Gertrudis calves bring out this point clearly. The results in Table 4–3 show how the growth of the Shorthorn was superior to that of the Brahman at 10° C but inferior at 27° C. The Santa Gertrudis combined the growth rate of the Shorthorn with the heat tolerance of the Brahman. In general, fattening cattle of the European beef breeds grow uniformly within the temperature range −5° C to +20° C when they are protected from extremes of wind and precipitation; for younger growing animals, a lower limit of 0° C is considered preferable. It is generally agreed that young calves require a higher temperature, 10° C to 25° C.

B. Development: Change in Body Form

There are characteristic differences in appearance between animals which have grown up in different climates. Those in the cold typically appear short and thickset, with rough coats, while those in the warm are long, lean, and sleek-coated. Other visible changes invole the size of the ears, tail, and other appendages. Some of these differences are of measurable value to the animal in conserving or dissipating heat, and it can be argued that they represent adaptive responses to an unfavorable environment. This aspect will be considered in more detail later. Here, changes in the relative sizes of body parts are considered as manifestations of the influence of temperature on animal development, a slightly different viewpoint.

1. Body Composition: Lean and Fat

The changes in the rates of protein and fat deposition which occur as a consequence of changes of environmental temperature are naturally reflected in the body composition of the animal. As a corollary of the generalization made earlier, that the deposition of protein is proportiontately less retarded by environmental stress than that of fat, it is to be expected that animals will be fattest in a thermoneutral environment and that departure from the zone of thermoneutrality will result in leaner animals. This thesis has been repeatedly confirmed in work with small laboratory animals. Table 4–4 gives results from an experiment of mine

Table 4–4. The Amounts of Fat and Fat-free Tissue in Pigs 10 Weeks Old, after 8 Weeks at Different Temperatures

(*Data from Fuller, 1965. Brit. J. Nutr. 19:531.*)

Temperature (° C)	10	15	20	25	30
Fat (kg)	7.5	9.4	10.2	10.9	7.3
Fat-free tissue (kg)	24.1	26.2	28.2	27.9	22.6
Total (kg)	31.6	35.6	38.4	38.8	29.9
Fat as % of total	24	26	27	28	24

which demonstrate the same effect with pigs. Unfortunately, there is no comparable information for sheep or cattle.

It is convenient at this point to mention the interesting observation that the body fat of a variety of animals living in cold climates has a lower melting point and contains a higher proportion of the liquid and unsaturated fatty acids. The mechanism involved is unknown, but in view of the fact that thermoregulatory heat is produced largely by the oxidation of body fat, it may be that the mobilization of triglycerides for this purpose is at least somewhat selective.

2. WEIGHTS OF ORGANS

It has frequently been reported that the visceral organs are enlarged as a result of cold exposure. Hypertrophy of the lungs, heart, spleen, liver, and kidneys has been noted in several species. Here again it may be reasonable to argue for the adaptive significance of these effects of cold stress. A chronically elevated metabolic rate means more work for the organs of respiration and circulation. The effect on heart, lungs, and spleen may be explained as a form of work hypertrophy, the organ growing to meet the demands made of it. Furthermore, since food intake is normally increased by cold exposure, the increased sizes of the liver and kidneys may be similarly viewed since there is evidence also for the adjustment of the sizes of these organs in response to the load put upon them.

There is, however, a difficulty in describing all such changes as these, namely, that of deciding what reference base should be used to compare the sizes of the organs. The point is that in some degree organ size can be expected to be related allometrically to the size of the body as a whole. The effect of temperature can then be measured as a deviation from allometric growth. (This is the assumption made in covariance analysis.) However, if the growth of an organ can be specifically accelerated or retarded by temperature, cannot also the growth of, say, the muscles? Since the muscles com-

prise the largest single component of the reference base against which the growth of individual organs is compared, the assumption of allometry becomes questionable. On the other hand, to compare the sizes of organs directly, without reference to the sizes of other parts of the body, is to deny that the growth of the separate parts is interrelated.

The solution, insofar as one exists, is probably that allometric considerations should be taken into account, but with caution. The assumptions of allometry can be checked by comparing the components of the reference base, *e.g.*, muscle:bone ratio, muscles + bone:gut, and so on. If there is no effect of temperature on these, then it is reasonable to regard a change of, say, the ratio of kidney:muscle + bone as a real effect of temperature on the kidney.

IV. ADAPTATION TO THE CLIMATIC ENVIRONMENT

We can assume that for any given animal there is a climatic environment for which it is physiologically best suited. Indeed, in the course of evolution those kinds of animal have survived which were best fitted to the environment—in the widest sense of the term—in which they found themselves. The development of a species by genetic selection, natural or artificial, represents one kind of adaptation. In addition, we have to consider the adaptation of the individual to the climatic environment into which it is born. This involves the development, during its growth, of mechanisms which fit it better for an environment to which its species is unaccustomed. We will discuss this latter aspect first.

A. Adaptation of Individuals

Three basic kinds of adaptation are open to the animal living in an adverse climate. The first possibility involves a change of physical regulation; the animal may develop more effective means of conserving or dissipating heat. Second, by means of a change of chemical regulation, the animal's resting metabolic rate may be altered to more nearly suit the environmental demand. Third, and this adaptation applies only in cold weather, the animal may increase its food intake so that, in spite of an increased energy expenditure, growth may be unimpaired. This latter aspect has already been discussed.

1. CHANGE OF PHYSICAL REGULATION

In Section IIA, the three components of the animal's insulation were described and estimates of their magnitude given. Can these insulations be varied so as to permit more comfortable life in a hostile environment? The insulation of the boundary layer of air is a physical, aerodynamic phenomenon over which the animal has no control. The insulation provided by the coat increases with its depth, so that one possible adaptation lies in growing more or less hair, fur, or wool, according to the climatic circumstances. The coats of many large mammals have been observed to thicken as winter approaches. The seasonal growth and shedding of the coat by cattle is controlled to some extent by temperature, but far more by changing day length. The same is true of sheep. This may simply be a quirk of evolution; on the other hand, the fact that change of day length tends to foreshadow change of temperature gives the coat time to respond. Young calves which are reared in the cold grow more hair than those kept in warm conditions: even pigs, a species in which the insulation of the coat is slight, grow more hair when raised in a cold environment.

An increased tissue insulation can be achieved by absorbing more of the temperature gradient between central tissues and environment within the body, that is, by lowering skin temperature by means of reduced cutaneous blood flow. There is some evidence that prolonged exposure to cold can lead to the toleration of lower skin temperature, and thereby, enhanced tissue insulation.

In environments above the zone of thermoneutrality, problems of heat dissipation arise. In some circumstances full vasodilation may not completely short-circuit the insulation of the subcutaneous tissues; there may then be some slight advantage to being lean—having little subcutaneous fat—but in general full vasodilation brings skin temperature very close to deep body temperature. Any improvement in heat tolerance by removal of insulation must be effected by a reduction of the insulation of the coat. The range of insulative adaptation to heat is therefore slight. Vaporization is the most important form of physical regulation in heat; there is, however, no evidence that by prolonged exposure to heat the efficiency of vaporization by farm animals is improved although this kind of acclimatization is known to occur in man.

A further aspect of adaptation of physical regulation is by the development of regional body cooling. The coat insulation over an animal's surface is not uniformly distributed; it is thickest on the dorsal trunk and thinnest on the lower legs, tail, face, and ears. In hot environments these sparsely covered areas dissipate a disproportionately large fraction of the total heat production. They are to the animal what a radiator is to a car. In animals kept chronically in a hot environment, some or all of these areas become larger than they do in animals kept in neutral or cool conditions. Such *trophic adaptation* has been extensively investigated in laboratory animals; it has even been suggested that the length of a mouse's tail could be used as a new kind of thermometer. Similarly, Missouri workers found that in calves reared at 26° C, a higher proportion of the surface area was formed by the legs and tail than in others kept at 15° C. The same phenomenon has been observed in pigs, which grew larger ears in a hot environment.

2. METABOLIC ADAPTATION

Heat stress is aggravated by a high resting metabolic rate. Any reduction in resting metabolism is therefore an aid to thermal stability in heat; of primary importance in this regard is the reduction in food intake that was discussed in Section IIB; the decline in heat production is, of course, equal to the heat increment of the feed refused. The resting metabolism of animals exposed to heat may, in addition, diminish more than can be accounted for by a reduced food intake. This is a form of metabolic acclimatization associated with a lowering of thyroid activity.

The converse of this, metabolic acclimatization to cold, independent of increased food intake, was described by Gelineo in his classic researches (see review by Hart, 1957). The value of an elevated metabolism at all temperatures is simply that it enhances summit metabolism and extends the lower limit of the animal's range of survival; there is no change of critical temperature. Although metabolic adaptation is of major importance in small animals, large domestic animals probably rely entirely on insulative adaptation.

3. Nonshivering Thermogenesis

In unacclimatized animals exposed to cold, thermoregulatory heat is provided largely by shivering. As the exposure to cold is continued, however, shivering gradually disappears although the elevation of metabolism persists. This is *nonshivering thermogenesis*. It has the advantage that it avoids the increase in convective heat loss associated with the movement of the skin in shivering. Nonshivering thermogenesis in the newborn of several species is associated with a particular kind of fatty tissue, called brown adipose tissue, located at strategic points in the body. Nonshivering thermogenesis is stimulated by norepinephrine, which, in the course of cold acclimatization, becomes strongly calorigenic. The heat is

thought to be produced by the cyclic esterification and hydrolysis of triglycerides.

B. Differences Between Species and Breeds in Their Resistance to Climatic Stress

In the early history of nomadic civilization, man began to hunt, and later, to domesticate certain of the wild animals which shared his habitat. He chose to husband those animals which possessed in greatest measure the qualities he valued. Thus began the subordination of natural to artificial selection in the evolution of domestic livestock. As a consequence of the two processes, there exist today profound differences between species, and between breeds within species, in their resistance to climatic stress. These differences may be traced to the same physical factors that distinguish acclimatized from unacclimatized animals. Amongst these, insulation, the distribution of insulation, the sizes of the appendages (Table 4–5), and the metabolic rate may all be considered as manifestations of interspecies and interbreed differences in heat tolerance. In this way, species and breeds can be classified as naturally resistant or susceptible to heat or cold. It can be argued, for example, that since the interspecies norm of basal metabolism is 70 kcal/$kg^{0.75}$ day, sheep, with a metabolic rate of about 55 kcal per $kg^{0.75}$ day, are inherently more heat tolerant than are cattle, whose basal metabolism typically amounts to some 88 kcal/$kg^{0.75}$ day, 15 percent above the interspecific mean. Differences between species in such factors as size, shape, coat, color, and so on, are too obvious to require mention, but they

Table 4–5. The Contribution of Ears and Dewlap to the Total Surface Area of Calves of Different Breeds

(*Data from Johnson et al., 1961. Mo. Agric. Exp. Sta. Res. Bull. 770.*)

Type	Beef			Dairy		
Breed	Brahman	Santa Gertrudis	Shorthorn	Holstein	Brown Swiss	Jersey
Area of the ears and dewlap as % of total surface area	9.6	7.6	5.0	3.4	5.9	3.3

PLATE 5

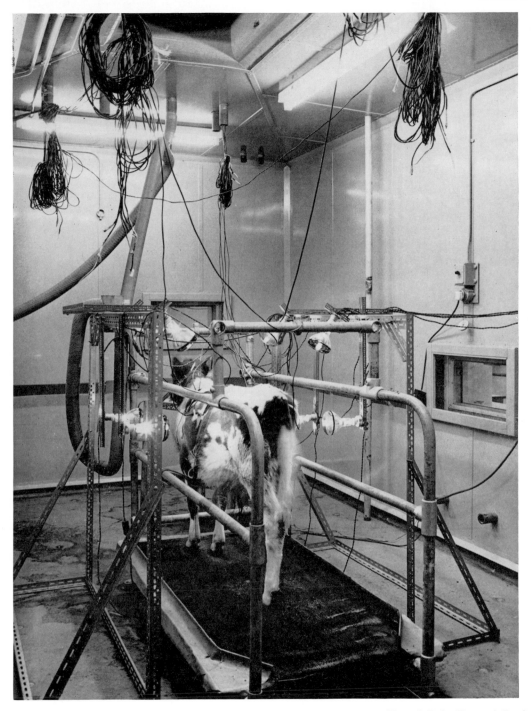

A calf being subjected to infrared radiation in the precision climatic room of the Hannah Dairy Research Institute, Ayr, Scotland. The wires are thermocouples for measuring temperature. (*Photograph by courtesy of McCallum of Ayr, Scotland.*)

PLATE 6

The exterior instrumentation of the climatic laboratory shown in Plate 5. For further explanation, see the text. (*Photograph by courtesy of McCallum of Ayr, Scotland.*)

have a profound influence on the physics of heat exchange.

Differences between breeds are scarcely less pronounced. The review by Wright (1954) contains many illustrations which highlight more clearly than words the variety of external forms within a given species. Compare, for example, the smooth-coated, loose-skinned Sindhi cow with the stocky, shaggy, Highland cattle of Scotland; or the short legs, little ears and heavy coat of the Shetland pony with the large ears, long, thin limbs and sleek hair of the Abyssinian jackass. These breeds, representative of different parts of the world, are clearly adapted in their external features to their local environments. Even within the British Isles, different breeds vary widely with locality and its associated climate.

It is generally true that those breeds of cattle developed in temperate regions are more productive than their tropical counter-parts. They are, however, poorly tolerant of heat, and if they are imported into the tropics, the performance may be no better, or frequently worse, than that of native breeds. This is largely because they rely for their productivity on a high rate of food intake, which, in a hot climate, they would be unable to sustain.

In recent years, attempts have been made to combine the heat tolerance characteristic of tropical breeds with the high productive capacity of European stock. Perhaps the most famous of these new breeds is the Santa Gertrudis, developed from the Brahman and beef Shorthorn. The research team at the Missouri Agricultural Experiment Station has made extensive comparisons of the physiological reactions to climatic stress of several breeds of cattle, including the Indian Brahman.

Some of the salient facts emerging from this work which are relevant to growth are:

a. Food intake of European breeds (Holstein, Jersey, Shorthorn) is greater in neutral and cold environments than that of the Brahman, but declines more rapidly with increasing temperature.

b. Because of its relatively low food intake, the Brahman cow has a critical temperature 10 to 15° C above that of a comparable Holstein or Jersey cow.

c. At high temperatures (35° C) the respiration rate and respiratory vaporization of the Brahman are increased much less than that of European breeds.

d. The cutaneous vaporization, although initiated at a higher temperature in the Brahman, is no more effective than that of European breeds.

e. The superior heat tolerance of the Brahman is probably explained largely by its greater surface area and lower metabolic rate.

V. RESEARCH TECHNIQUES

In his investigation of the effects of climate on animals, the research worker has two distinct, alternate approaches open to him. One, perhaps the more fundamental, is concerned with isolating the effects of individual climatic factors by exposing the animals to carefully controlled conditions in a climatic chamber. Here, all components of the climate except that which is deliberately being varied may be held constant. This method lends itself primarily to the elucidation of such basic questions as the mechanism of body temperature regulation and the control of heat loss.

The second approach is at once more empirical and more complex. One observes, using whatever measurements are appropriate, the reactions of animals to such natural variations of climate as may occur, recognizing that in this situation all components of the environment may be varying simultaneously.

The method of choice depends in the first place on the objective of the investigation. The first produces results which are relatively simple to interpret, but which may have little immediate practical application; its principle value is that the animal is exposed to a completely defined set of climatic factors which can be recreated at will. The second method relies on retrospective analy-

8

sis of a climatic situation which may never be repeated exactly, but which, taken over a sufficiently long period, may approximate that very situation which may be expected to arise in practice, and to which the results are to be applied. Furthermore, in studying large populations or groups of animals in natural surroundings, it is frequently impossible to work in a climatic laboratory without disturbing the social or ecological situation; the approach is then dictated by the prerequisites of the experiment.

Modern air-conditioning technology has provided the experimental worker with some sophisticated tools for his research. Plate 5 shows the interior of a modern climatic chamber for farm animals at the Hannah Dairy Research Institute in Scotland. The room can be maintained at any temperature between $-5°C$ and $+60°C \pm 0.2°C$. Humidity can be varied from near zero to saturation. The chamber is constructed within an outer room, so that the walls can be maintained at the same temperature as the air. There are 120 thermocouples and provisions for the measurement of respiratory rate, sweat rate, blood pressure, and electrocardiogram. In Plate 6 the exterior instrumentation of the room can be seen. Installations of this size are nearly always designed and built by professional air-conditioning engineers. More modest requirements can frequently be satisfied by mass-produced equipment such as domestic air-conditioning units which are modified and installed by the laboratory staff. The basic object in designing a system for climatic control is to create, within specified limits of accuracy, a given set of conditions which shall be uniform throughout the volume of a confined space.

For many purposes it is not necessary to control all the components of the environment; in studies of the effect of cold, for example, the influence of such variations in humidity as are likely to be encountered can frequently be considered negligible. In certain cases all that is required is a range of environments with different air tempera-

tures, since it is recognized that while variations in other factors would affect the result, air temperature is the most important, and most simply controlled, single variable.

For the most part, control equipment acts on the environment from a point source, so that gradients of environmental conditions are inevitable. These can be minimized by adequate insulation and minimal ventilation coupled with a sufficient rate of air movement within the room.

To make separate estimates of each component of the climatic environment and then to summarize the separate effects of each in order to form a prediction of the overall effect of the environment is a lengthy and difficult procedure. The channels of heat loss from the animal are not independent and their interactions are complex.

For these reasons, the idea of making a single measurement which would characterize the effect of the environment on an animal is an attractive one. The globe thermometer is an instrument that integrates the effects of air temperature, radiant temperature, and air movement. It is not, however, affected by the interaction of the various climatic factors in the same way as an animal.

Other instruments have sought to imitate animals more closely, and model pigs, sheep, and even men have been constructed. These are models more in the abstract than visual sense. A typical model of this sort consists of a hollow metal cylinder painted black so as to have an emissivity equal to that of the animal's surface, and containing a heater and thermostat which maintain the interior at a temperature of $38°C$. The measure of the net thermal effect of the environment is made by recording the electric current demanded by the thermostat in order to maintain a constant temperature. The shape and size of the instrument are chosen so that although it may bear no obvious resemblance to the animal in question, its physical characteristics are comparable. The model may be provided with internal and external insulation which can also be made to vary at a certain temperature and thus mimic the animal's physical temperature regulation. Unlike an animal, however, a model does not lose heat by evaporation, and is thus immune to changes of humidity. This restricts its usefulness at high temperatures. Instruments of this kind, simple enough to be used by unskilled persons, have been employed to measure environmental coldness in the field.

REFERENCES

Bianca, W., & Blaxter, K. L. (1961). The influence of the environment on animal production aud health under housing conditions. In: *Festschr. zum Intern. Tierzuchtkongress in Hamburg.* p. 113, Stuttgart, Eugen Ulmer.

Blaxter, K. L. (1962). *The Energy Metabolism of Ruminants.* London, Hutchinson.

Brody, S. (1945). *Bioenergetics and Growth.* New York, Reinhold.

Brody, S. (1948). Physiological backgrounds. Mo. Agric. Exp. Sta. Res. Bull. No. 423.

Burton, A. C., & Edholm, O. (1955). *Man in a Cold Environment.* London, Edward Arnold.

Dill, D. B., Adolph, E. F., & Wilber, E. G. (1964) Adaptation to the environment. Section 4. In: *Handbook of Physiology* Washington, D.C., Amer. Physiol. Soc.

Findlay, J. D. (1950). The effects of temperature, humidity air movement and solar radiation on the behaviour and physiology of cattle and other farm animals. Hannah Dairy Res. Inst. Bull. No. 9.

Hambridge, G. (Ed.) (1941). *Climate and Man.* Yb. U.S. Dept. Agric.

Hart, J. S. (1957). Climatic and temperature induced changes in the energetics of homeotherms. Rev. Canad. Biol. 16:133.

Herzfeld, C. M. (1963). *Temperature, Its Measurement and Control in Science and Industry.* Hardy, J. D. (Ed.), Vol. III, part 3. New York, Reinhold.

Kleiber, M. (1961). *The Fire of Life.* New York, John Wiley and Sons.

Schmidt-Nielsen, K. (1964). *Desert Animals.* Oxford, Clarendon Press.

Smith, R. E., & Hoijev, D. J. (1962). Metabolism and cellular function in cold acclimation. Physiol. Rev. 42:60.

Wright, N. C. (1954). The ecology of domestic animals. In: *Progress in the Physiology of Farm Animals.* J. Hammond (Ed.), London, Butterworth.

Chapter 5

Social Environment and Growth

By J. L. Albright

MAN began the domestication process by observing the behavior of animals. He started by domesticating dogs, reindeer, sheep, and goats sometime prior to 6700 B.C. (Zeuner, 1963). In retrospect it seems unfortunate that only a few of the large number of vertebrate species have been domesticated. Once man's food requirements were satisfied there was little incentive to domesticate other animals. Also, domestication of some species had been successful because the habits of the animals fitted in with the habits of man. No small part of man's success on earth is due to the animals that have clothed him, carried him, and cultivated his fields.

From the viewpoint of the animal scientist, the important aspects of behavior are those that directly influence survival, reproduction, growth, and productive capacity. This chapter presents the importance of the emerging field of social environment and its relationship to domestic animals under both natural conditions and intensive husbandry systems. Also considered are animal densities, space, and complete confinement systems. Behavioral aspects including social orders, sensory capacities, "nutritional wisdom," and their measurement are discussed.

I. NATURAL CONDITIONS

A. Wild State

The wild state refers to living under natural conditions, not tamed or domesticated. From the standpoint of ecology and resource management, a wild or natural community is likely to be more productive than any unfertilized pasture we may substitute for it. Short-grass prairies in the United States produced more meat when it was cropped by bison than it does today, when it is cropped by cows. Similarly, African game is more productive than the cattle replacing it. This conclusion appeals to the investigator who has compared the lush forest of the Amazon with the farms trying to take its place (Leigh, 1968).

B. Pasture

The behavior of animals on pasture is conditioned by such factors as climatic conditions, season, age and condition of the livestock, species and quality of forage, and manuring (since herbage near dung is rejected). The primary concern of all animals is the gathering of food. Essential features of grass-covered land upon which domestic animals are placed to graze are an adequate supply of forage, water, and shade. The best pastures are formed from well-drained, loam soils which should be rich in plant food, especially nitrogen and organic matter. It is advisable to pasture the various domestic animals separately, since horses and cattle may injure each other, and sheep and goats graze so closely to the ground that they deprive larger animals of their share of the forage. Care must be taken in selective grazing. The motivation and innate reactions of grazing animals have developed during the evolution of each species, but they are not necessarily linked to survival under man's husbandry conditions (McBride *et al.*, 1967).

1. GRAZING BEHAVIOR

From the standpoint of social behavior, cattle and sheep can be kept together. Strong social relationships develop between these species and time adds to the strength of the association (Bond *et al.*, 1967). In pastures containing more than one lamb, lambs remain together but in pastures with one lamb and one steer, the lamb will stay with the steer. In one case, when it was necessary to separate the pair, the steer attacked the man holding the lamb. It is unusual for a steer to attack man unless he is confined to a small area and becomes excited, but in this instance the constant bleating of the lamb after it was separated from the steer may have served as a distress call to which the steer responded. In other trials, sheep paired with steers followed and stayed with the steers with which they had been paired. Steers displayed no tendency to follow their paired sheep. The relationship is of a leadership-followership type, with the steer always being the leader and protector. This type of behavior (leadership-followership) is important within species, especially sheep and cattle, since the young animals follow their mothers and later generalize to all larger individuals.

In sheep, distinct seasonal changes have been found in patterns of grazing (Arnold, 1962). Grazing usually starts during the dawn and decreases at dusk. In spring and summer, there are two peaks of grazing, but there is only one in fall and winter. The percentage of grazing taking place between 6 a.m. and 6 p.m. varies with season and is highest in spring and early summer. In the spring 87 percent of grazing takes place during 6 a.m. and 6 p.m. and very little takes place during darkness. There are fewer but longer grazing periods in spring than in the other three seasons, in which only 68 per cent of the grazing takes place between 6 a.m. and 6 p.m. There is a considerable amount of grazing during darkness. The seasonal changes in the pattern of grazing appear to be independent of the grazing pressure or of the nutritional requirements of the sheep. There is no evidence that the behavior of one sheep influenced that of another, even a twin, grazing nearby.

Continuous surveillance of beef cows was made over a 12-month period which was divided into a grazing period (February to July) and a supplemental grain period (July to February). Since rainfall and forage production were below average, the beef heifer weight showed similar growth curves for the two seasons in spite of differences in climatic conditions (Fig. 5–1).

Like sheep, cattle feed at all hours of the day and night with peak periods from daybreak until 6 hours later and from late afternoon until darkness. Cows move great distances each day to secure feed, salt, and water, to search for their calves, and to pursue miscellaneous activities. The results of a study of distances traveled in six range pastures are shown in Table 5–1. As could have been expected, feeding accounted for the greatest amount of distance covered and time spent daily. Most of the movement took place during daylight hours.

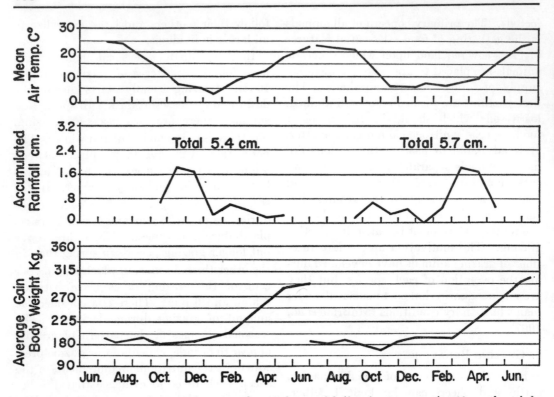

Fig. 5–1. Average growth curves for unsupplemented weaned heifers for 2 consecutive 12-month periods. Rainfall was unusually low during observation period February to February. (*From Wagnon, 1963. Calif. Agric. Exp. Sta. Bull. 799.*)

Table 5–1. Distances Traveled by Range Cows on Pasture While Engaged in Various Activities

(*Wagnon, 1963. Calif. Agric. Exp. Sta. Bull. 799.*)

Activity[1]	Meters/day	%
Feeding	2765	63
Day	1996	(72)
Night	769	(28)
Hunting calf	133	3
Running from heel fly . .	75	2
Water	395	9
Salt	115	3
Supplement	140	3
Other[2]	768	17
Day	574	(75)
Night	194	(25)
TOTAL: all activites . .	4391	100

[1] 24 hour observation periods on 6 different range pastures.

[2] Resting, idling (grooming self or other animals, pawing dirt, rubbing, scratching, waiting for supplement, movement during estrus).

A grazing dairy animal also walks a considerable distance to harvest feed. In the process she damages a noticeable amount of forage through defecation and urination and physically harms the growing plants and moist earth by trampling. In a single season of intensive grazing during which a field would be grazed four times, the hooves of the grazing animal would cover each square inch of ground 3.2 times. Larsen (1963) measured the actual distance traveled by the grazing animal. By means of a bunk feeding scheme, he also characterized the behavior of cows while grazing and being fed green feed. During intensive grazing in which the cows were given a new area to graze twice daily, each animal walked from 1,280 to 2,048 meters per day to obtain necessary forage. The cow's filling time on pasture averaged about $2\frac{1}{3}$ hours. With the normal four major fillings each day a cow must spend about $9\frac{1}{2}$ hours daily eating even where there is abundant forage. All cows take over 50 bites per minute at the beginning of the grazing period and by the end of the cycle the bites per minute are reduced by about 50 percent. Limited observations imply that the more productive animals are more aggressive and active in their eating habits than low producers and dry cows. In a bunk feeding system which provides adequate bunk space for all cows to eat simultaneously, the distance traveled by one cow will vary from 427 to 1,158 meters per day. In other words the cow covers 40 percent of the distance traveled by an intensively grazed cow. With cut green feed the time required for filling averaged 1½ hours.

Observation of a group of cattle reveals

the presence of a "boss cow." Closer observation over any extended period of time reveals a "dominance" or "peck order" throughout the group. Characteristic feeding patterns are associated with various positions within the "peck order."

Figure 5–2A shows the feeding pattern of an animal on green feed. This animal occupies position one or two in the dominance order. She moved only 107 meters and filled herself with feed in about 1 hour. She encountered no resistance to feeding wherever or whenever she desired. This characteristic feeding pattern is much different from that of a herdmate occupying a low position in the dominance order.

Figure 5–2B shows the feeding pattern of

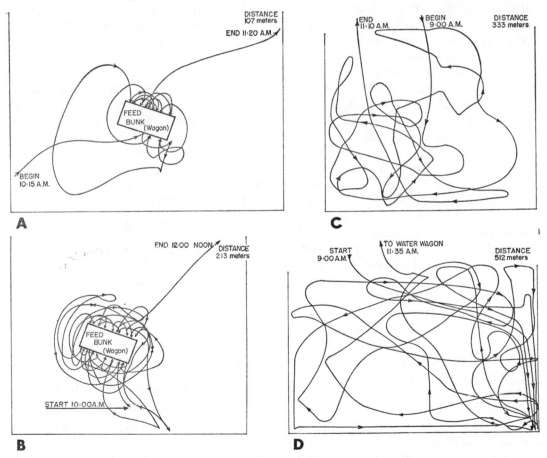

Fig. 5–2. A. Feeding pattern at a bunk. Animal occupies high position in dominance order. B. High producing small cow—low position in dominance order. She takes every opportunity afforded her to eat. C. Dominant cow able to graze unmolested. D. Grazing behavior of a cow at the low end of the peck order. Note the wandering pattern of grazing searching for a spot to graze unmolested. (*From Larsen, 1963. J. Anim. Sci. 22:1134.*)

a high-producing small cow occupying a very low spot in the peck order. This cow was knocked about continuously, required 2 hours to fill, and walked twice as far as the animal in Figure 5–2A. She showed a strong impetus to eat by taking every chance and opportunity afforded her to occupy, even for brief moments, abandoned positions at the feeding wagon.

Feeding patterns based on intensive grazing schemes show sharper characteristic differences between animals than do patterns based on bunk feeding schemes. An established dominance order is as much in evidence on pasture as on confined-lot feeding. Peck order position determines to a large extent the grazing pattern exhibited by the individual animal. Figure 5–2D shows the grazing pattern of a cow on the lower end of the order. It is characterized by wandering indicative of an animal searching for a spot to graze unmolested. In contrast to this highly active roaming pattern, Figure 5–2C shows a slow and deliberate pattern characteristic of a cow at the head of the peck order. The pattern, although a wandering one, differs from that of the animal on a lower peck order. This dominant cow could graze unmolested wherever she chose and, as a result, she walked only 333 meters during a cycle. Between the upper and lower extremes of the peck order the group members seem to "stake out a claim" which is protected from members of the group lower in the peck order. These claims seem to be grazed intensively for about the first $1\frac{1}{2}$ hours of the feeding cycle. The claims are then relinquished and a smaller area is sought in which the filling is then completed. There is considerable overlapping of the claims; this results in butting and pushing.

Among a given group of animals of this size there often seems to be a "public servant" who licks the head, face, or neck of any other animal in the group upon request in the form of a gentle nudge that is quite unlike the movement made to define a claim. The other members of the group seem to be aware of the service since they will travel some distance to request it. Position in the social order does not seem to determine who receives the service.

When the group enters a new ungrazed area, the members appear to go searching. In 10 to 15 minutes they settle down to a grazing pattern which is held for $1\frac{1}{2}$ to 2 hours. There then seems to be a rest period. When the weather is warm or flies are bothersome, they will crowd together to face the wind. This period, during which grazing comes almost to a standstill, will last from 10 to 15 minutes. No ruminating appears to take place during the rest period; only an occasional belch is noted. Grazing is then resumed in a wandering fashion and continues for about 30 minutes. Then as one cow leaves the grazing area and heads for the water, all the others follow in less than 5 minutes.

Dairy bulls in a group are quite different. They form a restless community. Apart from disturbances within the group, the presence of men, dogs, vehicles, and bulls from other groups disturbs the group's equilibrium. The signs of disturbance are bellowing, pawing, horning the ground, fighting, and outbursts of sexual activity (Dalton *et al.*, 1967).

II. INTENSIVE HUSBANDRY SYSTEMS

Marked changes have occurred in the behavior of farm animals. The results of domestication are influenced by both genetic selection and economics. Management of herds and flocks has undergone considerable change in respect to feeding and housing. During the past quarter of a century, labor-saving devices have replaced manual labor.

A. Living Space and Animal Density

Floor space requirements for maximum growth rates and efficiency of feed utilization vary with the species. Density and population size determine how frequently any two individuals meet, and factors like these affect social relationships. Overcrowding causes conflict which results in reactions frequently assigned to other causes. For example, the availability of feeding and watering space is more critical than the quantity of feed available, especially in such competitive species as swine and poultry. When animals are under social stress and in undue competition for feed space, they tend to eat very rapidly, thus affecting proper digestion and feed efficiency (Hafez & Lindsay, 1965).

While hens in large groups and with low social pressure require 0.3 to 0.4 square meter per bird, only 0.093 square meter is needed by single birds or pairs. Four birds housed in a cage lay fewer eggs, have greater mortality, and eat less food than do 3 birds housed in the same cage (Bramhall & Little, 1967). Although 4 birds per cage give greater economic returns, it is important to know the reason for the reduction in feed consumption which contributes to the decline in production. Four birds to a cage spend less time feeding because they have greater difficulty in moving about the cage and in reaching the troughs than does the 3-bird group. Although members of both groups show a preference for feeding together, this is more marked in the 3-bird cages. Some birds spend two or three times as long as others at the trough. In every settled community of grown birds, each has an almost fixed social status. In large flocks there is good evidence that subflocks of about 100 exist. Therefore, feed, water, and nests should be evenly distributed in a large pen so that members of one subgroup will not have to cross into another subgroup's area.

A dominance order may be a stabilizing feature of the group in that, once established, it tends to prevent the outbreak of serious fighting. For instance, organized groups of animals exhibit greater growth than do those maintained in a continuous state of disorganization.

Cannibalism is one of the reactions of flocks to the crowding of cage living. Early debeaking has provided a practical control of cannibalism. In addition to increased egg production, advantages of early precision debeaking by machine at 1 week of age include (a) easier handling of chicks, (b) reduction of labor and use of unskilled labor, (c) consistent accuracy even though the operator may tire, (d) consumption of less feed during rearing, and (e) reduction of cannibalism (Bramhall & Little, 1967).

When adequate feeder space is provided for pregnant gilts, there is no evidence of a dominance order and the position of penmates at the feeder space fails to show a regular pattern. With inadequate feeder space, a dominance order is exhibited, with aggressive behavior such as biting, particularly around the ears, and pushing. Body weight does not appear to be a significant factor in the establishment of dominance or rank order. Management techniques such as providing individual feeding stalls should minimize the disruptive effects of aggressive behavior (Rasmussen *et al.*, 1962).

In checking growth patterns of swine, behavioral information is of value in assessing the use of shades and wallows in summer experiments and in observing activity in space allotment studies. Under hot-weather conditions, continuous observation revealed that swine spent 44 percent of their lives lying in shade, 26 percent eating, 20 percent wallowing, 9 percent standing and walking, and 1 percent lying in the sun. Observation intervals up to but not beyond 60 minutes apart showed 42, 27, 21, 9 and 1 percent of their time devoted to the aforementioned activities (Heitman *et al.*, 1962).

Data on optimum space requirements for farm species are lacking. Recommendations suggesting space for the routine housing of domestic (laboratory) animals have been assembled (Table 5-2).

Evidence with laboratory animals indicates that initially aggressiveness and reproductive capacity of social groups can be increased proportional to population density (Hafez & Lindsay, 1965). In addition, mice kept in groups had larger adrenals than isolated controls. As their numbers increased there was an increase in adreno-

Table 5–2. Suggested Space for the Routine Housing of Laboratory Animals

(After Guide for Laboratory Animal Facilities and Care. 1968. USDHEW,
Public Health Services, Publication No. 1024.)

Species	Weight or Age	Type of Housing	Number of Animals	Housing Area per Animal Sq. feet
Cattle: adult	350–750 kg	Stanchion	1	16–27
	550–750 kg	Pen	1	120–150
calves	50–75 kg	Pen	1	24
	1½–10 mo	Group pens	up–10	20–25
Chicken: adult	3 kg	Single cage	1	1
	3 kg	Group cage	2–4	1.5–3.0
Horses	500–750 kg	Tie stall	1	44
	500–750 kg	Pen	1	144
Rabbit	2–4 kg	Cage	1–2	3
Rat	150–250 g	Single cage	1–3	0.2–0.7
	150–250 g	Group cage	4–10	0.2–0.5
Sheep and goats		Pen	–	15–22
Female with young		Pen	–	20–30
Adult male		Pen	1	20–30
Swine: adult sow		Pen	1	25–40
Sow with young		Pen	1	48–88
Adult boars	18–45 kg	Pen	1	6–12
	45–100 kg	Pen	1	12–16

corticotropic activity and a decrease in gonadotropin output. It appears that behavior operates as a homeostatic mechanism for size of populations. The hypothesis has been advanced by Wynne-Edwards (1962) that, unlike man, most animals maintain fairly constant population levels. They do so by forms of social behavior (individual territories) that limit reproduction to avoid overexploitation of food resources. Before the classical theory of Wynne-Edwards was advanced, the prevailing belief had been that population is regulated by a set of negative natural controls. It was accepted that animals produce young as fast as they can, and that the main factors that keep population density within fixed limits are predators, starvation, accidents, and parasites causing disease.

In livestock production, oftentimes individuals must be rotated and moved from one group to another. Dairymen have separated cows into several production categories— high, medium, low and nonlactating (dry). As one cow moves to a new area, cows in the new group often resent and butt her. Several approaches have been tried to elim-

inate the problem. One method is separation of lots by an electric fence in close proximity to the cows so that they can see one another and even reach through the fence to touch one another. Experimental introduction of unfamiliar animals to groups of dairy cows gives rise to increased agonistic (combative) behavior (Hafez & Schein, 1969). New herd members should be integrated with the herd by degrees in an adjoining field or pen.

An attempt was made to determine how early in life such agonistic behavior occurs. Four sets of male chicks were used. One set was placed in isolation and the other 3 were placed in pens. Every other day one chick from each of two groups was interchanged. Similarly, a chick from isolation was placed into the third penned group as one was removed and placed in isolation. The two rotation patterns were continued for 8 weeks after hatching (Fig. 5–3). All forms of agonistic behavior were at a higher frequency in the pen containing chicks from isolation. Chicks introduced from isolation exhibited a stronger stimulus for attack than those rotated between groups.

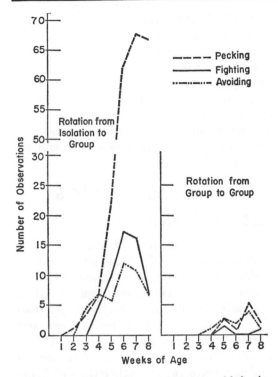

Fig. 5–3. Differences in the frequency of behavior patterns of male chicks rotated between isolation and a group and those rotated from group to group. The latter show social inertia. (*From Guhl, 1958. Anim. Behav. 6:92.*)

A corollary exists with dairy heifers. Rank order is not stable and changes frequently and drastically with the removal of a subject or introduction of a new animal into the group. Heifers with dominating tendencies had a threefold greater increase in total conflicts with other animals than did a random group of heifers. Body weight was significantly related to dominance in one third of the heifers tested (Hook *et al.*, 1965).

B. Complete Confinement

A trend different from the traditional pasture scene has emerged which favors keeping animals on concrete throughout their lifetime. This trend was accepted rapidly at the expense of not having available data on the long-term effects of such confinement. Concern has been expressed over the long-range effects of this environment upon the foot and limb structure of cattle and swine. Because of the lack of data and as a safeguard, many dairymen remove their cows from concrete at least during the dry period. Others remove their cows to exercise or pasture areas whenever the opportunity presents itself.

Further concern has been expressed over the unsolved problem of managing the excreta from large aggregations of animals. Removing animal wastes by hauling, anaerobic or aerobic digestion, incineration, or recycling is expensive and perhaps explains why, to date, the large commercial complete-confinement operations are to be found in the warmer climates where the sun dries the manure and barnlots.

Livestock wastes (manure and bedding) were once considered advantageous by-products of dairy farms. Commercial fertilizer manufacturers have become quite competitive and the cost of the labor and equipment to handle manure in confinement is expensive. In an attempt to reduce both the volume and number of times manure is handled, farmers have adopted new management practices and housing systems. Slotted floors represent one attempt to reduce the cost and time of handling manure. European farmers were using slotted floors as early as 1950. About 15 years later the first slats were installed and used for dairy cattle in the United States in New York (steel) and Wisconsin (concrete). Cattle walk differently when placed upon slats. There is an absence of aggressive behavior, possibly because animals have much less confidence on slotted floors than on other types of flooring (Lees, 1962). The average daily gain of feeder cattle on slats with a cleaner surface was superior to a flat concrete floor (Morrison *et al.*, 1966). Animals with sheltered slotted floors and 58 ft² gained as much as unsheltered animals in a dirt corral with 355 ft² (Givens *et al.*, 1967).

C. Housing and Facilities

Domestic animals respond differently to various types of housing. Herdsmen study the behavior of their animals and use this knowledge to increase production. For instance, feeders and watering systems must be placed where the young or inexperienced animal can find them.

Given the opportunity, swine and cattle avoid lying in their excreta. In particular, if dairy cows are kept in free stalls (individual stalls within a loose housing barn), they void excreta in the passageway and remain much cleaner than with other systems of housing. After overcoming the apprehension of lying down, beef cattle prefer sloping floors that drain and stay cleaner than flat ones. A slope of 5 to 7 degrees affects gain and efficiency more than a nearly flat concrete floor with its accumulation of manure, which actually depresses gains (Morrison *et al.*, 1966).

Except when air-conditioning is used, housed cattle tend to be restless and do not lie down for very long. This activity may be due to conditions other than housing (Crowl, 1952). Over a two-year period, Crowl observed differential responses to the type of bedding used in a stanchion barn. Employing the number of times observed standing as an index of comfort and observing cows at 15-minute intervals for a total of 34 times during the night, he found that cows were least comfortable when ground or crushed corncobs were used and most comfortable when chopped wheat straw was used. Wood shavings and sawdust gave intermediate responses.

Behavioral responses of cows to loose- and free-stall housing under the same roof in a complete confinement system were observed. Cows in loose housing exhibited more group action while animals in free stall were more individualistic. About 40 percent showed a significant preference for individual stalls, while 60 percent used the 20 stalls at random (Schmisseur *et al.*, 1966).

Man determines the space and facilities available for domestic animals. Management methods based on cost per man-hour rather than the optimal productive efficiency of animals often dictate practices. Accessibility of feed may be more important to animals than the actual amount of food present. A predetermined minimum amount of feeding space per animals in several areas of the barnlot is best for optimum production. Efforts must be made for reducing competition for water, minerals, and shelter. Also, space, density, and distribution of feed are closely associated factors. Some competition is essential since individually fed animals do not eat as much or gain as well as animals in a group.

III. BEHAVIORAL ASPECTS

A. Social Behavior

With the exception of cats, domestic animals form herds, flocks, or packs. Strong social relationships are frequently found under grazing and range conditions. Further, domestic animals have a very close relationship to man because of frequent contacts during feeding, herding, and milking. Unfortunately, many animals are often treated like inanimate objects rather than as living habit-forming organisms. Possibly due to haste, ignorance, or fatigue, man considers only the work to be done rather than the needs and responses of his animals. To be successful with livestock, he should be able to understand and anticipate their needs.

1. SOCIAL ORDER

A herd of cows behaves as a unit in which all members engage in the same behavior at the same time. Within the unit, several social hierarchies which function to define priority rights to food, water, and leadership for each member are known to develop. At least 3 (dominance, leadership-followership, and entrance order) and perhaps 4 (submission) distinct social orders evolve among cattle. Milking order and leadership are related in dairy cows and the dominance hierarchy is so stable that observation for a single day can determine the order (Dickson *et al.*, 1965). Social order is a function of the test to which animals are subjected. The dominance-submission tendencies of calves can be altered by manipulating early feeding and rearing conditions (Donaldson *et al.*,

1966). The factors which normally determine the ultimate dominance pattern in a herd of cattle are weight, age, and breed. Range cows low in the social order do not gain weight as well as those of high rank since the low-ranking ones do not have ready access to feed. Low-ranking members wait several hours in order to get food or water.

A dominance order for cattle exists in the stockyard when the animals come to be weighed. Animals pass over the scales in a consistent order. Neither weight, sex, horned condition, nor weaning was related to weighing order (Tulloh, 1961). A relationship between temperament and live weight exists, indicating that animals which are generally docile grow better than animals that are restless, nervous, wild, or aggressive. Dietrich *et al.* (1965) and Albright *et al.* (1966) observed that there was a consistent entrance order into the milking parlor. They attempted to alter it by training each cow to come in when her number was called. By the end of 15 days training, 70 percent of the cows came when their number was called. When the calling was discontinued, however, the cows reverted to their former entrance order. Injections of a tranquilizer did not change the routine milking order. Body weight and age were not significantly correlated with entrance rank nor was entrance order significantly correlated with milk production.

Preweaning. The neonate is equipped with well-developed behavioral patterns such as suckling and play. Other types of behavior are developed under the influence of stimulation from the mother and from subsequent learning, training, and experience. Ungulates are very precocious. They stand and walk shortly after birth and learn the source of food rapidly. Domestic animals have a brief period early in life when their first social attachments develop. If the neonate is exposed only to human beings during the first 2 or 3 days of life, it becomes attached to its handler in spite of subsequent opportunities to become attached to its own mother. This phenomenon is called "socialization" and in ungulates this critical period occurs within the first week of life.

Handling young animals in early life will generally cause them to be easier to handle later on. In some cases in which the man-animal interaction is high because of too few animals per caretaker, certain problems develop such as: pets that anticipate and expect a great deal of attention, difficulty as breeders, and excessive fleshing. Calves that are reared in group pens and bucket-fed tend to suckle each other after drinking. This tendency carries through into their adult life and poses a problem for herdsmen. The solution is to feed and rear calves individually.

Postweaning. Following weaning, most animals are raised in groups with the exception of the breeding male or the female about due to give birth to her young. The social orders that are formed at the feed bunk, gate, and milking parlor are quite constant. Unfamiliar or new animals entering the group give rise to instability and conflict. As a result, certain animals eat less and lose weight.

Several behavioral differences exist between normal adult pigs and pigs fed a diet deficient in protein for eight weeks immediately after early weaning for experimental purposes. Tests revealed a change in emotionality, motivation, and a more limited mental capacity in the protein-deficient group. When the experimental pigs were put on an *ad libitum* diet, they ate more at more frequent intervals than the controls, particularly at night (Moore *et al.*, 1967).

B. Sensory Capacities

The recognition and study of sensory differences in domestic animals have had their beginning only recently. Data are therefore limited.

1. GUSTATORY

In cattle, sweet taste is most acceptable and bitter least acceptable, with salt and sour intermediate. Kare (1966) specified that cows find sucrose appealing but are indifferent to saccharine. Cows can distinguish tastes as well as different concentrations of each. Ambient temperature modifies taste reception. Domestic chickens are so sensitive to water temperature that when it is raised to slightly above their body temperature (40° C), they will suffer from acute thirst rather than drink. In dogs, cats, rabbits, and chickens, unlike in man, water conveys a taste stimulus.

Self-regulation of intake of specific nutrients or drugs is desirable. Knowledge of the sensory mechanism can facilitate these objectives. For example, animals (including fowl), stop eating but continue to drink when they are ill. For this reason, drugs are administered to them in water (Kare, 1966).

Nutritional Wisdom. Animals are commonly thought to possess some degree of nutritional wisdom as a guide in the selection of their food, but grazing sheep and cattle fail to correct a phosphate deficiency when presented the opportunity. Also cattle on phosphorus-deficient pastures exhibit no preference between a calcium supplement and a calcium-phosphorus supplement. Still, cattle on calcium-phosphorus deficient diets eat cinders, bones, and sacking materials that are largely ignored by normal cattle. On the other hand, sodium deficiency in sheep that were confined or on pasture was alleviated when sodium (solutions or forage) was available. Sheep on a self-selection feeding system do not choose feed to meet their nutritional requirements when they are bearing or rearing a lamb or twins or when they are suffering from mineral deficiencies (Gordon & Tribe, 1951).

Ruminants evolved from environments in which survival depended upon selection of plant materials containing adequate energy, protein, and minerals. They show an extreme preference for green forage instead of dry stored forage. This is nutritionally wise in many situations. However, on improved pastures, this preference means that rumi-

nants acquire a diet that is high in protein and low in fermentable carbohydrates. On the other hand, when feeding on dry forage, ruminants, in particular sheep, may not select the legume component when this choice could be nutritionally wise (Arnold *et al.*, 1966; Hafez, 1968).

Beef and dairy calves show some nutritional wisdom regarding the composition of the milk they consume. Less time was spent suckling and more dry matter consumed on a 19.5 percent total solids ration as compared to 13 percent (cow's milk) or 6.5 percent dry matter (milk replacer). As might be expected, calves fed a 19.5 percent total solids ration by machine (Nurs-ette) grew faster than calves fed a ration of 6.5 percent total solids (Hafez & Lineweaver, 1968).

Although domestic chickens reject xylose solutions which impair vision, they also reject nutritious alfalfa. Chickens will also consume sodium tungstate solutions in choice situations until they die, or, if they are vitamin deficient, they will select thiamine-enriched feed over one devoid of the vitamin. In chick experiments in which feed constituents were replaced with alternatives of lesser or greater nutritional value, chick preference reactions were not consistent as an index of the feed material. In fact, chickens died of protein deficiency with a dilute casein solution placed in front of them. Similar experiences were found with rats regarding discrimination between essential and nonessential amino acids. The animals selected and preferred the nonessential amino acids (glycine and alanine) and avoided the essential amino acids (methionine, tryptophane, and valine). In summary, whether or not an animal prefers a food is not a reliable guide to its nutritional value (Kare, 1966).

2. AUDITORY

Cattle have well-developed hearing (Baryshnikov & Kokorina, 1964). Over an 11-year period, Wagnon (1965) found there was an average response to vocal calls of

95 percent during the hand feeding of supplements to range cattle. When a battery-operated electronic device with a prerecorded message (herdsman's voice) was fastened to a normal cow-halter, it was effective in that the bovine responded to and could be herded from the pasture with a mechanical auditory stimulus (Albright et al., 1966).

Behavior During Sound Exposure. Swine can tolerate for short periods and become conditioned over long periods to sound intensities up to 120 decibels with no effect upon growth (Table 5–3). When a nursing sow was subjected to loud sounds she initially exhibited alarm. She rose to her feet and appeared to search for the source of sound, but then her young resumed suckling and she was apparently indifferent to it. When pigs were exposed to the absence of their mother, however, they did appear to be alarmed and crowded together. No differences were detected in the responses to the various sounds used: frequencies ranging from 200 to 5,000 cycles per second at 100 to 120 decibels of intensity and a recorded squeal of a pig reproduced at 100 decibels all elicited like responses. Animals were almost invariably alarmed on initial exposure to loud sounds, but quickly became conditioned to them.

There was no effect on weight gain but hatchability and setting tendency of hens were affected when chickens were subjected to aircraft sounds in excess of 110 decibels. In Japan, motorboat racing noises caused significant decreases in breeding efficiency and milk production of dairy cows. The growth of calves and heifers was not affected.

Table 5–3. Effects of Chronic Exposure to Aircraft Sound[1]

(Bond, et al., 1963. USDA Tech. Bull. 1280.)

Group	Gain Daily kg	Feed per Pig Daily, kg	Feed per kg of Gain
Control	0.8	3.1	1.8
Exposed	0.8	3.0	1.8

[1] Intensity of sound was 120 to 135 decibels. A sound intensity of 90 decibels is roughly equivalent to that heard near an elevated train or a farm tractor.

3. VISUAL, OLFACTORY, CUTANEOUS

Cattle are able to distinguish the colors of the spectrum and differentiate the intensity of light. An assessment of the role of vision in the development of body growth was made with twin dairy bulls (Table 5–4). The eyes

Table 5–4. Growth in Normal and Blind Twin Bulls

(Hale, 1966. J. Anim. Sci. 25, Suppl. 36.)

Pair and Breed	Body Weight (Kg) Normal	Blind
Brown Swiss	500	441
Holstein	536	455
Holstein	484	509
Ayrshire[1]	632	586
Holstein	477	473
Holstein[1]	514	484
AVERAGE	524	491

[1] Identical twins

of one of each pair were enucleated prior to 6 weeks of age. The animals were reared in separate, nonadjoining stalls in the same building. At 18 months of age, the bulls with sight weighed 524 kg as compared to their blind twins, which weighed 491 kg. Histological comparisons indicated that absence of vision had no adverse effect on gonadal structure and development.

High sensitivity of the olfactory bulbs has been noted. Cattle are able to differentiate between the following smells: ammonia 1:10,000 and 5 percent acetic acid, amyloacetate and camphor, fresh and stale corn silage, bergamot, lavender and cinnamon oils. Absolute skin sensitivity fluctuated between the pressure range of 0.1 to 2.5 cm water.

The functional development of the sensory organs is arranged in the following order: olfactory, cutaneous, auditory, visual, and gustatory (Baryshnikov & Kokorina, 1964).

C. Abnormal Behavior

Most abnormal behavior is associated with some physical or physiological aberration. Cattle with cerebral lesions or brain tumors will often revert to abnormal gaits and the head pushing behavior exhibited by calves. They characteristically lose weight and must be isolated from the herd. Overcrowding can cause physical injury such as lameness and undesirable (abnormal) behavior such as tail-biting in swine. To circumvent this latter problem, many swine herdsmen routinely remove the tails from their pigs.

Abnormal patterns of maternal behavior have been reported in sheep (Alexander, 1960). In some ewes that had lambed previously, maternal activity was first seen hours or days before parturition and occasionally led to attachment of a lamb to a ewe that was not its mother. Complete desertion at birth and delays in the onset of grooming were associated with ewes undergoing long and exhaustive parturition. Similar responses were found with cattle. Heifer calves were removed from their dams and reared from birth under experimental conditions of minimal and maximum conflict. (Donaldson *et al.*, 1966). Normal growth was apparent in all groups. Two years later, all heifers that had been fed and reared separately accepted their calves, cleaned them, protected them, and allowed them to nurse. Of the animals reared under one of three competitive systems, 50 percent rejected their calves, and neither cleaned them nor allowed them to nurse. Heifers from the noncompetitive group (fed and reared separately) made little or no noise when caring for their calves. Heifers from the competitive group who accepted their calves were quite vocal. When these same animals calved for the second time, all cows accepted their offspring.

Stockmen have observed abnormal behavior such as cows who develop the habit of sucking themselves as well as other cows. Some animals are consistent in terms of licking their housing facilities, watering troughs, and the earth when presented with the opportunity. Licking is confounded by the interaction of appetite, nutritional need, and curiosity. Whether this activity is abnormal or due to some dietary deficiency remains to be resolved.

IV. RESEARCH TECHNIQUES

A. Analysis

Shepherds and herdsmen have analyzed their flocks and herds and made decisions for the growth and development of the animals of the next generation from the acquired information. Although much of the behavioral information handed down from one generation to the next contains a fabric of folklore intermixed with superstition, students of growth should use this storehouse of knowledge regarding social behavior.

Domesticated species have been selected to include man as a part of their environment. Man, being an inconsistent member of the environment, however, can influence the behavior of the observed. The response of the animals to the observer varies according to the species, housing or natural environment, parturition, time of day (*e.g.*, mice are more active at night), and the extent of previous contact between the man and his animals.

Analysis of collected data is more accurate and sophisticated since the advent of electronic computers. Data comparing individual rank, teat order, and dominance-subordination patterns with growth rates would be helpful in assessing their relative importance in husbandry.

B. Observations

Methods which can be used to eliminate, at least partially, the effect of the observer upon the observed include (Ewbank, 1966): the use of binoculars to watch from a distance; the use of cinematography with an automatically controlled camera or closed-circuit television; time-lapse photography (a movie camera with single frame exposures at predetermined intervals); the use of animal hides in hiding; and the use of "one-way" windows.

The ideal procedure is one which considers all behavior, but continuous observations are fatiguing, exasperating, and time-consuming. More animals can be observed with interval observation techniques for periods of 5, 15, 30 and 60 minutes. Major activities (grazing, ruminating, idling) can be predicted with 15 and 30 minute intervals, but continuous observations are needed to witness the total behavioral pattern and such minor activities as urination, defecation, drinking, and walking. Several animals need to be observed since highly significant differences in behavior exist among the individual animals.

The use of a light-weight, battery-operated tape recorder can solve the difficulty of observing and recording the behavior of the subjects by one person. The mere act of writing down comments in a notebook inevitably means a break in visual contact. The use of a recording code can simplify and improve data collection. Symbols and check sheets aid in developing a recording code. Timing devices, in particular a clock which records only the hour, minute, and seconds, help to minimize mistakes that could be made by the observer. The exact time is recorded without having to interpret a face clock.

The problem of individual identification occurs when dealing with strange animals and large groups. Marking crayons, colored neck tags, conspicuous ear tag numbers, and numbers that are glued to the rump area of the hide (used at sales and auction yards) are useful and helpful. Freeze-branding methods facilitate permanent identification. Freeze-branding with cattle is done by applying a precooled ($-70°$ C) copper or bronze branding iron to a hide which has been clipped and is wet with alcohol. The branding instrument should be applied for 30 seconds for maximum results. Because the branding destroys melanocytes, the subsequent hair growth is white and remains so. With white cattle the hair regrowth follow-ing freeze-branding is slightly discolored (Farrell *et al.*, 1966).

C. Measurement

Compared to behavior, there are simpler and more convenient characteristics of animals which can be measured such as body weight, growth, litter size, milking time, and so on. In the final analysis, animal scientists are ultimately concerned with the the intact organism, of which behavior is only a part.

Two fundamental objectives of any scientific study are to describe and predict. Describing behavior is a very broad goal, ranging from brief verbal statements to sophisticated quantitative statements. Predictions of behavior range from complete inability to make estimates of future performance to more exact statements. Adequate description and prediction can be achieved only when techniques have been developed for measuring behavior.

Devices have been developed (Hafez & Schein, 1969) such as implanted transmitters in cows to record their heart rates; one which records jaw movements and bites per minute; rangemeter harness to determine distances travelled; an electronic counter activated when an animal licks a tube to obtain water; pedometer attached to the leg to record steps taken prior, during, and following estrus; photoelectric relays; and a battery-powered electronic device with a prerecorded message fastened to a cow halter to herd cattle with a prerecorded familiar voice stimulus.

Such items as procedure and reliability of measurement; methods of behavior study including naturalistic, experimental, and statistical analysis; special methods—comparative, age-behavior relationships, using each animal as its own control, matching, split-litter techniques; the relation of animal growth to major factors such as the nervous system, hormones, deprivation, ingestion, and parental, sexual, and social behavior are suggested for further study (Denenberg & Banks, 1969).

REFERENCES

Albright, J. L., Gordon, W. P., Black, W. C., Dietrich, J. P., Snyder, W. W., & Meadows, C. E. (1966). Behavioral responses of cows to auditory training. J. Dairy Sci. 49:104–106.

Alexander, G. (1960). Maternal behavior in the Merino ewe. Proc. Aust. Soc. Anim. Prod. 3:105–114.

Arnold, G. W. (1962). The influence of several factors in determining the grazing behavior of Border Leicester X Merino sheep. J. Brit. Grassland Soc. 17:41–51.

Arnold, G. W., Ball, J., McManus, W. R., & Bush, I. G. (1966). Studies on the diet of the grazing animal. I. Seasonal changes in the diet of sheep grazing on pastures of different availability and composition. Aust. J. Agric. Res. 17:503–556.

Baryshnikov, I. A., & Kokorina, E. P. (1964). Higher nervous activity of cattle. Dairy Sci. Abstr. 26:97–115.

Bond, J., Carlson, G. E., Jackson, C., Jr., & Curry, W. A. (1967). Social cohesion of steers and sheep as a possible variable in grazing studies. Agron. J. 59:481–482.

Bramhall, E. L., & Little, T. M. (1967). Effects of de-beaking and cage density on egg production. Calif. Agric. 21:2–4.

Crowl, B. W. (1952). Bedding cows with different materials. Hoard's Dairyman 98:879, 893.

Dalton, D. C., Pearson, M. E., & Sheard, M. (1967). The behavior of dairy bulls kept in groups. Anim. Prod. 9:1–5.

Denenberg, V. H., & Banks, E. M. (1969). Techniques of measurement and evaluation. Chap. 8. In: The Behavior of Domestic Animals. E. S. E. Hafez (Ed.), 2nd ed., London, Balliere, Tindall & Cox Ltd.

Dickson, D. P., Barr, G. R., & Wiechert, D. A. (1965). Social relationship of dairy cows in a feed lot. J. Dairy Sci. 48:795.

Dietrich, J. P., Snyder, W. W., Meadows, C. E., & Albright, J. L. (1965). Rank order in cows. Amer. Zool. 5:713.

Donaldson, S. L., Black, W. C., & Albright, J. L. (1966). The effect of early feeding and rearing experiences on dominance, aggression and submission behavior in young heifer calves. Amer. Zool. 6:559.

Ewbank, R. (1966). Observational techniques in the study of the behavior of the domesticated animals. Proc. Soc. Vet. Ethology 1:12–14.

Farrell, R. K., Koger, L. M., & Winward, L. D. (1966). Freeze-branding of cattle, dogs and cats. J. Amer. Vet. Med. Ass. 149:745–752.

Givens, R. L., Morrison, S. R., Garrett, W. N., & Hight, W. B. (1967). Influence of pen design and winter shelter on beef performance. Univ. Calif. Beef Day pp. 64–71.

Gordon, J. G., & Tribe, D. E. (1951). Self selection of diet by pregnant ewes. J. Agric. Sci. 41:187–190.

Hafez, E. S. E. (1968). Behavioral adaptation. Chap. 15. In: Adaptation of Domestic Animals. E. S. E. Hafez (Ed.), pp. 202–214. Philadelphia, Lea & Febiger.

Hafez, E. S. E., & Lindsay, D. R. (1965). Behavioral responses in farm animals and their relevance to research techniques. Anim. Breed. Abstr. 33:1–16.

Hafez, E. S. E., & Lineweaver, J. A. (1968). Suckling behavior in natural and artificially fed neonate calves. Z. Tierpsychol. 25:187–198.

Hafez, E. S. E., & Schein, M. W. (1969). The behavior of cattle. Chap. 9. In: The Behavior of Domestic Animals. E. S. E. Hafez (Ed.), 2nd ed. London, Balliere, Tindall and Cox Ltd.

Heitman, H., Jr., Hahn, L., Bond, T. E., & Kelly, C. F. (1962). Continuous versus periodic observations in behavior studies with swine raised in confinement. Anim. Behav. 10:165–167.

Hook, S. L., Donaldson, S. L., & Albright, J. L. (1965). A study of social dominance behavior in young cattle. Amer. Zool. 5:714.

Kare, M. R. (1966). Taste perception in animals. Agric. Sci. Rev. 4:10–15.

Lees, J. L. (1962). Dairy cows on slats. Agriculture 69:226–229.

Leigh, E. G. (1968). Making ecology an applied science. Science 160:1326–1327.

McBride, G., Arnold, G. W., Alexander, G. & Lynch, J. J. (1967). Ecological aspects of the behavior of domestic animals. Proc. Ecol. Soc. Austr. 2:133–165.

Moore, A. U., Barnes, R. H., Pond, G. W., & Reid, I. M. (1967). Behavioral characteristics of adult pigs after recovery from early malnutrition. Amer. Zool. 7:217.

Morrison, S. R., Mendel, V. E., & Bond, T. E. (1966). Sloping floors for beef cattle feedlots. Management of farm animal wastes. ASAE Publication No. 3P-0366, p. 41–43, St. Joseph, Mich.

Rasmussen, O. B., Banks, E. M., Berry, T. H., & Becker, D. E. (1962). Social dominance in gilts. J. Anim. Sci. 21:520–522.

Schmisseur, W. E., Albright, J. L., Dillon, W. M., Kehrberg, E. W., & Morris, W. H. M. (1966). Animal behavior responses to loose and free stall housing. J. Dairy Sci. 49:102–104.

Tulloh, N. M. (1961). Behavior of cattle in yards. I. Weighing order and behavior before entering scales. Anim. Behav. 9:20–24.

Wagnon, K. A. (1965). Social dominance and its effects on supplementation of range cows. Univ. Calif. Beef Day. pp. 38–43.

Wynne-Edwards, V. C. (1962). Animal Dispersion in Relation to Social Behavior. Edinburgh, Oliver & Boyd, Ltd.

Zeuner, F. E. (1963). A History of Domesticated Animals. London, Hutchinson.

Chapter 6

Voluntary Feed Intake

By B. R. Baumgardt

I. FEED INTAKE AS A HOMEOSTATIC MECHANISM

A. Relation of Voluntary Intake to Energy Requirements

THE goal of animal nutritionists is to provide nutrients to the animal at levels, and in such a balance, so as to maintain the animal in a homeostatic condition while allowing productive processes to proceed at a high rate. *Homeostasis* refers to the tendency of an organism to maintain a uniform and beneficial physiological stability within and among its parts. The American physiologist, Walter B. Cannon, is generally credited with first using the term "homeostasis" to describe such mechanisms (Cannon, 1929). He recognized, however, that the concept of homeostasis had been alluded to by the German physiologist, Pfluger, and the Belgian physiologist, Fredericq, in the last half of the 19th century.

Negative feedback is a concept of more recent origin which implies that a deviation in one direction is followed by a reaction in the opposite direction. Thus, an updated definition of homeostasis is "the self-regulating negative feedback systems which serve to maintain the constancy of the internal environment." The regulation of feed intake is one example of such a homeostatic system (Fig. 6–1). Energy balance of the animal body is determined by the difference between energy input (feed) and energy output in the form of feces, urine, and heat plus the energy expended for maintenance, milk production, reproduction, and activity. A positive energy balance is the result of an increase in feed-energy input over energy output, a decreased output, or combination of these. Similarly, a decrease in feed-energy input or an increased energy output results in negative energy balance. The tendency of an adult animal to maintain zero energy balance (*i.e.*, to maintain a constant body weight) over rather long periods of time will be examined before proceeding to the components of the regulating system.

Most adult humans and animals maintain body weight at the same level, often for years, in spite of great variations in energy expenditure. This homeostatic principle can also be seen in children and young animals which grow at well-defined rates even while exhibiting marked variation in activity and energy expenditure. This implies that feed

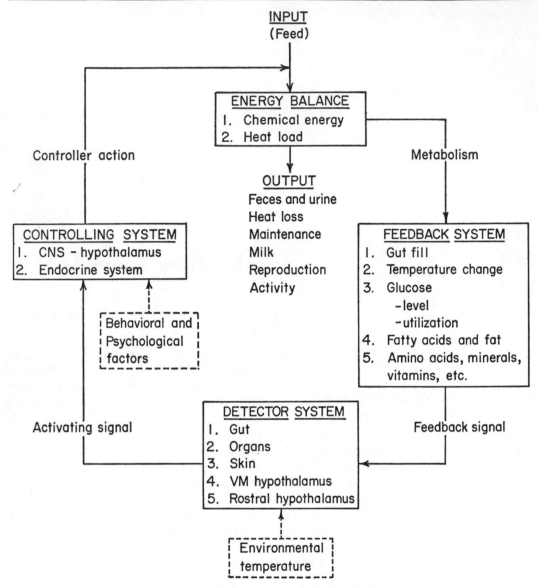

Fig. 6–1. A diagram of the components of the system regulating food intake.

intake must be adapted to energy expenditure. An excellent demonstration of the maintenance of body weight in the rat, with variations in exercise, is shown in Figure 6–2. The precision of the regulation is good within a wide range of activity. It is apparent, however, that below this range (in the "sedentary" range) and above it (in the "exhaustion" range) energy homeostasis is not maintained.

If the energy content of a ration is decreased by dilution with an indigestible material, the animal will adjust the amount of food eaten so that caloric intake remains fairly stable. Such a response was demonstrated by Cowgill with dogs as early as 1928 and has since been shown in many species including chickens, rats, swine, sheep, and cattle. Over a rather wide range of energy concentration in the ration, animals are able

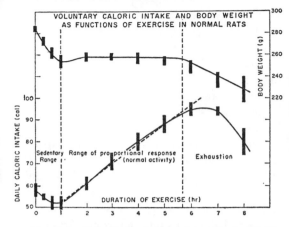

VOLUNTARY CALORIC INTAKE AND BODY WEIGHT
AS FUNCTIONS OF EXERCISE IN NORMAL RATS

FIG. 6-2. Voluntary caloric intake and body weight as functions of exercise in normal animals. (*By permission from Mayer et al., 1954. Amer. J. Physiol. 177:544.*)

to adjust the amount of feed voluntarily consumed so as to maintain equal caloric intakes. Under a variety of circumstances and on a variety of rations, "animals eat for calories."

At first glance, ruminants appear to be exceptions to the generalization concerning energy homeostasis. The feed intake situation seems to operate in reverse on many roughage feeding programs. For example, ruminants consume more early-cut immature forage than late-cut mature forage. Since the digestible energy content of the early-cut forage is higher, the animal consumes much more energy from the early-cut forage; energy intake differs between these two forages. This is an example of a breakdown in a homeostatic system due to interference by a secondary but potent force. The very low energy concentration in the late-cut forage coupled with its bulky nature results in a filling or saturation of digestive tract capacity at a level of energy intake below that which is called for by the homeostatic mechanism. This phenomenon can be demonstrated in nonruminant species as well as in ruminants if the ration is diluted to a very large extent with indigestible, bulky material. The example with ruminants is more obvious because their normal ration is often high in roughage.

B. The Central Nervous System and Feed Intake

The sequence of events in the homeostatic system regulating feed intake and energy balance indicates the feedback or signal system to be next in line for consideration. However, it will be more convenient and intelligible to consider next the means by which signals are detected and integrated (detector and controlling systems, Fig. 6-1). Ultimately, the brain is the prime controller of feeding (Anand, 1961). The basic control systems involving the central nervous system have facilitory and inhibitory mechanisms stemming from the higher nervous levels which are superimposed on reflex actions operating through lower levels. Most of the homeostatic mechanisms in the animal body operate in similar ways. For example, the regulation of respiration and the pattern of nervous regulation of most other visceral and autonomic activities, including those of the cardiovascular system, digestive system, and body temperature, are similar. Activities of the visceral and autonomic nervous system along with those of the endocrine system are ultimately responsible for homeostasis. The arrangements within the central nervous system for the regulation of feed intake follow a similar pattern. The search for localization of these functions in the central nervous system has involved destruction, damage, or stimulation of particular structure by electrical or chemical means. The cerebral cortex and limbic system are involved in feeding behavior, prejudices, selection preferences, and other complex integrations which finally determine what and how much an animal will eat. However, conclusive information is not available to permit a clear-cut description of the exact mechanisms.

The hypothalamus is very directly involved in the regulation of feed intake and energy balance and much is known about its mode of operation. "Centers" involved in feeding and in satiety have been attributed to the hypothalamus. In this context, the term "centers" refers to areas of greater density of cells with a given function (in this case regulation of feed intake) rather than areas of exclusive localization of cells of a certain type. Reference will be made to the following areas which exist as symmetrical, bilateral structures:

> Feeding center: lateral areas
> Satiety center: ventromedial areas

The ventromedial nuclei or satiety center may be regarded as an integrating relay station for satiety information (feedback signal, Fig. 6–1). If this region is destroyed by electrical or chemical means, the animal no longer is able to receive satiety signals. This results in cumulative overeating and, finally, obesity, as was first demonstrated in rats by Hetherington and Ranson (1939). Since these lesions were rather large, and may have had diminished spontaneous motor activity, it was suggested that decreased exercise may have caused the obesity. However, it is now well established that controlled lesions in the ventromedial areas of the hypothalamus result in overeating. The effects of hypothalamic lesions on food intake have been demonstrated in rats, mice, cats, monkeys, and, very recently, in goats (Baile *et al.*, 1967). Obesity can also be produced in mice by injecting them with gold thioglucose at an LD_{50} level. This compound destroys certain small areas of the brain, including the ventromedial nuclei, and thus produces a chemical lesion with results similar to the electrically induced lesions.

An opposite effect on feed intake is produced by placing lesions in the outer part of the lateral hypothalamus. Animals treated in this way show a complete lack of feed intake behavior and will die unless they are force-fed. Some of these animals will swallow food placed in the mouth, indicating that some eating reflexes remain operative although they will not voluntarily eat. If kept alive by forced-feeding, some lateral-lesioned animals may recover and resume eating, especially if offered a very palatable feed.

Activity of the lateral areas of the hypothalamus is necessary for feeding behavior, and thus, these areas are called "the feeding centers." The feeding center serves to integrate all the complex visual, auditory, olfactory, tactile, gustatory, and digestive tract reflexes associated with feed intake behavior.

Effects opposite to those produced by forming lesions can be achieved by electrical stimulation. Thus, stimulation of the lateral area induces eating and stimulation of the ventromedial area produces satiety. In goats and sheep, rumination results from the stimulation of the same regions that produce hyperphagia, indicating the complexity of these nervous regions.

When feed is eaten by an animal, certain changes (signals) are produced in the body which stimulate the activity in the satiety center which, in turn, suppresses the activity of the feeding center. After a period of time these signals are diminished and the inhibitory action of the satiety center on the feeding center is removed. This results in re-activating the feeding center and the animal again exhibits a normal feeding behavior (Fig. 6–3).

C. Signals to the Regulating System

A change in the energy balance or status of an animal produces a feedback signal which ultimately is integrated in the hypothalamus. A long and continuing search for such signals has led to the conclusion that there are a variety of signals, any one of which may be predominant at a given time. Two general classes of signals can be discerned (Fig. 6–4). There are signals triggered by distension or filling of the digestive tract and others more closely related to metabolism which can be classed as either chemostatic or thermostatic. Rations that are low in "nutritive value" (due to either low digestibility or high bulkiness) are consumed at a low level because the digestive tract becomes distended and thus dry matter intake is inhibited before the complete demand for energy has been satisfied. As the

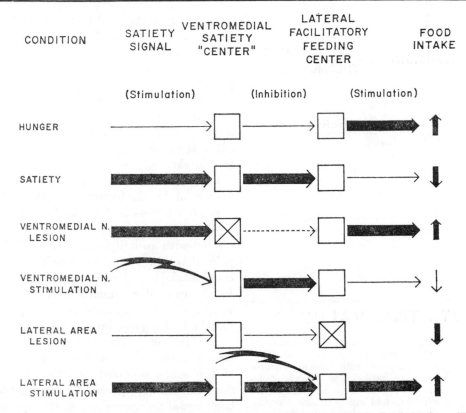

| CONDITION | SATIETY SIGNAL | VENTROMEDIAL SATIETY "CENTER" | LATERAL FACILITATORY FEEDING CENTER | FOOD INTAKE |

Fig. 6–3. Diagram of the role of the hypothalamus in the determination of feeding behavior. (*From Jay Tepperman: Metabolic and Endocrine Physiology. 1962, Year Book Medical Publishers, Inc. Used by permission of Year Book Medical Publishers. [See review by Anand, B. K. 1961, Physiol. Rev. 41:677–708, for documentation.]*)

nutritive value of the ration is increased, both feed and energy intake increase until energy intake reaches the point set by the physiological demands of the particular animal. Further increases in the nutritive value of the ration are accompanied by a decrease in feed intake of a magnitude to allow approximately stable energy intake. Most nonruminant rations fall in the category of stable energy intake, whereas most ruminant rations are of the type in which energy intake is limited by distension.

Suitable units for "nutritive value," as used in Figure 6–4, are lacking. Studies with several species fed a variety of rations show that expressions of digestibility such as total digestible nutrients (TDN), digestible dry matter (DDM), or digestible energy (DE) are not sufficiently descriptive for these purposes. Although TDN and DDM

give an approximation of the weight of indigestible material left in the digestive tract, they do not allow for differences in the volume occupied or for differences in the rate of removal. Expressions such as DDM × density (Montgomery & Baumgardt, 1965) or the ratio between the energy content and volume of the ration (Mraz *et al.*, 1957) are more reliable units of measure. If DE is substituted for DDM in the first expression, the two expressions become identical and may be referred to as "caloric density."

The localization and mode of action of the important signal systems will now be considered.

1. METERING IN THE MOUTH, PHARYNX, AND DIGESTIVE TRACT

Metering of feed in the mouth, pharynx, or esophagus is not an important signal in

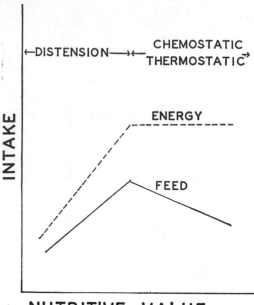

FIG. 6–4. Diagram of the relationship between nutritive value of rations and feed dry matter and energy intake. Feed intake increases with increased nutritive value to the point where distension no longer limits intake. Further increases in nutritive value are accompanied by decreases in feed intake to adjust energy intake to the level set by physiological mechanisms.

the normal regulation of feed intake. Animals retain their ability to meter calories under experimental conditions in which the oropharyngeal region is bypassed. Dogs that are pre-fed all or a part of their normal meal by stomach tube make fairly accurate adjustments in the voluntary intake of the meal when it is offered in the usual way. Rats can be trained to feed themselves food that does not pass through the oropharynx. They eat by pressing a bar with their paws to inject a liquid diet directly into their stomachs through a permanently implanted gastric tube. Rats equipped in this way continue to ingest normal amounts of food, maintain a relative constancy of food intake and body weight, adjust intake to dilution rapidly and precisely, and continue to ingest normal meals on a normal diurnal cycle.

Experiments of approximately the reverse type show that the passage of food through the oropharynx and esophagus will not pro-

duce satiety if the food is not allowed to be digested. When boli of normally swallowed food are removed from the rumen of cows, the cows will continue to eat for extended periods of time in an attempt to gain the energy metabolites which normally result from the ingestion of a meal. Such experiments also suggest that tiring of the jaws of cattle is not an important factor in limiting feed intake.

Gastric distension, on the other hand, is known to be an important satiety signal (Figs. 6–1 & 6–4) under some feeding conditions. Even nonruminant rations can be diluted with an indigestible material to the point that energy intake will be impaired because intake is limited by gastric distension. In the case of ruminants being fed high-roughage rations, distension is the primary signal inhibiting feed intake. The placement of air-filled or water-filled balloons in the rumen decreases hay intake of cattle because of the space occupied by these balloons. Such distension responses are a normal feedback signal working through the hypothalamus. Distension and tension receptors have been localized in the esophagus, stomach, duodenum, jejunum, and small intestine. Distension in these areas increases the electrical activity in the vagus nerve and in the satiety center of the hypothalamus in proportion to the amount of distension imposed. Neural impulses arising because of gut filling are afferent to the central nervous system and thus serve to inform the controlling system of the degree of filling (Fig. 6–1). Under most conditions these signals prevent overloading of the gut and therefore play the role of a safety device.

In addition to distension receptors, there are chemoreceptors (special nervous regions which are activated by one or more chemical entities) distributed in the gastrointestinal tract. Their role in the regulation of feed intake is yet to be determined.

2. CHEMOSTATIC REGULATION

Glucostatic Regulation. Feed intake and blood glucose concentration are closely related, especially in nonruminants. Detailed investigations of this relationship led Mayer and his co-workers (1964) to identify a glucostatic signal as one component in the regulation of feed intake. The hypothalamic satiety center, and possibly other central and peripheral areas as well, contain glucoreceptors which are specialized areas of cells sensitive to blood glucose to the extent that they utilize it. Utilization of glucose in all or a part of the body can be estimated by determining the arterial and venous blood glucose levels. When the peripheral arteriovenous blood glucose difference is small (*i.e.*, when glucose utilization is low), hunger is experienced. On the other hand, when the peripheral arteriovenous glucose difference is high (*i.e.*, when glucose utilization is high), a subjective feeling of satiety is reported by human subjects. It should be stressed that it is the *rate of glucose utilization* and not simply the level of blood glucose which is considered to be the signal. Glucose occupies a key role in animal metabolism and it is reasonable that blood glucose should, in some way, be related to energy homeostasis. Glucose is the prime, if not the only, source of energy for the central nervous system. During the interval between meals, body content of fat and protein, which is quite large, decreases very little, while body stores of carbohydrate (*i.e.*, glucose, glycogen), which are limited, decrease proportionately more so. The depleted body stores of carbohydrate are replenished largely by intake of feed. Furthermore, carbohydrate metabolism is interrelated with protein and fat metabolism and all are influenced by the endocrine system. Thus, glucose should be a sensitive indicator of energy status.

The metabolic effects of at least two hormones, insulin and glucagon, can be integrated with glucostatic regulation of feed intake. Insulin, produced by the beta cells in the pancreas, enhances glucose utilization and lowers blood sugar. After a brief latency period during which the readily available glucose is utilized, the administration of insulin results in an increased feed intake.

Glucagon, on the other hand, produces hyperglycemia, apparently by stimulating breakdown of glycogen in the liver; the administration of glucagon usually results in an inhibition of feed intake.

Additional support for the involvement of a glucostatic signal can be drawn from studies in which the electrical activity of the satiety and feeding centers has been followed in relation to changes in blood glucose. Intravenous infusion of glucose results in the satiety center becoming more active electrically and the lateral areas less active. Conversely, the hypoglycemia produced by insulin injection can be correlated with a low activity in the satiety center and an increased activity in the feeding center. Such alterations in electrical activity of the hypothalamus are not seen after infusing protein hydrolysate or fat emulsions and hence seem to be intricately related to glucose metabolism.

The exact locality of glucoreceptors (Detector system, Fig. 6–1) is yet to be determined conclusively. The detector is assumed to be a neural mechanism that is sensitive to glucose utilization or that is affected by neural impulses from peripheral glucoreceptors.

Available evidence indicates that the glucostatic system applies only to nonruminants and is not operative in ruminants. This can be related to the comparatively low blood sugar level of ruminants and to their increased dependence on volatile fatty acids (made available from rumen fermentation) as energy sources. Neither the intravenous infusion of glucose nor intraperitoneal infusion, to resemble absorption from the digestive tract, produces satiety in cattle or sheep.

Lipostatic Regulation. Another chemostatic system with some support is the lipostatic theory proposed by Kennedy (1966), who visualized communication between the aggregate fat depot and the regulating structures of the central nervous system by way of the blood stream. Animals mobilize each day a quantity of fat proportional to or at least increasing with, the total fat content of the body. This could provide an inventory statement of the size of the total fat deposit to be kept by the brain if some critical metabolite were released when fat was mobilized. It can be visualized in this hypothesis that body fat is the parallel of glycogen in the glucostatic hypothesis; both are forms of stored energy. Fat is continually being mobilized from fat stores for use as an important energy source. It is released from fatty tissue as free fatty acid (FFA) bound to albumin. The metabolite, or perhaps hormone, that serves as the feedback signal is unspecified at this time.

The lipostatic theory can help to explain some experimental observations made by Hervey on parabiotic (surgically produced Siamese twin) rats. The hypothalamic satiety center of one member of the pair was destroyed by electrolytic lesioning. This member developed hyperphagia and became obese while its parabiotic partner lost weight due to hypophagia. The undamaged hypothalamic centers of the "normal" partner appeared to respond to some change taking place in the lesioned animal as a result of hyperphagia. This change and "feedback signal" is capable of crossing through the parabiotic union. The normal partner decreased its feed intake to the greatest extent when the lesioned partner became very fat. Such observations kindle interest in the lipostatic hypothesis and raise questions about hormonal involvement. It may be that the hypothalamus receives a postmeal signal of satiety associated with lipogenesis (or insulin), followed later by a hunger signal of lipolysis (or growth hormone).

Volatile Fatty Acids as Signals in the Ruminant. As was pointed out earlier, the ruminant does not appear to possess a glucostatic signal system. Suggestions for a possible chemostat in ruminants have come from investigations of blood metabolite changes after a meal, the idea being that a metabolite serving as a satiety signal would

increase in concentration after feeding. Studies with cattle show that blood glucose and FFA concentrations decrease for one to two hours after feeding. Both blood acetate and ketone concentrations increase after feeding; acetate levels peak with the cessation of eating, whereas ketone levels continue to rise for several hours. Simultaneous sampling techniques show that blood acetate concentration parallels ruminal levels of total volatile fatty acids (VFA) and acetate during and following a meal.

Intraruminal infusions of VFA's can reduce feed intake in ruminants. Acetate has received more attention as a possible satiety signal than the other VFA's because of the patterns described in the preceding paragraph and because acetate is the only VFA which reaches peripheral blood in significant concentrations. Propionate is converted to glucose in the liver while butyrate is converted to beta-hydroxy butyrate in the rumen wall and liver. However, intravenous *and intraruminal* infusion of acetic, butyric and propionic acids at fairly high levels will reduce voluntary feed intake in sheep, goats, and cattle. These results must be reckoned with since the propionic acid and butyric acid infused intraruminally should give the same response as acids produced in the ruminal fermentation of feedstuffs.

Since butyric acid and propionic acid will reduce feed intake and yet only very small quantities of these acids get past the liver, it is difficult to visualize that their effect on feeding is due to detection in the hypothalamic satiety center. Direct attempts have failed to identify the receptors in the hypothalamus of goats that are responsive to either glucose or acetate. Thus, if the feed intake response to VFA's is due to a chemoreceptor system, the receptor site must not be located in the hypothalamus. The ruminal wall and the portal-hepatic areas are good possibilities.

3. THERMOSTATIC SIGNALS

Environmental temperature has profound and predictable results on feed intake. Homeotherms increase feed intake in the cold and decrease it in the heat. Feed intake falls with rising temperature to a point where animals will not eat if the environment is so warm that they already experience hyperthermia (abnormally high body temperature). The environmental temperature at which this occurs is not always the same for the same animal, and it varies from one species to another. Also, there is a difference between younger and older animals, and nonlactating and lactating animals. In rabbits and rats, feed intake essentially stops at an environmental temperature of 36° C, at which time the rectal temperature is about 40° C. The feed intake of cattle is not completely inhibited until the environmental temperature reaches 41° C, although there is a marked drop at 37° C. Lactating cows show this drop in intake at about 28°C. Goats show a similar reduction in feed intake when the temperature is above 35°. Considering the data from all species, there is a gradual fall in feed intake as the *environmental temperature* approaches 32° C, and a more abrupt fall as the *body temperature* approaches 40° C. A body temperature of 40° C seems to have serious import since we usually associate it with a severe infection. However, this temperature itself is not hazardous as it is a normal one for a man undergoing vigorous exercise.

Brobeck (1960) proposed that feed intake is controlled as if it were an integral part of the systems regulating body temperature when he suggested that "animals eat to keep warm, and stop eating to prevent hyperthermia." This concept is referred to as the thermostatic theory. Eating feed increases the body's heat production in at least three ways. The first and most apparent is the specific dynamic action (SDA), or the increase in heat production occurring within a few hours after feeding. The second is the increase in metabolic rate (heat production) with increased level of nutrition, and the third is the increase in heat production observed as body weight increases. In the thermostatic hypothesis of feed intake regulation, no distinction is made between the heat produced as SDA and that obtained by oxidation of tissues. This theory has the advantage of integrating the effects of protein, fat, and carbohydrate as contributors to the satiety signal. Differences in SDA among these nutrients offer explanations for apparent errors in caloric regulation on some rations. Fat has a low SDA and animals often "overconsume" calories on a high-fat ration. Conversely protein has a high SDA and the caloric intake is often reduced on a high-protein ration.

This theory has received special consideration in ruminants. Artificially induced temperature changes in the hypothalamus of goats altered feeding patterns (Balch & Campling, 1962). Cooling induced eating in satiated goats while heating inhibited eating in hungry goats. Feed intake can be inhibited in ruminants by infusions of acetate, propionate, or butyrate. Each of these metabolites could be exerting a satiety response in proportion to the heat produced during metabolism. It is interesting that acetate has the most marked effect on feed intake and also has the highest heat increment or SDA.

Although a precise relationship between metabolic heat production and feed intake regulation can be argued, there is no doubt that body temperature regulation and feed intake regulation are interrelated. Deep body temperature alone cannot be the parameter that influences feed intake. The most consistent relationship is between skin temperature and feed intake, thus, temperature regulation and feed intake are linked most clearly through mechanisms of peripheral sensation. Current concepts of the regulation of body temperature visualize the temperature which is regulated to be an average body temperature, taking into account skin, subdermal tissue—and deep core temperatures. It is likely that the regulation of feed intake to body temperature likewise depends upon some integrated value representing mean temperature.

D. Varying Levels of Regulation

The level of energy intake at which feed intake is regulated can, and in fact does, vary with the physiological state of the animal. Even the animal with medial hypothalamic lesions, which overeats and becomes obese, will eventually reach a stable phase and begin to regulate caloric intake at a new level. Variation in the level of regulation is quite noticeable in domestic livestock. The lactating animal regulates feed intake at a higher level than one that is not lactating. A similar situation exists when comparing pregnant and nonpregnant animals. The growing animal has a higher feed intake per unit of metabolic weight ($W_{kg}^{.75}$) than the nonlactating adult. This is another way of saying that nutrient requirements change and feed intake is regulated in relation to requirements.

It is not at all clear whether these varying stages of regulation are achieved by changing the set-point on the regulator or simply by increasing the demand. One way to visualize this problem is to think of a thermostat which regulates the temperature in a house. Fuel consumption by the furnace will be increased by raising the setting on the thermostat or by a marked drop in the outside temperature, while the thermostat is set at the same point. But, regardless of the mechanism, the level of regulation does change and thus has an important influence on the type of ration required for various animals. This is shown in Figure 6–5, which provides an extension of the concept presented in Figure 6–4. A ration only moderately high in caloric density will enable the low-producing cow or the slow-gaining steer to meet its energy demand within its capacity to hold feed (junction of diagonal and horizontal lines). As the production level or rate of gain increases, the ration must be made higher and higher in caloric density if the animal is to be able to consume sufficient feed to meet its energy requirement. At the very high levels of milk production, it is nearly impossible to formulate a ration high enough in caloric density to permit the lactating cow to meet energy needs and regulate calories rather than have intake limited by distension.

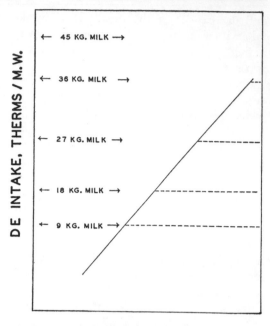

CALORIC DENSITY

FIG. 6–5. Diagram of the relationship between caloric density of the ration and digestible energy (DE) intake per unit of metabolic weight (MW). The horizontal lines represent the level of energy required per unit of MW for various levels of milk production by a dairy cow. In the case of a beef animal being fed for meat production, the 9, 18, and 27 kg of milk production would be equivalent to 0.6, 1.1, and 1.9 kg of gain daily. The diagonal line shows the increase in energy intake which accompanies increased caloric density when distension is limiting intake. Each horizontal line and the diagonal line up to the point of intersection represents the relationship between voluntary energy intake and ration caloric density. Note that the caloric density of a ration necessary to allow energy homeostasis varies with the level of energy demand.

II. PALATABILITY AND FEED PREFERENCE

A. Perception of Taste and Odor

Of all the terms employed in discussions of feed intake, the word "palatability" is the one most often misused. Palatability may be defined as the degree of readiness with which a particular feed is selected and eaten. It should be clear that palatability and level of feed intake are *not* synonymous. Palatability involves the senses of taste and smell, which are somewhat susceptible to modification by training and various psychological factors. Most experimental studies of palatability involve cafeteria-type situations in which the animal is allowed to exhibit a preference in selecting different rations or ingredients. In most practical feeding situations the animal is not allowed to choose among several rations but is offered only one. A ration that ranks low in a cafeteria study may be consumed in amounts equal to one that ranked higher when one ration is offered at a time and no choice is allowed. The former is an indication of palatability while the latter is a measure of voluntary feed intake.

Another fallacy that must be overcome is the idea that taste and flavors which appeal to man will also appeal to his livestock. Dr. Morley Kare, who has conducted extensive studies on taste perception in animals, concludes that domestic animals *do not* share man's taste world. All air-breathing vertebrate animals have their taste buds concentrated in the mouth. The number of taste buds varies widely, as the following examples indicate: chicken, 24; dog, 1706; man, 9000; pig and goat, 15,000; cattle, 25,000. The meaning of the number of taste buds is not clear and the number does not necessarily reflect the taste sensitivity of a species. As an example, the chicken, which has 24 taste buds, will reject certain flavored solutions which apparently are imperceptible to the cow with 25,000 taste buds.

Man is inclined to assume that all animals will find a sweet taste to be appealing. However, only the rat gives a response to sweet taste similar to that shown by man. Man would describe both a 10 percent solution of sucrose and an equally sweet solution of saccharin as sweet and pleasant. Among the animals that have been studied, only the rat would find both of these equally acceptable. Cattle find sucrose more appealing than man does, but are indifferent to saccharin. Birds are totally indifferent to sugar solutions and indicate some rejection of saccharin. Dogs find sugar appealing but tend to reject saccharin.

In addition to species differences there are differences in taste preferences within a species. For example, when pigs from the same litter were tested with saccharin solutions, some were offended by saccharin and some were indifferent, while the majority found the solutions appealing.

Sex differences in taste preference for glucose and saccharin solutions have been demonstrated in rats. Female rats prefer sweeter solutions than males do and the females retain their preference for a very sweet saccharin solution while males change their initial preference for a saccharin solution to a preference for a glucose solution after several days.

The importance of palatability to the rat is intensified if either the satiety center or the feeding center is damaged. If the palatability of a standard diet is reduced by the addition of quinine or by presenting the diet in powder rather than in pellet form, hypothalamic-hyperphagic animals reduce their food intake markedly while normal animals maintain a relatively constant food intake. If the lateral feeding area is destroyed the animal fails to eat; however, if such an animal is kept alive by force-feeding, eventually it will begin to eat very palatable food such as moistened cookies.

Ruminants and nonruminants can discriminate among the basic tastes but all animals within a species do not respond alike. Individual differences exist among calves, for example, in their acceptance and rejection of the usual four classes of taste: sweet (glucose), bitter (quinine), salt (sodium chloride) and sour (acetic acid). There are good indications that differences in taste discrimination are largely genetic in origin.

Factors such as color, texture, and temperature of food are of much less importance to animals than to man. The feeding practices which suggest otherwise have been developed largely to please the feeder rather than the animals he is feeding. Cattle and sheep cannot distinguish between blue and red, and in general appear to be color blind. The temperature of feedstuff is important only insofar as it affects the chemical and nutritive stability of the feed. Similarly, combinations of moisture and heat can cause problems in stability and enhance growth of microorganisms. Any effects on feed intake are not due to the temperature and moisture *per se*, but to the deteriorative changes they permit to take place in the feed.

B. Palatability versus Caloric Regulation

A distinction must be made between palatability (eating for taste) and caloric regulations (eating for calories). In the majority of practical circumstances, cattle, sheep, swine, and poultry eat for reasons other than flavor. Two examples of feed ingredients used in ruminant feeds which are unpalatable in significant concentrations are urea (used as a protein substitute) and sodium propionate (used as a ketosis preventative). Cattle begin to detect urea when incorporated in grain mixtures at the 1.5 percent level and feed intake may be depressed at the 2.5 percent level. Another common feed discrimination situation can be seen in cattle and sheep in pasture; these animal generally avoid forage contaminated with feces. Even here, however, it has been demonstrated that if sheep are accustomed to the smell of feces they will eat feed contaminated with them.

Pigs offered a choice between sugar-coated pellets and plain pellets will eat more of the sugar-coated pellets. If no choice is allowed, however, they will eat about as much of one as of the other. If growing cattle are given hay pellets coated with sugar they show a transient rise in voluntary intake, but after three or four weeks the animals given coated pellets eat no more than animals given untreated pellets.

Molasses is often credited with increasing the palatability of feeds. It no doubt has some positive effect on palatability. Most of this effect, however, is due to a decrease in the dustiness of the feed and to an increased rate of breakdown, which, in the case of ruminants, lessens the fill and distension. For the most part, molasses simply overcomes a negative palatability factor rather than exerts a consistent positive effect on feed intake by its own flavor.

Feed intake is remarkably stable despite changes in palatability. For example, feed made totally unacceptable in a situation where the animal has a choice among feed must have the offensive quality increased tenfold to reduce caloric intake over a period of time in a no-choice situation. However, when feed intake is depressed below normal because of such factors as the weaning of domestic animals or a drastic change in environment, a highly palatable feed can reduce this transitory depression. This improvement may justify some special cost and effort in the formulation of highly palatable rations for baby pigs and calves, for example, especially if early weaning is practiced. Even with adult animals, it is advisable to make changes in the ration as gradual as possible This permits the animal to become accustomed to a strange flavor and, in the case of ruminants, allows time for the rumen microbiol population to adjust to the new substrates. Maintenance of a constancy in digestive ability promotes a constancy in feed intake.

There is little that can be done by way of palatability and flavor to increase feed intake above normal. This is in spite of some claims for "broad-spectrum," "miracle-flavor" additives. A trace of aromatic essence simply cannot shift the basic nervous regulation of feed intake. In many feeding situations, palatability may entice an animal to the feed trough and thus play a role in the *initiation* of eating, but it has little to do with the *amount* eaten.

Experience gained with rats substantiates the minor role played by palatability in meeting energy needs. Taste and smell sensations are important for finding and identifying feed and fluid. They aid survival in the wild and make an important contribution to the motivation to work for food. When a single food is freely available and motivational demands are slight, however, the sensations are not necessary for the control of the amount of food eaten. Rats that cannot taste or smell the food they are eating maintain constancy of intake and eat normal meals on a diurnal cycle. Other sensations previously discussed as feedback signals are sufficient for the central neural control of feed intake.

C. Appetite for Specific Substances

Animals are credited with varying degrees of nutritional wisdom for the selection of necessary components of a diet. In some cases they make very wise choices and in other cases their choice leads to disastrous results. For example, if chickens are thiamine-deficient, they will select thiamine-enriched feed over one devoid of the vitamin. On the other hand, they will consume sodium tungstate solutions from a choice situation until they kill themselves.

Extremes in protein level of the ration have marked effects on feed intake. Feed intake is depressed if animals are fed either a very low- or a very high-protein ration. The low feed intake on a low-protein diet is related to the inability of such a diet to support normal growth and milk production. Very low protein in ruminant rations decreases intake also by retarding microbial growth and activity in the rumen. Intake of a poor quality roughage such as straw can be increased by infusing urea directly into the rumen. The urea nitrogen can be used by ruminal microorganisms to promote growth and enzyme synthesis. Increased microbial activity results in a greater and more rapid fermentation of cellulose in the straw, which results in less fill or distension per pound of straw consumed so the animal can eat an additional amount of straw until the fill limit is once again reached.

Depressed feed intake on very high-protein rations is due, in part, to the high SDA associated with protein metabolism (thermostatic regulation) and, in part, to a deficiency of enzymes involved in amino acid catabolism. The latter explanation accounts for the transient nature of the depressed feed intake on high-protein rations since the activity of many of the enzymes involved in amino acid catabolism increases as the animal adapts to the ration. Increased activity of these enzymes contributes to the maintenance of homeostasis by facilitating removal of the amino acids from the excess protein. Animals which are severely deficient in protein will select a protein supplement or a high-protein ration in an attempt to correct the deficiency. Mechanisms involved in detection of the deficiency and selection of protein are unknown.

Feed intake in rats and other nonruminants can also be decreased by the dietary pattern of amino acids. Rats will greatly decrease their intake of a diet that is *deficient* in one of the essential amino acids. Force-feeding of such a diet results in pathological lesions such as fatty infiltration of the liver and atrophy of pancreatic and intestinal cells. The reduced intake of the deficient diet by rats fed *ad libitum* protects them for some time from the adverse effects associated with the ingestion of a large amount of an amino acid-deficient diet.

Rats also exhibit a marked ability to detect an amino acid *imbalance* in their diet (Harper, 1967). The state of imbalance should be distinguished from the deficient state. A deficient diet is devoid or at least much less than adequate in the absolute amount of one or more amino acids. The imbalance state is created by starting with a basal diet that is balanced with respect to its amino acid pattern. The imbalance can be imposed in two ways: (1) by adding to the basal diet a small amount of the one or more amino acids that are second limiting for growth, or (2) by adding to the basal diet a mixture of all but one of the indispensable amino acids. If the imbalance is created by adding to the basal diet an amino acid mixture devoid of histidine, for example, a rat eats little and consequently gains little weight (*e.g.*, 2 gm in 2 wks.) whereas this same diet including histidine (a "corrected" diet) will result in normal feed intake and normal growth (*e.g.*, 33 gm in 2 wks.). The feed intake depression occurs within one day after switching the rats to the imbalanced diet.

Nutritional wisdom is lacking in rats offered a choice between a protein-free diet and an imbalanced diet devoid of histidine. In such an experiment rats ate gradually increasing amounts of the protein-free diet and their intake of the imbalanced diet fell until, toward the end of the experiment, they were eating almost exclusively the protein-free diet, which will not support life. They were rejecting the imbalanced diet, which will support a moderate rate of growth. On the other hand, when rats were offered a corrected diet (well balanced) together with a protein-free diet, they ate very little of the protein-free diet and selected the corrected diet almost exclusively.

All animals require salt and have been credited with the ability to select and regulate salt intake. Free-choice feeding of salt either in block form or loose form is common practice with cattle and sheep. These animals will travel great distances to find "salt licks" if they are otherwise deprived of salt.

More recent investigations have dealt with the mechanisms involved in appetite for salt and with the accuracy of its regulation. Large amounts of sodium are conserved and recycled in ruminants via the saliva. If one of the parotid glands of the sheep is fistulated and the secretion of this gland removed, the sheep will lose 300 to 600 mEq of sodium per day. If this deficit is not corrected the condition of the animal deteriorates very rapidly. Such animals will select solutions containing sodium to correct this deficit. The advent of drinking such solutions is associated with restlessness and exploratory behavior. If various concentrations of sodium (as sodium chloride or sodium bicarbonate) in water are made available, the depleted sheep will vary the quantity drunk with the concentration present, and the deficit will be corrected within 24 to 48 hours. Sheep that have undisturbed salivary glands and are not depleted

of sodium show very little interest in sodium solutions.

Exact mechanisms governing salt appetite and sodium homeostasis are yet to be discovered. Current information does not suggest that variation in the concentration of Na^+ in blood plasma determines the appetite specific for sodium. This does not preclude the possibility that changes of Na^+ content within specific cells of the nervous system determine appetite, if it were assumed that the rate of penetration of Na^+ into cells with change of plasma Na^+ concentration is slow. The hormone aldosterone, secreted by the adrenal gland, is intimately involved in electrolyte balance and homeostasis. There is a hypersecretion of aldosterone in sheep under field conditions of sodium deficiency. Best evidence indicates a connection between a fall of salivary Na^+/K^+ ratio in deficient animals and the aldosterone secretion of the adrenal gland. This striking endocrine control of a digestive secretory mechanism is an important adaptation with considerable survival value in the evolution of the capacity of ruminants to adapt to stringent ecological conditions involving pasture change and periodic feeding.

Adrenalectomized rats show a spontaneous appetite for salt when given a choice between water and 0.15 M sodium chloride solution to drink. The appetite is relatively specific for sodium solutions, and the preference (taste) threshold for sodium chloride is reduced. The blood level of aldosterone influences spontaneous intake of sodium chloride; either a reduction or an increase in aldosterone secretion rate increases sodium chloride intake. The increase in sodium chloride intake resulting in decreased aldosterone secretion rate has survival value, whereas the increase in sodium chloride intake resulting in increased aldosterone secretion rate may have pathological consequences. There must be a factor to turn off sodium ingestion as well as to turn it on. This factor resides in the mouth, pharynx, and esophagus, or, possibly, in the stomach.

Very little is known of animals' ability to select micronutrients such as vitamins. Chickens and rats which are thiamine deficient show a very clear selection for thiamine. The situation with vitamin A is less clear; this deficiency responds slowly compared with thiamine deficiency.

III. MANAGEMENT AND FEEDING PRACTICES

A. Feeding System

Although the animal may possess physiological mechanisms for regulating feed intake in relation to energy demand, domestic animals are not often allowed to make use of this mechanism because they are offered a limited amount of feed. This limitation is, of course, desirable when feeding adult beef cattle and sheep for maintenance needs only. On the other hand, lactating dairy cows or finishing cattle and lambs often have difficulty meeting their energy need even when fed a high-grain ration.

Of the producing animals, dairy cows are the ones most often limit-fed. Forage is usually fed *ad libitum* but the concentrate part of the ration is fed in some ratio to the amount of milk produced. If an acceptable grain-to-milk ratio is selected on the basis of forage quality and size of cow, this system works well with cows beyond the peak of lactation. In early lactation, however, the situation is complicated, especially for the very high producer. Milk production increases are very rapid in the first week or two of lactation and a grain-to-milk feeding system is always behind needs. One attempt to solve this problem is embodied in the challenge-feeding or lead-feeding concept. Under this system, cows are accustomed to significant amounts of grain before parturition and then the level of feeding is increased rapidly after parturition until the cow is fed *ad libitum*. For a period of time high-producing cows regulate energy intake in a very acceptable manner. However, some cows with a lower genetic ability for milk production, and even the high-ability

cows later in lactation, will deposit body fat at an increasing rate rather than convert the extra energy into milk. Obviously, this can become uneconomical but it does *not* necessarily mean that the cow lacks the ability to regulate energy intake! At least two other explanations are possible.

The first is that the set-point on the regulator has been raised to an unusually high level. This can be explained on the basis of evolution and selection since farm livestock have been selected for high production. Such selection may have resulted in animals that would be considered, if they were rats, or especially, humans, to have "pathological appetites." Since lactation has a higher biological priority than fattening, obesity is not observed as often in dairy cows when limit feeding is practiced.

The second explanation involves no change in the set-point on the regulator. The main difference is that energy status is monitored on the basis of a circulating metabolite pool or undissipated heat load rather than on the basis of the energy balance of the animal body *per se*. In such a system, fat which is deposited can be considered more as an "output" than as a part of energy balance insofar as a feedback signal is concerned. Thus, increased enzymatic potential (the level of which is under genetic control) can can be visualized as removing metabolites from the circulating pool at an accelerated rate in the fattening animal. Intake is related to energy output but all functions using energy, including fattening, are considered as drains on the energy metabolite pool which is the parameter being monitored.

Individual versus group feeding can cause variation in feed intake. Cattle, sheep, and swine consume slightly more feed and make more rapid gains when fed in groups than when fed individually. This increase is credited to the behavioral effects of competition. Although the increase in intake is slight, it is persistent and is another demonstration that energy intake is regulated within a range, but not at a very precise level.

The frequency of feeding can influence

feed intake and metabolism. If animals are offered discrete meals frequently (*e.g.*, 8 or 10 times per day), they eat slightly more than when they are fed once or twice each day. Psychological and behavioral factors are involved. Growing lambs and heifers exhibit a slightly greater efficiency in the use of feed when it is fed several times a day than when it is fed only once or twice. Part of this response is still apparent if the total feed intake is held constant, and can be attributed to a more even pattern of rumen fermentation and more efficient nitrogen utilization. Responses to feeding frequency are less apparent with mature ruminants in which nitrogen demands are not so high.

Quite different responses are noted with rats, in which the metabolic effects appear to be similar to those in man. The rat is normally a "nibbler," *i.e.*, he eats many small meals throughout a 24-hour period. It is likely that primitive man followed a similar schedule before the advent of work schedules and food preservation. When the rat is trained to eat his food in one meal per day he deposits more of the food energy in his tissues as fat than he does when allowed to nibble many meals each day, even if the total amount of feed is held at the same level. Metabolic studies show that fat-synthesizing enzymes are more active in the meal-fed rat than in the nibbler. The large load of metabolites from one big meal must be handled by the body in some way, and since only a very limited amount of energy can be stored as carbohydrate (glycogen), the extra calories are deposited as fat. Although fat can be mobilized later as an energy source, the rat "prefers" to consume additional feed at the next meal if the quantity is not controlled. (It is interesting to compare this situation to the regulation of feed intake in fattening livestock discussed in the preceding section.) The meal-fed rat reduces his energy expenditure, and thus, his need to mobilize the extra fat deposited, by markedly reducing his exercise or physical activity. Doesn't this sequence of events sound remarkably similar to that developed

by many white-collar workers eating one or two *major* meals a day in an affluent society? In fact, medical research confirms many of these findings and some clinicians are recommending that we eat a larger number of smaller meals a day and force ourselves to exercise.

B. Feed Processing

Traditionally, farm animals have been fed roughage in the form of hay and grains in the whole and ground form. Many hay-crop forages are preserved as silage and grains are cracked, ground, crushed, crimped, dried and steamrolled, flaked, pelleted, pressure-cooked, toasted, and even popped.

Pelleting complete rations for swine, lambs, and cattle usually results in a slightly higher feed intake. This is partly due to compensation for a negative palatability factor, dustiness of a finely ground feed, and to an increased rate of passage through the digestive tract. This greater intake results in an increase in rate of body weight gain. The magnitude of response to pelleting is not constant, however, but varies with the quality of the ration. Large responses are usually obtained by pelleting poor-quality rations, and only small improvements are found with good rations. This is because fill or distension is the factor limiting intake of poor-quality rations and pelleting increases the rate of removal, and hence decreases fill.

Cracking, grinding, and cold rolling show an improvement over whole grain in feed intake and performance. Greater improvement is obtained with corn and milo through flaking after moist heat treatment. Treating barley by steaming and rolling will usually increase gains and feed intake, but the feed requirements per unit of gain are not affected. On the other hand, steam processing and flaking of corn and milo appear to increase the digestibility of these two grains.

If hay and high-moisture silage made from the same crop are fed to cattle and sheep, less dry matter will be consumed in the form of silage. An apparent explanation would be that the silage required more space in the digestive tract because it contained so much water. However, in experiments in which the silage was dried and the hay was soaked in water, the trend was not reversed; less dry matter was still consumed in the form of the dried silage. Another logical explanation is that silage might have an objectional odor and must, therefore, be unpalatable to the animal. If the juice from a wet silage is poured over hay, the voluntary intake of the hay is reduced. Pumping silage juice directly into the rumen via a fistula also causes the animal to decrease its voluntary intake of hay, silage, grain, or whatever it might be eating. As little as 100 ml of some silage juice put in the rumen can have drastic effects on feed intake in sheep. The tentative conclusion is that during the undesirable fermentation with many high-moisture, high-nitrogen forages, some compound or compounds are produced which function pharmacologically as appetite depressants. Exact identification of these is lacking but the activity is associated with some low molecular weight nitrogen-containing compound. Amines such as tryptamine, serotonin, tyramine and histamine have been isolated from so-called unpalatable silages. Such amines can be produced by amino acid decarboxylations in wet, cold fermentations. It may be that none of the amines thus far isolated is directly responsible for the reduced feed intake, but another similar compound produced during the fermentation may be involved. Because feed intake can be reduced by intraruminal infusion of acetic, butyric, and propionic acids, and since these acids are produced in significant quantities in wet silages, it is possible that they also contribute to the lowered intake. Thus, the intake problem with direct-cut silage can be attributed largely to factors other than palatability. Heavily wilted silage does not possess this factor since such silages are consumed (on a dry basis) as well or better than hay. A crop that is greatly reduced in moisture content before ensiling undergoes a much less extensive fermentation than a wet forage.

Maximum feed intake can be achieved only by feeding nutritionally balanced rations to animals that are in good health. Imbalanced rations are consumed at submaximal levels either because digestion is impaired or because tissue metabolism is upset. Unhealthy animals do not possess the feeding drive to consume at the maximum rate. For example, ruminants often go "off-feed" on an all-concentrate or high-concentrate ration. This can usually be related to an abnormally low rumen pH, reduced motility of the digestive tract, and general acidosis or breakdown in the acid-base homeostasis of the animal body. The common denominator in animal nutrition and physiology is homeostasis. If the ration being fed is not compatible with homeostasis, the animal will voluntarily reduce its intake in an attempt to return to the homeostatic state.

REFERENCES

Anand, B. K. (1961). Nervous regulation of food intake. Physiol. Rev. *41*:677–708.

Baile, C. A., Mahoney, A. W., & Mayer, J. (1967). Preliminary report on hypothalamic hyperphagia in ruminants. J. Dairy Sci. *50*:1851–1854.

Balch, C. C., & Campling, R. C. (1962). Regulation of voluntary food intake in ruminants. Nutr. Abstr. Rev. *32*:669–686.

Brobeck, J. R. (1960). Food and temperature. Recent Progr. Hormone Res. *16*:439–466.

Cannon, W. B. (1929). Organization for physiological homeostasis. Physiol. Rev. *9*:399–431.

Harper, A. E. (1967). Dietary protein and amino acids in food intake regulation. Chap. 11. In: *The Chemical Senses and Nutrition*. M. R. Kare & O. Maller (Eds.), pp. 155–167, Baltimore, The Johns Hopkins Press.

Hetherington, A. W., & Ranson, S. W. (1939). Experimental hypothalamico-hypophyseal obesity in the rat. Proc. Soc. Exp. Biol. Med. *42*:465–466.

Kennedy, G. C. (1966). Food intake, energy balance and growth. Brit. Med. Bull. *22*:216–220.

Mayer, J. (1964). Regulation of food intake. Chap. 1. In: *Nutrition—A Comprehensive Treatise*. Vol. I, pp. 1–40, New York, Academic Press.

Montgomery, M. J., & Baumgardt, B. R. (1965). Regulation of food intake in ruminants. 2. Rations varying in energy concentration and physical form. J. Dairy Sci. *48*:1623–1628.

Mraz, F. R., Boucher, R. V., & McCartney, M. G. (1957). The influence of the energy : volume ratio on growth response in chickens. Poult. Sci. *36*: 1217–1221.

Chapter 7

Growth Regulators

By J. R. Carlson

ANIMAL growth is affected by genetics, nutrition, disease, hormones, tissue-specific regulatory factors, and nearly all aspects of the animal's environment. All chemical compounds, even those indigenous to the animal, would probably affect growth when they are deficient or administered in sufficient quantity. Proper nutrition provides a balanced supply of calories, amino acids, lipids, vitamins, minerals, and water which furnishes energy and raw materials for growth; but it does not regulate growth except when deficiencies or imbalances cause abrupt growth disturbances.

Animal growth involves a correlated increase in total body mass resulting from an increase in the size of the individual tissues and organs. Changes in body weight are closely related to the production and degradation of tissue mass and these are intimately associated with alterations in protein metabolism.

Substances which have a substantial and direct effect on growth may be classified as growth regulators and include tissue-specific growth factors and certain hormones. Tissue-specific growth factors act as cellular growth regulators by affecting the proliferation and

development of specific types of cells. These include epidermal growth factor, nerve growth factor, erythropoietin, promine, retine, and tissue-specific chalones.

I. CELLULAR GROWTH REGULATORS

A. Epidermal Growth Factor

Injection of salivary gland extracts from male mice into newborn mice results in gross anatomical changes including: (a) precocious opening of the eyelids as early as 7 days instead of the usual 12 to 14 days; (b) precocious eruption of the incisors at 6 to 7 days instead of 8 to 10 days; and (c) stunting of the animals and inhibition of hair growth. The precocious opening of the eyelids and eruption of the incisors are a consequence of an increased proliferation and enhanced keratinization of the epidermis. The factor in salivary gland extracts which increases epidermal growth is called the *epidermal growth factor* (EGF). The growth-stunting effect is not associated with purified fractions of EGF.

Epidermal growth factor is found in significant amounts only in the submaxillary gland of male mice. Attempts to detect

EGF in salivary glands of female mice and other species as well as in a variety of mouse tissues have been unsuccessful because of the relatively insensitive *in vivo* assay system. It is unknown whether EGF is synthesized in the mouse submaxillary gland or only stored there.

Purified EGF is a homogeneous protein containing all of the amino acids except phenylalanine and lysine. During purification the absence of phenylalanine and lysine is a criterion of purity since contaminating proteins would be expected to contain these amino acids. Biological activity of EGF is associated with the protein structure since denaturation or precipitation with antibodies destroys the activity.

Epidermal growth factor stimulates cell proliferation *in vivo* and *in vitro*. Injection of purified EGF into mice, chicks, rats, and some other animals increases the dry weight, total nitrogen, DNA, RNA, and some enzymes of the epidermis. This change in chemical composition is associated with an increased thickness of the keratin and cellular layers of the epidermis. The thickness of chick epidermis is also increased during *in vitro* incubation with low concentrations of EGF (Plate 7 A and B). Epidermal growth factor acts directly on the epidermis rather than through secondary control mechanisms as shown by its potent activity *in vitro* and by the fact that it stimulates mitosis in the epidermis in the absence of the dermis layer (Cohen, 1964). In addition to stimulating epidermal growth in mice and chicks, EGF is also effective on cultures of rat epidermis, esophagus, ureter, ductus deferens, trachea, uterus, vagina, prostate, submandibular, parotid, and exorbital lachrymal gland, but not on cultures of liver, thyroid, adrenal, anterior pituitary, pineal, thymus, or lymph node (Jones, 1966).

B. Nerve Growth Factor

In 1951, *nerve growth factor* (NGF) was discovered in two types of mouse tumors (sarcoma 180 and 37) which enhanced the growth and differentiation of sympathetic ganglia in the chick embryo. More potent sources of NGF were later found to be mouse submaxillary gland and snake venom. These sources have provided material for experimental investigation. Even though high concentrations of NGF are found in the submaxillary gland, it is not produced there. Evidence supporting this conclusion includes: (1) Removal of the salivary gland does not affect the sympathetic ganglia. (2) Sympathetic nerve cells are approximately the same size in male and female mice, but the male gland has at least 10 times as much NGF. (3) NGF is not detected in salivary gland of other animals, but trace amounts are present in serum. (4) Some tumors produce NGF. The nerve growth factor is also found in small amounts in tumor tissue of numerous vertebrates, in mouse and human serum, and in sympathetic ganglia of several mammals.

Injection of antiserum to NGF into chick embryos prevents the development of sympathetic nerve cells and kills the differentiated neurons. In several animals, administration of antiserum causes complete atrophy of sympathetic ganglia without affecting other nerves or organs. Isolated nerve cells can live in culture only with added NGF. These facts indicate that NGF plays a role in normal development and life of sympathetic nerve cells.

Nerve growth factor is a small protein that can be separated into two components by column chromatography. It has little species specificity, but is highly specific for the type of nerve tissue that it stimulates. Injections of NGF into chick embryos cause massive invasion of viscera and blood vessels by sympathetic nerve fibers. These fibers branch profusely and disrupt the functional and structural integrity of the embryo. In 6- to 12-day-old chick embryos both sensory and sympathetic ganglia respond to NGF, but the sensitivity of the sensory ganglia is soon lost. A dense halo of neurites appears on sensory ganglia of young chick embryos when they are cultured with NGF *in vitro* (Plate 7 C and D). In newborn mice and adult animals, only sympathetic ganglia retain the ability to respond to NGF. Enlargement is caused by both hyperplasia and by hypertrophy in mice less than 9 days old. In mice over 9 days old and adult animals, sympathetic ganglia respond to NGF by hypertrophy only. Nerve growth factor stimulates amino acid incorporation and RNA synthesis (Levi-Montalcini, 1964, 1966). It is active at extremely low concentrations; about 10 molecules of it are required to evoke the appearance of new neurites from a ganglion (Schenkein *et al.*, 1968).

C. Erythropoietin

Erythropoietin (EP) is a hormonal substance circulating in the blood which stimulates the formation of red blood cells. The production of EP increases in response to anemia, hypoxia, or cobaltous ions, and results in accelerated erythropoiesis. The kidney is important in the production of EP, but synthesis continues in nephrectomized animals, showing that other tissues also contribute to EP production. Erythropoietin is present in all species that have been investigated, and its action is not species specific. Erythropoietin activity in serum increases in direct relation to the severity of anemia and represents a response to stimulate erythropoiesis under these conditions.

Erythropoietin is assayed in rats or mice by measuring the rate of uptake of radioactive iron by erythrocytes or by counting the number of reticulocytes in the blood after injection of a sample containing EP. With these biological assays, EP is measurable in the kidney, plasma, and urine of anemic, hypoxic, or cobalt-treated animals, and in plasma and urine of normal animals. Injection of antiserum to EP into normal animals decreases erythropoiesis for as long as antibodies are given, but erythrocyte production rapidly returns to normal after antibody administration is discontinued. This shows that EP functions in the control of erythropoiesis in normal animals.

The best source of EP for chemical characterization is the plasma of anemic animals.

Erythropoietin has not been completely purified, but it is probably a glycoprotein containing hexosamine and sialic acid as the carbohydrate moiety. It is a poor antigen which makes antibody production difficult (Lange & Pavlovic-Kentera, 1964).

Erythropoietin stimulates the synthesis of RNA, hemoglobin, and several enzymes in erythroid cells, and it causes an increased uptake of glucosamine into the stroma. Nuclear RNA synthesis is increased soon after exposing marrow cells to the hormone, indicating that EP may act at the transcription level to induce specific synthesis characteristic of erythroid cells. This occurs before DNA synthesis is affected. Erythropoietin apparently acts on a primitive stem cell by causing differentiation, but the exact nature of the target cell remains in question. It facilitates the entry of iron into bone marrow before the stimulation of hemoglobin synthesis is detected, and it increases the release of erythroid cells into the circulation. In contrast to EGF and NGF, EP does not dramatically stimulate the mitotic rate of erythroid cells (Goldwasser, 1966).

D. Promine and Retine

Certain extracts of calf thymus glands inhibit malignant growth in tumor-bearing mice while other extracts promote growth. The growth inhibitor is called *retine* and the growth promoter is called *promine*. These materials are widely distributed in nature. Retine is present in the thymus gland, muscle, tendon, urine, and in large blood vessels. Especially high concentrations of retine are found in clams (*Mercenaria mercenaria*) and mushrooms (*Agaricus campestris*).

Neither promine nor retine has been isolated in pure form, but they share most physical and chemical characteristics. Promine and retine are low molecular weight compounds which are normally bound to a colloidal carrier. Retine has been studied more extensively than promine, and is probably a methylglyoxal derivative. Promine and retine are present together in tissues that have been studied, but the mechanism by which they influence the growth of cancer cells is unknown. They presumably elicit their individual stimulatory and inhibitory effects on the cells, and the ratio of these compounds and their absolute concentration are important in the regulation of cell division (Szent-Györgyi, 1965, 1966).

E. Chalones

Substances which selectively inhibit the mitotic activity of specific tissues are called *chalones*. They have been extracted from liver, kidney, epidermis, and granulocytes. These mitotic inhibitors are tissue specific, but not species specific. Most available information concerns the epidermal chalone, which has been extracted from mouse, rat, guinea pig, rabbit, and fish skin (Bullough & Laurence in Teir & Rytömaa, 1967). The epidermal chalone inhibits the mitotic activity of the tissue in which it is produced. This effectively maintains the functional activities of the cells since they are not required to

direct their synthetic capabilities toward preparations for mitosis. Since the mitotic activity of epidermal tissue is inversely proportional to chalone concentration, Bullough concluded that the epidermal chalone is responsible for normal mitotic control in the epidermis (Fig. 7-1).

In vitro experiments have demonstrated that chalones lack their full potency for mitotic inhibition unless epinephrine is present. Epinephrine increases mitotic inhibition in the epidermis in the presence of chalones. The active mitotic inhibitor is an unstable chalone-epinephrine complex, and within each cell the concentration of this complex is in equilibrium with free chalone and epinephrine. For example:

Chalone + epinephrine = chalone-epinephrine complex

The addition of either chalone or epinephrine drives the reaction to the right and increases mitotic inhibition. A decrease in the cellular epinephrine concentration by diffusion and/or oxidation reverses the reaction, causing less mitotic inhibition. This response has been confirmed by studying the diurnal mitotic rhythm in mouse ear epithelium. With chalones present, the mitotic rate is lower when epinephrine secretion is increased and higher when epinephrine secretion is depressed. Adrenalectomy increases the mitotic rate and destroys the mitotic rhythm.

The chalone-epinephrine complex blocks the cell cycle at the prophase stage before the division process begins by inhibiting the mechanisms which trigger mitosis. The period of DNA synthesis and the actual mitotic process are the least sensitive to inhibition by the chalone-epinephrine complex (Bullough, 1964, 1965).

F. Growth Control

Growth may occur by both hyperplasia and hypertrophy. In embryos and young animals, growth occurs primarily by hyperplasia and later in life it continues by hypertrophy in many tissues. In animals, cell division is regulated by sensitive control mechanisms resulting in harmonious growth

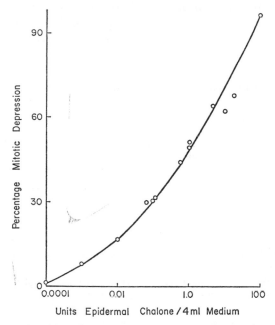

Fig. 7-1. The relationship between concentration of epidermal chalone and mitotic depression *in vitro*. Ear epidermis of adult male mice was incubated in the presence of chalone. Colcemid was added to arrest mitosis in metaphase for microscopic counting. (*Bullough et al., 1964. Exp. Cell Res. 36:192.*)

of the body and organs of young animals and the maintenance of body and organ size within narrow limits in adults. It is clear from biological activity studies and chemical characterization that tissue-specific growth regulatory substances are present in animal tissues. Each factor has its own characteristic chemical properties, tissue distribution, and mode of action; but they are all important in regulating growth of cells and tissues. Their mode of action varies greatly and includes stimulation of cellular proliferation (EGF, NGF, promine), inhibition of cellular proliferation (retine, chalones), and cellular differentiation (EP) (Table 7–1).

The mechanism by which cellular growth is regulated and the role of these growth factors in an integrated scheme of growth control is not known, but the possibility remains that they act in combination with other compounds, including hormones. Indeed, the epidermal chalone is an example of an active complex with a hormone. Future work with these and other tissue-specific growth factors will be important for a complete understanding of growth control and may be essential for an understanding of how cancer cells have escaped from these control mechanisms.

Partial hepatectomy results in rapid regeneration of liver to the original size, and a burst of mitotic activity is responsible for the rapid wound healing process. In neither case does growth continue appreciably beyond normal limits. Experiments on regenerating liver tissue have established the existence of other humoral growth regulators which have not been characterized (Wrba & Rabes in Teir & Rytömaa, 1967). These humoral growth factors are produced in relation to the amount of liver present and are important in controlling the overall size of liver in the organism. From studies in liver and other organs, it is clear that humoral growth regulators are important in autoregulation of tissue growth and function by some feedback mechanism.

The nature of these growth regulatory mechanisms is under intensive investigation, and several theories on the specific control of organ growth have been proposed. The fundamental difference between these theories has been whether cell division is regulated by inhibition, stimulation, or by a combination of both. As a result of his work on chalones, Bullough (1964) has proposed that the mitotic rate in organs is controlled only by tissue-specific inhibitors. He stated that it is the nature of cells to prepare for and to undergo mitosis whenever possible, and that it is opportunity, not stimulus, that is needed for cell division. This theory suggests that high concentrations of chalones suppress the genes responsible for mitosis

Table 7–1. Cellular Growth Regulators

Name	Chemical Properties	Function
EGF	Heat stable protein, lack phe and lys, MW - 14,600	Stimulates epidermal cell proliferation and keratinization
NGF	Protein with 2 subunits, MW - 20,000	Stimulates sympathetic and sensory nerves by hypertrophy and hyperplasia in embryos; stimulates sympathetic nerves by hypertrophy in growing and adult animals
EP	Glycoprotein containing hexosamine and sialic acid	Promotes differentiation of stem cells into erythroid cells; stimulates erythroid maturation and release from bone marrow
Retine	Methylglyoxal derivative	Inhibits mitosis in tumor cells
Promine	Uncharacterized	Stimulates mitosis in tumor cells
Chalones	Basic glycoprotein, MW - 40,000	Inhibits mitosis in specific tissues; inhibitory effect increased by epinephrine in epidermis

PLATE 7

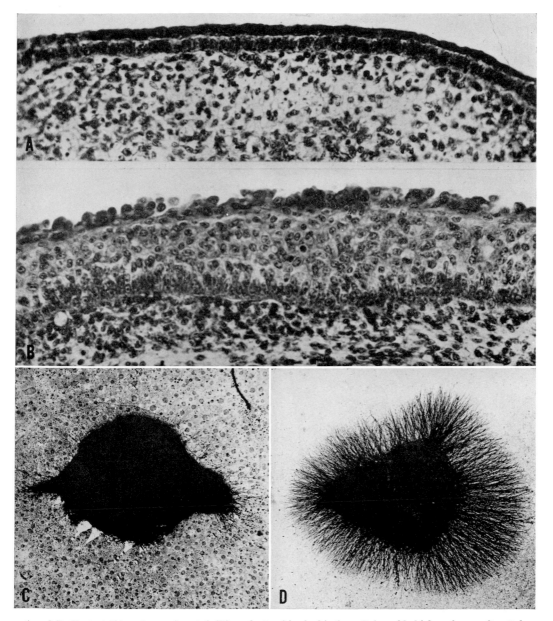

A and B, Control (A) and experimental (B) explants of back skin from 7-day-old chick embryos after 3 days incubation with 5 µg/ml EGF. Note increased cellular proliferation and thickness of epidermal layer in experimental explant. (× 380.) (*Cohen, 1965. Develop. Biol. 12:394.*)

C and D, Sensory ganglia from 8-day-old chick embryo after 24 hours of culture *in vitro.* (C) Ganglion in control medium. (D) Ganglion in medium supplemented with NGF. This treatment stimulated profuse growth of neurites from experimental ganglion. (× 50.) (*Levi-Montalcini, 1964. Ann. N. Y. Acad. Sci. 118:149.*)

and permit the genes for tissue function to be active. Decreased chalone concentrations release the mitotic inhibition allowing mitosis to proceed. According to this theory, regulation is by inhibition only.

The theory of Weiss and Kavanau (1957) differs fundamentally from that of Bullough since it proposes that organ growth is regulated by a balance of stimulatory and inhibitory factors (Fig. 7–2). According to this theory, each cell type produces key compounds characteristic of itself called "templates" which stimulate cell division in a particular organ. Growth rate is proportional to the concentration of these intracellular "templates" which remain in the cell under normal conditions. Each cell also produces characteristic "antitemplates" which inhibit the templates by forming inactive complexes. The antitemplates may diffuse out of the cell into the circulation. Their total concentration is proportional to the amount of organ mass. As the concentration of antitemplates in the circulation increases, intracellular concentration likewise increases, inactivating the templates and causing a decreased growth rate. A decrease in circulating antitemplates caused by a reduction in tissue mass will stimulate organ growth. When a stationary equilibrium between intracellular and extracellular concentrations of antitemplates is reached, organ growth will cease, but it will maintain its size through a balance of these stimulatory and inhibitory factors.

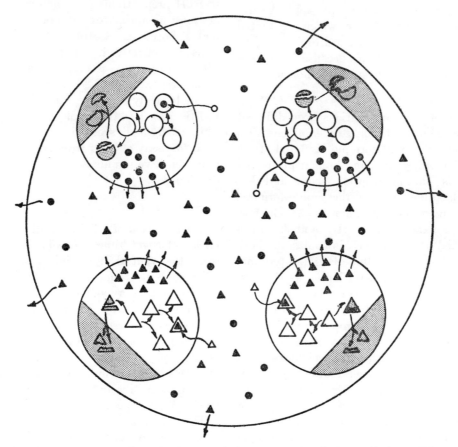

FIG. 7–2. Model of organ-specific molecular control of organ growth. Organ-specific intracellular templates (*large circles and triangles*) are produced by the cells and stimulate mitosis. Cells also produce diffusable antitemplates (*small full circles and triangles*) which inhibit mitosis by forming inactive complexes with the templates. Growth is regulated by the relative concentrations of templates and antitemplates. (*Weiss & Kavanau, 1957. J. Gen. Physiol. 41:1.*)

II. HORMONES AND GROWTH

Probably all hormones directly or indirectly influence growth by altering biochemical reactions; many of them influence the size of specific tissues and organs. However, only somatotropin, thyroxine, androgens, estrogens, and glucocorticoids are generally considered to have a direct effect on whole body growth. These hormones affect body mass and dimensions primarily by altering skeletal and/or nitrogen metabolism.

A. Somatotropin (Growth Hormone)

Somatotropin (STH) is secreted by the anterior pituitary, and is essential for normal growth in young animals and for the maintenance of normal metabolism in adults. The name "growth hormone" was applied early to the pituitary factor which promoted growth. The term "somatotropin" is preferred by some authors because it denotes several metabolic effects of the hormone.

1. CHEMISTRY AND SPECIES SPECIFICITY

Somatotropin preparations from different species exhibit strikingly different chemical properties. Somatotropin isolated from whale, monkey, pig, and human pituitaries contains one polypeptide chain and has phenylalanine as both the amino-terminal and carboxy-terminal amino acid. The hormone preparation from cattle and sheep is a branched polypeptide chain containing both phenylalanine and alanine as amino-terminal amino acids and phenylalanine as the carboxy-terminal amino acid. The iso-electric point ranges from pH 6.8 in cattle to 4.9 in man. Human somatotropin contains 188 amino acid residues with a molecular weight of 20,000. The molecular weight of STH preparations from several other species is about the same; but species differences exist in the amino acid composition. The most definitive data on species specificity comes from immunological studies. Somatotropin forms species-specific antibodies and antiserum to cattle STH will cross react with sheep STH but not with whale, pig, monkey, or human STH. Antiserum to human STH

will react only with human and monkey STH. Among these 6 species there are at least three pairs of antigenically similar hormones: (a) cattle, sheep (b) pig, whale, and (c) man, monkey. It follows that structural similarities exist between these pairs of antigenically similar hormones. The biological activity of STH is retained after partial hydrolysis with proteolytic enzymes, and it is possible that the "active core" of STH is similar for many species but that parts of the amino acid chain are different, allowing species-specific recognition (Evans *et al.*, 1966).

Chemical and immunological differences among species result in strict species-specific biological activity. Primates do not respond to STH preparations from nonprimates nor do most nonprimates respond to the primates' hormone. Cattle and pig preparations are ineffective in chickens, but they are active in hypophysectomized fish. In contrast, preparations of STH from all species are biologically active in rats. The characteristic growth in the epiphyseal plate of the tibia of hypophysectomized rats was the basis for assay of STH from all species until more sensitive immunological methods were developed.

2. SECRETION

The secretion of STH is regulated by the release of *somatotropin-releasing factor* (SRF) from the ventral hypothalamus. Plasma STH concentrations are also increased by insulin-induced hypoglycemia, fasting, and exercise. In addition, some nonspecific stimuli such as hemorrhage, histamine, epinephrine, and pain cause STH release. The effect of these physiological and stress conditions on SRF secretion is unknown and will remain so until a sensitive and specific assay for SRF has been developed.

The secretion of STH varies with age. There is no evidence that STH is produced in the fetus or is necessary for fetal growth, but the amount of STH in the pituitary increases after birth. As the animal grows older, the concentration of STH in the pituitary remains fairly constant. Since the relative size of the pituitary decreases with age, the total amount of STH in the gland decreases in relation to body weight in rats, pigs, cattle, and other animals. The variation in actual secretion rate with age is not known in detail since the glandular content of STH cannot be related to the secretion rate.

3. BIOLOGICAL EFFECTS

Somatotropin causes a general increase in body mass and affects the body composition of treated animals. Rats treated with STH grow to a larger size and store more protein and water, but less fat and total calories, than do nontreated rats (Table 7–2). In 8 weeks, STH-treated rats gained approximately 100 gm as opposed to 60 gm for pair-fed controls. During the feeding period the STH-treated rats maintained a body composition characteristic of younger animals since it was identical to that of the rats killed at the beginning of the experiment (Gaebler, 1965; Gaunt, 1954).

Table 7–2. Composition of Body Gain in Pair-Fed Control and STH-Treated Rats*
(*From Table 5, M. O. Lee & N. K. Schaffer. 1934. J. Nutr. 7:337.*)

	Water	Ether Extract	Total Nitrogen	Calories
Control	45	39	2.2	4.4
Treated	63	13	3.1	2.3

* Values in percent of empty carcass weight and kcal per gm of empty carcass weight.

Somatotropin affects many aspects of metabolism, but the two primary effects in promoting growth involve skeletal and protein metabolism. Somatotropin stimulates the growth of the epiphyseal plates of long bones and of endochondral bone in growing animals. It does not alter the time of closure of the epiphyseal plate. In adult animals, a condition known as acromegaly results from acidophilic tumors of the adenohypophysis causing hypersecretion of STH. This disease

is characterized by disproportionate overgrowth of the skeleton and some soft tissues. Only the long bones in which the epiphyseal plate has not closed, and to some extent periosteal and endochondral bone respond to hypersecretion of STH. Disproportionate growth results from differences in the time of epiphyseal closure in different bones.

Somatotropin administration promotes nitrogen retention and growth in animals that are fed a diet which does not normally permit growth. Under these conditions, fat is mobilized to provide energy for protein synthesis and growth. Hormone-induced decreases in plasma amino acid levels and urea excretion result from an increased transport of amino acids into skeletal muscle, heart, kidney, and liver. Somatotropin elicits a direct and unequivocal stimulation of amino acid transport into these tissues and this effect may be considered a site of action of STH. Transport of the nonmetabolizable amino acid, α-aminoisobutyric acid (AIB), is not affected by blocking protein synthesis with puromycin, indicating that STH may act directly on the amino acid transport system. Somatotropin also stimulates protein synthesis by directly increasing the ability of the ribosomes to incorporate amino acids into protein. Somatotropin can act alone to promote growth and nitrogen retention, but its full effects are observed only in the presence of insulin and thyroxine.

Insulin and STH have similar effects on protein metabolism, but STH has an anti-insulin effect on carbohydrate metabolism. Hypophysectomy results in hypoglycemia, decreased liver and muscle glycogen, and increased sensitivity to insulin. Somatotropin administration causes hyperglycemia, glucosuria, and ketonuria under some conditions. This diabetogenic effect may be related to increased fatty acid mobilization and oxidation. Direct evidence concerning the mechanism of action of STH on carbohydrate metabolism is unavailable.

Somatotropin increases lipid mobilization from adipose tissue and reduces lipid synthesis. These changes result in greater fatty acid oxidation, but these responses are probably not direct effects of STH and the mechanisms that are involved remain obscure (Knobil & Hotchkiss, 1964). Somatotropin has not been used to a significant extent in animal production because of difficulties in obtaining sufficient quantities and the need for injection as the method of administration.

B. Thyroxine

Thyroxine is an iodinated tyrosine compound that is secreted by the thyroid gland. It is synthesized in the epithelium of the thyroid follicles after iodide is transported into the gland from the circulation. Iodide is converted to monoiodotyrosine and diiodotyrosine. Triiodothyronine is formed by coupling one molecule each of mono- and diiodotyrosine:

$$\text{HO}-\overset{\overset{\displaystyle I}{|}}{\underset{\underset{\displaystyle I}{|}}{\bigcirc}}-\text{O}-\overset{\overset{\displaystyle I}{|}}{\underset{\underset{\displaystyle I}{|}}{\bigcirc}}-\text{CH}_2-\underset{\underset{\displaystyle NH_2}{|}}{\text{CH}}-\text{COOH}$$

3,5,3',5' – Tetraiodothyronine

(Thyroxine)

$$\text{HO}-\overset{\overset{\displaystyle I}{|}}{\bigcirc}-\text{O}-\overset{\overset{\displaystyle I}{|}}{\underset{\underset{\displaystyle I}{|}}{\bigcirc}}-\text{CH}_2-\underset{\underset{\displaystyle NH_2}{|}}{\text{CH}}-\text{COOH}$$

3,5,3' – Triiodothyronine

Both T_4 and T_3 are biologically active and, in certain species, T_3 is several times more active than T_4. These iodinated compounds are bound to a colloidal glycoprotein (thyroglobulin) in the follicles of the thyroid gland. Enzymatic hydrolysis of thyroglobulin liberates thyroxine, and the secretion of thyroid-stimulating hormone (TSH) from the pituitary causes the release of thyroid hormones into the circulation (Turner, 1961).

1. BIOLOGICAL EFFECTS

The best known example of the effect of thyroid hormones on growth and development is the stimulation of precocious metamorphosis in tadpoles. In animals, normal growth and development occur only in the presence of thyroid hormones, indicating that they play a "permissive" role in growth regulation. Since they are necessary for normal growth, thyroid hormones are classified as growth stimulators when present in optimal amounts.

It is well known that thyroid hormones are essential for maintenance of the basal metabolic rate (BMR); consequently, they must be involved in growth processes. Hypothyroidism decreases the BMR, and the ability of the animal to respond to STH secretion. It also decreases appetite and the reduced food intake slows growth. Force-fed hypothyroid rats, however, are able to maintain their body weight and nitrogen balance. These force-fed rats have a drastically altered body composition. Total body water and protein content are lowered while the fat content is extremely high and virtually no skeletal growth occurs after thyroidectomy. In spite of the ability to maintain a positive nitrogen balance, hypothyroidism results in a lower rate of protein synthesis and increased nitrogen excretion. Hyperthyroidism resembles a deficiency resulting in weight loss and general catabolic changes, including greater nitrogen excretion. The metabolic rate and related oxygen consumption are stimulated, but the energy utilization is less efficient. Thyroxine is a nonspecific stimulant of corticosteroid secretion which may be responsible for the catabolic effects (Gaunt, 1954). Although the effects of thyroxine on BMR and many aspects of metabolism are well known, the precise mechanism of action remains in question. Rosenberg and Bastomsky (1965) suggest that the fundamental action of optimal amounts of thyroid hormones is to stimulate energy-requiring reactions such as protein synthesis, and that the hormone-induced increase in BMR is a secondary response to the need for more energy. T_4 stimulates protein synthesis *in vitro* in the presence of mitochondria by increasing the transfer of amino-acyl sRNA to ribosomal protein. The role of mitochondria is unclear, but it is not to generate ATP.

2. Effects on Animal Production

Thyroid hormones and goitrogens, such as thiouracil or its derivatives, have been given to swine, sheep, poultry, beef, and dairy cattle in an attempt to improve production. In swine, feeding of thyroprotein (iodinated casein) sometimes increases body weight gain, but it is difficult to use under practical production conditions. Beneficial effects are limited to young growing pigs, but after fattening begins and they weigh over 100 lbs thyroprotein retards the rate of gain and decreases feed efficiency. Goitrogens in moderate doses have a limited usefulness and sometimes increase the feed efficiency during the last 3 to 4 weeks of fattening in swine. Higher doses decrease feed efficiency and weight gain.

Feeding thyroprotein to poultry causes a striking increase in feathering, but its advantages for stimulating other aspects of poultry production are not clear. Variation in the response of egg production, growth, hatchability, and fertility to thyroprotein feeding is caused by difficulties in obtaining the proper dose, and by differences in the strain of chicks, type of ration, and season of year. These variables usually result in deleterious effects. Goitrogens used in moderate doses slightly increase feed efficiency during the last part of the fattening period, but in general they have not been used to a significant extent except in combination with estrogens. Carcass qualities are improved when a combination of goitrogens and estrogens are fed to poultry.

Thyroprotein stimulates wool growth in sheep and milk production in ewes, resulting in heavier lambs. There is no advantage in feeding goitrogens to growing-fattening lambs. Feeding thyroprotein to dairy cattle stimulates milk production by 10 to 25 percent and fat content by 0.2 to 0.3 percent. Thyroprotein should be fed during the first 3 to 4 months of the lactation and additional feed is necessary to support increased production. Goitrogens have given deleterious results and are not used extensively for beef cattle (Blaxter *et al.*, 1949; N.R.C., 1959).

C. Corticosteroids

Hormones of the adrenal cortex have such far-reaching effects that there are few physiological processes which escape their influence. In addition to androgenic and estrogenic products, the adrenal cortex elaborates other substances which may be divided into three groups on the basis of chemical structure and physiological function: (a) steroids with oxygen at carbon 11 such as cortisol, cortisone, and corticosterone; (b) steroids without oxygen at carbon 11 such as deoxycorticosterone; (c) steroids with an aldehyde group at carbon 18 instead of a methyl group. Aldosterone is an example. The 11-oxygenated corticosteroids are especially potent in affecting protein and carbohydrate metabolism, and they are sometimes called *glucocorticoids*. These compounds are discussed in relation to their effects on growth. The *mineralocorticoids* in group b and c mainly affect water and electrolyte metabolism.

a. **Cortisol**   b. **Deoxycorticosterone**   c. **Aldosterone**

The width of the epiphyseal plate decreases in growing adrenalectomized rats, indicating that adrenal corticosteroids are necessary for maintenance of bone growth in young animals. They also decrease the stimulatory effects of STH on bone, depending upon the relative amounts of glucocorticoids and STH that are present (Drill, 1961). Excessive amounts of corticosteroids, especially mineralocorticoids, are catabolic to bone and produce a negative calcium balance.

Removal of the adrenal glands almost completely inhibits growth and will cause death, but the cortical hormones do not have a direct effect on growth. A high sodium chloride diet or 1 percent NaCl in the drinking water maintains life and supports appreciable growth (Fig. 7–3). In adrenalectomized rats, life is maintained and some growth occurs with the administration of cortisone, but the best results are obtained with deoxycorticosterone acetate (DCA), which causes the retention of electrolytes. Cortisone in combination with DCA reduces the growth rate. Intact rats show little growth response to NaCl and DCA administration, but corti-

sone significantly decreases growth. These results show that glucocorticoids are growth inhibitors (Gaunt, 1954).

An important site of action of glucocorticoids is on protein and carbohydrate metabolism. Glucocorticoids cause an increase in urinary nitrogen excretion and a stimulation of carbohydrate synthesis from amino acids (gluconeogenesis). The primary source of amino acids is muscle protein, but another source is lymphatic tissue, resulting in pronounced atrophy of the thymus and circulating lymphatic cells. Even though growth of most tissues and the total body weight are depressed by glucocorticoid administration, the liver does not lose weight and may even increase in size. In fact, many enzymes in the liver that are involved in amino acid degradation and gluconeogenesis are substantially increased by the administration of glucocorticoids. Glucocorticoids affect protein metabolism by specifically decreasing protein synthesis in skeletal muscle, thymus, appendix, spleen, lymphocytes, and reticulocytes. They also increase the transport of amino acids into the liver and decrease trans-

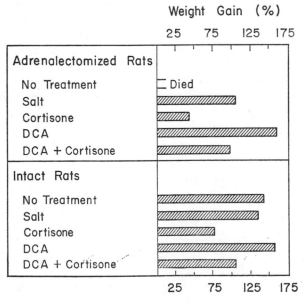

FIG. 7–3. Effect of adrenal hormones on growth of adrenalectomized and intact rats. Nontreated rats are compared to rats treated with 1 percent NaCl in the drinking water, injections of cortisone (1 mg/day), deoxycorticosterone acetate (DCA) (1 mg/day), or a combination of cortisone and DCA (1 mg each/day). Growth is compared at the end of 20 days. (*Gaunt, 1954. Dynamics of Growth Processes. Boell (Ed.), Princeton, N. J., Princeton University Press.*)

port into skeletal muscle, thymus, and other tissues. With decreased protein synthesis, degradation continues and the overall effect is increased protein catabolism in extrahepatic tissue.

In the liver, cortisol stimulates synthesis of specific enzymes by causing the synthesis of mRNA. This is consistent with the idea that cortisol stimulates protein synthesis by activating specific genes, but the question of whether cortisol has a direct or secondary effect on the gene locus remains unanswered. In the thymus, catabolic effects of cortisol are apparently not related to changes in gene activity (Hechter & Halkerston, 1965).

D. Androgens

The common observation that mature male animals are larger than females of the same age shows that sex hormones are important in growth regulation. Male and female rats grow at approximately equal rates until the beginning of sexual maturity on about the thirtieth day of age. The males then begin to grow faster and are thus larger than females when growth ceases in both sexes (Fig. 7–4). Associated with this difference in growth rate is the appearance of secondary sex characteristics and a slower

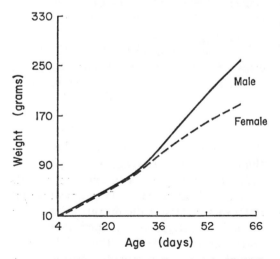

Fig. 7–4. Effect of sex on the increase in body weight of Sprague-Dawley rats. (From Fig. 1 "Steroids and Growth" by *Victor A. Drill, in Growth in Living Systems, Ed. M. X. Zarrow, Basic Books Inc., New York, 1961.*)

ossification of the epiphyseal plate in males. These changes are directly related to androgen secretion, so that androgens can be classified as "anabolic" or growth stimulatory compounds.

The chief sources of androgens are the interstitial cells of the testes and the adrenal gland. Testosterone is the chief androgenic steroid secreted from the testes. A closely related androgen with lower biological activity is secreted from the adrenal cortex. An increase in androgen secretion is observed in both male and female animals that are approaching maturity; this explains the burst of growth seen in both sexes at this time. The secretion of testosterone from the testes causes males to have a larger total secretion of androgens and a more rapid growth rate than females.

The effect of testosterone on the epiphyseal plate of rats varies considerably, depending upon the dose, duration of treatment, and the criteria used for observation. Castration of males slightly inhibits skeletal growth, indicating that testosterone has a slight stimulatory effect on growth and an inhibitory effect on epiphyseal fusion. Small doses of testosterone increase the width of the epiphyseal cartilage and augment the effect of low doses of STH on the epiphysis. Large doses of testosterone hasten the closure of the epiphysis and block the effects of STH. Androgens are more effective in stimulating skeletal growth in males than in females because estrogens, which promote epiphyseal closure, are present in smaller quantities in males.

In addition to skeletal effects, androgens also affect protein metabolism. They decrease urinary nitrogen excretion with no change in circulating nonprotein nitrogen, and increase nitrogen retention. The primary effect of androgens on protein metabolism is to stimulate protein synthesis rather than to decrease catabolism or decrease the conversion of amino acids to urea. This may involve stimulation of cellular RNA synthesis, but the precise mechanism of action is not established. The nitrogen-retaining effect of testosterone results in an increased body weight, and, especially, in greater skeletal muscle growth (Drill, 1961). Figure 7–5 shows a slight stimulatory effect of testosterone on the growth of young male rats. Although cortisone administration severely inhibits the normal growth rate, testosterone in combination with cortisone relieves a significant portion of this inhibition.

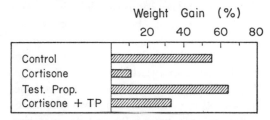

Fig. 7–5. Effect of testosterone propionate on growth of normal and cortisone-inhibited rats. Testosterone propionate (0.25 mg/day), cortisone (1 mg/day), or testosterone and cortisone was injected daily for 7 days into growing rats. (*Gaunt, 1954. In: Dynamics of Growth Processes. Boell, (Ed.), Princeton, N. J. Princeton Univ. Press.*)

In some respects the anabolic action of testosterone is similar to that of STH, but important differences exist. Androgens have relatively little effect on males producing adequate amounts of testosterone, and they will not stimulate somatic overgrowth. The mild growth stimulation caused by androgens does not exceed that of normal males, while STH stimulates growth in proportion to dose in young animals and sometimes produces gigantism. Prolonged administration of androgens requires increasing doses in order to maintain an anabolic response, but constant doses of STH are continually effective. Testosterone causes masculinization and growth of sex accessory organs, but STH has no effect on these parameters (Gaunt, 1954). Some nortestosterone derivatives possess normal anabolic activity but very few of the masculinizing properties of testosterone. Attempts to utilize the anabolic properties of androgens in feeding livestock have met with only limited success. The effects of anabolic steroids on growth rate of swine, poultry, and ruminants have been small and variable. Large doses of testosterone produce decreases in rate of gain, but feed efficiency is not altered. Androgenic compounds improve the carcass quality by decreasing the amount of fat and increasing the proportion of edible tissue, especially in swine. They have no beneficial effects in poultry except for a slight stimulation of gain in growing capons and sometimes in growing females. In beef cattle, limited beneficial effects of moderate doses of androgens include increased nitrogen retention and a lower carcass fat content. Females respond more readily than males to androgen administration (N.R.C., 1959).

E. Estrogens

Estrogens affect somatic growth of animals in addition to having the well-documented endocrinological effects on sex organs. The major estrogens which are produced by the ovary, placenta, and to a smaller extent, by the adrenal cortex of both sexes are estradiol, estrone, and estriol. Gaunt (1954) classified estrogens as somatic growth inhibitors because castration of the female results in greater growth than would otherwise occur and because sizable doses of estrogens depress the growth of male animals. This effect is in contrast to the most important endocrinological effects of estrogens which result in substantial growth stimulation in sex organs.

Estrogens inhibit skeletal growth by decreasing growth of the long bones. Increased secretion of estrogens during sexual maturity is associated with a narrowing of the epiphyseal cartilage and a more rapid fusion of the epiphysis. Ovariectomy of female rats just prior to the beginning of sexual maturity eliminates some of the decrease in epiphyseal width, slows epiphyseal fusion, and permits further skeletal growth. This indicates that estrogen secretion is important in inhibiting the epiphyseal growth in normal animals. Estrogens do not affect the epiphyses of hypophysectomized rats, but estradiol depresses the response to a standard STH dose, implying that an interrelationship exists between estrogens and STH (Drill, 1961).

The effects of estrogens on nitrogen retention are variable, depending upon the dose, age, sex, and species of the animal. At reasonably high doses, estrogens increase nitrogen retention in ruminants and fat deposition in poultry, but negative results are usually obtained in other animals. A contributing factor to the catabolic effects of estrogens is that they stimulate secretion of corticosteroids which have catabolic properties (N.R.C., 1959).

Growth of sex organs, especially the uterus, and other specific tissues such as the deep layers of the skin is stimulated by estrogens. Estrogens trigger some physiological mechanism which stimulates growth by both hypertrophy and hyperplasia in uterine tissues and increases the mitotic rate in the skin. The target organs for estrogens contain a specific estrogen-binding protein which forms hormone-protein complexes that in some way initiate anabolic activity in these tissues at the chromosomal level. Thus, in the target organs, estrogens presumably stimulate the synthesis of enzyme systems needed for growth by causing the activation of specific genes.

The effects of several hormones on skeletal and protein metabolism are summarized in Table 7–3. Somatotropin, thyroxine, and androgens are anabolic hormones while estrogens and glucocorticoids are generally catabolic. The growth-inhibiting effect of estrogens is most pronounced on the skeleton, but when administered in high doses, estrogens also reduce nitrogen retention in some animals. Since estrogens stimulate growth of sex organs and skin, and growth of ruminants under some conditions, the decision of whether estrogens are classified as inhibitory or stimulatory agents depends upon the sex, species of animal, and criteria of observation.

III. FEED ADDITIVES AND OTHER GROWTH FACTORS

A. Synthetic Hormone-like Compounds

The use of natural and synthetic estrogens and progestogens in animal feeding has contributed substantially to increased production, especially in fattening ruminants. Implantation of a synthetic estrogen-like compound, diethylstilbestrol (DES), increases the rate of gain by about 0.3 to 0.5 lb/day in beef cattle. Diethystilbestrol is effective in females, castrated males, and in growing bulls, but it is more effective when the rate of gain is high. Feed efficiency is generally increased by DES administration. High doses of DES sometimes adversely affect

Table 7–3. Effects of Hormones on Growth

General Effects	Hormone	Effects on Skeleton	Effects on Protein Metabolism
Anabolic	Somatotropin	Stimulates growth of endochondral bone and epiphysis of long bones	Increases N-retention and protein synthesis; thyroxine and insulin stimulate effect
Anabolic	Thyroxine	Stimulates growth of long bones; essential for STH effect	Stimulates protein synthesis and BMR; deficiency or excess is catabolic
Anabolic	Testosterone	High dose is weak stimulator of epiphyseal closure and antagonistic to STH; low dose increases width of epiphysis and augments effect of STH	Increases N-retention; promotes muscle growth and development of sex characteristics
Catabolic	Estrogens	Inhibits skeletal growth; promotes epiphyseal closure	Increases N-retention in ruminants, high doses catabolic in other animals; promotes growth of sex organs
Catabolic	Glucocorticoids	Decreases growth of epiphysis; decreases stimulation on epiphysis by STH	Increases protein and amino acid degradation; inhibits protein synthesis in extrahepatic tissue

cattle by stimulating growth of urogenital tissues and lowering carcass grades. Lower carcass grades are associated with a reduced fat content and an increase in both moisture content and weight of visceral organs in the live animal.

Estrogen administration improves performance of fattening lambs in a manner similar to that in beef cattle. Experimental administration of DES and other natural and synthetic estrogens to poultry effectively stimulates weight gain and increases the fat content of the carcasses. This effect is in contrast to the slight reduction in fat content in ruminants. Estrogen administration has not proved to be beneficial for swine.

The increased weight gain stimulated by DES in ruminants is associated with a greater nitrogen retention, muscle growth, and water uptake. It also causes an increase in the weight of the adrenal cortex and pituitary gland as well as some hypertrophy of the male secondary sex tissue. Diethylstilbestrol presumably stimulates ACTH secretion from the pituitary, and this in turn elevates adrenal androgen production, which stimulates growth. Diethylstilbestrol might also stimulate growth by increasing STH secretion. It does increase STH release from the pituitary in intact ruminants but probably not as a direct effect since DES has no effect n STH release from the pituitary *in vitro* N.R.C., 1959).

A summary of some hormone and hormone-like compounds which are used as feed additives to stimulate growth in farm animals is shown in Table 7–4. Many other hormone-like compounds have been used experimentally in an attempt to find more satisfactory growth-promoting substances.

B. Naturally Occurring Growth Factors in Feeds

At least 50 plant materials produce estrogenic effects in animals. The amount of estrogenic substances present in the plants varies considerably, but they are important in livestock production because they sometimes reduce fertility and cause genital and mammary development. Under some conditions, positive growth responses occur in ruminants. Alfalfa (*Medicago sativa*) and clover (*Trifolium pratense*) are the feed plants most likely to contain appreciable estrogenic compounds. In addition, alfalfa seeds, soybean (*Glycine max*), cottonseed (*Gossypium hirsutum*), and linseed (*Linum usitatissimum*) oil meals contain estrogenic compounds which are not usually detectable in grasses.

Many naturally occurring growth inhibitors are also present in livestock feeds. The trypsin inhibitors of soybeans and the gossypol content of cottonseeds are well-established examples of substances which cause growth inhibition. These compounds act by

Table 7–4. Hormones and Hormone-Like Compounds Used in Animal Production for Growth Promotion

Product	Animal	Dose
Diethylstilbestrol	Beef cattle	10 mg/head/day[1]
	Beef cattle	24 mg/head[2]
	Sheep	2 mg/head/day[1]
Melengestrol acetate	Beef heifers	0.25 to 0.5 mg/head/day[1]
Progesterone + estradiol benzoate . . .	Steers	200 mg + 2.5 mg[2]
	Lambs	25 mg + 2.5 mg[2]
Testosterone prop. + estradiol benzoate .	Beef heifers	200 mg + 20 mg[2]
Testosterone prop. + diethylstilbestrol .	Beef cattle	120 mg + 24 mg[2]
Thyroprotein (iodinated casein)	Beef, dairy and sheep	0.5 to 1.5 gm/100 lb body wt[1]
	Swine	100 mg/lb feed[1]
	Poultry	50–100 mg/lb feed[1]

[1] In feed.   [2] Subcutaneous.

interfering with the availability of other nutrients. Many other deleterious compounds in plant material, including alkaloids and specific toxins, reduce growth of farm animals under certain conditions.

C. Antibiotics

Antibiotics and arsenicals are used extensively as growth stimulants, especially in poultry and swine production (see Table 7–5). These substances increase growth and feed efficiency, but they do not alter the mature body size of the animals. In healthy animals there is usually little improvement in gain, but antibiotics reduce the number of unthrifty, slow-growing animals. Antibiotics are effective in eliminating subclinical infections and diseases and thus create a better environment for rapid growth. It is generally accepted that they modify the intestinal flora, but how this affects growth is not completely understood. Antibiotics could act by: (a) increasing bacterial synthesis of essential and stimulatory growth factors, (b) inhibiting bacteria which produce deleterious compounds, (c) inhibiting bacteria which compete for nutrients, and (d) inhibiting organisms which damage intestinal tissues (Robinson, 1962). Some antibiotics have a sparing effect on B vitamins in rats and poultry, and act by increasing the absorption of these vitamins.

Table 7–5. Antibiotics Used as Feed Additives for Growth Promotion

Product	Animal	Dose (mg/lb feed)
Bacitracin and derivatives	Poultry	2–25
	Swine	5–25
	Beef	17–35[1]
Arsanilic acid, sodium arsanilate . . .	Poultry	22–45
	Swine	22–45
Chlortetracycline	Poultry	5–25
	Swine	5–25
	Beef	25–70[1]
	Sheep	10–25
Dynafac	Poultry	45–90
	Swine	200[1]
	Sheep	200[1]
	Beef	300–400[1]
Erythromycin	Growing chicks and turkeys	2.3–9
	Beef	37[1]
Furazolidone	Poultry	3.5–5.0
	Swine	75
Oleandomycin	Broilers and turkeys	0.5–1.0
Oxytetracycline	Poultry	25
	Swine	
	(10– 30 lb)	12–25
	(30–200 lb)	3.5–5
	Beef	25–75[1]
	Sheep	5–10
Penicillin	Poultry	1.2–25
	Swine	5–25
Roxarsone	Poultry	11–22
	Swine	11–33
Tylosin	Poultry	2–25
	Swine	5–10

[1] mg/head/day

D. Anthelmintics

Anthelmintics have been used as promoters of growth under practical feeding conditions. Beneficial effects of anthelmintics are generally related to the control of parasite infestations. A possible exception is an organic phosphate, DDVP (2,2-dichlorovinyl dimethyl phosphate), which is used to stimulate growth in fattening cattle. DDVP has both an anthelmintic and a separate growth stimulatory effect in cattle. The mechanism of growth stimulation has not been elucidated (I. A. Dyer, personal communication).

E. Tranquilizers

Feeding of tranquilizers sometimes causes a slight increase in weight gain and feed efficiency in fattening beef and sheep, but there are usually no beneficial effects in other animals. Tranquilizers used in feeding experiments include reserpine, hydroxyzine, perphenazine, and chloropromazine.

IV. RESEARCH TECHNIQUES

Experimental investigations of growth regulators require the availability of adequate methods of growth measurement. Because of the broad interpretations of growth, many different methods are used to quantitate the growth of cells, tissues, organs, and whole animals.

Cellular growth is often expressed in terms of the number of cells undergoing mitosis. This is determined by measuring the mitotic index and mitotic rate. The mitotic index is the proportion of cells undergoing division at a particular time; it is determined by microscopic examination. The mitotic rate is measured by administering colchicine or colcemid to arrest mitosis, and then microscopically counting the number of arrested cells which accumulate in a specific period of time. A relative change in either of these parameters indicates a change in the rate of cell proliferation. The rate of mitosis is also measured by determining the number of cells incorporating tritiated thymidine into newly synthesized DNA. A dose of ^{3}H-thymidine is administered to growing cells, and, after a suitable interval, the number of labeled cells is counted using autoradiography. When dispersed cells or uniform suspensions are available, cell proliferation may be determined by microscopic or electronic counting.

Cellular growth is not discernible by the methods described above if it involves an increase in cell size rather than cell number. Growth can still be measured by determining the total cell mass or by making a chemical analysis for a cellular constituent such as protein. Even microscopic observations of changes in cell size and development are used to measure cell growth (Iversen, in Teir & Rytömaa, 1967).

Growth of tissues and organs can be measured by observing changes in tissue or organ weight. The degree of hyperplasia is determined by chemically analyzing for DNA and relating it to tissue mass or protein content. Growth by hyperplasia may also be estimated with the methods used for measuring cellular proliferation.

Body weight, nitrogen retention, and changes in linear dimensions are used to measure growth in whole animals. Increased body weight can result from retention of electrolytes and water, fat deposition, or an increase in lean body mass. Nitrogen retention is intimately associated with increased protein synthesis and growth of lean body mass, and it may be used as a reliable index of growth. In feeding experiments with farm animals, an increase in body weight is usually the only criteria of growth. This broad interpretation ignores the relative contributions of water, fat, and protein to changes in body weight. It also fails to detect abnormal growth responses which may result from an imbalanced diet, pathological conditions, or other factors. Even though under normal conditions the growth of young animals is primarily associated with increased protein deposition and skeletal growth, body weight alone should not be used as a reliable index of true growth.

Skeletal growth in the long bones causes a linear increase in body size. This continues until maturity, when the epiphyseal cartilage calcifies. In growing animals, the width of the epiphyseal cartilage and the time of fusion are criteria for assaying the effectiveness of skeletal growth regulators.

REFERENCES

Blaxter, K. L., Reineke, E. P., Crampton, E. W., & Peterson, W. E. (1949). The role of thyroidal materials and of synthetic goitrogens in animal production and an appraisal of their practical use. J. Anim. Sci. *8*:307–352.

Bullough, W. S. (1964). Growth regulation by tissue-specific factors or chalones. In: *Cellular Control Mechanisms and Cancer*. P. Emmelot & O. Muhlback (Eds.), p. 124, New York, Elsevier.

Bullough, W. S. (1965). Mitotic and functional homeostasis. A speculative review. Cancer Res. *25*: 1683–1727.

Cohen, S. (1964). Isolation and biological effects of an epidermal growth-stimulating protein. Nat. Cancer. Inst. Monogr. *13*:3–38.

Drill, V. A. (1961). Steroids and growth. In: *Growth in Living Systems*. M. X. Zarrow (Ed.), pp. 383–405, New York, Basic Books.

Evans, H. M., Briggs, J. H., & Dixon, J. S. (1966). The physiology and chemistry of growth hormone. In: *The Pituitary Gland*. G. W. Harris & B. T. Donovan (Eds.), Vol. I, pp. 439–491, Berkeley, University of California Press.

Gaebler, O. H. (1965). Growth and pituitary hormones. In: *Newer Methods of Nutritional Biochemistry*. A. A. Albanese (Ed.), Vol. II, pp. 85–121, New York, Academic Press.

Gaunt, R. (1954). Chemical control of growth in animals. In: *Dynamics of Growth Processes*. E. J. Boell (Ed.), pp. 183–211, Princeton, New Jersey, Princeton University Press.

Goldwasser, E. (1966). Biochemical control of erythroid cell development. In: *Current Topics in Developmental Biology*. A. A. Mascona & A. Monroy (Eds.), Vol. I pp. 173–211, New York, Academic Press.

Hechter, O., & Halkerston, I. D. K. (1965). Effects of steroid hormones on gene regulation and cell metabolism. Ann. Rev. Physiol. *27*:133–162.

Jones, R. O. (1966). The *in vitro* effect of epithelial growth factor on rat organ cultures. Exp. Cell Res. *43*:645–656.

Knobil, E. K., & Hotchkiss, J. (1964). Growth hormone. Ann. Rev. Physiol. *26*:47–74.

Lange, R. D., & Pavlovic-Kentera, V. (1964). Erythropoietin. Progr. Hemat. *4*:72–96.

Levi-Montalcini, R. (1964). The nerve growth factor. Ann. N. Y. Acad. Sci. *118*:149–170.

Levi-Montalcini, R. (1966). The nerve growth factor: its mode of action on sensory and sympathetic nerve cells. Harvey Lect. *60*:217–259.

National Research Council (1959). Hormonal relationships and applications in the production of meats, milk, and eggs. Publication 714, Washington, D. C., National Academy of Science.

Robinson, K. L. (1962). Value of antibiotics for growth of pigs. In: *Antibiotics in Agriculture*. M. Woodbine (Ed.), pp. 185–202, London, Butterworth.

Rosenberg, I. N., & Bastomsky, C. H. (1965). The thyroid. Ann. Rev. Physiol. *27*:71–106.

Schenkein, I., Levy, M., Bueker, E. D., & Tokarsky, E. (1968). Nerve growth factor of very high specific activity. Science *159*:640–643.

Szent-Györgyi, A. (1965). Cell division and cancer. Science *149*:34–37.

Szent-Györgyi, A. (1966). Growth and organization. Biochem. J. *98*:641–644.

Teir, H. & Rytömaa, T. (Eds.) (1967). *Control of Cellular Growth in Adult Organisms*. New York, Academic Press.

Turner, C. D. (1961). *General Endocrinology*. 3rd ed., pp. 107–146, Philadelphia, Saunders.

Chapter 8

Developmental Anomalies

By W. Binns, L. F. James, R. F. Keeler, and K. R. Van Kampen*

ANATOMICAL ANOMALIES

A. Introduction

THE subject of anatomical anomalies is an extremely large one, involving a study of all the possible errors which may occur during the complex process of development. Genetic causes, deficiencies or excesses of certain nutritional elements, and ingestion of certain poisonous plants or chemical compounds at specific periods of embryonic or fetal development have been associated with congenital deformities.

Developmental anomalies ordinarily occur before birth, in the embryonic or fetal stage of life, when the body organs or the supporting structures are in the process of development. There are exceptions. If the epiphyseal cartilage of a long bone is severely injured in a young growing animal, as often occurs in chronic osteomyelitis, for instance, the bone stops growing and the animal develops a shortened and deformed leg. The exact incidence of prenatal malformations in livestock is difficult to determine, but may reach 5 percent or more annually. The wide variety of congenital malformations in livestock is astounding.

B. Description

The various types of deformities indicate that they depend on a variety of errors in the developmental mechanisms of the embryo. The mechanisms of some of the most commonly understood deformities are as follows:

1. Arrest of a specific part during the development of the embryo. This arrest results in the absence (aplasia) or decrease (hypoplasia) in the size of the specific structure or organ (Fig. 8–1).

2. Failure of certain embryological openings, fissures, or grooves to close properly as the development of the embryo proceeds in its early life. Many such developmental failures occur from an incomplete closure of the neural groove which results in a congenital fissure of the cranium (cranioschisis) or a congenital fissure of the spinal column (rachischisis). A patent foramen ovale in the heart and a persistent cloaca in which

* U.S. Department of Agriculture, Agricultural Research Service, Animal Disease & Parasite Research Division, 1150 East 14th North, Logan, Utah 84321.

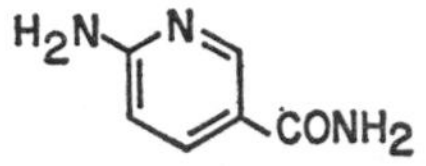

6-Aminonicotinamide 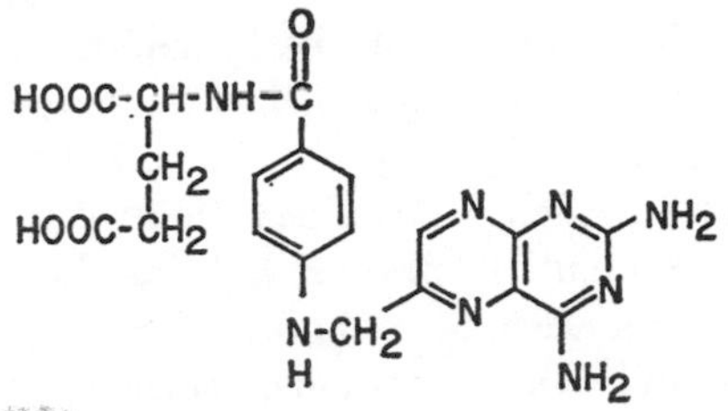   Aminopterin   β Aminopropionitrile

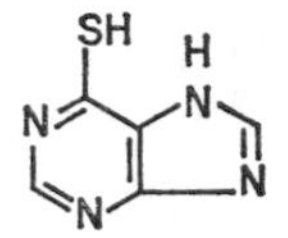

6-Mercaptopurine

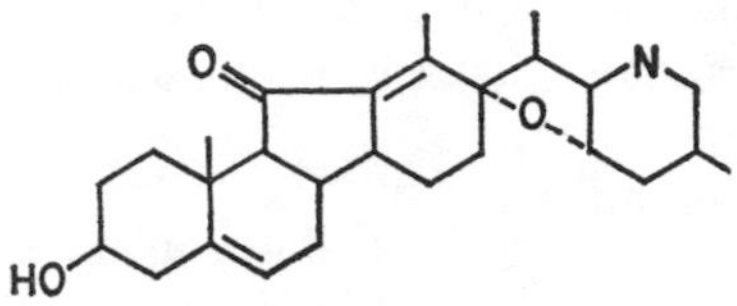

Jervine 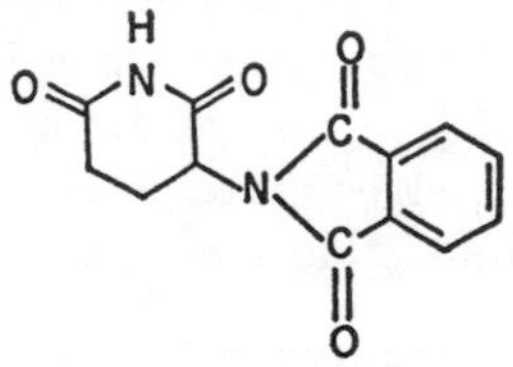   Thalidomide

FIG. 8–1. Structures of certain teratogenic compounds. Aminopterin, ethionine, 6-aminonicotinamide, and 6-mercaptopurine are examples of antagonists of dietary factors and metabolic intermediates. β-Aminopropionitrile is a typical lathyrogen, thalidomide, a tranquilizer responsible for limb deformities in man, and jervine is one one of the teratogenic alkaloids from *Veratrum* producing cyclopia in sheep.

the rectum and external genital openings are not separated can also be included in this developmental failure.

3. Failure of certain embryonic or fetal structures to disappear as the embryo develops. Examples are persistence of the ductus arteriosus, the thyroglossal duct, and patent urachus. The common malformation of atresia ani results from failure of the overlying skin to disappear from the anal opening.

4. Tissue cells that wander or deviate from the usual or normal course (aberrant, ectopic or heterotopic structures). This deformity is more common than most livestock men realize because it is seen more often at postmortem than at clinical examination.

5. Duplication. In early embryonic life, each specific cell or group of cells is destined to produce a specific structure in the adult. If such a cell or group of cells were to divide without further differentiation for a specific structure of an organ or tissue, two cells or groups of cells with identical potentialities would result and identical organs or body structures would be formed so that there would be two or more organs instead of the usual one. If the two-cell stage embryo makes a complete division, there will be two complete identical twins (monozygotic type). Experimental studies with certain amphibian species have shown that a chemical hormone called an "organizer" is formed in the appropriate regions of the embryo to produce the necessary stimuli for the cells to differentiate into the specific organs or structures. Twin monsters and many other malformations have been experimentally produced by depriving the embryo in its early life of sufficient oxygen, by the maternal ingestion of certain toxic agents, or by unduly decreasing or increasing the body temperature.

There may be many types and degrees of duplication. Any part of the body may be duplicated, but the duplicated parts may not necessarily be equal in size and function. Calves are occasionally born with 2 mouths and 4 eyes of equal size. The heads are a common attachment of the cranium. One mouth may have a cleft palate while the other is normal. The four eyes and two mouths may move simultaneously.

Many types of supernumerary parts may be present, but are usually minor in nature. For example, in animals extra phalanges are usually bilaterally symmetrical, imperfect, and accessory. Supernumerary teats are common in cows. In some instances, the duplicated part may attain only a shapeless mass and be recognized as an accessory organ only by its histological structure.

II. GENETIC ANOMALIES

A. Predisposition

One of the prime factors in embryogenesis is the early determinative effect present in the fertilized ovum. This effect is known as "predetermination" and is thought to be a function of cytoplasm carried with each new cell during cleavage since the genome is identical.

During development, after organal differentiation has occurred, the cells of an organ differentiate from one another. A second factor, besides predetermination is thought to be present. This is called "embryonic induction" and is due to an interaction of inductor tissue and responding tissue. The beginning and conclusion of induction are different for each developing organ. Little is known about the inducer substances although the mechanisms are being studied.

There are many well-documeted abnormalities of growth and development which arise from genetic abnormalities.* These

* For additional information the reader is referred to Fishbein (1963); Crew (1923); Detlefsen & Carmichael (1921); DiPaolo & Kotin (1966); Eldridge *et al.* (1951); Fischer (1953); Grunberg (1963); Jacobson (1966); Kalter & Warkany (1959); Pruzansky & Thomas (1961); Shrode & Lush (1947); Shupe *et al.* (1967); and Weber & Ibsen (1934).

may be "predetermined" or may arise from defects in the "induction mechanism." Both types of defective factors may be inherited. Manifestations of these anomalies may vary from a slight malformation of a specific organ to abnormalities or absence of several organ systems. If the latter occurs and vital organ systems are affected, the developing embryo or fetus may die *in utero* and be absorbed or aborted, or, if carried to term, may not be compatible with extrauterine life. If this occurs the genetic makeup is said to be lethal. Offspring with this genetic composition will not live to reproduce; hence, the genes must be carried as recessive in a heterogeneous population.

B. Lethal Genes

A well-documented genetic lethality is found in Dexter cattle. The calves are chondrodystrophic. The limbs are severely shortened and brachycephaly is present. These "bulldog" type calves are usually aborted at 4 months gestation or later. Occasionally hydrops amnii will be present in the dam.

Another lethal genetic abnormality is described in Hereford cattle. The axial and appendicular skeletons are severely deformed. Cleft palate and hydrocephalus are present. These calves are dead or die at birth. The cows carry the calves approximately 2 weeks beyond term and show no signs of impending parturition.

C. Dwarfism

One of the most prevalent genetic abnormalities is the chondrodystrophic "dwarf," found in all breeds of cattle. These brachycephalic cattle have bulging foreheads, retruded maxillas, protruded mandibles and bulging eyes. The vertebrae are compressed longitudinally and there is premature closure of the synchondroses of the cranium. Chronic bloating and nasal dyspnea are commonly associated with this condition. A few years ago, this deformity was widespread within beef breeds in the United States because of selection for a very short-legged, compact type of beef animal.

PLATE 8

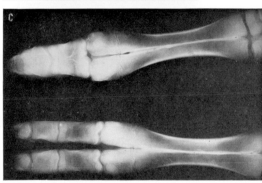
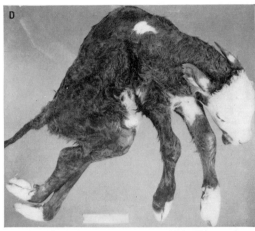
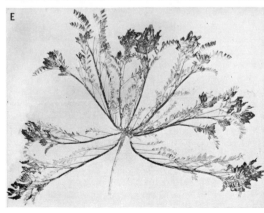

A. Crooked calf disease induced by maternal ingestion of *Lupinus sericeus.*
B. Duplication of a calf's head with a common cranial attachment.
C. Radiograph of fusion of the phalanges in the foot of a Holstein calf (syndactylism).
D. Hereditary skeletal deformities in Hereford cattle. *(Courtesy James L. Shupe.)*
E. Locoweed, *Astragalus lentiginosus.*
F. *Veratrum californicum* (false hellebore) in early bud stage.

PLATE 9

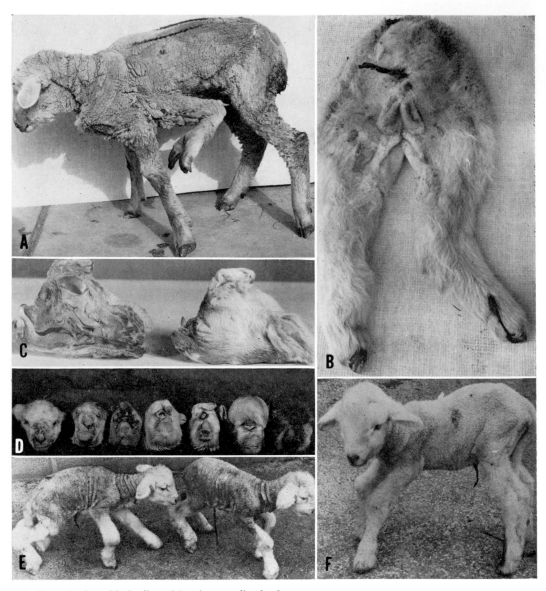

A. Extra foreleg with duplicated feet in a yearling lamb.

B. Congenital malformation resulting in development of only the posterior portion of a lamb's body. Born at full-term gestation.

C. The malformed head of a lamb (cyclopic type) induced by ingestion of *Veratrum californicum* (medical sagittal section).

D. Graduation of cyclopian type of malformed lambs produced by maternal ingestion of *Veratrum californicum*.

E. Osteolathyrism in lambs from maternal aminoacetonitrile ingestion.

F. Deformity in a lamb induced by maternal ingestion of locoweed.

D. Musculoskeletal Deformities

The musculoskeletal system is most often affected by known genetically controlled anomalous development. Ancon or otter sheep are characterized by a rather normal axial skeleton with shortening of the bones of the appendicular skeleton. The bones located in the distal portions of the extremities are shortened much more than are the bones proximally. Syndactylism or "mule foot," which is due to a recessive gene, occurs in Holstein-Friesian cattle and in some breeds of hogs. A single third phalanx is usually present. The condition occurs more commonly in the forelimbs than the hindlimbs. Occasionally, only a single limb will be affected.

Double muscling is a recessive factor of variable expression found in most milking and beef breeds of cattle. There is hyperplasia of the muscle fibers of the thigh, rump, loin, and shoulders. The increased muscle bulk is accentuated by the very thin skin associated with the condition. Affected animals may be infertile because of genital hypoplasia.

E. Others

Many organ systems are affected with some of the inherited abnormalities. Adenohypophyseal aplasia is described as a genetic trait in Guernsey cattle. The multiplicity of organ systems affected is striking. Although the fetus may be carried alive *in utero* 100 days or more beyond term, its size remains that of a 7-month fetus. In this syndrome, the genetically defective fetus is unable to initiate parturition and gestation is interrupted only by the death of the fetus. The fetal anomaly is characterized by craniofacial deformities with an associated aplasia of the adenohypophysis. Concomitant endocrine changes are present. The thyroid glands are hypoplastic and inactive, the testicles are devoid of Leydig's cells, and the zona glomerulosa of the adrenal cortex is undifferentiated. The ovaries and pancreas appear normal.

Other genetic abnormalities of animals affect the eyes, the ears, the hematopoietic tissue and certain enzyme systems. Many of these are related to corresponding diseases in man and offer a practical biomedical model for further study of these conditions.

III. NUTRITIONAL

A. Toxic Plants

The maternal ingestion of *Veratrum californicum* (false hellebore) (Plate 8) on the 14th day of gestation (neural plate stage of the embryological development) causes developmental anomalies restricted primarily to the head and central nervous system. All grades of disorder may occur, ranging from defective development to complete absence of areas, parts, or regions, and these can be unilateral or bilateral. These include absence of the brain (anencephaly), incomplete closure of the skull (cranioschisis), various degrees of cerebral herniations, abnormally large size or smallness of brain (macrocephaly or microcephaly), absence of eyes (anophthalmia), presence of a single eye in centrally located orbit (cyclopia), various degrees of gross dilation of cerebral ventricles with fused hemispheres (hydrocephalus) (Plate 9), and shortening of maxillae causing a marked foreshortened appearance of the face which emphasizes the domed appearance of the cranium. The latter probably inspired the name "monkey-faced" by which lambs with these anomalies are commonly known among sheep ranchers. Usually, in the cyclopic type of deformity, the olfactory bulbs, the optic chiasm, and the pituitary are absent. Only a single optic nerve is present. The mandible is unchanged except for an exaggerated upward curvature. The tongue of normal size protrudes from the oral cavity because of the lack of space in the mouth.

Maternal ingestion of *Lupinus caudatus* (tailcup lupine) or *Lupinus sericeus* (silky lupine) by cows during pregnancy causes a congenital developmental deformity in calves commonly called "crooked calf disease." The deformity is characterized by a persistent flexure or contracture with twisting of the legs (arthrogryposis), twisting of the neck and unnatural position of the head (torticollis), and abnormal curvature of the vertebral column (scoliosis). In field cases, congenital fissure of the palate and roof of the mouth (cleft palate) may be present. One or more of these characteristic deformities may occur simultaneously in the affected calf.

Lupine plants grow abundantly on all ranges where "crooked calf" deformities have occurred, and the malformations can be produced by experimental feeding of the plant. Because of hybridization, identification of specific species is sometimes very difficult. All species of lupine are not toxic or teratogenic. Many species are considered valuable forage for livestock.

Locoweeds (species of *Astragalus* and *Oxytropis* genera) are poisonous to livestock (Plate 8). They produce a teratogenic effect in lambs and calves through maternal ingestion. The deformity is characterized by contracted tendons and flexure of the forelimbs, anterior flexure with hypermobility of the hock and stifle joints, and abnormal shortness of the under jaw (brachygnathia).

The dosage level of the teratogenic species is less than the amount necessary to cause clinical toxic signs of maternal poisoning. The amount of toxic and teratogenic agent present in the plants may vary from one year to the next because of certain unknown environmental factors.

Hybrid sudan grasses are known to cause urinary incontinence and fetal abnormalities in horses. Only the green plant is toxic. The active principles have not been isolated.

Many postnatal anomalies are produced directly in animals ingesting toxic plants. Examples include neuro- and osteolathyrism from ingestion of members of the *Lathyrus* and *Vicia* spp., liver tumors from ingestion of *Senecio*, and certain skeletal changes from ingestion of molbydenum and selenium accumulating plants.

Thus, it is quite clear that ingestion of range plants can produce developmental anomalies in the offspring of livestock that graze on these plants. Little, however, is known concerning the chemical agents in the plants responsible for the teratogenesis or the mechanisms by which they act. A few notable exceptions exist.

The compounds from *Lathyrus odoratus* (common garden sweet pea) and possibly other *Lathyrus* and *Vicia* spp. responsible for the developmental anomalies of osteolathyrism have been extensively investigated.

The compound β-aminopropionitrile (Fig. 8–1) and the glutamyl derivative are responsible for some of the prenatal and postnatal skeletal deformities of lathyrism in man and laboratory and farm animals. A closely similar nitrile, aminoacetonitrile, is extensively used as an experimental tool in studies of lathyrism.

The mechanism by which the lathyrogen exerts its effect is by preventing normal cross-linking of connective tissue. This suppression results in structurally defective connective tissue which allows the abnormal skeletal characteristic of the disease to develop.

The prenatal malformations resulting from ingestion of locoweed and lupine plants in sheep and cattle are clinically similar to osteolathyrism. The question of whether all three abnormalities are produced by similar compounds and therefore arise by similar mechanisms is being investigated.

The cyclopian malformation of sheep, which results from ingestion of *Veratrum californicum* (false hellebore), is due to certain steroidal alkaloids elaborated by the plant. One of these compounds whose structure is known is jervine (Fig. 8–1). A related compound, cyclopamine, appears to be the naturally occurring compound in *Veratrum californicum* (false hellebore) responsible for the developmental anomalies.

B. Nutritional Imbalances

1. VITAMINS AND ANTAGONISTS

A variety of prenatal anomalies results from nutritional deficiencies, excesses, and antimetabolites. Indeed, the work of Hale on teratogenic effects in pigs from maternal vitamin A deficiency is acknowledged to be the beginning of teratology. Hale demonstrated that anophthalmia, ectopic kidneys, harelip, cleft palate, malformed limbs, and other defects in piglets resulted from vitamin A deficiency in the sow.

The teratogenic effect of riboflavin deficiency is being studied extensively in rats. Malformations include short mandible, cleft palate, syndactylism, and short tail as well as various skeletal defects including the absence of the tibia or fibula and fused ribs. Folic acid deficiency, also studied extensively in rats, gives rise to hydrocephalus and various abnormalities of the eyes. Rats deficient in pantothenic acid give birth to offspring with anencephaly, anophthalmia, localized hemorrhages at the extremities of the limb, deformed digits, malformations of the thorax, and edema. Other vitamin deficiencies including vitamins C, D, E, thiamine and niacin give rise to malformations in offspring. The experimental work is being done on laboratory animals primarily, but it seems probable that domestic animals could also be affected providing conditions which reduce ruminal availability of any of the vitamins normally produced intraruminally existed.

Congenital anomalies resulting from dietary excesses are less common, but are well documented in limited instances. For example, hypervitaminosis A in pregnant rats caused malformations in offspring. Cleft palate, cleft lip, brachygnathia, shortening of the maxilla, open eye, exophthalmos, cataract, anophthalmia and microphthalmia were among the defects produced.

Antagonists of required dietary factors apparently produce malformations through their ability to block the normal function of the dietary factor competitively. Such antagonists produce the same type of malformations as are produced by deficiency of the dietary factor. Examples of antagonists in this category include 6-aminonicotin-amide, a nicotinic acid antagonist; aminopterin, antagonist of folic acid; and ethionine, an antagonist of methionine (Fig. 8–1).

Such antagonists are useful experimental tools, but are seldom encountered under natural conditions. Antagonists of metabolites involved in intermediary metabolism also are known teratogens. Various nucleic acid antagonists such as the purine analog, 6-mercaptopurine, are good examples (Fig. 8–1).

2. MINERALS

Developmental anomalies are generally associated with the prenatal development of an animal. However, many anomalies can occur during postnatal development. Some of these anomalies occur because of various nutritional excesses, deficiencies, or imbalances. Many of these deviations can be corrected through diet while others leave the animal permanently deformed in spite of improvement in the diet.

Rickets is a classic example of a dietary-induced postnatal anomaly. Rickets is due to a derangement of the mineral metabolism of developing bone in a manner that prevents proper calcification of growing bones. The ends of the long bones and costochondral junctions become enlarged and in severe cases the limbs may become permanently bent (bow legs) from the weight of the animal on the poorly calcified bone. Fractures and lameness are common in rachitic animals.

Big head or nutritional secondary hyperparathyroidism in horses is induced by feeding rations high in phosphorus with a low or normal amount of calcium. The high dietary phosphorus results in a hyperphosphatemia followed by a hypocalcemia. The hypocalcemia induces a hyperparathyroidism, and, subsequently, a generalized osteitis fibrosis. This condition is characterized by lameness, a hollow sound over the sinuses on percussion, and a mild enlargement of the jaw.

Selenium toxicity can result in both pre- and postnatal anomalies in lambs and colts born to dams that have grazed seleniferous range areas. The eyes of some lambs develop various degrees of deformity: microphthalmia, microcornea and coloboma. Most of the lambs died soon after birth. Many of those that survived developed deformed legs.

Chronic selenium poisoning causes hoof lesions, deformities, and loss of the long hair. The hoof lesions are manifest by a circular break on the wall of the hoof below the coronary band. As the hoof grows the break in the wall of the hoof moves downward with the growth of new hoof and is gradually sluffed off. The hoof may grow excessively long.

Although molybdenum poisoning does not cause any permanent structural changes in the animal's body, it does produce abnormalities that should be noted. Diets high in molybdenum can result in unthriftiness, rough hair coat, hair color change (red to straw color, and black to mouse grey), arching of the back, and brittle bones. The hair color and brittleness of the bones can be changed by improvement of the diet.

Excess fluorine in the diet of animals can result in mottling of the teeth and varying bone changes. The bone changes associated with osteofluorosis are osteoporosis, osteosclerosis, periostial hyperostosis, and osteomalacia. In some instances, cows with marked osteofluorosis also develop malalignment of the limbs.

3. PROTEIN AND ENERGY

Stunting of animals from lack of sufficient total food or of an individual nutrient is not to be considered by some investigators as anomalous development. However, the condition is a deviation from the normal during the postnatal development of an animal. An animal can compensate for the lack of early development; however, if the dietary deprivation is continued over a long time, the animal may become permanently stunted.*

IV. INFECTIOUS DISEASES

Most of the anomalies in domestic animals that are caused by infectious agents are due to viruses (Table 8–1) (cf. Marsh, 1965; Smith & Jones, 1961).

V. OTHER ANOMALIES

A. Radiation-induced Anomalies

Whole body gamma irradiation has induced prenatal deformities in swine, sheep, and cattle. Ewes irradiated between the 22nd and 25th days of gestation gave birth to lambs with deformities of the thoracic limbs, while irradiation between the 23rd and 27th days resulted in malformed pelvic limbs. Limb deformities were induced in cattle by gamma irradiation on the 32nd day of gestation and in swine by neutron exposures on the 21st day of pregnancy. The minimum dose that produced anomalies in sheep was 200 rads and in cattle, 300 rads.

B. Chemically-induced Anomalies

A variety of prenatal malformations are produced by a miscellaneous group of chem-

* For additional information the reader is referred to Binns et al. (1963); Dye & O'Hara (1959); Greenwood et al. (1964); James et al. (1967); Keeler & Binns (1968); Keeler et al. (1967); Krook & Lowe (1964); Rosenfeld & Beath (1964); Shupe et al. (1967); and Stout & Koestner (1964).

Table 8–1.　Virally-Induced Anomalies

Virus	Natural Host	Viral State	Induced Anomalies
Rat	Rat	Natural	Cerebellar hypoplasia
Panleukopenia	Cat	Natural	Cerebellar hypoplasia
Hog cholera	Swine	Vaccine strain Modified Live Virus	Nasal and kidney deformities and edema
Bluetongue	Sheep	Vaccine strain Modified Live Virus	Encephalopathy

ical compounds including drugs derived from both organic synthesis and natural sources. Again, much of the information comes from trials in laboratory animals and accidental exposure in man. A parital list of these miscellaneous compounds includes nitrogen mustard, chlorambucil, busulfan, dactinomycin, thalidomide, benzopyrene, urethan, colchicine, estrogens, testosterone, trypan blue, nicotine, quinine, vinca alkaloids, insulin, cortisone, terramycin, penicillin, streptomycin, actinomycin D, and sodium salicylate.

Possibly, the most startling of all teratogenic effects is the very high incidence of limb deformities in children that occurred in the late 50's in West Germany. The abnormality is traced to the tranquilizer thalidomide (Fig. 8–1), which was in widespread use at that time. The drug was thought to be relatively free of undesirable side effects and indeed, doses 20 to 40 times the normal hypnotic dose produced only sleep and severe headache. The unfortunate teratogenic effect was learned only in retrospect. It is estimated that over 6,000 deformed babies were born whose deformities resulted from maternal ingestion of that drug. The period of fetal insult producing deformed limbs or absence of limbs is between the 31st and 50th day after the last menstrual period.

REFERENCES

Binns, W., James, L. F., Shupe, J. L., & Everett, G. (1963). A congenital cyclopian-type malformation in lambs induced by maternal ingestion of a range plant, *Veratrum californicum*. Amer. J. Vet. Res. *24*:1164–1175.

Crew, F. A. E. (1923). The significance of an achondroplasia-like condition met within cattle. Proc. Roy. Soc. (Biol.) *95*:228–255.

Detlefsen, J. A., & Carmichael, W. J. (1921). Inheritance of syndactylism, black and dilution in swine. J. Agric. Res. *20*:595–604.

Di Paolo, J. A., & Kotin, P. (1966). Teratogenesis-oncogenesis: A study of possible relationships. Arch. Path. *81*:3–23.

Dye, W. B., & O'Hara, L. J. (1959). Molybdenosis. Nevada Agric. Exp. Sta. Bull. No. 208. Reno, Nevada, University of Nevada.

Eldridge, F. E., Smith, W. H., & McLeod, W. M. (1951). Syndactylism in Holstein-Friesian cattle:

Its inheritance, description, and occurrence. J. Hered. *42*:241–250.

Fischer, H. (1953). Die Genetik der Doppelendigkeit beim Rind. Fortpfl. Besamung Haustiere *3*:25–27.

Fishbein, M. (Ed.) (1963). *Congenital Malformations*. International Conference, Philadephia, J. B. Lippincott.

Greenwood, D. A., Shupe, J. L., Stoddard, G. E., Harris, L. E., Nielson, H. M., & Olsen, L. E. (1964). Fluorosis in cattle. Special Report #17. Utah Agric. Exp. Sta., Logan, Utah.

Grunberg, H. (1963). *The Pathology of Development. A Study of Inherited Skeletal Disorders in Animals.* Oxford, England, Blackwell Scientific Publications.

Jacobson, A. G. (1966). Inductive processes in embryonic development. Science *152*:25.

James, L. F., Shupe, J. L., Binns, W., & Keeler, R. F. (1967). Abortive and teratogenic effects of locoweed on sheep and cattle. Amer. J. Vet. Res. *28*:1379–1388.

Kalter, H., & Warkany, J. (1959). Experimental production of congenital malformations in mammals by metabolic procedure. Physiol. Rev. *39*:69–115.

Keeler, R. F., & Binns, W. (1968). Teratogenic compounds of *Veratrum californicum* (Durand). V. Comparison of cyclopian effects of steroidal alkaloids from the plant and structurally related compounds from other sources. Teratology *1*:5–10.

Keeler, R. F., James, L. F., Binns, W., & Shupe, J. L. (1967). An apparent relationship between locoism and lathyrism. Canad. J. Comp. Med. *12*:334–341.

Krook, L., & Lowe, J. E. (1964). Nutritional secondary hyperparathyroidism in the horse. Path. Vet. *1*: 1–98.

Marsh, H. (1965). *Newsom's Sheep Diseases*. 3rd ed., Baltimore, The Williams and Wilkins Co.

Pruzansky, S., & Thomas, C. C (Eds.) (1961). *Congenital Anomalies of the Face and Associated Structures.* Proceedings of an International Symposium. Springfield, Ill.

Rosenfeld, I., & Beath, O. A. (1964). *Selenium Geobotany, Biochemistry. Toxicity and Nutrition.* New York, Academic Press.

Shrode, R. R., & Lush, J. L. (1947). The genetics of cattle. Advances Genet. *1*:209–261.

Shupe, J. L., Binns, W., James, L. F., & Keeler, R. F. (1967). Lupine, a cause of crooked calf disease. J. Amer. Vet. Med. Ass. *151*:198–203.

Shupe, J. L., James, L. F., Balls, L. D., Binns, W., & Keeler, R. F. (1967). A probable hereditary skeletal deformity in Hereford cattle. J. Hered. *58*: 311–313.

Smith, H. A., & Jones, T. C. (1961). *Veterinary Pathology*. 2nd ed., pp. 725–738. Philadelphia, Lea & Febiger.

Stout, R. W., & Koestner, A. (1964). Skeletal lesions associated with a dietary calcium and phosphorus imbalance in the pig. Amer. J. Vet. Res. *26*: 280–294.

Weber, A. D., & Ibsen, H. L. (1934). The occurrence of the double muscled character in purebred beef cattle in all European beef breeds. Proc. Amer. Soc. Anim. Prod. *27*:228–232.

Chapter 9

Aging

By H. J. Curtis

ALL animals have a definite life cycle which includes birth, a period of rapid growth, puberty, a gradual cessation of growth, senescence, and, finally, death. The pattern is rather rigidly fixed by the genetic background of the organism, and can be varied only by rather drastic measures. The developmental aspects of the problem are considered in other chapters, and only that part of the life cycle involved with degeneration and senescence will be considered in this chapter. Animals have very efficient repair mechanisms which allow them to recover from almost any type of damage or disease, and yet, in the absence of demonstrable malfunction, the organism degenerates and finally succumbs to one of the degenerative diseases.

This chapter will describe the anatomical and physiological changes which take place in mammals as they age, discuss briefly the basic biological processes concerned with aging, and, finally, review some of the environmental conditions which may alter the process. No attempts will be made to discuss lower forms of animals such as insects or fishes, since this is a separate study and the

subject as developed for mammals is applicable only in part to other species.

The longevity of the different mammals is surprisingly difficult to ascertain, and is known reasonably well only for man, laboratory rats, mice, dogs, and race horses. The most accurate measure of longevity is the median life span of a fairly large population of the animals in question, but such data is unknown except for the animals just noted. However, maximum life spans seem to be more interesting, if less meaningful. These have been recorded for a few species (Table 9–1).

There seems to be very little correlation between the anatomical or physiological characteristics of the different animals and the characteristics of the aging process of each group. In general, the large animals live longer than the small ones. Man, however, apparently outlives both the elephant and the whale, and small dogs outlive large ones.

In 1909 Rubner measured the metabolism of a large number of animals and concluded that the total metabolism of an animal summed over his life span is about the same

Table 9–1. Approximate Maximum Life Spans
of Some Mammals

(*Data from Comfort, A. 1964. In: Aging, The Biology of Senescence. Holt, Rinehart & Winston, New York.*)

Mammals	Years
Man	90
Elephant	70
Horse	50
Hippopotamus	50
Rhinoceros	50
Donkey	50
Whale	50
Chimpanzee	40
Baboon	30
Monkey	30
Domestic cat	30
Goat	20
Domestic dog	20
Cow	20
Bat	16
Fox	15
Rabbit	15
Guinea pig	8
Laboratory rat	5
Laboratory mouse	$3\frac{1}{2}$
Shrew	$1-1\frac{1}{2}$

for all mammals. The large animals have a small surface-to-volume ratio so the metabolism per gram of tissue required to maintain body temperature is quite low. The shrew, for example, has the highest metabolism per gram body weight of all the mammals, and by far the shortest life span. This is a very interesting observation and should be kept in mind. There are so many exceptions to the general rule, however, that one must infer that metabolism is only one of many factors which must be taken into account.

About the same time, Friedenthal noted a correlation between life span and the size of the brain relative to the body weight. There is better correlation in this parameter than in any of the others, but even so, the correspondence is poor. For example, on this basis, the median life span of man should be 150 years. Furthermore, it is difficult to see why brain size should be correlated with longevity, especially for man and domestic animals.

The average life span of each animal is, to a large extent, determined by its genetic

background. However, there are several factors which alter this genetically programmed aging considerably. It is thus important to understand something of the basic biological mechanisms involved.

I. ANATOMICAL CHANGES

In most animals, growth virtually stops soon after puberty and body weight remains relatively constant for most of the rest of the life span. An exception is the rat, whose skeleton continues to grow at a slow rate throughout life, and whose weight increases accordingly. Each animal has a preferred body size at any age, and will attempt to attain this size if proper food is available. This size is determined by genetic factors. If adult animals are restricted in their food intake for a period of time, they will lose weight in proportion to the dietary restriction. If they are then allowed to eat all they want, they will gain weight, not to the original weight, but to the weight they would have attained if they had been allowed to eat *ad libitum*. The nutrition of the very young animal will, of course, greatly influence his eventual size as an adult.

One of the most common observations with regard to aging is the decline in muscular performance. This is especially noticeable in athletes and race horses, who must function to the peak of perfection. Several factors contribute to this gradual decline. At birth, each of the muscles contain a definite number of muscle fibers or cells. As the animal grows and increases the size and strength of the muscles, the size of the individual fibers increases but not their number. In senescence, the number of fibers decreases, although exact cell counts have not been made. The muscle mass tends to decrease with age, but the amount of decrease is quite variable. Korenchevsky (1961) estimated an average decrease in muscle mass in man of 15 percent between the ages of 25 and 70, and about the same percentage change in rats for a comparable time. Another factor is a similar loss which takes place in the central nervous system. It has been estimated by actual cell counts (Brody, 1955) that the average man loses about 10,000 brain cells per day out of about 10 billion. It is not surprising, then, that muscular control is not as precise in older animals and that reflexes are slowed.

Another factor is the accumulation of collagen. This protein is the chief constituent of scar tissue and is very important in maintaining the structural integrity of the soft tissues. Once it is deposited it is very insoluble and tends to remain in the tissues for years. Many of the connective tissue cells die and are replaced by collagen fibers, so that some organs such as skin become infiltrated with collagen fibers. These fibers tend to shrink and join together in the course of time, forming a dense network. This accounts for the wrinkled appearance of the skin of old persons. In addition, the dense network tends to impede normal function. In the lung, it interferes with the oxygenation of the blood and in the muscles, with the flow of oxygen and nutrients from the blood capillaries to the muscle cells. It can even shrink and restrict some of the blood capillaries and thus cause further impairment of function (Kohn, 1963).

Most of the other organs of the body tend to decrease somewhat in size as the individual ages. The liver and kidneys decrease by about 20 percent in man between 25 and 70 years of age, and the spleen by as much as 40 percent. The endocrine glands may undergo rather profound changes, depending on the function performed.

Another tendency in both animals and man is to accumulate fat as they grow older. This inclination is probably largely a product of civilization since there are few fat animals in the wild. They never live long enough or well enough to become fat. In domestic animals, however, as the muscle mass declines with age, there is usually at least a corresponding gain in fat tissue, and often, a very large gain. This obesity varies greatly among different animals and even among individuals of the same species.

II. PHYSIOLOGICAL CHANGES

In their internal environment, animals maintain a constancy within quite narrow limits into extreme old age. This includes the body temperature (for warm-blooded animals), the concentration of blood cells in the blood, the chemical composition of the blood, the oxygenation of the blood, and many other factors. Indeed, it is very hard to find something of a physiological nature which will characterize an old animal. The reason for this is not difficult to understand. The normal animal has a great deal of reserve capacity when it comes to maintaining the constancy of the chemical composition of the blood, for example. Much of this constancy is regulated by the function of the kidneys. It is well known that an animal can function adequately on one of its two kidneys. In fact, if the animal is not subjected to very violent chemical stresses, it can probably live quite comfortably on 25 percent or less of its normal kidney function. Other bodily functions such as digestion, metabolism, and circulation of blood fall in this same category.

It is thus evident that as many of these organs become deficient in function because of loss of cells, a deficient circulation, or other reasons, they will still have enough function to maintain homeostasis. The situation is analogous to a furnace which heats a house to maintain a constant temperature. In a well-designed heating system, the furnace is only operating a small fraction of the time. When it becomes old and inefficient, it still maintains a constant temperature but does so by running longer. It is not until extremely cold weather occurs that an old furnace may eventually be unable to maintain the temperature.

It is often said that the most striking decrement with age is that of muscular performance. Animals customarily function close to the maximal performance, so any decline in functional capacity would be quickly noticed. Actually, the limiting factor is probably the cardiovascular-pulmonary function, and the decrement in the maximal

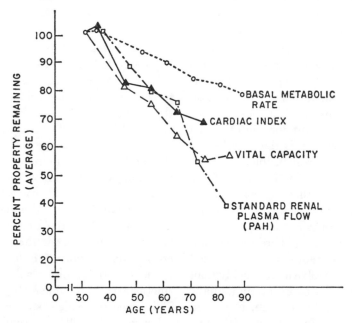

FIG. 9–1. Decline in various human functional capacities with age. They decline, but not seriously enough to jeopardize the life of the individual. (*From Shock, 1960.* In: *The Biology of Aging. Strehler, Ed., Amer. Inst. Biol. Sci., Washington, D. C.*)

function of this sysem is approximately the same as that of the kidney (Fig. 9–1).

The control systems of the body likewise suffer a small linear decline with age. When placed in a cold environment, an old rat takes much longer to adjust its metabolism to the changed conditions than does a young one. In response to exercise, an older person takes longer to increase the heart rate and respiration. If one injects glucose into the blood, a young animal can immediately mobilize the body defenses to bring the blood glucose levels to normal, but an older animal takes much longer.

The endocrine glands have been implicated in the aging process for many years. These glands control both development and function of many of the organ systems of the body. In general, they undergo important changes in function during growth, development, maturation, and aging. These changes are dictated by the genetic background of the individual and occur at the same times in all members of a species. For example, sexual maturity as well as the decline in reproductive capacity are genetically pro-

grammed. The other endocrine glands function in much the same way to regulate the function of the individual at a level appropriate to the developmental stage.

When the endocrine glands are examined anatomically or physiologically, they are found, in general, to be functioning quite normally. If the blood level of sex hormones, for example, is quite low in old age, it is not because the glands are incapable of maintaining higher levels, but because the functional state of the animal at that time requires those levels. Their function is part of the genetically programmed growth and development of the animal. This is not to say that the endocrine glands do not suffer some loss of efficiency and functional capacity with age. They probably do, but not enough to contribute materially to the aging syndrome in the vast majority of cases (Gitman, 1967).

Reaction times become progressively longer with age. The change is relatively small for simple reflexes, but for responses requiring several different senses and muscles, reactions can be very much slower in old age. This is especially true when some decision-making is required. The deficits are almost wholly attributable to the central nervous system. There is apparently some slowing of the conduction velocity of peripheral nerves and some deficiency of function of peripheral sense organs, but these are not nearly enough to account for the observed slowing. The conclusion is that the loss of brain cells is probably responsible.

When the more complex functions of the brain are considered, there is considerable question concerning the changes due to age. Old rats cannot learn to negotiate a maze as quickly as young ones. There are certain complex problems, however, which old monkeys can learn to solve more quickly than young ones. In man, many attempts have been made to assess the change, if any, in intelligence or reasoning power with age. The qualities depend heavily on experience as well as inherent ability. Few would doubt that intelligence declines in extreme old age; however, it seems to remain remarkably high in most persons even into the eighth decade. It appears that the loss of brain cells is largely offset by the accumulated experience (Birren, 1964).

III. DEGENERATIVE CHANGES

The decline in organ function is very important in relation to aging, but the end result of aging, death, is related to this decline in only a very peripheral way if at all. In populations of domestic or laboratory animals, very few deaths result from such a drastic reduction in the function of any one organ that the animal dies. There seems to be enough functional reserve in all vital organs to more than last out the maximum life span of the species.

In general, the cause of death of the older domestic animals today is degenerative diseases. These comprise the various forms of cancer, the cardiovascular diseases, and the autoimmune diseases. This latter class of disease has been receiving increasing attention in recent years, and the different diseases in this class are being recognized as

belonging to the same group. These are such diseases as arthritis, muscular dystrophy, diabetes mellitus, and many others (MacKay & Burdet, 1963).

The degenerative diseases can be characterized as ones arising primarily from within the body itself. For some reason, some of the cells of the body do not function properly and this may cause the entire machinery of the body to malfunction. Cancer is the common example. Here, certain cells break away from the constraints which limit them to a normal division rate, and divide much more often than the normal cells of that tissue. If unchecked, they will eventually crowd out the normal cells not only in the tissue of origin, but also in other organs (metastasize) to which they have spread. This process, which starts with a single somatic cell, will eventually cause such serious malfunction that the animal will die. The other degenerative diseases are caused by malfunctions of cells in other ways. In all cases, it is a process of self-destruction. The probability of death increases very rapidly with age (Fig. 9–2).

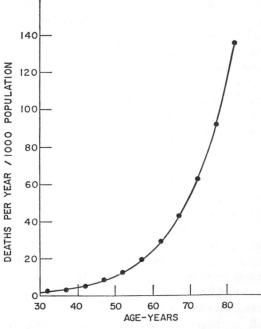

FIG. 9–2. Death rate versus age for United States males in 1949, showing that the probability of death increases very steeply with age.

When the death rates from individual degenerative diseases such as a particular form of cancer are examined, they all follow the same form of curve, suggesting that all the diseases have a common basis. Furthermore, those treatments which alter the life span of a population tend to do so by changing the incidence of all these diseases at all ages. One example of this is the delayed effects of a dose of ionizing radiation such as x rays. The entire incidence curve for each of the degenerative diseases is shifted to the left, *i.e.*, higher rates at every age. Also, it is well known that dietary restriction prolongs the life of rodents, and it does so by shifting the mortality curve to the right. These facts all argue for a common basis for all these diseases. Other examples could also be cited (Curtis, 1966a).

The general concept amounts to a different definition of aging, namely, that biological process which leads to an increasing probability of the induction of the degenerative diseases. It is as if the aging process makes the ground increasingly fertile for the germination of these seeds of self-destruction. This is a more limited definition of aging and considers only part of the problem, but it is a very important part for the individual.

IV. BIOLOGICAL MECHANISMS

The question of the nature of the process which can cause the initiation of the degenerative diseases then arises. This process has been the subject of speculation for centuries, but no attempt will be made here to present all the ideas which have been advanced. The most promising ones will be briefly discussed.

The first has been variously called the "stress theory" or the "wear and tear theory." According to this idea, a stress of any kind destroys tissue in the individual; recovery is never complete. As the unrecovered damage from more stresses accumulates, the organism is finally unable to resist one of the degenerative diseases. An extensive series of experiments has been per-

formed on mice using various noxious chemicals as stress agents. It was found that no matter how severe the nonspecific stresses, if the animal is allowed to recover it will have as long a life expectancy as if it had not been placed under stress (Curtis, 1966a). This does not apply, of course, to specific poisons such as mercury salts, which can produce irreparable kidney damage. Thus, stress *per se* does not contribute to natural aging.

The "somatic mutation theory" postulates that spontaneous mutations develop in the cells of the body, and, since mutations are irreversible, they may build up to rather high levels in the course of time. It is reasonably certain that virtually all such mutations would be deleterious to the function of the cells. If the mutation is in one of the genes controlling a vital function of the cell, the cell will die. If the mutation is in some other gene, it may change the function of the cell to make it less efficient or perhaps to make it grow in an abnormal, cancerous growth.

There is a great deal of direct evidence supporting this theory. First, ionizing radiation, which is an extremely potent mutagenic agent, also hastens the appearance of all the degenerative diseases. It is quite different from other stress agents, which do not cause mutations.

Next, a method has been developed to determine chromosome damage in liver cells of mice. The amount of damage is an index of the mutations present. The method involves inducing liver regeneration and examining the anaphase chromosomes in the dividing cells under a light microscope to judge whether they are normal figures. These cells are taken as typical of the somatic cells of the body. By means of this technique, many correlations have been made (Curtis, 1963). For example, the number of liver cells with chromosome aberrations increases steadily with the age of the animal and may be present in as many as 50 percent of these cells in old animals (Fig. 9–3). If all these cells have visible defects, many more must have invisible mutations.

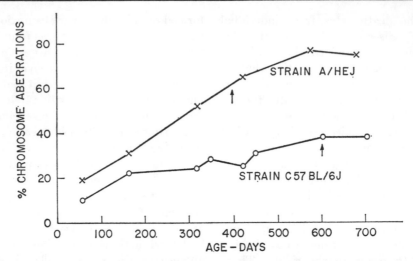

FIG. 9–3. Percent chromosome aberrations in liver cells of short-lived (AHe/J) strains versus long-lived (C57BL/6J) strains of inbred mice. The median life span is indicated by the arrows. This shows that short-lived mice develop mutations in their liver cells (and presumably all cells) at a faster rate than do long-lived mice. (*From Crowley & Curtis, 1963. Proc. Nat. Acad. Sci. 49:626.*)

In mice, a dose of ionizing radiation causes a dramatic increase in mutations, and the frequency of mutations may or may not return to control levels over a period of many months. Other stress agents which cause no shortening of the life span also cause no mutations. There are a number of different types of radiation and radiation regimes, each of which shortens the life span by a certain percentage. The amount of life shortening produced in each case is proportional to the amount of chromosomal damage which that treatment produces.

Some inbred strains of mice are rather long-lived, some are short-lived. The long-lived strains do not develop chromosomal defects nearly as rapidly as do the short-lived strains (Fig. 9–3). An experiment with dogs, which live six times longer than mice, indicates that they develop chromosomal defects much more slowly. These experiments indicate a close correlation between the stability of the chromosome structure of liver cells and longevity. It would be very strange indeed if there were no causal relation between them.

Since these are liver cells, which rarely, if ever, divide in the normal adult, the concept has developed that aging takes place in the nondividing cells of the mammal. For example, senility and muscular weakness, prominent features of the aging syndrome, occur in the neuromuscular system, whose cells never undergo division in the adult. If defects occur in the chromosome structure of these cells, the cells will either die and not be replaced, or function inefficiently. The dividing cells, such as those of the bone marrow, probably throw off defective cells by cell selection at the time of cell division. As a result, aging would not be expected to take place in these organs. However, it has recently been found that even bone marrow cells, which are continually dividing, eventually develop chromosomal defects (Chlebovsky *et al.*, 1966). This finding implies that the nuclei of all the somatic cells in the body develop defects with age and eventually either die or malfunction.

This latter finding is in complete accord with the research of Hayflick (1965) who studies cells in tissue culture. He finds that, in general, cells taken from a young human donor will undergo about 50 divisions and then develop chromosomal defects and the whole culture then dies. Occasionally such a mutation will occur in one of these cells that the cell will essentially be turned into a cancerous one, and the culture will grow indefinitely. It then seems that when cells differentiate in embryogenesis they somehow lose the ability to live and divide indefinitely.

Nuclear damage can be repaired, at least to some extent, even in highly differentiated cells. Spontaneous lesions occur in the DNA structure of all cells, and specific enzyme systems operate to repair the damage. Only very occasionally will the damage be such that it cannot be repaired, and in this case we say the cell has suffered a mutation (Howard-Flanders, 1965). The germ cells of the mammal are not subject to this constraint, but can continue to divide indefinitely.

It is concluded that mutations in the somatic cells of the body play a dominant role in the aging of the animal. However, it is now evident that there are other factors of importance. A number of instances have been found in which there is no correlation between something which changes the life expectancy and the change, if any, in chromosomal defects. A striking example of this is the fact that hybrids between two inbred strains live longer than either of the parents, but nevertheless develop chromosomal de-

fects at an intermediate rate (Curits *et al.*, 1966). Furthermore, if the somatic mutation theory is taken in its simplest form and the shape of the mortality curve for a population predicted from it, it does not fit the facts at all.

This does not negate the mutation concept, but rather, it indicates that there are other factors which also play an important role in the etiology of the degenerative diseases. A mutation is rare, so that one can assign a certain probability to its occurrence. We can assume that other chance events must also occur in a cell before one of the degenerative diseases is initiated. A schematic diagram of this concept is shown in Figure 9–4. It assumes that each of the many degenerative diseases is caused by a series of changes in an individual cell and its progeny. This gives rise to a new cell line which may be cancerous or may have altered immunological characteristics. This assumption and certain simplifying mathematical assumptions make the basis for an equation for the form of the mortality curve for a mammalian population (Curtis, 1966b). It is:

$$\frac{dN}{dt} = Kt^{(n-1)} \qquad (9–1)$$

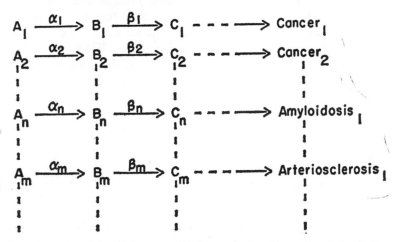

Fig. 9–4. Schematic representation of the composite theory of aging. It assumes that each degenerative disease arises from a normal somatic cell, represented here as types A_1, B_1, C_1, D_1, and so on. In order for one of these to become transformed into a cell whose daughter cells will give rise to the disease, a series of changes must take place in the genetic structure of the cell. Thus type A_1 must be transformed to type A_2, which must, in turn, change to type A_3. Such a scheme is necessary to explain the development of the degenerative diseases. (*From Curtis, 1966b. Gerontologist 6:143.*)

in which N is the number of individuals in the population at time, t; K is a constant dependent on the mutation rate and probability of occurrence of the other stops, and n the number of steps or events required to produce a cell line which will lead to the death of the individual.

When this equation is applied to mortality curves, it is found to fit the data very well and indicates that there is an average of six steps in the process of acquiring these diseases in man (Fig. 9–5). Taking data for the incidence of individual degenerative diseases, the data of each of these also fits very well. Each has a different exponent, indicating a different number of steps in the induction process for each.

There are exceptions to this general formulation. The simplifying assumptions entailed in the derivation are not valid for some diseases, so that a more exact derivation must be used. When this is done, all the degenerative diseases so far investigated

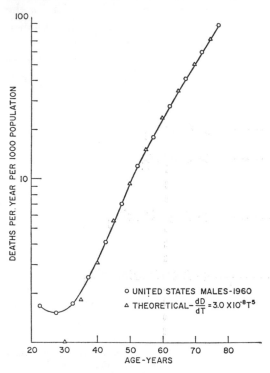

FIG. 9–5. Log death rate versus age for United States males as predicted from Equation 9–1. This shows that the composite theory (Fig. 9–4) leads mathematically to a prediction of the mortality rates. (*From Curtis, 1967. Sympos. Soc. Exp. Biol. 21:51.*)

follow this pattern very closely (Burch, 1966).

The fact that the concept leads logically to the correct form of mortality curve does not prove the correctness of the idea. However, if it did not lead to the curve, the concept would be untenable. In addition, there are a number of other facts which support this concept. For example, it is well known that cancer induction is a multistep process, and that there is often a very long delay between the application of a carcinogenic agent and the formation of a tumor. The nature of the different steps is largely unknown except for the fact that specific mutations constitute one or more of them.

Other symptoms of aging, such as muscular weakness, have been considered to be due to causes different from those for the degenerative diseases. They, too, however, can be fitted into the conceptual framework of Figure 9–4. Most of these effects can be explained by cell loss due to cell death and a lack of their replacement. Cell death requires a much less complex form of transformation than that required to initiate a degenerative disease. A one- or two-step process is all that is required.

In summary, aging appears due to the gradual change which takes place in the genetic structure of the individual somatic cells during the life span. The changes are discrete probabilistic events occurring in the genome. If the change results in the loss of the ability of a cell to synthesize an essential enzyme, the cell will die. Some of the more complex changes may lead to the development of one of the degenerative diseases.

V. NUTRITION AND AGING

One of the most striking experiments which can be performed in the laboratory in the field of aging is restriction of the diet of a group of animals. This is best demonstrated by feeding a group of mice only about 66 percent of the amount of food eaten by their controls when fed *ad libitum*. The animals must be caged individually and a measured amount fed each experimental animal every day. The experimentals will weigh about 66 percent of the weight of controls. Part of the weight difference is due to the fact that the restricted diet animals are small all over, and part to the fact that they have no fat. The striking features of the experiment, however, are that the restricted-diet animals live up to 50 percent longer than the controls, and are very much more active throughout their lives.

There were early reports that puberty was delayed in these animals, but this is true only when food restriction is held to starvation levels. Further, in the moderately restricted animals the reproductive period is extended by an even greater proportion than the increased life span.

This phenomenon, first discovered by McCay in 1939, is of very wide application in the animal kingdom. This principle applies to lower forms of animals and even plants. In the United States, persons 25 percent overweight have an average life expectancy 3.6 years shorter than normal, and those 67 percent overweight lose 15.1 years.

The specific elements of a diet responsible for decreasing the life span are not fully understood. Feeding experiments in animals have shown that if the diet is deficient in any essential element such as a vitamin or protein, the beneficial effects of calorie restriction can be nullified. This requires rather strict diet regulation, however, so for practical purposes only the number of calories counts (Ross, 1961). Most of the dietary recommendations for persons on a reducing diet have to do with psychological rather than physiological factors. However, there certainly are various kinds of diets legitimately designed to accomplish definite therapeutic purposes.

The reasons for the beneficial effects of caloric restriction are obscure. It is found that all of the degenerative diseases are retarded in their time of appearance. A striking example of this is the delay in appearance of virtually all forms of cancer (Tannenbaum & Silverstone, 1953). The mechanism is unknown. In order for any of the degenerative diseases to develop, there must be active cell division. Perhaps a restricted diet keeps the blood sugar levels so low that cell division is not stimulated as much as if the individual were on a non-restricted diet. However, this is pure speculation.

Obesity has just the opposite effect from caloric restriction. There is a recessive genetic character in mice which, when present in the homozygous condition, causes the individuals to eat excessively and become extremely obese. The mice become so fat they can hardly walk because their legs barely reach the ground and they are nearly spherical. In any litter carrying this trait, some mice will be obese and some quite normal. The obese mice have only about half the life expectancy of their normal litter mates. However, if their caloric intake is limited to that of their litter mates, they will appear normal in all respects and live just as long.

It is also possible to make animals obese by destroying the satiety center of the hypothalamus. This can be done either by surgical methods or by the injection of gold thioglucose. By either means, the animals get very obese, and often reach three times the weight of the controls. They have a correspondingly short life span. Again, if the food intake is restricted to that of the controls, they will live a normal life span. The genetically obese animals are exceedingly phlegmatic and almost never move except to eat. The mice made obese with gold thioglucose are very active and aggressive, but the life span seems to depend on the obesity, not the degree of activity.

VI. STRESS AND EXERCISE

Although stress *per se* does not seem to be a life-shortening agent, specific stresses may be quite detrimental to the life expectancy of an individual. For example, excessive mental stress can cause gastric ulcers. Excessive stress on the heart can cause permanent damage. These are likely to be rather extreme causes, however.

On the other hand, there is a very strong subjective impression among gerontologists that the best way to retard aging is to keep active mentally and physically. However, the positive evidence bearing on this assertion has been very fragmentary. Pearle tried to show from epidemiological studies that hard work (exercise) is equivalent to stress, which shortens the life span. He failed to obtain conclusive evidence because in the populations he studied stress was linked to living standards. More recently, studies in which the controls were more carefully selected have almost universally shown that exercise and hard work prolongs, not shortens, the life span (Ordy *et al.*, 1967). These studies have been well reviewed recently by Fox and Skinner (1964). While none of these is completely conclusive in itself, together they constitute quite an impressive body of evidence to indicate the beneficial effects of exercise. For example, McDonough *et al.* (1965) showed that in a group of carefully paired farmers 60 years of age, those who did their own farm work had only one sixth of the prevalence of coronary heart disease as those who managed their farms without doing any of the actual work themselves. Other studies have not shown such dramatic results as this one, but as methods of study have been refined, it has become quite clear that at least insofar as cardiovascular disease is indicative of aging, moderate exercise appears to retard the aging process.

There is apparently a link between exercise and obesity. Mayer (1967) has studied obese versus normal children and finds that they both eat about the same amount of food but the obese ones exercise much less. He finds almost the same to be true of the genetically obese mice. Also, when obese persons are placed on an exercise regime with an unrestricted diet, they lose weight. Farmers have known for centuries that animals could be fattened for market by confining them so that they could hardly move.

It is apparent that more research is needed in the fields of nutrition and exercise physiology before a clear picture of the aging process can be formulated. However, it is now clear that aging can be retarded very considerably by an optimum diet and an exercise regimen.

REFERENCES

Birren, J. E. (1964). *Relations of Development and Aging*. Springfield, Ill., Charles C Thomas.

Brody, H. (1955). Organization of the cerebral cortex. III. A study of aging in the human cerebral cortex. J. Comp. Neurol. *102*:511–556.

Burch, P. R. J. (1966). Age and sex distributions for some idiopathic non-malignant conditions in man. Some possible implications for growth-control and natural and radiation-induced aging. In: *Radiation and Ageing*. P. J. Lindop & G. A. Sacher (Eds.), pp. 117–157, London, Taylor & Francis, Ltd.

Chlebovsky, O., Praslicka, M., & Horak, J. (1966). Chromosome aberrations: Increased incidence in bone marrow of continuously irradiated rats. Science *153*:195–196.

Curtis, H. J. (1963). Biological mechanisms underlying the aging process. Science *141*:686–694.

Curtis, H. J. (1966a). *Biological Mechanisms of Aging*. Springfield, Ill., Charles C Thomas.

Curtis, H. J. (1966b). A composite theory of aging. Gerontologist *6*:143–149.

Curtis, H. J., Tilley, J., Crowley, C., & Fuller, M. (1966). The role of genetic factors in the aging process. J. Geront. *21*:365–368.

Fox, S. M., & Skinner, J. S. (1964). Physical activity and cardiovascular health. Amer. J. Cardiol. *14*: 731–746.

Gitman, L. (1967). *Endocrines and Aging*. Springfield, Ill., Charles C Thomas.

Hayflick, L. (1965). The limited *in vitro* lifetime of human diploid cell strains. Exp. Cell Res. *37*:614–636.

Howard-Flanders, P. (1965). Molecular mechanisms in the repair of irradiated DNA. Jap. J. Genet. *40*:256–263.

Kohn, R. R. (1963). Human aging and disease. J. Chronic Dis. *16*:5–21.

Korenchevsky, V. (1961). *Physiological and Pathological Aging*. New York, Hafner Publishing Co.

MacKay, I. R., & Burdet, F. M. (1963). *Autoimmune Diseases*. Springfield, Ill., Charles C Thomas.

Mayer, J. (1967). Nutrition, exercise and cardiovascular disease. Fed. Proc. *26*:1768–1771.

McDonough, J. R., Hames, C. G., Stuth, S. C., & Garrison, G. E. (1965). Coronary heart disease among Negroes and whites in Evans County, Georgia. J. Chronic Dis. *18*:443–461.

Ordy, J. M., Samorajski, T., Zeman, W., & Curtis, H. J. (1967). Interaction effects of environmental stress and deuteron irradiation of the brain on mortality and longevity of C57BL/10 mice. Proc. Soc. Exp. Biol. Med. *126*:184–190.

Ross, M. H. (1961). Length of life and nutrition in the rat. J. Nutr. *75*:197–210.

Tannenbaum, A., & Silverstone, H. (1953). Nutrition in relation to cancer. Advances Cancer Res. *1*: 451–502.

Chapter 10

Diseases and Parasites

By C. K. Whitehair

In a discussion of disease and nutrition the implication is that these are two separate disciplines. Disease may be defined as any divergence from normal. This would include disturbances due to the microbial infectious agents as well as toxicoses and nutritional deficiencies. Nutrition is a science concerned with "nourishment" to obtain growth and normality. There is obviously much over-lapping. Only at the extremes are these separate entitites. In the animal organism, they exist as a single dynamic entity.

Research in animal nutrition has been mainly concerned with determining the nutritive requirements for growth, reproduc-tion, and lactation for maintenance of health under favorable environmental con-ditions. To obtain reliable information on what these requirements are, special efforts have been made to eliminate the role of infectious disease factors. Likewise, in research on animal diseases, the emphasis has been on disease processes; little attention has been given to the role that nutrition may have in the cause, control, or prevention. Therefore, the subject of disease as related to nutrition and growth has been in the "twilight zone" between research on the nutritive requirements of normal animals and research on specific diseases. Only a meager amount of information is available about how infectious diseases modify nutri-tive requirements and influence growth. With current emphasis on efficient animal production and the maintenance of large numbers of animals in concentrated areas, integration of problems of nutrition, diseases and parasites will be essential, since that is how they exist under conditions of inten-sive animal production. In addition to feeding rations that will allow efficient growth and production, consideration must be given to the nutritive requirements of animals exposed to a variety of diseases, parasitisms, and mismanagement practices and maintained under unfavorable environ-mental conditions. These latter factors im-pede the normal assimilation and metabo-lism of food and therefore reduce the growth rate. There is a wide variability in the effect of diseases and parasites on animal growth. At one extreme are acute, severe diseases that result in the death of the animal, *e.g.*, hog cholera in swine. The fact that the disease exists is usually obvious, so that the cause can be determined and control measures applied. At the other extreme and of greater importance economically are chronic diseases and parasites. These may exist unrecognized in animals and not only interfere with the normal growth due to good nutrition and inherited capacity, but also influence lactation and reproduction. Specific clinical signs of chronic disease are often not evident. They become apparent by an improved growth rate and feed efficiency in response to chemical or nutritional additives to the ration. At times chronic diseases may be suspected from the poor growth rate of young animals. The improved growth response of animals to a specific drug or chemical that inhibits or corrects a chronic disease may essentially be a return to normal growth and therefore to attribute "growth-promoting properties" to some drugs and chemicals may at times be misleading. It is a growth promotion for the animals afflicted with a disease or parasitism but in healthy animals it would constitute essentially normal growth. Chronic, insidious diseases tend to exist in young animals maintained in concentrated areas and assembled from a wide variety of sources. These diseases, especially those involving the digestive system, influence growth and are in turn minimized by adequate nutrition and specific nonnutritive substances. The specific nutritional requirements may be altered to a minor or major degree, depending primarily on the nature, severity, and duration of the disease process.

I. INFECTIONS AND ENVIRONMENT

Animals reared in one type of environment often grow better than comparable animals fed essentially the same ration in other types of environment. While many environmental factors could account for this difference in growth, experimental work indicated that microbial contamination has a major role (Coates *et al.*, 1963). In new or clean facilities, maximum growth was achieved and antibiotic supplementation resulted in no additional growth response in chicks. Exposure of the clean facilities to material

Table 10–1. Growth Rate, Feed Efficiency and Response to Chlortetracycline, 10 mg/lb Ration in Unthrifty Pigs in a Contaminated Environment in Comparison to Healthy (Cesarotomy Derived) Pigs in a Clean Environment. Both Groups were Fed Essentially the Same Ration.

(*Data adapted from Whitehair & Thomas, 1956. J. Amer. Vet. Med. Ass. 128:94, and Whitehair & Hillier, 1952, Okla. Agri. Exp. Sta. MP–27, 77.*)

	Basal Ration	Basal + Chlortetracycline	Improvement %
Unthrifty pigs, contaminated environment			
Av. initial wt, lb	35	35	—
Av. daily gain, lb	0.8	1.2	45
Feed per 100 lb gain	493	370	25
Cesarotomy-derived pigs, clean environment			
Av. initial wt, lb	54	51	—
Av. daily gain, lb	1.4	1.4	0
Feed per 100 lb gain	288	278	0

from contaminated environments reduced the growth rate, and these effects were reversed by antibiotic feeding. This same phenomenon exists in swine (Table 10–1). In some environments normal growth and production are achieved while rather simple rations are fed and in other environments slow growth results even though excellent rations are fed. Likewise, diet is often erroneously blamed for poor performance. Infectious agents apparently exist in the environment, and are associated especially with animal wastes or exist in the animals themselves as subclinical infections. While the influence of some pathogens in obstructing efficient production is known, the influence of others alone or in various combinations is not established. These pathogens associated with certain environments are usually not a problem when the facilities are first used but tend to increase in virulence and in numbers as use of the facilities continues. Since these pathogens can be either in the facilities or in the animals, attempts to eliminate their influence on growth and production require not only a cleansing and disinfecting of the facilities and equipment but also a restocking with animals free of the pathogens that reduce normal growth. These infectious agents have a marked influence on the requirements for certain nutrients and must be recognized as partly responsible for the difference in response to the same ration in different environments. Animals tend to become adapted to certain environments. This adaptation is largely by the indigenous flora of the gastrointestinal tract (Schaedler *et al.*, 1965). The microflora, therefore, has an influence on growth, feed efficiency, and resistance to infectious agents and toxic substances.

II. INFECTIONS AND METABOLISM

Pathogens influence the metabolism of nutrients in a great variety of ways. Only some of the general aspects will be discussed here.

Appetite. The appetite is often referred to as a "barometer" of health. The extent to which it is reduced may be an indication of the severity of infection. Likewise, an improvement in appetite is associated with recovery from disease. While it is discussed more in detail elsewhere, appetite is a complex conscious activity dependent on many factors ranging from habit to neurologic, physiologic, and biochemical aspects. Even mild or subclinical infections may reduce or completely inhibit appetite. Since lack of appetite is a sign common to many infections, it is probably the single most important factor initiating nutritional deficiencies and malnutrition. In addition to an insufficient consumption of food as a consequence of infection, at times a perverted or depraved appetite may develop and result in the consumption of foreign objects such as dirt, hair, and wood which cause tissue injury or abnormal metabolism of nutrients.

Absorption. The importance of the digestive tract in the conversion of food into living tissue and vital substances becomes more apparent as the knowledge of nutrition and pathology increases. Of all the tissues concerned in the metabolism of food, the digestive tract must be considered as the most important. Infections that directly or indirectly affect the digestive system cause nutrient losses and poor growth. Reduced absorption and diarrhea result in the loss of water, protein, and electrolytes (Table 10–2).

Table 10–2.   Effect of Transmissible Gastroenteritis Infection in Pigs on Food Consumption and Excretion of Nutrients
(*Adapted from Reber & Whitehair, 1955. Amer. J. Vet. Res. 16:116.*)

Nutrients	Controls	Infected
Daily food consumption (gm)	160	94
Daily fecal excretion:		
Water (gm)	1.3	40.7
Nitrogen (gm)	0.03	0.56
Sodium (mg)	1.1	49.8
Potassium (mg)	5.0	231.0

The large loss of water is accompanied by the loss of water soluble vitamins and unidentified growth factors. Injury to the mucosal cells reduces their metabolic activity, such as the conversion of carotene to vitamin A, and also reduces the production of specific enzymes essential to the digestion of food (Maronpot & Whitehair, 1967). Gastrointestinal infections that impede absorption and metabolism of nutrients are an important factor in precipitating malnutrition. In man it is estimated that on a worldwide basis, diseases and disorders in which diarrhea is an outstanding manifestation account for the deaths of 5 million infants and children each year (Munro, 1964). In animal production, gastrointestinal infections are a common problem in young animals. Even if mild in their manifestations, they greatly influence nutritional requirements and growth.

Intestinal Flora. Most animals normally harbor an extensive bacterial flora, not only in the rumen and large intestine, but also in the stomach and small intestine. This flora has an influential effect on the nutritive requirements and general well-being of its host

(Dubos & Schaedler, 1960). In the animal with a simple stomach, the microbial flora is responsible to an extent for the synthesis of some vitamins, especially K and B_{12} and unidentified nutrients. A desired flora contributes to the formation of essential chemical and physical factors and has a tendency to suppress microbial species undesirable for good nutrition and growth.

Biochemical Lesions. The metabolism of energy, protein, and many specific nutrients requires the active participation of numerous other nutrients. Thus, a deficiency of one nutrient can lead to inappetence, and, eventually, to a complex deficiency. A deficiency or excess of a specific biochemical, which would be indicative of an alteration or lesion, is the first event that leads to morphologic changes and the malfunctioning of specific organs and tissues. Early recognition and correction of biochemical changes, whether of primary or secondary origin (*e.g.*, precipitated by infection), would do much to minimize the influence of these changes on the proper functioning of tissues and impairment of the health of animals.

III. ANTIMICROBIAL AGENTS

The benefits of small amounts of antibiotics and similar drugs in livestock rations have done much to emphasize the interrelationship between nutrition and infection. The more noticeable effect has been in the improved growth of the young in a variety of species (Whitehair & Pomeroy, 1967).

Basis for Use. The inclusion of low levels (5 to 20 mg/lb ration) of antibiotics in the rations of experimental animals, especially young ones, improves their growth rate and feed efficiency when compared to that of untreated controls. The improved growth rate is a manifestation of increased feed consumption and absorption of nutrients. At higher levels (25 to 50 mg/lb ration), antibiotics are used to treat specific infections of the digestive tract. These infections are indicated primarily by clinical signs of diarrhea, dehydration, inappetence, and poor health. At still higher levels (75 to 250 mg/lb ration), the antibiotics are used in the feed or water to treat specific acute systemic infections, especially those infections which cause limited impairment of the digestive system.

The choice of antibiotic will depend on the species, severity of infection, environment, and sensitivity of the pathogen to the antibiotic. Feeding the antibiotics even at low levels has a "sparing effect" for it improves the utilization of nutrients such as protein and vitamin B_{12}.

Mode of Action. Research using germ-free animals suggests that the antibiotics function by inhibiting the growth, activity, or multiplication of harmful microorganisms (Forbes & Park, 1959). Any nutrient-sparing effect would be an indirect effect of either a more desirable flora of the digestive tract for synthetizing nutrients for use by the host animal or simply a better absorption of nutrients in the diet. Thus, the use of feed as a vehicle for the administration of antibiotics and other drugs has labor-saving advantages in the mass medication of large numbers of animals maintained in concentrated areas. However, there is usually less control over the dosage administered and the safety factors than when they are used to treat a specific disease in an individual animal or herd.

Drug-resistant Microorganisms. The problem of resistant strains of microorganisms that had previously been susceptible to antibiotic therapy has been clearly demonstrated in the laboratory. The magnitude of this problem under field conditions is not well established and will be difficult to determine. The initial response to the inclusion of antibiotics in rations was rather dramatic, yet enteric infections in calves, pigs, lambs, and poultry are still a major problem today. What appears to be of even greater importance than the development of resistance to antibiotics by specific microorganisms is the ability of some microorganisms to transfer this antibiotic resistance to other species of microorganisms (Anderson, 1968). This development of microbial resistance and transfer to other organisms is of concern to public health officials in that the routine use of antibiotics in livestock rations to improve growth could jeopardize the value of antibiotics in the treatment of specific infections in man.

Residues. The problem of minimizing antibiotic or drug residues in animal products used in human food has received much attention. The manner in which a drug is metabolized has an important influence on its effectiveness in treating an infection. The drug must come in contact with the specific pathogen. Those that are degraded during metabolism to harmless products that are eliminated from the animal body do not present a problem. For those drugs that persist in edible foods after being fed to animals, withdrawal from the ration for definite periods is recommended before processing for human food.

IV. GERM-FREE TECHNIQUES

Growth and health are closely correlated. The mechanics and symbiotic relations of microbes in the production of disease and slow growth are very complicated. Likewise, the desirable effects of microbial life such as the synthesis of vitamins in the gastrointestinal tract are poorly understood. Therefore, much interest has been manifested in the germ-free animal as a technique to obtain more precise information about the role of microbial life on absorption of nutrients, growth, and health. A variety of plastic isolators, equipment, and techniques have been devised during the past decade so that animals can be procured and maintained either free of microbial life or associated with specific microorganisms. Germ-free techniques also can be used to obtain information about the interrelationship of microorganisms, about relationships between the microbial flora of animals, and about requirements for specific nutrients. It might be considered analogous to the techniques of using purified diets in nutrition studies. An example might be given of how the germ-free technique was used to demonstrate the infectious etiology of wasting disease in mice. Germ-free mice thymectomized at birth grow normally. However, exposure of these mice to a conventional flora as late as 8 months after thymectomy results in stunted growth (Coates, 1968). Thus, the germ-free technique has contributed information concerning the role of the thymus in impaired growth and the immune response.

The early work was done mainly in the rat and mouse and was stimulated by the need for a more uniform and reliable experimental animal. The same general procedures may be used for larger species. Plastic film isolators can be made in a wide variety of designs and can thus accommodate most animals that are used routinely in research. The larger species may actually have some advantages over the smaller laboratory species in that they are mature enough at birth to consume various experimental rations readily. They also are of a size that permits handling and obtaining more precise information about factors that inhibit or improve growth. In addition, tissue samples of sufficient size for analytical work may be obtained.

In procuring and maintaining germ-free animals the "closed system" as described for laboratory animals is preferred (Trexler & Reynolds, 1957). This technique allows complete control of the environment in which the animals are maintained (Plate 10, *Top*). If the environment is not completely controlled, then this technique has no advantage over the conventional methods of rearing animals. If suitable tests fail to demonstrate the presence of microorganisms, then these animals are referred to as "germ-free" or "axenic" (without strangers). The term "gnotobiotic" (known life) also has been widely used for animals maintained in the "closed system." Since the gnotobiotic animals may be purposely contaminated with a specific organism they may be germ-free or may have one or more known species of microorganisms in their environment.

Technology. The only initial source of germ-free animals is the sterile fetus. The fetus must be delivered without contamination into a sterile isolator. Contamination of germ-free animals may come from two sources, the uterus or the environment. When an animal is placed in a closed system it is sealed into that system with whatever contamination it possessed initially. It is important, therefore, that the dams from which young are to be collected are free of as much contamination as possible, especially any pathogen that might be transmitted *in utero* to the young. In addition to checking the herd and the individual health history and conducting appropriate clinical testing for infection, some suggest the administration of antibiotics to the dam to reduce contamination possibilities. Meticulous planning is required since neglect of any detail will negate the whole procedure. Once mastered, however, the procedures become routine.

Morphological Characteristics. In general, the germ-free animal is similar to its healthy, conventionally reared counterpart. Both the amount and activity of lymphoid tissue appear less in germ-free animals. The intestinal tract appears thinner and devoid of lymphatic tissue. Germ-free animals have a decreased amount of immune globulins, but they are not completely devoid of them. It should be remembered that while diets are sterile they are not free of antigens. Low antigen rations are currently being investigated. Germ-free animals, when introduced to conventional animal rearing environments, are generally more susceptible to infections, which result in reduced growth rates. After they have adapted to conventional procedures of rearing and if they have been maintained free of disease, they grow more uniformly and efficiently than conventional animals (Dubos & Schaedler, 1960).

They not only grow rapidly on complete diets but are more resistant to poor growth on deficient diets than conventional animals. Cecal distention is a perplexing problem in germ-free mice, rats, and guinea pigs; however, in the larger germ-free animals it has not been a problem.

Applications. Germ-free animals are expensive to produce and maintain yet they are invaluable for studies of the beneficial or harmful influence of microorganisms on growth. It has been concluded that implications from studies concerned with germ-free animals compel a reinterpretation of the meaning of nutritional requirements for growth (Dubos & Schaedler, 1960). In addition many of the characteristics, such as growth that have been assumed to be inherited, can in reality be determined by the microbial flora of the intestinal tract. One practical advantage is establishing herds or flocks free of the infectious diseases that impede growth. The pathogenesis of specific infections can be determined without the influence of other organisms or extraneous factors. Experiments may be conducted with the control and experimental units adjacent to each other so that virulent pathogens can be transmitted from one to the other with a minimum of danger to other animals or man. The technique would seem useful to exploit the growth of the ruminant animal, and especially, to elucidate the role of the symbiotic microflora in the synthesis of protein from nonprotein nitrogen, cellulose digestion, and water-soluble vitamin synthesis.

V. NUTRITION—INFECTIOUS DISEASE INTERRELATIONSHIPS

Well-fed animals are more resistant to infection than those poorly fed. At times this statement has been exploited and used without scientific evidence to support it. In many instances, it is probably more an association of good nutrition and feeding practices and a minimal amount of infectious disease problems. Thus, the animal producer that feeds well formulated rations also employs good disease prevention practices. Nutrition-infection interrelationships are very complex and will depend on both the specific type of infection and class of nutrients involved. Information about the relationship of different types of infectious diseases and nutrition has been summarized (Miner, 1955).

A. Viral Diseases

Well-fed animals are more susceptible to viral infections than those poorly fed. The viral pathogens apparently require a well-nourished cell for growth and multiplication. In general, viral infections initiate inappetence, significant nitrogen losses, and enhanced requirements by increasing basal metabolism (fever). Thus, if extended for a sufficient length of time, malnutrition becomes evident. Some examples of diseases that are more severe in well-fed animals are foot-and-mouth disease, the avian leukosis complex, Newcastle disease, Rous sarcoma virus, and canine distemper. The research on the influence of nutrition on canine distemper virus provides an illustration of this interrelationship (Newberne, 1966). Eighty pure-bred beagles were divided into groups and fed a balanced diet which supplied 90 to 100 (high level), 70 to 75 (normal), or 40 to 45 (low level) kcal/kg of body weight per day. After an adjustment period of 6 weeks, the dogs were exposed to the distemper virus by intracerebral injection. The mortality rate was as follows: high level group, 87 percent; normal, 74 percent; and low level group, 31 percent. In addition to a higher mortality rate in the obese dogs, the clinical signs were also more severe. It is also of interest to point out that in recent years, concurrent with improved rations and feeding practices in livestock, we also have witnessed an increased incidence and severity of viral disease problems. This has been true particularly in poultry production.

B. Bacterial Diseases

A large volume of experimental data indicates that good nutrition is associated with a reduction in the incidence and severity of bacterial infections. As to susceptibility to infection, this is opposite to viral infections. As an example, the incidence and severity of tuberculosis is less in well-fed animals than in those poorly fed. Similar trends are noted for infections caused by salmonella, shigella, clostridia and streptococci. The methods by which nutrition aids in protection against bacterial diseases are multiple. The more important are:

1. *Tissue Integrity.* A reduction due to malnutrition would increase permeability of intestinal and other mucosal surfaces; decrease of mucous secretions; decrease of normal tissue repair; loss of ciliated epithelium and presence of edema and cellular debris to give a favorable culture medium. Thus, a deficiency, such as vitamin A, enhances permeability of mucosal surfaces to pathogenic bacteria.

2. *Antibody Production.* Immune substances are protein in nature, and quality and quantity of dietary proteins as well as vitamins are vital for their synthesis.

3. *Ability of Liver to Detoxify Bacteria.* In malnourished animals the ability of the liver to detoxify toxic products from bacteria is reduced.

4. *Maintenance of Reticuloendothelial Activity.* In nutritional deficiencies, an atrophy of bone marrow, spleen, liver or lymphoid tissue reduces reticuloendothelial activity, particularly phagocytosis and antibody production.

C. Parasites

Parasitic infections are similar to bacterial diseases. Much of the previous information about bacterial infections and nutrition likewise applies to parasitic infections. Internal parasites are a prevalent cause of malnutrition and it is the young of many species that are afflicted. Ascarid and hookworm infections have long been associated with malnourished children and puppies. Parasitic infections are usually a chronic disease and may exist unknown in the young for a long period of time while causing poor growth and unthriftiness. In general, parasites cause malnutrition by injury to the tissues of the host. They utilize the host's food and blood, cause mechanical obstruction, produce wounds for entrance of other pathogens, and secrete or excrete hemolytic toxins and antienzymes. In the ruminant, internal parasites cause anorexia, which initiates malnutrition. Cattle and sheep having even a mild infection do not go on feed as readily as nonparasitized animals. The detailed effects of parasites on the absorption of specific nutrients have been of interest to investigators. In chicks infected with *Eimeria necatrix* at first there was an increased absorption of zinc-65 due to mild intestinal injury and congestion. Later, after severe injury, absorption and intestinal motility decreased or stopped (Turk & Stephens, 1966).

When parasitized animals are fed adequate rations there is usually a loss of worms and a reduction in egg production. Anthelmintic compounds are more effective in well-nourished animals and may actually be hazardous to use in heavily parasitized, malnourished animals.

D. Rickettsial and Protozoal Infections

While experimental data are more limiting, they do suggest that a high plane of nutrition reduces the problems of rickettsial infections. Under practical conditions epidemics of typhus have been associated with malnutrition more than any other disease.

Nutrition seems to influence some protozoal diseases more than others. Especially those involved in the digestive tract such as amebiasis and coccidiosis are reduced by adequate nutrition. In man and monkeys

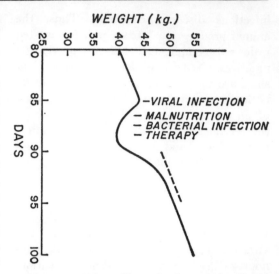

Fig. 10–1. Scheme illustrating interrelationships of infection and nutrition on reduced growth and enhanced growth following therapy.

fed inadequate diets amebic dysentery is a serious problem and usually responds promptly to an improved nutritional regimen. The malaria parasites of the blood appear to have rather specific nutritional requirements and are enhanced by adequate nutrition.

The many interrelationships among infectious agents, nutrition, therapy, and growth become apparent and might be summarized (Fig. 10–1). Thus a growing pig might be fed an adequate ration and be susceptible to a viral infection. The resulting inappetence, diarrhea, and increased nutritive requirements (due to increased metabolic rate) would precipitate malnutrition. This latter condition would enhance the possibility of bacterial or parasitic infection. Finally, depending on the nature of the disease, the animal either dies, recovers spontaneously, or responds to therapy. Therapy must consider not only the nutritional deficiencies but also the underlying disease. Children, though underweight, dwarfed, and dying because of malnutrition, are often suffering mainly from untreated diseases (Williams, 1965).

Under field conditions nutritional supplements may at times produce rather dramatic improvement in animal health. This re-

Plate 10

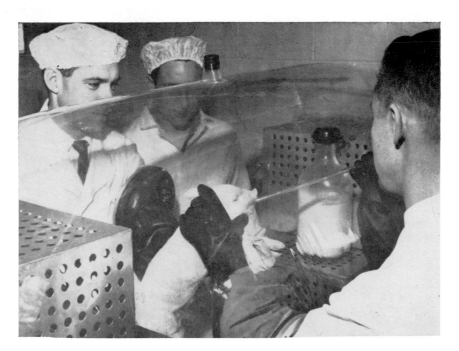

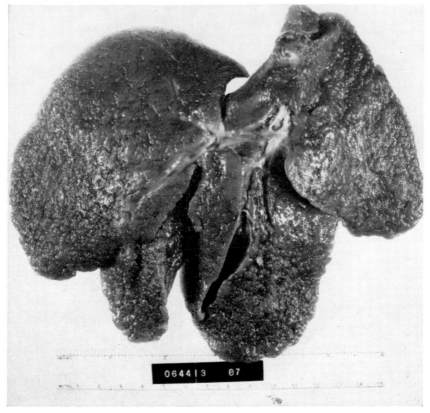

Top, Plastic film isolator for maintaining germ-free animals. Arm-length rubber gloves are on both sides of the isolator for handling the animals and equipment. The animals are maintained in stainless steel cages.

Bottom, Dietary liver necrosis in swine due to a deficiency of selenium–vitamin E.

sponse to nutritional therapy is often due to a deficiency precipitated some time previously by an infection. This response could erroneously imply that a nutritional deficiency initiated the disease when actually, an infectious agent was responsible. Caution must therefore be exercised in ascertaining the role of nutrition in animal health problems. An accurate diagnosis with a reliable health history will do much to minimize misunderstanding in handling animal health problems.

VI. TOXICANTS

A. Food

The presence of both artificial and naturally occurring toxic substances has long been recognized in animal food. As in the case with the infectious agents, there is variability in the role of nutrition in preventing or treating these diseases. Reduced growth in young animals is a clinical sign common to most of these disturbances. In some instances rations apparently have no influence; in other cases, rations have a marked influence in reducing toxicosis. Well-nourished animals will usually not consume these feeds; and if they do, they are more resistant to the resulting toxicosis. These problems are, therefore, associated more with droughts, limited feeding programs, and actual mismangement practices. In most instances, feed is an innocent vehicle for a noxious substance. There are some instances of toxic substances associated with the composition of food. While there is some overlapping, toxicants in food may be divided into (1) pathogens, (2) specific chemical poisons, and (3) naturally occurring toxicants in foods.

1. PATHOGENS

While there are over 100 pathogens shared by man and animals, only a few are associated with animal feeds and are deleterious to the health of animals (Sadler, 1965). Salmonella species, especially in products of animal origin, are perhaps of most concern since they produce specific infections in animals and are of importance to public health. The toxin produced by *Clostridium botulinum* is responsible for outbreaks of food poisoning, primarily in carnivorous animals. The toxin is usually ingested in foods in which this organism has grown. Pathogens in animal feeds are primarily due to inadequate sanitation and improper processing of animal food products.

2. CHEMICAL POISONS

A voluminous amount of information is available on the deleterious effects of poisonous plants, chemicals, drugs, insecticides, and herbicides to animal health. In most instances the feed is accidentally contaminated and, therefore, an innocent vehicle to poison animals sporadically. Many of the poisons cause liver injury in animals. Diets high in protein and carbohydrate protect against liver injury, while fat diets tend to enhance the injury. Well-fed animals are not only more resistant to the toxicosis, but are also less likely to consume feed containing poisons. In addition, malnourished animals with a gluttonous—often depraved—appetite will consume a larger total quantity of the poison. Consumption of even small amounts of poisonous substances, and the ensuing illness, results in refusal to consume that particular feed later unless there is nothing else to eat.

A problem more important to animal health than specific poisons is the contamination of feeds during processing. These substances have not only caused heavy animal production losses; in addition, the losses appear insidiously in herds and flocks; and the cause is extremely difficult and confusing to determine and prevent. While the examples to be mentioned are perhaps only of historical interest they do illustrate animal health problems to guard against in the future. Chlorinated naphthalenes in lubricants used in feed manufacturing were responsible for hyperkeratosis in cattle during the late 1940's and early 1950's (Hansel & McEntee, 1955). The disease afflicted hundreds of thousands of cattle and is probably the most disastrous chemical poisoning in the history of cattle production. Of interest to the subject of nutrition was the fact that these chemicals interfered with vitamin A metabolism. Afflicted animals had low values of vitamin A in the blood and liver. Supplements of vitamin A for adequate nutrition were of definite therapeutic value. After prolonged feeding, trichloroethylene extracted soybean meal produced an aplastic anemia in cattle, and was toxic to sheep and horses. Nitrogen trichloride (Agene) was once used as a bleaching agent for flour, and small amounts produced specific neurologic disturbances in dogs. This substance acted as a specific antimetabolite in that it interfered with glutamic acid metabolism in brain tissue. Another serious disease due to feed contamination was the toxic fat syndrome that appeared in broilers in 1957. The disease was characterized by ascites, hydropericardium, and subcutaneous edema. The disease was caused by a chlorine containing compound in processed animal fats. Chlorinated compounds (including insecticides or their derivatives) seem especially deleterious to animal health, and extreme caution should be practiced in the use of these compounds around feed manufacturing operations.

3. Naturally Occurring Toxicants

This group of substances is closely associated with the composition of feeds and is known to have toxic properties (or produce toxic effects) in animals. Many of the substances either directly or indirectly have an influence on nutritive requirements (N.A.S.–N.R.C., 1966). Most of them have a chronic effect, reducing optimum growth and production, which may be corrected, at least partially, by proper nutrition. There are even some substances such as selenium that are essential nutrients in small amounts but toxic in larger amounts. Some of the more important toxicants associated with feeds are:

Aflatoxins. Mold growth on feeds has been long recognized to produce sporadic outbreaks of toxicosis in animals. The widespread toxicosis in animals resulting from *Aspergillus flavus* in peanut meal initiated much research on this subject. It is now apparent that a variety of molds plays an important part in diseases of livestock. Molds have been incriminated in silage, corn, and most cereals, concentrates, hay, and grasses. Most of the molds produce specific liver injury. Improper harvesting, drying and storage are important factors contributing contamination and toxin production to feeds. Isolation of the specific toxic chemicals has allowed experiments to be conducted on the role of nutrition in some of these problems (Madhavan & Gopalan, 1968). In short-term experiments, low-protein diets markedly enhanced aflatoxin injury. In more long-term experiments, rats fed the low-protein rations were more resistant to the development of hepatomas. This illustrates the variability of the role of nutrition in toxicoses. The active principle of moldy sweet clover hay is dicumarol, a true antimetabolite of vitamin K.

Nitrates and Nitrites. The occurrence of inorganic nitrates and nitrites in animal forage and drinking water is recognized as causing chronic toxicosis, especially in cattle. The acute toxicosis resulting from methemoglobin formation has perhaps been given more emphasis. This problem has been associated with the increased use of nitrogenous fertilizers and improper maturing of forage. It appears to be a more serious problem in the ruminant since nitrates are reduced to the more toxic nitrites in the rumen. A chronic toxicity, apparently due to nitrates, nitrites or some other toxic reduction product, has been observed, especially in cattle. Concentrate type rations and adequate vitamin A have a protective effect. They are also of value in the treatment of suspected nitrite poisoning.

Gossypol. The harmful effects of cottonseed meal, especially to nonruminant animals, are due to intoxication with gossypol. The adverse effects of gossypol are depression of appetite, reduced weight gains, lesions of gastroenteritis, hepatic, renal and pulmonary congestion, and anemia. The toxicity of gossypol can be reduced by adding iron salts to the rations containing cottonseed meal. This will decrease the absorption of gossypol from the intestinal tract. In recent years, varieties of cotton and methods of preparation have reduced the gossypol content in seeds, so the processed meal for feed is not the problem it once was.

Selenium. This compound is a dietary requirement in small amounts and a potent toxicant in larger amounts. It is required to prevent myodystrophic disturbances in many species, exudative diathesis in poultry and swine, and liver necrosis in rats and swine (Plate 10, *Bottom*). Chronic toxicosis (alkali disease) has long been recognized in range animals, especially cattle, in the seleniferous area of the midwest. In chronic toxicosis in cattle there is a loss of hair from the tail, the hoofs slough off, lameness occurs, food consumption decreases, and general malnutrition results. Acute toxicity (blind staggers) may occur in animals unaccustomed to selenium forages or when toxic amounts are added to the rations. In acute toxicity, selenium produces a severe gastroenteritis, hepatitis, and nephritis. Toxic amounts of selenium are believed to be antagonistic to sulfur metabolism. Inorganic and organic forms of arsenic are of some help in reducing the harmful effects of selenium.

Raw Soybeans. Extensive studies have been made on both the deleterious effect of feeding raw soybeans to rats and chicks, and how this protein may be corrected to one of high biological value by proper heat treatment. Many substances, including a trypsin inhibitor and hemagglutinin, have been incriminated as the toxic substances responsible for reduced growth rate and pancreatic hyperplasia.

There are additional substances associated with animal feeds that are toxic and influence growth and nutrition under somewhat isolated and special feeding conditions. Some examples are the estrogens, molybdenum, and oxalates. There are still other substances that are certainly not toxicants, even in large amounts, yet they may influence the metabolism of specific nutrients. For example, phytic acid (in soybean protein) and to a lesser extent calcium have a binding effect on zinc, precipitating a zinc deficiency in swine and poultry. In swine, parakeratosis was a chronic, insidious problem in many herds before it was discovered that zinc would prevent or cure it.

B. Environment

While polluted air has attracted attention as a factor in the health of man, it has received little attention regarding its influence on the health and growth of animals. With the trend toward high concentrations of animals in restricted areas, it may be a problem in the future. In poultry and laboratory animals, 20 ppm ammonia in the environment reduced performance and damaged the respiratory epithelium. Ammonia at these levels also increased the susceptibility of poultry to Newcastle disease (Anderson et al., 1964). Arsenic and fluorine are consumed with feed contaminated by air pollution from various industrial processes and causes serious losses.

VII. NUTRITIONAL DISEASES

Anatomical lesions resulting in malfunction and even death are the common terminal events of a disease. It is evident that disease in animals may be initiated, modified, and complicated by numerous factors (Williams, 1965). The clinical signs of inappetence, retarded growth, unthriftiness, and diarrhea are so general and common to both nutritional deficiencies and infections that they may be of limited diagnostic value.

The nature of the anatomical changes are usually helpful in diagnosis. There is some overlapping, however, and for this reason, a combination of the history of the disease, the clinical signs, and the gross and microscopic examination of tissues will suffice to make a diagnosis and formulate the treatment or preventive program. Information about the composition of rations fed, and especially the amount consumed daily, is also helpful in determining the role of nutrition in disease outbreaks.

In some deficiencies, characteristic tissue changes are pathognomonic or indicative for that deficiency. For example, parotid duct metaplasia is pathognomonic for a vitamin A deficiency in calves (Nielsen *et al.*, 1966). In other nutritional diseases, similar tissue changes may be produced by more than one deficiency. An example is that almost identical lesions are characteristic for selenium and vitamin E deficiencies.

The use of purified types of rations has been of value to characterize tissue changes associated with the specific deficiencies. This information is then used to establish the role of nutrients in diseased animals. While nutritional deficiencies usually exist as a complex deficiency, tissue changes often will suggest the primary cause. Pathology, especially microscopic, also supplies clues as to the function of nutrients in the living organism. Only the more important lesions associated with nutritional deficiencies will be discussed since the pathology is described in detail in other texts (Follis, 1958).

A. Energy

Caloric malnutrition is an important factor in initiating specific nutritional deficiencies and in reducing resistance to parasites and bacterial diseases. With a deficiency of dietary carbohydrates the animal organism uses available protein to satisfy energy requirements. Thus, an energy deficiency is expressed as a protein-calorie malnutrition (PCM). In children, this is the world's principal nutritional problem. It impairs physical growth and may cause irreversible mental injury. Most often it is not a total lack of energy, but rather a lack of enough food for the rate of growth and production expected. In animals, caloric undernutrition is not easily detected by clinical signs and thus, may exist insidiously unless reliable weight and age records are maintained. Of course, in more severe cases retarded growth, emaciation, weakness, and lowered production are obvious. After a period of time undernourished animals manifest a complex deficiency and have a reduced resistance to infection. On examination of adult animals at necropsy there is a conspicuous decrease of adipose tissue not only in subcutaneous areas, but in the mesentery and around the kidneys. A low fat content of the marrow of long bones is a common finding of prolonged starvation.

B. Protein

A protein deficiency can be expected to be manifested in a variety of ways because of the diverse functions it has in the living organism. In animals it most often occurs simultaneously with a deficiency of energy and is interrelated with energy as to the lesions manifested. Clinical signs are a loss of weight in the adult or retarded growth in the young, weakness, anemia, and a poor haircoat. The tissue changes associated with a protein deficiency are atrophy (a decrease in size) of the liver, thymus, lymph nodes, and spleen (Cannon, 1948). Proliferative activity is decreased in most tissues, including bones.

C. Minerals and Vitamins

Calcium, phosphorus, and vitamin D are important to the development and maintenance of the skeletal system. A deficiency of iron and copper causes a hypochromic, microcytic anemia in rapidly growing young animals fed primarily milk. Iodine is required for the proper development and functioning of the thyroid gland. Deficiencies of the other mineral elements required by animals have been characterized more by biochemical lesions and clinical signs than by histopathologic means. A deficiency of some minerals such as magnesium and potassium produces acute metabolic disturbances with limited necropsy changes. Other minerals, such as cobalt, are specific requirements for the ruminant; deficiency results in poor microbial fermentation with symptoms of inappetence, anemia, and general malnutrition (for details see Chapter 17).

Vitamin A deficiency is characterized by an atrophy of the epithelial cells, followed by reparative proliferation of the basal cells with growth and differentiation into a stratified keratinizing epithelium. Vitamin E and selenium have been implicated in a number of disease entities in animals. These include myopathies in cattle, sheep, swine, poultry, and mink; liver necrosis in rats and swine; "ill thrift" in cattle and sheep; exudative diathesis in poultry; "yellow fat" disease in swine and cats and acute circulatory failure in swine (Muth *et al.*, 1967). These are all primarily diseases of young animals and are associated with the feeding of polyunsaturated fats, fish products, or improperly harvested or stored cereal grains (spoiled). The incidence of vitamin E-selenium responsive diseases appears greater during years when the weather is unfavorable (rainy and cold) for harvesting cereals and forages. In the Midwest, the incidence of hepatic necrosis in swine is associated with the trend to harvesting high moisture corn and reliance on artificial heat for drying. The presence of linoleic acid and rancidity are believed critical for the development of the disease complex. Nutritional muscular dystrophy is characterized by the symmetrical distribution of the degenerative lesions. The muscle is edematous and has a white-waxy appearance (white muscle disease). Hepatic necrosis appears as scattered red spots on the surface, an irregular shape, and rough areas due to scar tissue (Plate 10).

Much research is in progress for a common physiological role to explain the diverse lesions associated with a deficiency of vitamin E and selenium. A belief long held by many workers, and still valid in many respects, is that they both function by their antioxidant properties. A more recent view suggests that they function in maintaining the integrity of membranes (Porta *et al.*, 1968). Altered membrane permeability would account for the observed transudation such as ascites and exudative diathesis in poultry and swine.

Vitamin K is required in blood coagulation.

The importance of the water-soluble vitamins in animal nutrition has been well established. Most practical rations are adequately fortified, since these nutrients are available from synthetic sources and are economical to use. Deficiencies are only rarely observed in animals fed practical rations. Nutritional diseases associated with an inadequacy of these nutrients are primarily of historical interest (for details see Chapter 18).

VIII. NUTRITION AND TREATMENT OF DISEASES

Nutrients and special rations have an important role in the treatment of diseased animals so they may return to normal growth and efficient production (Fig. 10–1). If the illness is primarily due to a nutritional deficiency it is a simple matter to provide a supplement of the deficient nutrient to allow a return to normal health. The use of nutrition as an aid in treatment of an infectious disease is a far more common problem, and it is also more complicated. In children, malnutrition often exists because of untreated disease (Williams, 1965). The response will depend to a degree on the nature and duration of the infection. Two considerations must be made: (1) determining and eliminating the cause of the infection and (2) providing the necessary nutrients which will aid in the return to health and efficient production. Adequate nutrition during the recovery phase of an infection also will insure the effectiveness of therapeutic drugs. The use of nutrients in the treatment of infectious diseases requires clinical judgment, knowledge of infectious processes, and applications of the principles of medicine and nutrition.

A. General Principles

While there is variability as to the role of nutrition in the susceptibility to infection, proper nutrition is of value in the recovery from an infectious disease regardless of cause. Infections that initiate inappetence, diarrhea, and high fever cause the greatest loss of nutrients from the body. Efforts should be made to minimize nutritional depletion. Of most importance are rations and nutrients that improve the appetite and restore losses from the digestive tract. Specific therapeutic rations such as a low-sodium ration in cardiovascular disease, a low-fat ration in pancreatic disturbances, and the use of lipotrophic nutrients, such as choline and methionine in liver disturbances, are considered in distinct disease entities. The therapy would consist essentially of providing a normal ration modified to insure an adequacy of the nutrients lost during the course of the infection. The oral route is usually the method of choice for administering therapeutic rations. They may be given in warm water by stomach tube if there is complete anorexia. In prostrate animals, parenteral therapy provides immediate nourishment.

B. Correcting Inappetence

A poor appetite accompanies most infectious diseases. The palatability of the ration is important since even the best-formulated ration is useless unless eaten. Ruminants will usually eat high-quality roughage, especially green forage, when they will not consume concentrates or other feed. Rations high in urea tend to be unpalatable. In ruminants, factors required in rumen fermentation, especially such minerals as cobalt, are also helpful in improving a poor appetite. Inoculation of the rumen with cud material from healthy animals may be helpful in chronic disturbances of the rumen. Small amounts of fresh feeds that animals are accustomed to, fed often, are preferable to a large amount at one feeding.

C. Energy Requirements

Energy is required not only to meet the normal requirements of body functions, but also in case of fever. The requirements are estimated to be enhanced 5 percent to 7 percent with each increase of $1°$ F in body temperature. Low-blood glucose values are indicative of weakness and may produce serious neurologic disturbances. Glucose administration not only supplies immediate energy but also aids the liver in detoxifying harmful substances, especially toxins. It also minimizes tissue nitrogen catabolism. In cold environments, body heat loss should be prevented by providing heated stabling or covering with blankets for sick animals.

D. Protein and Fat

Good quality and increased (4 percent to 6 percent more than recommendations) quantity of protein is required in animals recovering from an infection. Animal proteins of high biological value such as milk, milk products, and eggs may be used. Not only must vital tissue protein be replaced, but in addition food protein is needed to form enzymes, hormones, and specific immune substances that are essential in recovering from an infection. High levels of protein are contraindicated in kidney disturbances due to the threat of accumulating toxic nitrogenous end products in the blood.

A small amount (2 percent to 3 percent) of high quality fat has merit in a therapeutic ration. It adds palatability (avoids dustiness), is a concentrated source of energy, and aids in the metabolism of the fat soluble vitamins. It would also supply the essential fatty acids that may aid in skin disorders and general well-being.

E. Fluids, Electrolytes and Minerals

Fluid and electrolyte imbalance due to infectious diseases seriously impede normal organ function. In therapy, proper extra-

and intracellular electrolyte composition must be maintained. In gastrointestinal infections, large amounts of water and electrolytes, especially sodium and potassium, are excreted. The presence of alkalosis or acidosis dictates the need for administration of electrolytes. Caution should be used in potassium administration, since indiscriminate use may be deleterious. Fresh, clean water should always be readily available. Anemia is present in many diseases due to either increased loss of blood or deficient synthesis of hemoglobin. Factors such as iron and vitamin B_{12} required in hemoglobin formation should be included in rations for anemic animals.

F. Vitamins and Unidentified Growth Factors

During disease the requirement for vitamins and unknown factors is increased. The depletion of water-soluble vitamins is relatively rapid, but that of the fat-soluble vitamins is much slower. For therapy, a 5- to 10-fold increase in the intake of vitamins above the normal requirements is suggested. Lesions of the gastrointestinal tract and liver also interfere with the synthesis and metabolism of vitamins. Liver and milk are usually good sources of unidentified growth factors.

G. Balanced Ration

In animals recovering from an infection, as in normal animals, a balanced ration is essential. The vitamins and minerals are required, not only in the proper combination, but at the same time in order to metabolize the energy and protein portion of the ration. The aftermath of an infectious disease is usually a complex deficiency. The correction of a single specific deficiency is usually only temporarily helpful. All the nutrients should be supplied in their proper proportions at the same time.

REFERENCES

Anderson, D. P., Beard, C. W., & Hanson, R. P. (1964). The adverse effects of ammonia on chickens including resistance to infection with Newcastle disease virus. Avian Dis. *8*:369–379.

Anderson, E. S. (1968). Middlesbrough outbreak of infantile enteritis and transferable drug resistance. Brit. Med. J. *1*:293–304.

Cannon, P. R. (1948). *Some Pathologic Consequences of Protein and Amino Acid Deficiencies.* Springfield, Ill., C. C Thomas.

Coates, M. E., Fuller, R., Harrison, G. F., Levi, M., & Suffolk, S. F. (1963). A comparison of the growth of chicks in the Gustafsson germfree apparatus and in a conventional environment, with and without dietary supplements of penicillin. Brit. J. Nutr. *17*:141–150.

Coates, M. E. (Ed.) (1968). *The Germ-Free Animal in Research.* London, Academic Press.

Dubos, R. J., & Schaedler, R. W. (1960). The effect of the intestinal flora on the growth rate of mice and on their susceptibility to experimental infections. J. Exper. Med. *111*:407–417.

Follis, R. H., Jr. (1958). *Deficiency Disease.* Springfield, Ill., C. C Thomas.

Forbes, M., & Park, J. T. (1959). Growth of germ-free and conventional chicks: effects of diet, dietary penicillin and bacterial environment. J. Nutr. *67*: 69–84.

Hansel, W., & McEntee, K. (1955). Bovine hyperkeratosis (X-disease). J. Dairy Sci. *38*:875–882.

Madhavan, T. V., & Gopalan, C. (1968). The effect of dietary protein on carcinogenesis of aflatoxin. Arch. Path. *85*:133–137.

Maronpot, R. R., & Whitehair, C. K. (1967). Experimental sprue-like small intestinal lesions in pigs. Canad. J. Comp. Med. *31*:309–316.

Miner, R. W. (Ed.) (1955). *Nutrition in Infection.* N. Y. Acad. Sci. New York.

Munro, H. W. (Ed.) (1964). *The Role of the Gastrointestinal Tract in Protein Metabolism.* Philadelphia, F. A. Davis Co.

Muth, O. H., Oldfield, J. E., & Weswig, P. H. (Eds.) (1967.) *Symposium Selenium in Biomedicine.* The AVI Co., Westport, Conn.

National Academy of Science–National Research Council. (1966). *Toxicants Occurring Naturally in Foods.* Publication 1354. Washington, D. C.

Newberne, P. E. (1966). Overnutrition on resistance of dogs to distemper virus. Fed. Proc. *25*:1701–1710.

Nielsen, S. W., Mills, J. H. L., Rousseau, J. E., & Woelfel, C. G. (1966). Parotid duct metaplasia in marginal bovine vitamin A deficiency. Amer. J. Vet. Res. *27*:223–233.

Porta, E. A., Delaiglesia, F. A., & Hartroft, W. S. (1968). Studies on dietary liver necrosis. Lab. Invest. *18*:283–297.

Sadler, W. W. (1965). Animal-borne infections. Proc. West. Hem. Nutr. Cong., pp. 143–145. Chicago, Ill.

Schaedler, R. W., Dubos, R., & Costello, R. (1965). The development of the bacterial flora in the gastrointestinal tract of mice. J. Exper. Med. *122*: 59–66.

Trexler, P. C., & Reynolds, L. I. (1957). Flexible film apparatus for the rearing and use of germfree animals. Appl. Microbiol. *5*:406–412.

Turk, D. E., & Stephens, J. F. (1966). Effect of intestinal damage produced by *Eimeria necatrix* infection in chicks upon absorption of orally administered zinc-65. J. Nutr. *88*:2 1–266.

Whitehair, C. K., & Pomeroy, B. S. (1967). Veterinary Medical Basis for the Use of Antibiotics in Feeds.

Proc. Sympos. on The Use of Drugs in Animal Feeds. Washington, D. C. National Academy of Science–National Research Council.

Williams, C. D. (1965). Factors in the ecology of malnutrition. Proc. West. Hem. Nutr. Cong., pp. 20–24. Chicago, Ill.

III. Body Composition

Chapter 11

Muscle

By E. J. Briskey

ALTHOUGH muscle tissue generally develops more slowly than bone and faster than adipose tissue, muscles vary widely in their growth patterns (Butterfield & Berg, 1966). Certain muscles are "early-developing"; others are "average" or "late-developing," depending on when they increase in weight in relation to total muscle weight. With the exception of some animals such as mice (Goldspink, 1962), which are born very immature, there are no new muscle fibers formed after birth (Bendall & Voyle, 1967); subsequent muscle growth occurs by hypertrophy of existing fibers. Furthermore, muscle fiber numbers may actually decrease during advanced stages of maturity.

Connective tissues represent a very important part of muscle structure. In fact, no muscle fiber exists without some part of its surface being in contact with this tissue. All cells are contained in and held together by connective tissues. Connective tissue content decreases from the first to the sixth month after birth, and then remains constant at the lower value during the next 18 months. Collagen, the main structural protein of connective tissues, shows increases in cross-linking with advances in the chrono-

logical age of the animal. This situation may have some relationship to function. Tendons have considerable covalent cross-linking of the collagen molecules, while certain soft tissues have little cross-linking. Elastin, another connective tissue protein, exists in variable amounts in different muscles and is the principal structural protein in the walls of arteries. After maturity is reached and growth ceases there is very little turnover of elastin. Various other characteristics of connective tissues show obvious changes during growth, development, and aging (Goll *et al.*, 1964a, b; Herring *et al.*, 1967; Carmichael & Lawrie, 1967). The muscle fiber consists of a sarcolemmal membrane, sarcoplasmic proteins, nuclei, Golgi bodies, mitochondria, sarcoplasmic reticulum, T-system and contractile elements, the myofibrils. The sarcoplasm is the cytoplasm of the muscle fiber, within which, as in cells elsewhere, are two kinds of formed elements—organelles and inclusions. The term "organelle" is given to those bodies which are structurally specialized parts of the cytoplasm and could be regarded as part of the living substance of the cell. Conversely, the term "inclusion" embraces particles such

193

as glycogen and lipid droplets. Variable deposits of particulate glycogen, small lipid droplets, and occasional free ribosomes as well as certain quantities of minerals and nucleotides also comprise the main regions of the sarcoplasm; from a percentage standpoint, water is the main constituent of muscle. During prenatal and postnatal development, muscles change in composition and character of constituents as well as in their physiological responses to stimuli.

I. MORPHOLOGY

A. Macrostructure

Apart from species inheritance, the form of a muscle is determined by its function and, to a large extent, by its need for power and range of movement. A large number of muscles are placed on the skeleton (Plate 11 A) in such a way as to give direction, power, and speed to a particular movement, through complementary or cooperative action as well as noninterference. To avoid interference with movement, the muscle bellies are specially placed and tapered to tendons or blended with aponeurotic sheets where limbs must be free or where muscle is unnecessary. A tendon may run along or into the body of a muscle. In this case the muscle fibers are attached to one side of the tendon in a unipennate form of attachment. Bipennate attachments occur where both sides of the tendon are attached to the muscle surface and multipennate forms where several tendons invade the muscle. Muscles have been named according to their size (magnus), location (subscapularis), function (adductor), shape (trapezius), and structure (semimembranous).

Normally the percentage of muscle tissue in a carcass increases following birth, then decreases by fattening. The actual distribution of muscle mass over the carcass of an animal appears to be the result of functional demands by the animal to meet the challenges of its environment. There is more scope for efforts to increase the total amount of muscle per animal than for attempts to upset its proportionate distribution and thereby influence the intrinsic relations of this functional locomotor system.

The main factor in the genetic determination of muscle size appears to be the total number of fibers the muscle possesses rather than the size of the fibers. Obviously, genetic background is an important factor governing muscle growth. It is for this reason that the use of the longissimus area as a measure of muscling may have potential in the selection or evaluation of a sire's influence on the muscle mass of his offspring. Furthermore there are sex differences in regard to muscle growth (Suess, 1968).

1. CONNECTIVE TISSUES

A muscle is surrounded by a connective tissue layer, the epimysium, from the deep surface of which septa pass into the muscle at irregular intervals (Fig. 11–1). These septa, which are fine collagen fibers of the perimysium about 2 to 3 microns in thickness, invest 30 to 40 of these fibers and collect them into bundles or fascicles. Muscles used for power have larger fascicles with fewer fibers in each bundle than those which are placed for refined movement. The nerve bundles and small blood vessels run between the bundles of fibers. Very delicate extensions of fine connective tissue, the endomysium, pass from the perimysium inward to surround each muscle fiber (muscle fibers can be regarded as complete cells). Fine branches of nerve and vascular tissue are intricately associated with each individual fiber. The connective tissues are all continuous with each other as well as with the tendon of origin and the tendon of insertion. The term "origin" is usually applied to the proximal or less mobile point of a muscle's attachment and the term "insertion," to the distal and more mobile attachment. Most of the connective tissue in large muscles consists of collagen and, to a lesser degree, reticulin. Some bovine muscles such as the semitendinosus also contain large amounts of elastin in their epimysial and perimysial layers.

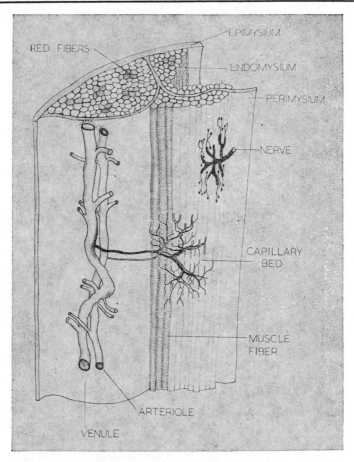

Fɪɢ. 11–1. Part of a muscle (*sketch*). This sketch shows connective tissue organization around bundles and fibers as well as a cross section of red and white fibers within bundles. (*Briskey & Fukazawa, 1969. Advances Food Res. 17.*)

Bovine intramuscular collagen solubility has been studied in regard to animal maturity. Young collagen solubilizes to gelatin far more easily than old collagen (Goll *et al.*, 1964a). Collagen solubility decreases with increasing chronological age of the animal (Hill, 1966). More specifically, the solubility declines precipitously and linearly from 1 month to 2 years of age, and then tends to level out. The decline in solubility reflects increasing covalent cross-linking in the collagen molecules as the animal ages.

2. Color

Muscles consist of red and white fibers. Red muscles contain a majority of red fibers while white muscles possess a majority of white fibers. This relationship between muscle color and fiber type can be seen very clearly. The gross effect of differences in red and white fibers can be seen in certain muscles of the pig which have light and dark portions (Plate 11 B). Researchers have concluded that muscle color in the pig is moderately heritable. The sex of the animal appears to have an influence on muscle color, with female pigs having the most intense color in their musculature. Some of the higher color intensity is obviously due to less fiber differentiation during growth. During later stages of maturation, however, an increase in color intensity can be due largely to greater deposition of myoglobin in the red fibers.

B. Microstructure

The essential structural unit of all muscles is the fiber. Fibers are long, narrow, tapered, cylindrical, multinucleated cells. They rarely anastomose with adjacent fibers—except perhaps in some facial muscles which are attached to the skin of the heads of large animals. Skeletal muscle fibers contain long, unbranching threads of protein. These threads are called myofibrils and run along the long axis of the fiber. Myofibrils are striated and adjacent myofibrils lie with their light and dark bands in register, conferring a striated structure upon the entire fiber. This striated appearance can also be seen with the light microscope.

Nuclei can also be seen with the light microscope and are located along the periphery of the cell. In the embryo, nuclei are located in the center of the cell but are translocated prenatally to the periphery and remain there throughout postnatal life, unless a dystrophic abnormality occurs which may result in a medial movement of the nuclei. In fetal muscle cells, nuclei are large in relation to cell size, and therefore, immature cells have a high nuclear-cytoplasmic ratio.

1. LENGTH OF FIBERS

Little is known about the lengths of individual fibers in the musculature of large animals. However, on the basis of work completed on dog and man, most of the fibers probably extend from one end of the muscle to the other. Some fibers, however, may have fine tapering extremities ending within the belly of the muscle. Additionally, some fibers may have their broad bases attached to a tendon while their other ends terminate in a fine attenuated extremity.

Animals vary in the manner in which their muscles grow in length. Postembryonic growth in length, however, is generally limited to small zones near the attachments of muscle to bone. In some muscles of the mouse, sarcomeres increase in length from 2.0 to 2.5 microns while the muscles lengthen from about 3.5 mm to their mature length

of about 7.00 mm (Goldspink, 1962). While most of the length change is due to an increase in sarcomere numbers, a significant proportion is attributed to an increase in the length of the individual sarcomeres. In the cow, however, no evidence has been found that the sarcomere length itself increases after the fourth week; the increase in muscle length appears to be due to the laying down of new sarcomeres.

2. DIAMETER OF FIBERS

Muscle fiber diameter varies with species, postnatal development in body weight, general body size, chronological age, nutrition, activity, breed, and color (red or white) of the fibers. It is difficult to assess accurately fiber diameter because of the shortening which might occur prior to or during excision. Shortening results in short sarcomeres and large fiber diameters.

Fiber diameter varies distinctly among muscles of the same animal. However, actual muscle size is influenced more by fiber number than by fiber diameter. This means that for the most part, the adult sizes of the various muscles are fixed by the beginning of the fetal stage (Hammond, 1961). The size of muscle bundles (fibers grouped by connective tissue) is positively associated with the visible coarseness of a cross-sectional area of the muscle and is known to increase somewhat with the chronological age of the animal.

3. FIBER TYPE

White muscle fibers of adult porcine, bovine, and ovine animals vary in diameter from approximately 50 to 100 microns, while red fibers are usually, but not always, smaller in diameter. The fibers are spherical and packed closely to one another if cut when fresh or frozen, but irregularly polyhedral if cut after fixation. White fibers are surrounded by fewer capillaries than are red fibers.

White muscle fibers have a punctate appearance and contain small myofibrils which are regularly spaced and surrounded with abundant sarcoplasm. Red muscle fibers contain areas of poorly delineated, large diameter myofibrils and little sarcoplasm (Hess, 1967). Histochemically, "red" muscle fibers develop pronounced reactions in media designed to demonstrate oxidative enzymes (such as diphosphopyridine nucleotide-tetrazolium reductase and reduced diphosphopyridine nucleotide-tetrazolium reductase) (Fig. 11–1) (Plate 11 C), but only weak reactions in media selected to detect adenosine triphosphatase and phosphorylase (Plate 11 D). Conversely, white fibers show a weak reaction for oxidative enzymes, but a very intense reaction in media selected to detect adenosine triphosphatase and phosphorylase. "Intermediate" fibers have been identified (Brooke, 1966). They give a moderate reaction to oxidative enzymes, but still show an intense reaction to phosphorylase and adenosine triphosphatase. Intermediate fibers are present in large numbers in the musculature of fast-gaining, stress-susceptible pigs (Cooper *et al.*, 1969) which develop pale, soft exudative characteristics after death (Briskey, 1964; Briskey *et al.*, 1966).

The differentiation of fiber types during growth is an area of considerable interest. The amount of fiber differentiation which will already have occurred at birth is related, in part, to the length of the gestation period and to the development of the animal at the time of birth (Dubowitz, 1963). Differentiation of muscle fibers in Rhesus monkeys is apparent, histochemically, at 120 days of fetal age. In the kitten, whether based on response to stimulation or intensity of staining for oxidative enzymes, the muscles are entirely red at the time of birth; in other words, no differentiation of fibers from red to white has occurred. At 10 days after birth, some differentiation between darkly staining and sparsely staining fibers can be noted, while at 15 days this differentiation is clearly evident. Enzymatic maturation has been studied in the muscles of man, the guinea pig, rabbit, rat, and hamster. Differ-

entiation is apparent at birth in the muscles of man, the rabbit, and hamster, but is most evident in guinea pigs, in whom muscles resemble the adult stage. Fiber differentiation in the muscle of the rat, however, has not been observed until it is at least 7 to 10 days of age. Biochemical studies on the muscle of the newborn rabbit show that its red and white muscles are not distinguishable until it develops the ability to move independently.

Muscle fibers are not differentiated in the longissimus muscle of the pig at any of the fetal stages. Only a very slight differentiation is apparent at 1 day of age. A further large increase in the area of white fibers occurs by 13 days of age. A large increase in the area of white fibers occurs as the animal attains an age of 200 days, but there is little additional change as the animal matures to 24 months of age (Plate 12). The red fibers form clumps within the bundles; however, most laboratory animals and man do not have any particular order of red fiber location within a particular bundle, but instead present a scattered appearance in fiber distribution.

4. Blood Supply

Muscle fibers depend on vascularization for nourishment. The larger branches of the arteries penetrate the muscle along the laminae of the perimysium. These arterioles give off capillaries at abrupt angles. A system of converging tubes, the veins, receive blood from the capillaries and unite to form fewer and larger vessels as they return blood to the heart. Oxygen, among other nutrients, is transported to the skeletal muscles by way of capillaries, while lactic acid, among other metabolic products, is transported out of the skeletal muscle by way of veins. Oxygen, reversibly bound to hemoglobin, is transported to the fiber via the capillaries; and by the process of diffusion passes through the capillary wall, the extracellular fluids, and finally, through the sarcolemma into the sarcoplasm. The removal of substances also occurs by diffusion, but in the opposite direction or from the muscle cell into the capillary lumen.

The capillaries (average diameter 7 microns) branch extensively without any change or with only a negligible change in caliber. These branches anastomose to form extensive networks with meshes which vary in size and shape. Capillary anastomoses are especially well developed at the motor end plates, sites which are especially active metabolically. In the capillary, a single layer of endothelial cells makes up the entire wall, but in arteries and veins, the endothelium is invested with accessory coats of muscle and connective tissue. The relationship of capillaries to arterioles and venules is shown in Figure 11–1. Lymphatic capillaries are not found between the individual muscle fibers. They are present, however, in the connective tissue of septa and along the blood vessels.

Capillary density is higher in red muscles than in white muscles (Smith & Giovacchini, 1956), and also in red fibers as compared to white fibers within the same muscle. Red muscle fibers are surrounded by approximately twice as many capillaries as white fibers. Many capillaries are closed to circulation in a resting skeletal muscle when it is under normal homeostasis. The number of open capillaries is found to increase after exercise, narcosis, shock, and other conditions (Carrow *et al.*, 1967). Additionally, muscles from trained animals have a larger number of open capillaries than muscles from untrained animals. Little is known about general changes in capillarization as a result of growth and nutrition; however, it is known that these factors can change fiber type (Beecher, 1966). Several authors are inclined to think that all capillaries are formed during early fetal life; therefore, the effect of any activity and growth may be reflected in the number of open capillaries.

5. NERVE

A neuron is defined as a nerve cell with all its processes. The protoplasmic processes of the neuron reach considerable distances from the body of the cell. The axons themselves form part of the ventral roots and efferent fibers of the peripheral nerves and terminate in the skeletal muscles of the body. Muscles concerned with delicate gradations of precise movement generally receive a more abundant supply of nerve fibers. In most muscles, however, the ratio is much lower and many muscle fibers are innervated by the terminal branches of the one nerve fiber. The nerve fibers enter the *perimysium externum* and *internum* in which they may bifurcate several times, thus permitting one neuron to innervate more than one muscle fiber. After repeated branchings within the perimysium, the nerve fibers pass to the individual muscle fibers where they terminate in structures known as motor end plates. At the motor end plate region the fiber is continous with the reticular fibers over the sarcolemma.

6. GROSS ASPECTS OF FAT CELLS

The architecture of the muscle influences the pattern of fat deposition. Looseness of fascicular organization generally parallels the quantity of interfascicular lipid. The capacity of muscles to accumulate increasing quantities of adipose tissue appears to be related to the changing skeletal proportions which accompany the enlargement of the body. If it is assumed that adipose tissue is most readily accommodated in body parts capable of the greatest enlargement, the more proximal muscles should be expected to reflect the trend more noticeably than those of the extremities, which have already undergone a considerable share of their growth prior to the period of maximal lipid deposition. Although there is a variation among species, there is a general tendency for the percentage of lipid content to increase with advancing age. In the pig, the lipid content of muscles continues to increase to 415 days of age; however, this tendency is more evident in the loosely organized muscles of the trunk and proximal segments of forelimb than in the tightly bound, consolidated, fusiform muscles nearer the extremi-

PLATE 11

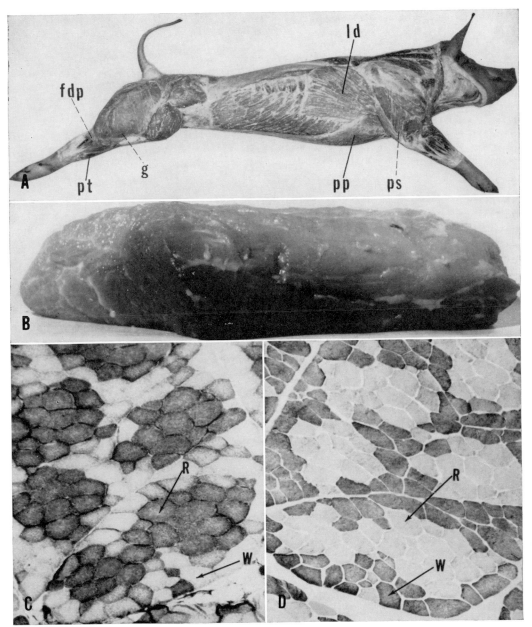

A, Lateral view of carcass to show placement and orientation (*Kauffman, 1965. Porcine Myology, Urbana, University of Illinois Press.*)

B, Red and white portions of the porcine semitendinosus.

C and D, Porcine trapezius muscle, fresh frozen section (magnification of 25× on 35 mm. negative). C represents a section reacted for DPNH-TR and D is the serial section reacted for amylophosphorylase. R denotes red fiber and W denotes white fiber. (*Moody & Cassens, 1958. J. Anim. Sci. 27, 961.*)

PLATE 12

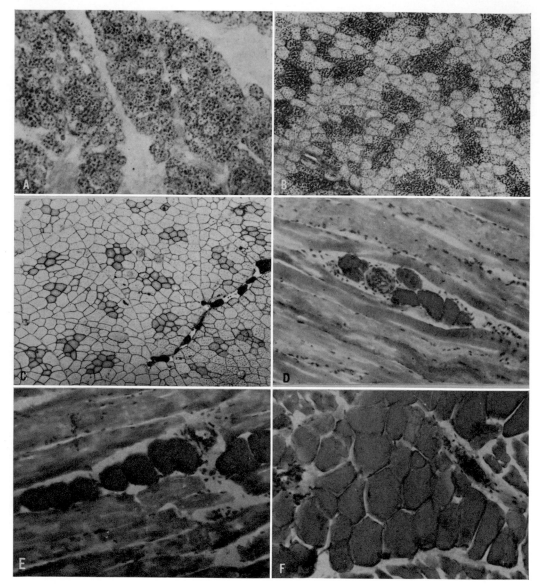

Histochemical changes in muscle fibers of longissimus dorsi of the growing pig.

A, 1 day old (430×). B, 13 day old (271×). C, 24 month old (43×). (*Cassens et al., 1968, J. Anim. Morphol. Physiol. 15.*)

D, E, F, Photomicrograph showing typical fat cell mass classifications. Fat cell size increases as the number of cells per mass increases. (*Moody & Cassens, 1968. J. Food Sci. 33:1, 47.*)

PLATE 13

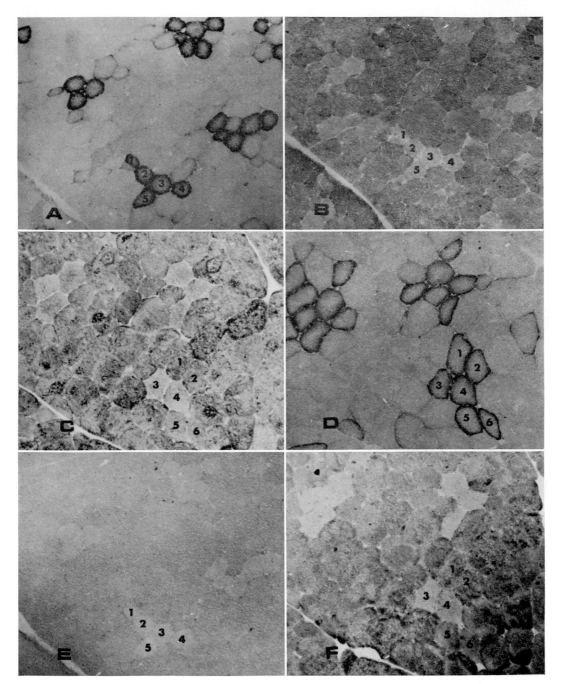

Histochemical characterization of red, white and intermediate porcine fibers. A, Diaphorase. B, Phosphorylase. C, ATPase. D, E, F, Different kinds of porcine animal but with stain as in A, B, C respectively. (*Cooper et al., 1969, J. Food Sci. 34.*)

PLATE 14

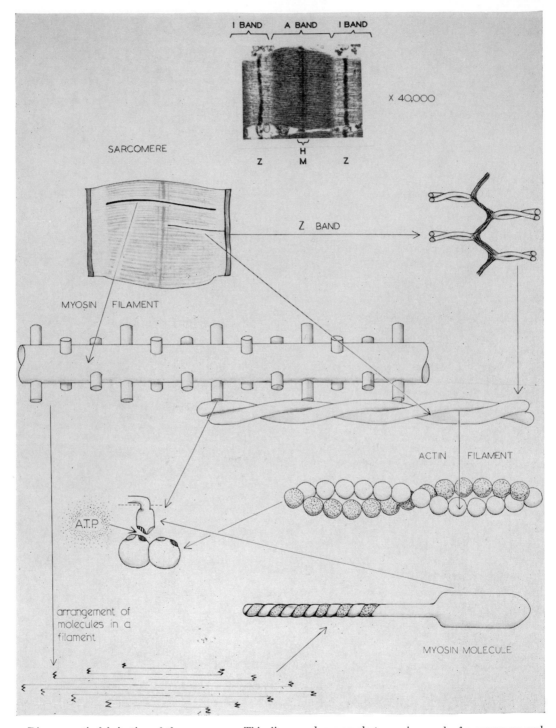

Diagrammatic fabrication of the sarcomere. This diagram shows an electron micrograph of a sarcomere and a sketch of the possible orientation and form of its principal components. (*Briskey & Fukazawa, 1969. Adv. Food Res.*)

ties. Fat deposition is in some way associated with the circulatory system. In a quantitative and morphological study of bovine longissimus fat cells, fat cell size was found to increase with increases in total chemical fat in the muscle (Plate 12 D, E, F). Fat cells accumulated and grew, however, in close proximity to portions of the circulatory system.

C. Ultrastructure

The muscle cell exists in a sleeve of collagenous connective tissue to which it is attached by an amorphous or faintly filamentous basement membrane of a few hundred Å in thickness. This basement membrane contains mucopolysaccharide and serves as a filtering barrier to unwanted substances or as a preferential diffusion pathway. It also contributes to the stability of the immediate cellular environment as well as "cements" the cell to its contiguous connective tissue (Slautterback, 1966).

1. Sarcolemma

Beneath the endomysium is a sheath called the "sarcolemma" which surrounds each fiber. The sarcolemma has a concentration gradient across its width keeping the K^+ concentration high in the interior of the cell and the Na^+ concentration high on the exterior of the cell. The sarcolemma has been shown by electron microscopy to consist of a double membrane of which the components are about 50 to 60 Å apart (Fig. 11–2). This is known as the unit membrane and is characteristic of many animal cell membranes as well as internal membranes such as those which are found in the mitochondria. The sarcolemma is about 0.1 micron in thickness and forms a resistant sheath enclosing the soft protoplasmic contents of the fiber.

2. Nuclei

The nuclei are of an ovoid form (Fig. 11–2). They are flattened, elongated (8 to 10 microns in length), and lie under the sarcolemma at the periphery of the fiber. In a

fiber some millimeters in length there may be several hundred nuclei. Their distribution is fairly regular along the length of the fiber, but toward the tendinous attachment they become more numerous and more irregularly distributed. The position of the nuclei beneath the sarcolemma, *i.e.*, hypolemmal, which is usual in adult muscle, differs from the embryo, in which the nuclei occupy a position in the middle of the fiber. Disease can cause shifts in the position of nuclei, as can activity and hypertrophy (Brooke, 1966).

The nucleus contains the genetic material, deoxyribonucleic acid (DNA), which determines the specificity of cellular behavior and controls its metabolic activities. The principal components of the interphase nucleus are the chromatin, the nucleolus, and the nuclear matrix. The term "chromatin" is used specifically for the DNA-containing, chromosomal substance of the nucleus. The nucleolus, rich in ribonucleic acid (RNA), is a rounded body that is usually basophilic and eccentrically placed in the nucleus. The chromatin and nucleolus are dispersed in the nuclear matrix. Evidence that the nucleolus is involved in the elaboration of the ribosomal RNA of the cytoplasm has come from many sources. Various analytical approaches indicate that the nucleolus plays a key role in nucleic acid metabolism and protein synthesis.

3. Golgi Apparatus

A small Golgi apparatus frequently can be seen in the cytoplasm at each pole of the nucleus (Fawcett & Selby, 1958) (Fig. 11–2). The arrangement of membranous structures comprising the Golgi apparatus is sufficiently distinctive and consistent from one cell type to another to permit identification of this organelle. It consists of aggregations of membrane-bounded elements of at least three kinds. The most conspicuous and characteristic of these are assemblages of cisternae piled one upon the other in close parallel array. The functions of this organelle have not been completely defined but it seems clear that it plays an important role in the secretory process.

4. MITOCHONDRIA

The fibrous matrix or the cristae components form the typical cross-strands which help to distinguish the mitochondria in ultra-structure studies. The sub-units of the mitochondria can also be clearly seen in preparations which have been negatively stained. The fundamental morphological structure of mitochondria in muscle is indistinguishable from the same type of particle seen in other cells (Fig. 11–2). The bulk of the electron transport through the Keilin-Hartree system, the linkage of this system with the citric acid cycle, and the site of oxidative phosphorylation are all contained

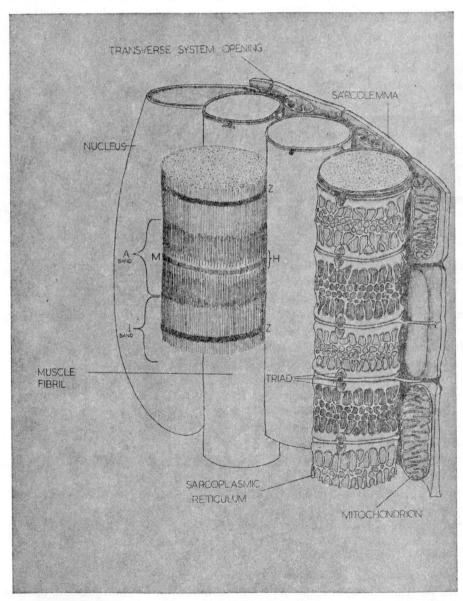

FIG. 11–2.   Part of a fiber (*sketch*).   This sketch shows the sarcoplasmic reticulum, transverse system, mitochondria, nucleus and myofibrils (muscle fibrils).   (*Briskey & Fukazawa, 1969.   Advances Food Res. 17.*)

within the mitochondria (Harmon, 1964). The mitochondrial population varies greatly from muscle to muscle, depending upon the intensity of metabolic activity. A high-frequency muscle contains many large mitochondria, and a low-frequency muscle contains few mitochondria. An adequate mitochondrial metabolism guarantees a consistent supply of energy to the contractile units.

In many muscles the mitochondria occupy exact positions in relation to the sarcomeres; they are situated against the I bands, near the triads, and close to the Z bands. These bodies are probably strategically situated to be close to these structures, which are so vitally in need of constant supplies of high-energy metabolites. At the site they occupy, such mitochondria afford ample high-energy materials for the immediately adjacent sarcotubular components. They may also satisfy some needs of the contractile units. In mammalian muscle they are most common in the sarcoplasm just opposite the I band and straddling the Z band and oriented with their long axes transverse so as to partially surround the myofibrils.

5. MYOFIBRILS

The word striated as applied to the muscle refers to crossbanding, which can be seen when its fibers are examined microscopically. Just as the muscle fibers show alternate light and dark bands, so does each myofibril. The appearance of cross-striation in whole fibers results from the fact that the light and dark bands of the myofibril are in register with each other. Myofibrils are about 1.0 micron in diameter and number about 2,000 in an average fiber of 50 microns in diameter (Bendall, 1966). These myofibrils are long, unbranching threads of protein which run along the long axis of the fiber. In the living cell the myofibrils, though not ordinarily sharply defined, are precisely aligned and so arranged as to present a repetitive pattern along their longitudinal facade. The bands which are dark are anisotropic and are called A bands, while the bands which are light are isotropic and are called I bands. This difference is due to the fact that the A bands are composed of material of a higher refractive index than the material of the I bands. The A band exhibits a strong positive birefringence while the I band is only weakly positively birefringent. Therefore, for purposes of nomenclature the latter is considered to be isotropic. These units of pattern and functionality are called sarcomeres (Fig. 11–2). Sarcomeres are bounded on each end by Z bands. Each Z band bisects the light I bands. The dense A band lies in the center of the sarcomere and is itself occupied by a central lighter zone, called H. Within this H zone is a thin line called the M band. The structure of the sarcomere reflects the intimate molecular arrangement of the special proteins responsible for the contractile process. In skeletal muscle the transverse dimension of a sarcomere is relatively constant from one sarcomere to the next in line. Thus, the consequence of end-to-end formation of sarcomeres is the production of uniform cylindrical structures running the full length of the myofibrils.

There is, however, considerable variation in the resting length of sarcomeres in different skeletal muscles. Sarcomeres are estimated to be around 2.3 microns in length in the rabbit, with the A band 1.5 microns and the I band 0.8 micron. The sarcomeres are thought to be considerably longer in bovine, porcine, and ovine muscles. In the *psoas major* of bovine animals, for example, the sarcomeres are approximately 3.5 microns in length. However, part of this length may be induced post mortem.

Most measurements of the diameter of myofibrils fall in the range of 0.5 to 2.0 microns irrespective of the source of the skeletal muscle. During fetal development, myofibrils split longitudinally and grow. Variations in diameter can occur within a given cell as well as between myofibrils of different sources. Differences in the reported values for the diameters of myofibrils are also partially due to variable amounts of shrinkage occurring during fixation. Additionally, postmortem shortening has a profound influence on diameter (Herring *et al.*, 1967).

6. SARCOPLASMIC RETICULUM

The myofibrils of most striated muscles are surrounded by a profuse and complex system of membrane-bounded tubules and vesicles (Fig. 11–2). These are known collectively as the sarcoplasmic reticulum and probably are continuous with the elements of the reticulum found at the poles of the nuclei and beneath the sarcolemma. The sarcoplasmic reticulum consists of two parts —one is known as the L longitudinal or sarcotubular system. It is made up of tubules which are parallel with the myofibril. At the A-1 junction in higher vertebrates, the L tubules meet the other part of the sarcoplasmic reticulum, the T-transverse or intermediate tubules which have direct access to the exterior of the cell. These tubules surround the myofibrils and are continuous from one fibril to the next transversely across the skeletal muscle cell (Slautterback, 1966). There is a vesicular expansion of the L tubules, where the L and T systems meet, and these three constituents —the two expanded parts of the L system and the T tubule between—are called the triad.

The sarcoplasmic reticulum has two functions: (a) to conduct impulses from the sarcolemma to the reactive bands, and (b) to act as a supply line for essential metabolites. Because of the special capacity of the sarcoplasmic reticulum for differential accumulation of calcium ions, it participates actively in contraction and relaxation. It seems clear that the sarcoplasmic reticulum represents the original relaxing factor. If the contractile protein system (actomyosin) contains bound Ca^{++}, it contracts with ATP; if part of this Ca^{++} is removed, it relaxes with ATP. It is the function of the sarcotubular system to control the amount of bound calcium in actomyosin by regulating the concentration of Ca^{++} in the environ-

ment. The quantity and character of sarcoplasmic reticulum changes with growth, development, and activity (Slautterback, 1966).

II. COMPOSITION

Muscle is a colloidal system of approximately 72 to 73 percent water, 18 percent protein, 1 to 2 percent soluble nonprotein substances, variable amounts of fat (1 to 20 percent) and small amounts of ash (1 percent) and carbohydrate (1 percent). All the essential amino acids are present in skeletal muscles. The ash content of muscle comprises various inorganic ions of which potassium, sodium, magnesium, phosphorus, iron, and zinc are most abundant. The carbohydrate is largely glycogen although other substances such as lactic acid also may be present. Certain nitrogenous constituents including amino acids, creatine, urea, carnosine, anserine, xanthine, adenine, and hypoxanthine also are found in muscle in small amounts at certain pre- and postmortem periods. Niacin, thiamine, and riboflavin are present in skeletal muscle in rather large quantities compared to the quantities in certain other tissues. These vitamins are especially high in porcine tissue.

Although proteins account for almost 20 percent of the wet weight of mammalian muscle (Perry, 1965), the actual quantity is nevertheless somewhat variable and is influenced by muscle type (fast or slow), training and stage of postnatal life. The muscle proteins may be categorized into sarcoplasmic (glycolytic enzymes and pigments), myofibrillar (contractile), and stroma (connective tissue) protein fractions.

A. Sarcoplasmic Protein

1. CHARACTERISTICS

Sarcoplasmic proteins are a heterogeneous mixture of proteins of the albumin type and are found in the cytoplasm of the cell. Proteins of the sarcoplasm are far too complex to review individually in this chapter; there are already voluminous reports on the subject of enzymes and muscle pigments. Consequently, this review, with a few exceptions, will be restricted to sarcoplasmic proteins as a more or less homogeneous group.

Sarcoplasmic proteins are globular in nature, low in viscosity, low in molecular weight, and soluble in water or low-ionic-strength solutions. These proteins are usually extracted in high yields by water or low-ionic-strength salt solutions. The ionic strength should not exceed that of muscle fluid (0.21 to 0.26) because higher salt concentrations will extract proteins from the myofibril. The protein which is soluble at low ionic strength varies from 28 to 44 percent of the total protein in muscle. These differences are ascribed partly to the fate of the muscle before and during the extraction procedure, and partly to the extraction procedure itself. The optimum conditions for quantitative extraction of muscle protein include the extraction of finely subdivided frozen muscle with 10 volumes of 0.03M potassium phosphate buffer at pH 7.4 ($=0.08$), changed twice, with each extraction permitted to continue for three hours (Helander, 1957).

2. Species Variations

The variation in sarcoplasmic proteins of different species became evident after the advent of moving boundary electrophoresis. With the development of zone electrophoresis techniques, additional positive species variations have been detected. Changes in electrophoretic patterns have been applied to detect adulteration of meat products from the horse, kangaroo, and deer (Thompson, 1961). Fish sarcoplasmic proteins also vary markedly with species. Since enzymes for anaerobic glycolysis comprise a large percentage of the sarcoplasmic proteins, it is understandable that the white muscle, which is geared for anaerobic metabolism, has more sarcoplasmic protein than does red muscle. Additionally, sarcoplasmic proteins of red and white muscles differ in electrophoretograms.

3. Relation to Growth and Activity

As fibers grow and differentiate they increase their total quantity of sarcoplasmic proteins (Helander, 1957). Although protein nitrogen increases from 9.7 to 18.5 g/kg during fetal life in man, the percentage of sarcoplasmic protein changes very little. In fact, in the fetal life of the pig there is actually a decrease in the concentration of this protein fraction (Dickerson & Widdowson, 1960). During the postnatal development of both man and pig, the concentrations of sarcoplasmic protein increase, the value of the concentration for the adult pig being a little higher than the value for the adult man. Also, with inactivity the muscles of animals increase their content of sarcoplasmic protein (Helander, 1966). However, when inactivity is accompanied by atrophy, the increase in sarcoplasmic protein appears to be only relative. The effect of age and activity on individual sarcoplasmic proteins is not known.

B. Myofibrillar Protein

The proteins of the myofibril that have been identified to date include myosin, actin, tropomyosin, an α-actinin-like compound, β-actinin, and troponin. These occur in the following percentage of the total protein: 55, 20, 7, 10, 2, and 2, respectively.

The dense or Λ bands (Plate 14; Fig. 11–3) consist of thick filaments (approximately 100 Å in diameter and 1.5 microns in length) which primarily contain myosin (Huxley, 1953, 1957). The cross bridges between the thick and thin filaments are also a part of the myosin molecule. Conversely, the light

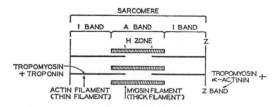

Fig. 11–3. Diagram of a sarcomere model indicating probable location of various myofibrillar proteins.

or I bands consist of thin filaments (approximately 80 Å in diameter and 1 or 2 microns in length, depending on whether filaments on both sides of the Z band are considered) which contain actin (Huxley & Hanson, 1957). Tropomyosin B (Bailey type) is present in the Z band and also in the thin I filament. The complex of tropomyosin and troponin (Ebashi & Kodama, 1966) which had earlier been termed "native tropomyosin" exists in the I bands. Alpha-actinin and β-actinin exist in the myofibril. Briskey and his associates (1967b) suggested that on the basis of cross linking, α-actinin probably exists in the Z band. Masaki and his co-investigators (1967), through fluorescent antibody experimentation, have shown that α-actinin is at least partially located in the Z band.

1. Myosin

The protein myosin represents a significant part of the protein of the myofibril (*i.e.*, 50 percent). The myosin molecule is an elongated structure (Fig. 11–4), with a length of about 1600 Å and a diameter of 15 to 40 Å. The molecule consists of (1) a rod shaped, 2-stranded, helical coil 15 Å in diameter and 12 to 1300 Å in length and (2) globular heads 40 Å in diameter and 250 to 350 Å in length. These workers envision the heads as consisting of polypeptide chains, each folded in tertiary structure to give the head region the characteristics of an elongated globular protein. The globular heads may contain two polypeptide chains. These globular heads possess the ability to split ATP, and also, to combine with actin, two properties which are of great physiological consequence (Driezen *et al.*, 1967).

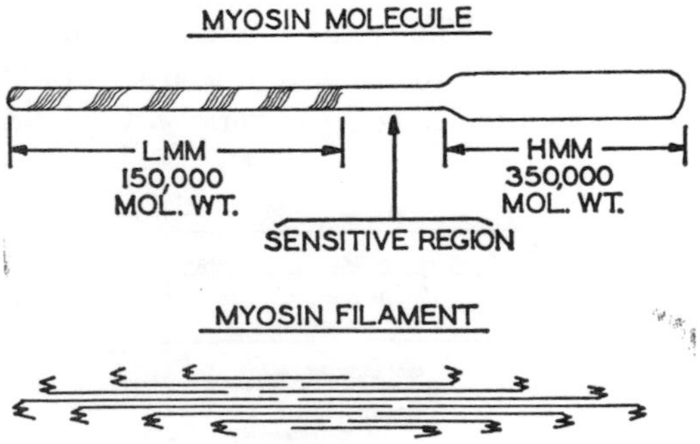

Proteolytic digestion of myosin with trypsin has been shown to split the globular heads from the rodlike tails. These two fragments have been called meromyosins. The head is called "heavy meromyosin" (HMM) and has a molecular weight of 350,000 and the rodlike tail is called "light meromyosin" (LMM) and has a molecular weight of 150,000. The ATPase of myosin is known to vary with muscle, species, and stage of postnatal growth.

2. Actin

The properties of actin which are of great biological concern are its capacity for polymerization and its interaction with myosin. Globular actin (G-actin) has a molecular weight of about 60,000 and recent hydrodynamic measurements indicate that its shape approximates a sphere. G-actin contains 1 molecule of ATP which is available to enzymes and which readily exchanges with the ATP of the medium. The transformation of G- to F-actin appears to take place by the polymerization of a globular molecule into a linear aggregate which may have a weight of many million (Fig. 11–5; Plate 14) (Mommaerts, 1966b). The polymer consists of a double helix, and each helix consists of globular monomers of actin about 55 Å in diameter. There are approximately 13 to 15 globular subunits per turn of the helix. Recent work shows that the strands are in a right-handed helix.

MYOSIN MOLECULE

LMM
150,000
MOL. WT.

HMM
350,000
MOL. WT.

SENSITIVE REGION

MYOSIN FILAMENT

Fig. 11–4. Sketch of myosin molecule and portion of thick filament. (*Briskey, 1967. Proc. Meat Ind. Conf., Chicago.*)

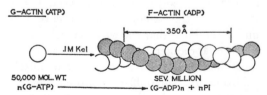

Fig. 11–5. Characteristics of G- and F-actin. (*Briskey, 1967. Proc. Meat Ind. Conf., Chicago.*)

3. Tropomyosin

Tropomyosin (Bailey, 1948) is characterized by the formation of highly polymerized viscous solutions at a neutral pH and in the absence of salts. The difference in the polymerization processes of actin and tropomyosin lies in the fact that the polymerization of actin is initiated by the addition of neutral salts and leads to the formation of a thixotropic structure, while the reverse is true for tropomyosin. Although more tropomyosin has been found in young animals than in older ones, little is known about the species and muscle variations of this protein. However, some workers found no differences between cardiac and skeletal tropomyosin.

4. Actomyosin

Some literature still refers to natural actomyosin as myosin B. However, the terms "natural actomyosin" and "synthetic actomyosin" (reconstituted from pure myosin and pure actin) seem to be much more appropriate in light of present day knowledge of these proteins. Natural actomyosin probably contains myosin A, actin, tropomyosin, troponin, α-actinin, and β-actinin. Conceivably, it might also contain many proteins still unknown. Apart from myosin A, which represents about 50 percent of the protein of the myofibril, there is still uncertainty regarding the amounts of the other proteins in the fibril. Several aspects of the interaction of actin and myosin, the exact nature of which is still unknown (Szent-Gyorgyi, 1966; Mommaerts, 1966a), are of fundamental importance to the function of muscle. First of all, the interaction between actin and myosin results in a viscosity change when the proteins are in solution. In the myofibril, this activity has been interpreted in terms of the formation of cross bridges between actin and myosin filaments. In the presence of a high level of ATP, actomyosin can be dissociated into actin and myosin (Szent-Gyorgyi, 1951).

5. Ca^{++}-sensitizing Protein (Tropomyosin-troponin)

When it is recombined with myosin, actin gives rise to either "relaxing" or "nonrelaxing" actomyosins. A "relaxing" actomyosin is inhibited by microsomes or the chemical chelation of Ca^{++} and is accelerated by the addition of Ca^{++}. The preparation of a Ca^{++}-sensitive actin has been described. Further, this actin has been shown to be contaminated with another protein or protein complex. This protein complex is thought to be present in natural actomyosin. For purposes of this chapter, the term for the Ca^{++}-sensitizing "regulatory proteins" will be accepted as "native tropomyosin" or a complex of tropomyosin and troponin. Fluorescent and fluorescent antibody techniques have been used to localize "native tropomyosin" along the entire length of the thin filaments.

6. β-Actinin

F-actin isolated directly from myofibrils without depolymerization or acetone treatment does not form large aggregates or undergo network formation in solution (Maruyama, 1965). This F-actin has a particle length of about 1 micron compared to ordinary Straub type of actin with a particle length of 3 to 4 microns. The presence of a protein factor (β-actinin) in muscle which inhibits the network formation of F-actin and serves to regulate the length of F-actin filaments (1 to 2 microns) *in vivo* is an extremely important discovery.

7. α-Actinin

Alpha-actinin is not a required factor in the contraction process (Briskey *et al.*, 1967a,b; Seraydarian *et al.*, 1967, 1968) although it does influence gelation and superprecipitation. Alpha-actinin (6-S component) is partially located in the Z band and partially (10-S component) in the M line.

8. Changes in Myofibrillar Proteins

The alterations which occur in certain properties of myosin during the growth and development of muscle have been delineated. Insufficient evidence exists about whether changes occur in actin, tropomyosin, α-actinin, and β-actinin. When considered collectively as myofibrillar proteins, it can be stated that they increase with hypertrophy, exercise, and during the early stages of growth (Helander, 1966). The myofibrillar protein in the muscle of the newborn bovine increases from 12.9 percent to 14.1 percent at 6 months and thereafter remains relatively constant (Bendall & Voyle, 1967). The concentration of fibrillar protein increases as development progresses and, at most stages of development, is higher than that of the sarcoplasmic protein (Dickerson & Widdowson, 1960). Changes in sarcoplasmic and myofibrillar proteins occur independently of each other, however. In the fetal period the muscles are high in connective tissue and low in sarcoplasmic and myofibrillar protein. Soon after birth both the sarcoplasmic and myofibrillar protein contents, especially the latter, rise. There is a small decrease in myofibrillar protein in old age.

From a structural point of view, the number of myofilaments increases with hypertrophy. If decreased activity sets in very rapidly and is severe, muscle atrophy ensues and selectively affects the myofilaments. If the reduction in activity is more gradual, however, there is no pronounced atrophy; sarcoplasm simply replaces myofilaments.

Postfetal growth occurs largely by addition of sarcoblasts to the ends of the myofibers (Slautterback, 1966). Little evidence exists concerning how actin and myosin filaments become arranged in position, but the following sequence probably occurs. Myosin and thin filaments are scattered randomly in the cytoplasm. The filaments become larger and longer and settle into position in the incipient A band region. In some muscles the Z band does not appear to take up its position before the A filaments have become oriented. After the myosin filaments become oriented, the actin filaments with their Z bands fall into place.

C. Stroma Protein

Connective tissue may be classified into three morphologically distinct classes of material: the cells, the ground substance, and the fibrous proteins. While connective tissue has very few cells, these possess a great variety. Among the cells found in connective tissue are the fibroblasts or the embryonic connective tissue cells responsible for the secretion of the precursors of the fibrous proteins, most cells which have an anticoagulant function, lymphoid wandering cells and macrophages (both of which have infection-resistance functions) and adipose cells.

Ground substance is the amorphous, structureless, jelly-like, homogeneous fluid in which connective tissue cells and fibrous protein fibers are embedded. It consists of complexes called mucoproteins composed of protein and carbohydrates of the mucopolysaccharide type. The mucopolysaccharides found are hyaluronic acid, chondroitin sulfates A, B, and C, keratosulfate, chondroitin, heparitin sulfate, and heparin. The nature of the linkage between protein and carbohydrate is not clear. An ester linkage is probably responsible for the strong binding. Ground substance probably acts to control diffusion of nutrients to the cell and to lubricate. The fibrous proteins of connective tissue in muscle have received most research attention. The three proteins are collagen, elastin, and reticulin.

1. Collagen

Collagen is a major constituent of the reticular sheath of muscle fibers and of septa between muscles. It is found in loose connective tissue adhering to the sarcolemma of muscle fibers as endomysium, separating bundles of muscle fibers as perimysium, and surrounding and connecting entire muscles as epimysium. Most obvious are the tendons which, by connecting muscle to bone, transmit the contracting force of a muscle to the movement of a part of the body. Macroscopically, collagen consists of white fibers 1 to 12 microns in diameter which frequently branch from one fiber to another. These fibers consist of smaller fibrils, 0.3 to 0.5 microns in diameter.

The basic feature of all collagens is the triple-chain backbone structure. The individual fibrils are composed of aggregates of collagen molecules in which the axes are parallel but the ends overlap in a regular manner to produce typical banded fibrils (Piez, 1966). All collagens have a high glycine content which represents close to one third of the total residues. Also present are two amino acids which are essentially unique to collagen—hydroxyproline and hydroxylysine. Six to seven residues of hydroxylysine exist in collagen per 1000 total amino acid residues. Collagen is the only protein known to contain large amounts of hydroxyproline. Over 70 percent of the amino acids in collagen are nonpolar in nature but a high dibasic acid and arginine content prevent it from being as nonpolar as elastin.

The molecule is triple-stranded, with each strand in a modified polyproline helix which has a repeat distance of 3 Å and three amino acids per repeat. The three polypeptide, left-handed helices are then wound together in a ropelike fashion to produce a right-handed super triple helix. This structure, with an approximate molecular weight of 300,000, is referred to as tropocollagen. The polypeptide backbones are tightly packed so that every third position must be a glycine residue with no room for a side chain.

Although the amino acid composition of mammalian collagen varies very little over a wide spectrum of species, wide differences are noted between the amino acid composition of mammalian, fish, and invertebrate collagens. The only exception appears to be that the glycine content remains relatively constant at one third of the total residues for all collagens.

The total content of collagen decreases with advancing age. The amount of covalent cross-linking is related to function of tissue and age of animal in terms of meat tenderness (Goll *et al.*, 1964a; Herring *et al.*, 1967). Mature collagens have more frequent and stronger cross-linkages between and among the tropocollagen molecules than do young collagens. Bovine collagen decreases in solubility after the animal exceeds 2 years of age. Five forms of collagen have been identified, which differ in solubility characteristics (Carmichael & Lawrie, 1967). Those forms which are extractable in neutral salt and dilute acid increase swiftly during the gestation period but fall to low levels between birth and 1 to 2 years of age. After birth there is a rapid rise in the concentrations of the more insoluble forms of collagen. These forms predominate by 1 to 2 years of age. The relative time periods for these changes were compatible with a transformation from the easily extractable forms into the more insoluble ones with increasing age.

2. Elastin

Elastin is normally present in animal connective tissue in rather small amounts. This protein may, however, make up 70 to 80 percent of the protein in the walls of arteries and in the elastic ligaments (*ligamentum nuchae*) which support the heads of large ruminants. In cattle there are considerable amounts of elastin in the muscles of the hind limbs, but in other muscles the amount may be quite small and associated mainly with the walls of blood vessels. Ordinary loose connective tissue contains only a fraction as much elastin as collagen. Elastin fibers are smaller than collagen fibers and branch freely to form networks.

The amino acid composition of elastin is unusual for an animal protein in that the content of hydrophilic side chains is very low. Ninety percent of the amino acid residues of elastin are nonpolar in nature. Elastin resembles collagen because its most abundant amino acid is glycine and it contains nearly 13 percent proline by weight. However, the hydroxyproline content of elastin is only 1.6 percent compared to 13.3 percent for collagen. Partridge and associates (1963) found a very high yield of two previously unknown amino acids in elastin. They called these substances desmosine and isodesmosine. Both amino acids are tetra-carboxylic-tetra-amino acids. On the basis of analysis of the products of enzyme hydrolysis, it is clear that both desmosines link together at least two independent chains in the elastin network. For more detailed information on elastin structures the reader is referred to another summary report (Partridge, 1966).

3. Reticulin

Although reticulin is distributed quite widely, it is not present in large amounts in any one tissue. More research is required to ascertain its importance in muscle tissue, although on the basis of its nature it could be exceedingly important to muscle structure and function as well as to meat tenderness.

D. Other Constituents

1. Glycogen

The principal carbohydrate found in skeletal muscle is glycogen. Its content is high (4 to 8 percent) in newborn pigs but generally represents about 1 percent (fresh weight) of mature mammalian skeletal muscle. The Hampshire pig, however, has close to 2 percent glycogen. Little else is known of other breed or species differences of this important constituent. The molecular weight of glycogen is about 3 to 5×10^6 gm/mole. Part of the variation in molecular weight is due to the method of preparation. Glycogen consists of D-glucopyranose with ether linkages. The principal linkage is α-1-4, with branching every 12 or 18 glucose units. Branching occurs through α-1-6 linkages. White muscles have higher initial levels of glycogen than red muscles; however, glycogen is broken down more rapidly in white than in red muscles. Both protein-bound and nonprotein-bound glycogen can

be metabolized. In the presence of oxygen, glycogen is metabolized to carbon dioxide and water, while in the absence of oxygen, it is metabolized to lactate.

2. Mucopolysaccharides

Mucopolysaccharides, which form the ground substance of connective tissue, make up the major portion of the extracellular carbohydrate. Chondroitin sulfates A, B, and C and hyaluronic acid make up the main part of the mucopolysaccharides in skeletal muscle. Marked changes occur in the mucopolysaccharides with increasing age, especially in arthritis and other connective tissue diseases. The amount of mucopolysaccharides in tissue decreases with increasing age. Some workers use the hexosamine/collagen ratio as a measure of physiological age. The growth hormone, when administered under controlled conditions, appears to promote a more rapid synthesis of mucopolysaccharides and gives the connective tissue the appearance of originating from an animal younger than its actual age.

3. Myoglobin

Although myoglobin was briefly discussed in regard to sarcoplasmic proteins, its importance in metabolism, association with muscle color, and extreme variability among species will be discussed. Its role in muscle is to acquire oxygen from circulating hemoglobin and to store it in the muscle for use in oxidative phosphorylation. Muscle needs an oxygen storehouse since there is not any steady state demand for O_2 in the muscle. Other species such as insects have a strong continuous demand for oxygen and possess no myoglobin but have very high concentrations of the cytochromes. Myoglobin increases markedly with advancing age and with exercise or activity. The increase in myoglobin concentration after maturity has been interpreted to mean that there is increasing difficulty in obtaining a sufficient oxygen supply in the older animals (Carmichael & Lawrie, 1967). The increase in fat deposition with age may also interfere with blood distribution and thereby create a need for greater quantities of myoglobin.

Species vary widely in myoglobin concentration. Bovine or horse muscle possesses 1 to 5 mg of myoglobin /gm fresh weight of muscle; porcine muscle, 1 to 3 mg/gm; chicken breast, 0.05 mg/gm; chicken leg, 0.5 mg/gm; and old bovine muscle, about 16 to 20 mg/gm. A positive correlation exists between myoglobin content and oxidative enzyme activity in several muscles of species varying widely in size and activity. Myoglobin concentrations are positively associated with the aerobic synthesis of ATP and negatively correlated with anaerobic glycolysis and energy-rich phosphate stores. As fibers differentiate or are altered by activity, there is an accompanying change in myoglobin concentration.

4. WATER AND INORGANIC CONSTITUENTS

During development, the percentage of water in skeletal muscle decreases from 91 percent in the early stages of fetal life to 74 percent in the adult animal (Dickerson & Widdowson, 1960). However, development is associated with an increase in the proportion of intracellular water. This change is accompanied by a fall in the concentrations of the extracellular sodium and chloride and a rise in intracellular potassium and phosphorus. Compared with the adults, the amount of calcium is also high, but magnesium low in the early fetal stages.

5. LIPIDS

Natural fats are composed of glycerol esters of the straight-chain carboxylic acids usually having an even number of carbon atoms. Triglycerides (glycerol with three fatty acid molecules) make up most of the lipid in muscle tissue, although small amounts of mono- and diglycerides and some free glycerol and fatty acid may exist in this tissue. Most of the triglycerides are mixed, indicating that the fatty acids are not identical.

The fatty acids found in the lipid of muscles differ in the length of the carbon chain and in the type of bonding between the carbon atoms. The fatty acids vary widely among animal species and are also influenced by diet, growth, and environment. The fat in muscle tissue contains large quantities of stearic, palmitic, and oleic acids. Adipose tissue has a number of enzyme systems involved in the deposition, synthesis, and mobilization of fat. It also has a comparatively dense capillary network and is innervated by the sympathetic nerve fibers.

The mechanisms responsible for the accumulation of lipid in muscle are not clear. Changes in the blood vessels and surrounding connective tissues are major factors associated with the lipomorphosis of human muscle with aging. Hormonal differences between animals have been implicated in lipomorphosis of muscle. Castration results in a greater accumulation of muscle lipid than occurs in normal, uncastrated males. The greater quantity of intramuscular lipid in the muscle of young uncastrated male porcines than in old ones has been partially attributed to the smaller percentage of fibers positive for enzymes (esterase and B-hydroxybutyric dehydrogenase) necessary for removing lipid from muscle. Fat cell size also increases with increases in total chemical fat in the muscle. Fat cells accumulate and grow, however, in close proximity to portions of the circulatory system (Moody & Cassens, 1968). In fact, some workers (Carmichael & Lawrie, 1967) have put forth the hypothesis that the increased deposition of fat in proximity with the circulatory system may interfere with blood distribution and thereby contribute to the need for increasing levels of myoglobin with advances in maturity. The increase in chemical fat which occurs in a carcass during growth and fattening involves an increase in the fat content of both fatty and muscular tissues. This increase, however, does not run parallel in the two tissues.

As the percentage of fatty tissue rises, a higher proportion of the extra fat is deposited in the fatty tissue and a smaller portion in the muscular tissue. There are major differences among breeds and species in the deposition of fat in muscular tissue. Additionally, muscles within a carcass vary widely in fat content. The percentage of intramuscular fat also shows a general increase with age. The effect of intramuscular fat deposition upon the growth coefficients of several large muscles has been studied. Coefficients of intramuscular fat deposition are almost twice as large as the muscle growth coefficients. It is inferred from this study that fat has an earlier relative maturity in the *gluteus medius*, *semimembranosus*, and *quadriceps* muscles than in the *biceps femoris* and *longissimus* muscles. Wide variations are found, however, in depositions of intramuscular fat within the length of the *longissimus*.

Generally, a decrease in activity, with other factors held constant, results in increasing quantities of lipid deposition in the muscle tissue. Further, intramuscular fat deposition is rather highly heritable. Some investigators (Suess, 1968) have found wide differences in intramuscular fat in progeny of different sires. The coarseness of intramuscular fat also varies according to the distribution and size of the blood vessels.

Muscular lipomorphosis is more prevalent in diseases in which the activity of muscles is abnormally restricted such as osteoarthritis and rheumatoid arthritis. Lipomorphosis induced by casting, which causes the muscle to atrophy, results in increasing quantities of fat being collected in the muscle tissue. Histologically, most of the fat seems to be deposited perivascularly and interstitially.

Deposition of lipid in adipose tissue is the result of two processes: incorporation of preformed lipid from the circulatory system, and *de novo* synthesis of lipid from precursors directly in the adipose cell itself. Adipose cells are the main site of fatty acid synthesis. The role of several of the adaptive enzyme systems has been described. These enzymes include: (a) glucokinase, (b) glucose-6-phosphate dehydrogenase, (c) citrate cleav-age enzyme, (d) acetyl CoA carboxylase, (e) fatty acid synthetase, and (f) the oxidative desaturase system. An essential requirement (Allen, 1968) for triglyceride synthesis is alpha glycerol phosphate, which is formed from glycerol and ATP by glycerokinase. Since the adipose tissues that have been studied in experimental animals indicate that there is little, if any, glycerokinase activity, all of the alpha glycerol phosphate must originate from the metabolism of glucose in other tissues.

Adipose tissue contains at least two lipolytic enzyme systems. The first, lipoprotein lipase, is located at or near the capillary endothelium and is involved in the deposition of preformed lipid in the adipose cell. Its function is to hydrolyze the free fatty acids (FFA) from the triglyceride protein complex, which may be either lipoproteins or chylomicrons, and thereby enable the FFA to enter the adipose cell. The activity of lipoprotein lipase is elevated by insulin or consumption of a fatty meal, and is a controlling factor in the deposition of fat at a particular site (Allen, 1968). After entering the cell, the FFA are esterified with alpha glycerol phosphate and deposited as triglycerides. During periods of energy need the adipose tissue triglyceride is broken down into glycerol and FFA, which are mobilized into the blood stream for transport to the tissues where energy is needed. Pigs that have not been fed or those placed on a severe dietary restriction for extended periods of time have a reduction in intramuscular fat. Another aspect of muscle which may influence its rate of lipid accumulation is its ability to oxidize fatty acids as a source of energy. Muscles can readily utilize variable amounts of free fatty acids.

III. METABOLISM

A. Integration

The term "metabolism" refers to the reactions in either the absorption and synthesis or breakdown and excretion of tissue components. Energy metabolism in skeletal muscle primarily involves these groups of fuels: fatty acids, glucose, glycogen, lactate and pyruvate.

Carbohydrate Metabolism: In muscle, carbohydrate metabolism involves: (a) reactions leading to glucose uptake, transport of glucose across the membrane, phosphorylation in the cell, and entrance into the hexosemonophosphate pool; (b) glycolysis leading to the production of lactate; and (c) the intramitochondrial formation of acetyl CoA from pyruvate, and the oxidation of the latter to carbon dioxide and water. Fatty acid and carbohydrate metabolism interact at this point because the oxidation of fatty acids and ketone bodies also yields acetyl CoA.

Deprivation of Oxygen: The effect of anaerobiosis on carbohydrate metabolism appears to be exerted primarily through relative concentrations of ATP and its breakdown products, ADP, inorganic phosphate, and AMP. This is due to the fact that the oxidation of glucose leads to the production of 20 times as much ATP by the citric acid cycle than does anaerobic metabolism through glycolysis.

Glucose Uptake: Glucose uptake and its glycolysis to pyruvate and lactate are accelerated by oxygen deprivation. Lack of oxygen increases the permeability of muscle to sugars by accelerating their transport. Much of this acceleration is probably mediated by the relative concentrations of ATP, AMP, and Pi.

ATP Synthesis: In muscle, ATP is synthesized by glycolysis and respiration or by breakdown of creatine phosphate to creatine:

$$ATP \rightarrow ADP + Pi \text{ (ATPase)}$$
$$2\ ADP \rightleftarrows ATP + AMP \text{ (myokinase)}$$
$$PC + ADP \rightleftarrows ATP + \text{creatine}$$
$$(Pi = \text{inorganic phosphate})$$

The overall glycolytic cycle, which operates in the absence of oxygen, splits glucose into two moles of lactic acid with a net synthesis of 2 molecules of ATP by substrate level phosphorylation. During aerobic respiration, these same 2 molecules of lactic acid would cause the synthesis of 36 molecules of ATP, since they continue to be involved in the TCA cycle and electron transport and oxidative phosphorylation. When muscle is deprived of oxygen, respiration ceases, and once the reserves of creatine phosphate have been used, synthesis of ATP is almost solely dependent upon glycolysis.

B. Adaptation and Function

Adaptation: In general, as muscle grows and develops with maturity or as it hypertrophies with activity, fiber differentiation is accelerated or altered. More mature fibers usually have more active phosphorylase and phosphofructokinase and have higher concentrations of ATP and phosphorylcreatine (PC). These fibers are more highly geared for anaerobic metabolism.

Glycogen in the skeletal muscle is metabolized readily during performance of work. Glycogen content of the skeletal muscle of man does not vary during the day or show diurnal variations. Even after 5 days of fasting, the glycogen content falls only 50 percent. It may be concluded that normal work does not require a high degree of muscle glycogen utilization. However, in some animals such as the pig, muscle glycogen is utilized quite extensively over a 48-hour fast.

Muscle contraction can take place even when the glycolytic chain is completely inhibited by iodoacetate. The necessary energy is derived from phosphorylcreatine; however, ATP is the initial source of energy in muscle contraction. Muscle contraction produces an increase in phosphorylase A activity. Thus, glycolysis is stimulated and more ATP will be formed. Phosphofructokinase is activated by AMP and ADP but inhibited by ATP. With a rising work load, it may be reasonable for an increasing part of ATP resynthesis to take place by means of myokinase. This might bring about some increase in the formation of AMP.

Sarcoplasmic proteins, primarily glycolytic enzymes, increase with growth or inactivity. Likewise, myosin adenosine triphosphatase (ATPase) increases with muscle growth and animal aging. Myosin ATPase activity is associated with speed of contraction (Barany, 1967).

Lipid depots also can be mobilized for energy in the muscle cell. Intramuscular fat deposition is altered by changing nutritional planes from high to medium without altering muscle growth. Merely restricting the nutritional level throughout growth is not as effective in regulating intramuscular fat deposition (Suess, 1968). Feeding frequency may have an influence on lipid deposition and subsequent metabolism in muscle (Gordon *et al.*, 1963). Opposite effects occur in human and pig experiments (Allen *et al.*, 1963). Single feeding reduces lipid in the muscle of the pig while multiple feeding is more effective in reducing lipid in the human.

Function: Some of the events surrounding muscle contraction may be described as follows: A nerve impulse to the muscle cell causes the external membrane to depolarize. This wave of depolarization goes along the transverse tubules to the interior of the cell. The influx of Na^+ possibly changes the permeability of the sarcoplasmic reticulum and causes the release of Ca^{++}. The free Ca^{++} concentration goes up to about 10^{-6} M. The Ca^{++} are then bound to troponin which has been held at the actin site by tropomyosin, causing a conformational change in the troponin molecule. This conformational change now permits the ATPase site on the myosin head, in proximity with a certain part of the actin monomer, to be activated by Mg^{++} and to split ATP. As ATP is split, the actin-combining site on myosin and the myosin-combining site on actin come closer together, generating a movement. These cross bridges between the thick and thin filaments generate a relative sliding movement of one set of filaments past the other. When the nerve impulse to the fiber ceases, the sodium-potassium pump re-establishes polarization, the sarcoplasmic reticulum reaccumulates Ca^{++} and maintains it at about 10^{-7} M or below 10^{-5} M. When Ca^{++} is pumped away the actomyosin ceases to split ATP. Upon cessation of the splitting, the ATP is reformed by the creatine phosphokinase system or by diffusion. In the presence of increased ATP, actin is then dissociated from myosin, the filaments slide, and the muscle relaxes.

IV. NECROBIOLOGY

When blood flow ceases, the metabolism quickly changes to an anaerobic nature. The oxygen retained with the myoglobin is used up by the metabolism associated with a few contractile movements. Red fibers, geared for aerobic metabolism during life, are greatly influenced by this shift to anaerobiosis. Nevertheless, these fibers are low in both glycolytic enzymes, particularly phosphorylase, and ATPase, namely myosin ATPase activity. Consequently, the postmortem metabolism is held somewhat in check by these inherent limitations. White fibers, on the other hand, are geared for anaerobic metabolism, and are not greatly influenced by anaerobiosis. Nevertheless, concurrent stimulation can activate the high ATPase. As phosphocreatine becomes depleted, a somewhat accelerated rate of glycolysis can result. Intermediate fibers, having both the machinery for aerobic and anaerobic metabolism, are greatly influenced by anaerobiosis. In this case, the fibers are dependent on oxygen during life, and when suddenly subjected to anaerobiosis, they have the high ATPase and phosphorylase activities which can result in very rapid rates of glycolysis. When ATP and adenosine diphosphate (ADP) levels are diminished, rigor mortis onset occurs. Rigor mortis is the loss of extensibility associated with at least a somewhat permanent interaction between the myosin and actin filaments (Fig. 11–6). The stiffness associated with the loss of extensibility lasts until there is some resolution of rigor mortis.

FIG. 11–6. Loss of extensibility with the onset of rigor mortis. (*Schmidt et al., 1968. J. Food Sci. 33:239.*)

The calcium uptake of the sarcoplasmic reticulum is low and diminishes rapidly in a quickly glycolyzing muscle and slowly in a slowly glycolyzing muscle (Greaser *et al.*, 1969a). In metabolism there seems to be some regulation at the phosphorylase site, some control of the rapid flux rate at the phosphofructokinase site, and some control

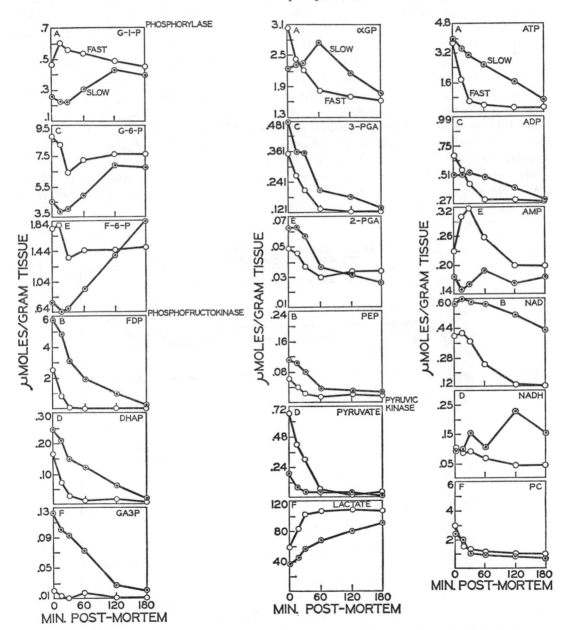

FIG. 11–7. Comparison of changes in glycolytic intermediates in "fast" and "slow-glycolyzing" muscles. (*Kastenschmidt et al., 1968. J. Food Sci. 33, 151.*)

at the pyruvic kinase site (Kastenschmidt *et al.*, 1968) (Fig. 11–7). As metabolism proceeds, the accumulation of lactic acid results in a lowering of pH. This will continue until the glycogen is completely depleted or until the enzyme phosphofructokinase is inactivated by the decreasing pH (Kastenschmidt *et al.*, 1968). This inactivation is thought to occur between the pH of 5.7 and 5.0, depending on the levels of ATP and fructose-diphosphate at the particular pH.

If the pH drops rapidly while the muscle temperature remains near *in vivo* body temperature, the myofibrillar proteins and the sarcoplasmic proteins are greatly reduced in solubility in high salt concentrations and water, respectively (Fig. 11–8). Simultaneously, the muscle loses its ability to bind its water and become pale, soft, and exudative. Acid rigor has developed in these muscles. Electron micrographs of these myofibrils

show a granularity of the A and I bands and an aggregation of protein in the Z band (Greaser *et al.*, 1969b).

If, however, glycogen is depleted before death, or if glycolysis is inhibited so that the pH will remain high, the muscle remains dark in color and holds its water very tenaciously. Alkaline rigor has developed in these latter muscles. They show virtually no change in the solubility of myofibrillar and sarcoplasmic proteins and exhibit a high degree of integrity in their ultrastructure.

Growth, development, nutrition, and handling, which influence any of the constituents mentioned, have a great effect on the necrobiology of the muscle tissue. The stress susceptibility of the animal is also highly associated with the nature of the necrobiology in the muscle tissue (Judge *et al.*, 1968; Lister *et al.*, 1969; Sair *et al.*, 1969).

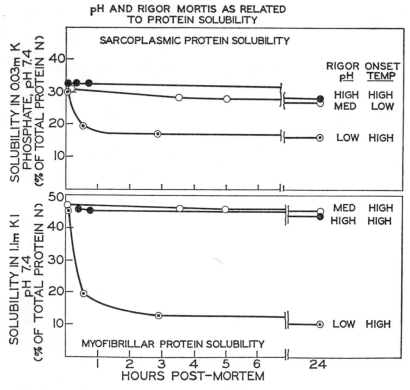

FIG. 11–8. Influence of pH and temperature on the solubility of sarcoplasmic and myofibrillar proteins.

REFERENCES

Allen, E. (1968). Fat deposition in the porcine animal. *Proc. Recip. Meat Conf.* Athens, Ga.

Allen, E., Cook, C. F., & Bray, R. W. (1963). A preliminary fat deposition study of pigs fed daily, once or at multiple intervals. J. Anim. Sci. *22*:825.

Bailey, K. (1948). Tropomyosin: a new asymmetric protein component of muscle. Biochem. J. *43*:271.

Barany, M. (1967). ATPase activity of myosin correlated with speed of muscle shortening. In: *The Contractile Process.* A. Stracher (Ed.), Boston, Little, Brown and Company.

Beecher, G. R. (1966). Biochemical Characteristics of Red and White Striated Muscles. Ph.D. Dissertation, University of Wisconsin, Madison.

Bendall, J. R. (1966). Muscle as a contractile machine. Chap. 2. In: *Physiology and Biochemistry of Muscle as a Food.* E. J. Briskey, R. G. Cassens, & J. C. Trautman (Eds.), pp. 7–16. Madison, University of Wisconsin Press.

Bendall, J. R. & Voyle, C. A. (1967). A study of the histological changes in the growing muscles of beef animals. J. Food Technol. *2*:259–283.

Briskey, E. J. (1964). Etiological status and associated studies of pale, soft, exudative porcine musculature. Advances Food Res. *13*:89–178.

Briskey, E. J., Kastenschmidt, L. L., Forrest, J. C., Beecher, R. G., Judge, M. D., Cassens, R. G., & Hoekstra, W. G. (1966). Biochemical aspects of post-mortem changes in porcine muscle. J. Agri. Food Chem. *14*:3.

Briskey, E. J., Seraydarian, K., & Mommaerts, W. F. H. M. (1967a). The modification of actomysin by α-actinin. II. The effect of α-actinin upon contractibility. Biochim. Biophys. Acta *133*:412–423.

Briskey, E. J., Seraydarian, K., & Mommaerts, W. F. H. M. (1967b). The modification of actomyosin by α-actinin. II. The interaction between α-actinin and actin. Biochim. Biophys. Acta *133*:424–434.

Brooke, M. H. (1966). The histological reaction of muscle to disease. Chap. 8. In: *Physiology and Biochemistry of Muscle as a Food.* E. J. Briskey, R. G. Cassens, & J. C. Trautman, (Eds.) pp. 113–135. Madison, University of Wisconsin Press.

Butterfield, R. M. & Berg, R. T. (1966). A classification of bovine muscles based on their relative growth patterns. Vet. Sci. *7*:326–332.

Carmichael, D. J. & Lawrie, R. A. (1967). Bovine collagen. I. Changes in collagen solubility with animal age. J. Food Technol. *2*:299–311.

Carrow, R. E., Brown, R. E., & Van Huss, W. D. (1967). Fiber sizes and capillary to fiber ratios in skeletal muscle of exercised rats. Anat. Rec. *159*:33.

Cooper, C. C., Cassens, R. G., & Briskey, E. J. (1969). Capillary distribution and liver characteristics in skeletal muscle of stress-susceptible animals. J. Food Sci. (In Press.)

Dickerson, J. W. T. & Widdowson, E. M. (1960). Chemical changes in skeletal muscle during development. Biochem. J. *74*:247.

Driezen, P., Gershman, L. C., Trotta, P. P., & Stracher, A. (1967). Myosin: Subunits and their interactions. In: *Contractile Process.* A. Stracher (Ed.), Boston, Little, Brown & Company.

Dubowitz, V. (1963). Enzymatic maturation of skeletal muscle. Nature (Lond.) *197*:1215–1218.

Ebashi, S., & Kodama, A. (1966). Interaction of troponin with F-actin in the presence of tropomyosin. J. Biochem. *59*:4, 425.

Fawcett, D. W. & Selby, C. C. (1958). Observations on the fine structure of the turtle atrium. J. Biophys. Biochem. Cytol. *4*:63.

Goldspink, G. (1962). Biochemical and physiological changes associated with the postnatal development of the biceps brachii. Comp. Biochem. Physiol. *7*:157–164.

Goll, D. E., Hoekstra, W. G., & Bray, R. W. (1964a). Age-associated changes in bovine muscle connective tissue. I. Rate of hydrolysis by collagenase. J. Food Sci. *29*:608–614.

Goll, D. E., Hoekstra, W. G., & Bray, R. W. (1964b). Age-associated changes in bovine muscle connective tissue. II. Exposure to increasing temperature. J. Food Sci. *29*:5, 615–621.

Gordon, E. S., Goldberg, M., & Chosy, G. J. (1963). A new concept in the treatment of obesity. J.A.M.A. *186*:50.

Greaser, M. L., Cassens, R. G., Briskey, E. J., & Hoekstra, W. H. (1969a). Post-mortem changes in subcellular fractions from normal and pale, soft, exudative porcine muscle. I. Calcium accumulation and adenosine triphosphatase activities. J. Food Sci. (In Press.)

Greaser, M. L., Cassens, R. G., Briskey, E. J., & Hoekstra, W. H. (1969b). Post-mortem changes in subcellular fractions from normal and pale, soft, exudative porcine muscle. 2. Electron microscopy. J. Food Sci. (In Press.)

Hammond, J. (1961). Growth in size and body proportions in farm animals. Chap. 16. In: *Growth in Living Systems.* M. X. Zarrow, (Ed.) pp. 321–335. New York, Basic Books, Inc.

Harmon, J. W. (1964). The ultrastructure of the muscle cell. Chap. 3. In: *Disorders of Voluntary Muscle.* J. N. Walton (Ed.), Boston, Little, Brown & Company.

Helander, E. (1957). On quantitative muscle protein determination. Acta Physiol. Scand. *41*:Suppl. 141.

Helander, E. (1966). General consideration of muscle development. In: *The Physiology and Biochemistry of Muscle as a Food.* E. J. Briskey, R. G. Cassens, & J. C. Trautman (Eds.), Madison, University of Wisconsin Press.

Herring, H. K., Cassens, R. G., & Briskey, E. J. (1967). Factors affecting collagen solubility in bovine muscles. J. Food Sci. *32*:534–538.

Hess, A. (1967). The structure of slow and fast extrafusal muscle fibers in the extraocular muscles and their nerve endings in guinea pigs. J. Cell. Comp. Physiol. *57*:63–76.

Hill, F. (1966). The solubility of intramuscular collagen in meat animals of various ages. J. Food Sci. *31*:161.

Huxley, H. E. (1953). Electron microscope studies of the organization of the filaments in striated muscle. Biochim. Biophys. Acta *12*:387.

Huxley, H. E. (1957). The double array of filaments in cross-striated muscle. J. Biophys. Biochem. Cytol. *3*:631.

Huxley, H. E. & Hanson, J. (1957). Quantitative studies on the structure of cross-striated myofibrils. I. Investigations by interference microscopy. Biochim. Biophys. Acta *23*:229.

Judge, M. D., Briskey, E. J., Cassens, R. G., Forrest, J. C., & Meyer, R. K. (1968). Adrenal and thyroid function in stress-susceptible pigs (*Sus domesticus*). Amer. J. Physiol. *214*:146.

Kastenschmidt, L. L., Hoekstra, W. G., & Briskey, E. J. (1968). Glycolytic intermediates and cofactors in "fast-" and "slow-glycolyzing" muscles of the pig. J. Food Sci. *33*:151–158.

Lister, D., Sair, R. A., Will, J. A., Schmidt, G. R., Cassens, R. G., Hoekstra, W. G., & Briskey, E. J. (1969). Metabolism of striated muscle of "stress-susceptible" pigs breathing oxygen or nitrogen. Amer. J. Physiol. (In Press.)

Maruyama, K. (1965). A new protein-factor inhibiting network formation of F-actin in solution. Biochim. Biophys. Acta *94*:208.

Masaki, T., Endo, M., & Ebashi, S. (1967). Localization of 6S component of α-actinin at z-band. J. Biochem. *62*:5, 630.

Mommaerts, W. F. H. M. (1966a). Molecular alterations in myofibrillar proteins. Chap. 18. In: *Physiology and Biochemistry of Muscle as a Food.* E. J. Briskey, R. G. Cassens, & J. C. Trautman (Eds.), Madison, University of Wisconsin Press.

Mommaerts, W. F. H. M. (1966b). *Laboratory Practice.* London, Partridge Printers, Ltd.

Moody, W. G. and Cassens, R. G. (1968). A quantitative and morphological study of bovine longissimus fat cells. J. Food Sci. *33*:1, 47–52.

Partridge, S. M. (1966). Elastin. Chap. 22. In: *Physiology and Biochemistry of Muscle as a Food.* E. J. Briskey, R. G. Cassens, & J. C. Trautman (Eds.), Madison, University of Wisconsin Press.

Partridge, S. M., Elsden, D. F., & Thomas, F. (1963).

Constitution of the cross linkages in elastin. Nature *197*:1297.

Perry, S. V. (1965). Muscle proteins in contraction. In: *Muscle.* W. M. Paul, E. E. Daniel, C. M. Kay, & G. Monckton (Eds.), New York, Pergamon Press.

Piez, K. (1966). Collagen. Chap. 21. In: *Physiology and Biochemistry of Muscle as a Food.* E. J. Briskey, R. G. Cassens, & J. C. Trautman (Eds.), Madison, University of Wisconsin Press.

Sair, R. A., Lister, D., Moody, W. G., Cassens, R. G., Hoekstra, W. G., & Briskey, E. J. (1969). Action of curare and magnesium on metabolism of striated muscle from stress-susceptible pigs. Amer. J. Physiol. (In Press.)

Seraydarian, K., Briskey, E. J., & Mommaerts, W. F. H. M. (1967). The modification of actomyosin by α-actinin. I. Survey of experimental conditions. Biochim. Biophys. Acta *133*:399–412.

Seraydarian, K., Briskey, E. J., & Mommaerts, W. F. H. M. (1968). The modification of actomyosin by α-actinin. IV. The role of sulfhydryl groups. Biochim. Biophys. Acta *162*:3, 424.

Slautterback, D. B. (1966). The ultrastructure of cardiac and skeletal muscle. In: *Physiology and Biochemistry of Muscle as a Food.* E. J. Briskey, R. G. Cassens, & J. C. Trautman (Eds.), p. 39–69, Madison, University of Wisconsin Press.

Smith, R. D. & Giovacchini, R. P. (1956). The vascularity of some red and white muscles of the rabbit. Acta Anat. *28*:342.

Suess, G. G. (1968). A Study of Certain Factors Influencing Beef Carcass Composition. Ph.D. Dissertation. University of Wisconsin, Madison.

Szent-Gyorgyi (1951). *Chemistry of Muscular Contraction.* New York, Academic Press.

Szent-Gyorgyi, A. G. (1966). Nature of actin-myosin complex and contraction. In: *Physiology and Biochemistry of Muscle as a Food.* E. J. Briskey, R. G. Cassens, & J. C. Trautman (Eds.), Madison, University of Wisconsin Press.

Thompson, R. R. (1961). Species identification by starch gel zone electrophoresis of protein extracts. II. Meat and eggs. J.A.O.A.C. *44*:787.

Chapter 12

Bone

By S. E. Zobrisky

A COMPREHENSIVE treatment of the development, growth, and composition of the skeleton (or bone) is far too extensive to be covered in a single chapter. Similarly, the information concerning the hereditary and nutrient influences on bone growth and composition could fill chapters. A complete treatise covering the techniques of the past and present, and those on the horizon for use in research on bone could also encompass volumes. Perhaps presumptuously, it is hoped that enough of the basic facts will be discussed here to engender in students an appreciation and understanding of the sequence of events and complexities involved in the development, growth, and composition of bone, and also the inherent hormonal and nutrient influences on bone per se of mammals.

All mammals tend to follow a common bone developmental pattern. The basic mode of development is essentially the same; structural details are also comparable. The pre- and postnatal variations that do occur are modifications or adaptations specific to the species. Differences in the developmental pattern of specific tissues and organs between animal species are minimized when

comparison is made on an equivalence of age basis (Brody, 1945). Yet the comparative viewpoint is mandatory for the attainment of a broad understanding of bone development during the embryonic, the fetal, and the postnatal periods.

A critical study of bone tissue quickly reveals a complicated structural and functional system with untold biochemical and physiological interrelationships. Bone is recognized not as a passive or inert tissue, but as a very complex tissue. Some of the most obvious differences between bone and other tissues are that bone is a dense, hard, mineralized, cellular tissue. The three cellular components of bone are of one cell type that changes in morphology in direct relation to specific functional needs of the tissue. This reversible interchangeability in morphology and function can be referred to as *cell-modulation ability*. These particular cells include the *osteoblasts*, charged with bone formation; the *osteoclasts*, charged with bone resorption; and the *osteocytes*, charged with the maintenance of living bone tissue. Thus, differential growth of bone, implying the development of a nonreversible specialized structure and function, is doubtful owing to cellular modulation coincident with the continuous exchange of ions, particularly during active growth, following injury, or when ionic prerequisites for other bodily physiological necessities are in demand.

Bone is further characterized by branching *lacunae* (cavities), *canaliculi* (fine canals), a dense matrix of collagenous, fibrous bundles in ground substance (cement) encased with calcium phosphate complexes (Plate 15). This dynamic living tissue is capable of structural alterations to accommodate mechanical stresses and biological demands incurred by pressure, and by vascular, nerve, endocrine, and nutritional influences. Bones of the skeletal structure support the body, aid in locomotion, protect important organs, and regulate part of the ionic environment of the body by their ability to incorporate and exchange various ions rapidly. The mineral microcrystalline structure of bone is remarkably constant (Robinson, 1951), but bone is not histologically homogeneous (Bell *et al.*, 1941). Most bones are *compact* in some areas and *cancellous* in others. Genetic factors that influence skeletal growth are assumed to effect some metabolic or enzymatic process, or the hormonal equilibrium in the mammal. Information concerning these genetic factors is incomplete. Bones are self-differentiating entities in which histogenesis and morphogenesis are not dependent on mechanical or environmental factors. For example, a suitable piece of a primordial femur from a chick embryo placed in culture media will differentiate, develop, maintain the typical form, and start to ossify (McLean & Urist, 1968). This does not mean, however, that mechanical forces are unimportant to normal development of the form "perfections" necessary to a functioning skeleton.

I. MORPHOGENESIS OF THE SKELETON

The skeleton or, more comprehensively, the endoskeleton as opposed to the exoskeleton of arthropods, consists of the *axial* portion (vertebrae, ribs, sternum, and skull), the *appendicular* portion (shoulder, pelvic girdles, and limb bones), and the *splanchnic* or visceral skeleton portion (bones that develop in the viscera or soft organs, *e.g.*, as penis of the dog and os cordis of the ox).

A. Axial Skeleton

The primitive axial support of vertebrates is the *notochord* (Fig. 12–1). Also known as *chorda dorsalis*, it is the cord of mesodermal cells that serves as the primitive backbone and is later replaced by vertebrae. The axial skeleton differentiates from the embryonic connective tissue, *i.e.*, mesenchyme, most of which traces to the mesodermal somites. Somites of the embryonic *mesoblast* break down into a mass of diffuse cells. This aggregate of mesenchymal cells, or *sclerotome*, migrates toward and surrounds the notochord. From these masses of cells the vertebrae and ribs are destined to evolve.

Vertebral Column. The masses of sclerotomic mesenchyme proliferate, divide, and reunite into new combinations. These recombinations of the primitive sclerotomes form the primordia of the definitive vertebrae. Growth of the primordia proceeds toward the mesial line to form the vertebral body, dorsally to form the vertebral *neural arch*, and ventrolaterally to form the costal processes, the rib rudiments. The spinous processes of the vertebrae are formed by a prolongation and fusion dorsal to the neural canal (Fig. 12–1). In most animals the spinous processes of the more anterior thoracic vertebrae are very long. During the recombination of the sclerotomic masses, intervertebral fissures are also established between the forming vertebrae. Mesenchymal tissue from the vertebrae primordia form into intervertebral disc within these interspaces. Ossification centers follow cartilage development, *i.e.*, *chondrification* in each half of the vertebral arch. These centers quickly follow in successive vertebrae posteriorly. The full union of these primary bony components is not completed until after birth. Disclike bony *epiphyses* form later from secondary centers of the vertebral

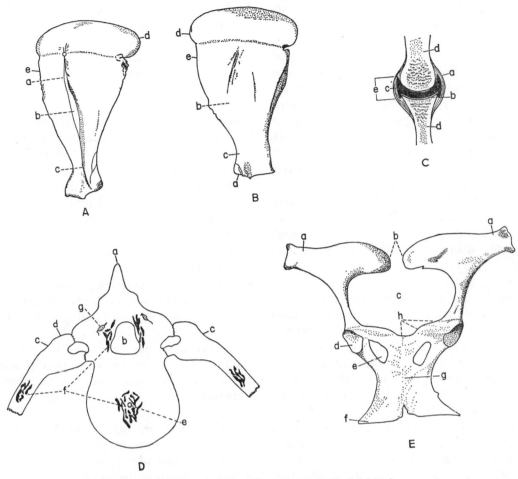

Fig. 12–1. Schematic drawing of different bones. (Drawings by S. K. Zobrisky.)
A, Scapula (lateral view: a, spine; b, acromium process (flat portion of spine); c, neck; d, cartilage; e, anterior edge. B, Scapula (medial view): a, coracoid process; b, body; c, neck; d, cartilage; e, anterior edge. C, Diarthrosal joint: a, fibrous tissues; b, articular cartilage; c, joint cavity; d, bone; e, joint capsule.
D, A developing vertebra and associated ribs: a, spinous process; b, neural canal (vertebral foramen); c, ribs; d, head of rib; e, notochord; f, ossification centers; g, primitive articular process. E, *Ossa coxarum* (equine) a, ilium; b, *tuber scarale* (scaral vertebral union area); c, pubic inlet; d, acetabulum cavity; e, *obturator foramen* f, ischium; g, *symphysis pubis;* h, dotted lines indicate where primitive bone portions unite.

body. These fuse with the rest of the vertebrae after puberty. With advancing age, the sacral vertebrae unite progressively into a single bony mass. Fusion of sacral vertebrae begins postnatally in most cattle prior to the age of 24 months. A similar fusion later occurs in the coccygeal vertebrae.

Costa or Ribs. The origin of the ribs begins with the primitive sclerotomes, which, among other entities, develop into the primordia of the ribs (costal processes). These mesenchymal rib tissues develop chondrification centers, the precursors of cartilage. Where the costal processes unite with the vertebrae, joints develop to receive the rib heads. Simultaneously, the transverse processes grow out and articulate with the developing elevation, *i.e.,* tubercles of the ribs (Fig. 12–1). The cartilaginous rib progressively ossifies. The distal ends of the thoracic ribs remain cartilaginous. In the thoracic region the ribs maintain a movable articulation with the vertebrae. These ribs follow the curvature of the body and join the sternum midventrally.

Sternum. The sternum (breast bone) originates from the mesenchymal tissue. Sternum rudiments are probably first observed as a pair of bands that lie ventrolaterally in the body wall before the end of the first third of embryonic development. Shortly after the ribs and sternal bars attach, the pairs of bands unite progressively from the front toward the caudal end. Ossification begins about midterm at the chondrification centers, some of which are not present until after birth. Sternum segmentation into sternebrae takes place later (Hanson, 1919).

Skull. At the cranial end of the notochord, the skull develops from a dense mass of mesenchymal tissue. The *blastematic* state of the skull development, the cranium-forming period, is the desmocranium. The chondrocranium period is the time of cartilaginous development of portions of the skull. The osteocranium, or ossification period of the skull, follows. Basal portions of the skull bones are of cartilaginous (endochondral) origin; the sides and top of the skull are of membranous origin. Ossification of the cartilage skull bone usually begins during the latter part of the first third of embryonic developmental period. Complete union and fusion of the bone components of the skull may not be attained until later during postnatal development.

B. Appendicular Skeleton

The appendicular skeleton consists of the shoulder and pelvic supports and the attached limb bones. Limb buds first appear as elevations on the body wall at the sites of shoulder and pelvic supports and the limbs, where dense somatic mesenchymal masses have formed. These primordia develop into temporary cartilaginous miniatures and later into bone. The differentiation proceeds proximodistally from the bud origins toward the distal end of the limbs. At first, the forelegs appear to develop faster than the hind legs, but development tends to equalize later on. Ossification proceeds in the cartilaginous bone precursors, such as in the *carpus* and *pedes* portions respectively of the fore and hind limbs, until after puberty.

Forelimbs or Legs. The clavicle or collar bone, not present in most mammals, is highly developed. It is one of the first bones to ossify in some climbing mammals.

An early primary cartilaginous center forms the *body, spine,* and *acromial* process of the scapula. The *coracoid* process of the scapula forms after birth and exists as a small elevation on the body of the scapula (Fig. 12–1). The humerus, radius, and ulna in cartilaginous form develop early and each ossifies from a primary center in the diaphysis (shaft of a long bone) and one or more epiphyseal centers at each end. Each of the *carpal* cartilages has an ossification center similar to each of the *phalanges.* Ossification of these cartilaginous tissues takes place postnatally.

Hind Limbs or Legs. The hip bone (*os coxae*) cartilaginous plate develops in relation to the sacral vertebrae. The *obturator foramen* forms in the lower portion of the cartilaginous os coxae precursor. Ossification centers appear as the ilium, pubis, and ischium develop. The *acetabulum* depression develops to accept the head of the femur in the region where the three hip bone portions unite. The two ilia articulate with the sacrum. The symphysis pubis is formed by the union of the two pubic bones (Fig. 12–1). Formative developmental processes of the femur, tibia, tarsus, metatarsus, and phalanges are not unlike those of the bones of the forelegs. The patella (knee cap) develops in the tendon of the quadriceps femoris muscle (Ruth, 1932).

C. Articulation of Joints

Immediately after the limb buds appear, cartilage development (*i.e.,* cartilaginification or chondrification) begins, and the precursors of future limb bones form. This cartilaginous tissue does not grow across the zones of future joints. Bones and associated joints very rapidly become established, and by the end of the embryonic stage, the joints have a form and arrangement characteristic of the species. *Cavitation, joint capsules, ligaments,* and *synovial tissue* develop as a continuation of the original proliferating limb bud blastemic cells.

Types of Joints. A joint or articulation is formed by two or more bones or cartilages. Bone is a part of most articulations; however, an articulation may be of a bone and a cartilage or two cartilages. Fibrous tissue or cartilage, or their mixture, is the principal uniting tissue. *Synsarcosis* is the union of part of the skeleton by muscle, as in the attachment of the forelimb in most domestic farm animals (Sisson, 1945). Joints may be classified by their developmental expression, type of uniting tissue, their form and amount, kind or absence of movement. The three subdivisions of joints generally referred to are *synarthroses, diarthroses,* and *amphiarthroses.*

In joints of the *synarthrodial* type, movement is practically precluded by the nature of the union. These joints do not have a cavity; most are temporary and later become ossified. Examples are the *sutura* of the skull, the symphysis pelvis, and the epiphyseal line, among others (Sisson, 1945). The mesenchyme that develops between these respective forming bones differentiates and forms a union of connective tissue, cartilage, or bone, known respectively as *syndesmosis, synchondrosis,* and *synostosis.*

Diarthrodial joints are true, movable joints of mobility, and are characterized by a joint cavity with a synovial membrane in the joint capsule (Fig. 12–1). The joint cavity forms from the encircling clefts in the mesenchyme between the developing bones. The capsule arises from the circumventing tissue of the cavity. Later, tissue fibers form a continuum with the periosteum, *i.e.,* the membrane covering the bone surface. The inner layer of the capsule develops into the synovial membrane, which forms a peripheral continuum to, but not into, bones or cartilages of the joint. In the case of an *articular disc,* the cavity forms as separate compartments and the intervening mesenchymal tissue differentiates into a fibrocartilaginous partition plate. Similar development is assumed for an *articular meniscus,* which partially subdivides a joint cavity (Haines, 1947). *Ligaments* are fibrous bands of mesenchyme originating in developing joints that interconnect bones to maintain their proper relationship.

In *amphiarthrodial* or cartilaginous articulations, the opposing bony surfaces, covered with hyaline cartilage, are united by fibrocartilaginous tissue that may have a slitlike space in its center, *e.g.,* pubic symphysis and sacro-iliac.

Sesamoid bones such as the patella develop in the tendon of the quadriceps femoris. The *digitals* in joints are also sesamoid bones of distal portions of the limbs that develop in relation to the joints and display an early cartilaginous origin. The fluidfilled sacs at areas of friction are known as *bursae.* They develop in late fetal life in connective tissue.

II. GROWTH OF BONE

Growth signifies the increase in mass acquired during the morphophysiological changes that occur from conception to time of death. Growth of bone can be loosely defined as the period when addition of ions predominates over withdrawal of ions, resulting in a net increase in size and possibly in number of bone crystals. The quantitative size and weight changes of the skeleton of animals have been recorded by many researchers. In contrast to the growth in length of cartilage tissue, the rigid, calcified nature of bone negates the possibility of proliferative *interstitial, i.e.,* intercellular growth (Plate 15). The increase in size (diameter, thickness) of bone is by *appositional* growth, the deposition of new tissue, *i.e.,* addition of minerals and matrix on the surfaces of preexisting bone. Thus, an increase in length of bone can occur only at the cartilaginous *epiphyseal plates*.

Bones increase in thickness by progressive development of new bone on the outer surface, not unlike the peripheral growth of trees. The length and thickness growth phenomena were originally documented by recording the stationary distance between marker holes drilled into growing bones, and later, by the static position of the red stain (ruberythric acid) incorporated into bones when growing pigs were fed *madder*. Similar pig madder-feeding experiments also partially elucidated the phenomenon of the change in size of marrow cavity of bone. Concurrently, the hypothesis that living bone is constantly changing "matter," *i.e.*, exchanging ions, was advanced. These fundamental processes involve the deposition of new bone in the *periosteal* (peripheral) region of the bone shaft, the absorption of preexisting bone on the surface of the marrow cavity, and also the absorption of existing bone on the external surfaces of the expanded *metaphyses* of long bones as length increases. The metaphyses are the extremities of the long bones where they join the epiphyses.

Bone formation (growth) is the specialized function of the osteoblast or bone-forming cells. In the periosteal region of a shaft of a long bone, the osteoblast cells form bone on the preexisting surface; on the shaft ends these cells form bone within erosions of the cartilaginous epiphyseal plates. Bone formation on the preexisting surface is referred to as *membranous* ossification or intramembranous bone formation; bone formation in eroded cartilage tissue is referred to as *endochondral* ossification. Endochondral or intracartilaginous ossification is dependent on a cartilaginous base and is therefore normally restricted to the period of active skeletal growth in length and mass when epiphyseal plates are present. It terminates when the skeleton attains "maturity," *i.e.*, epiphyseal closure. Membranous ossification also takes place during active growth, but continues after mature skeletal size is at-

tained; furthermore, it is independent of cartilage and is indispensable in the continuous structural remodeling process of bone. The cellular mechanism of bone destruction, referred to in the swine madder-feeding investigations, was possibly first documented with calves. Microscopic observations illustrate the distribution of large multinucleated bone-absorbing osteoclast cells on the surface of the metaphyses where bone absorption (destruction) occurs rapidly during growth.

"Bone maturity or age" can be determined by x-ray observation of the open and closed epiphyses of specific bones. Linear bone growth can occur as long as the epiphyses are separated from the bone shaft. Growth ceases after the epiphyses unite with the shaft, *i.e.*, after *epiphyseal closure* occurs. The epiphyses of the various bones close after puberty. The normal age at which each epiphyseal closure occurs is documented for man but not for many other mammals.

A. Types of Bones and Their Formation

Two types and formations of bones are generally recognized. Those that initiate development within the mesenchymal tissue, such as the skull, are known as *membrane bones*. Those that depend on a prior cartilaginous "scaffold," such as the deep bones, *e.g.*, appendicular bones and vertebrae of the body, are known as replacer or *cartilage bones*. The origin of deep bones is the mesenchymal cells which are collectively called sclerotomal tissue. Their mode of *histogenesis*—the actual method of bone formation—is the same. *Course-fibered* bone, *e.g.*, the skull, is initially formed in membrane tissue, and *fine-fibered* bone, *e.g.*, the femur, in cartilaginous tissue. As a result of remodeling, both bone types develop layers of *lamellae* between which are the small *lacunae* or *osteocyte* spaces (Plate 15).

Each lacuna has a bone cell, the osteocyte. The lacunae connect with one another by the *canaliculi*, small canals. The canaliculi communicate with the *Haversian canals* or larger spaces, the *marrow cavities*. The

PLATE 15

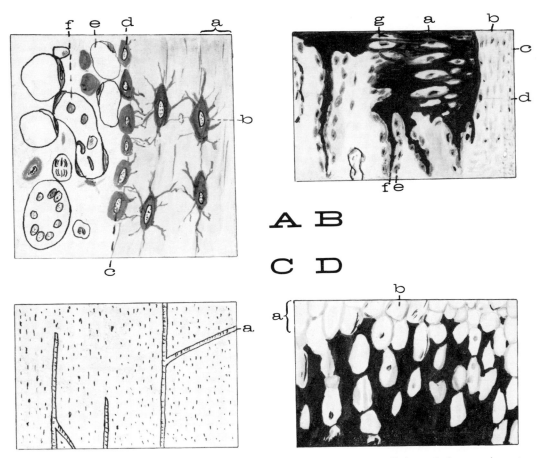

A, Schematic drawing of a section of bone and adjacent marrow: a, bone lamella layer; b, lacunae, *i.e.*, osteocyte space and cell; c, osteoblast cells; d, endosteum; e, fat cells in bone marrow; f, blood vessels and various other cells of bone marrow.

B, Schematic drawing of bone adjacent to spongiosa area: a, eroded cartilage; b, bone cortex; c, periosteum; d, osteocyte embedded; e, osteoblast; f, bone matrix and trabecula; g, cartilage cells.

C, Schematic drawing of compact bone: a, circulatory canal network of bone.

D, Schematic drawing of hypertrophic cartilage cells adjacent to the epiphyseal plate near the end of an immature long bone: a, proliferation zone of cartilage cells; b, cartilage cells.

(Drawings by S. K. Zobrisky.)

PLATE 16

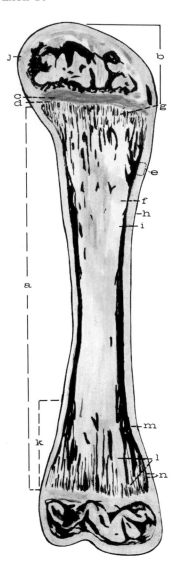

Semischematic drawing of sagittal section of a decalcified tibia: a, diaphysis; b, epiphysis; c, epiphyseal cartilage plate; d, proliferating zone of cartilage cells; e, compact bone (cortex) (Fig. 12–1C is an enlarged view of this area); f, marrow; g, spongy bone (trabeculae); h, periosteum; i, endosteal tissue; j, joint cartilage; k, metaphysis; l, metaphyseal spongiosa; m, Figure 12–1A is an enlarged view of this area; n, Figure 12–1B is an enlarged view of this area, and Figure 12–1D is an enlarged view immediately above "g." (Drawn by S. K. Zobrisky.)

Haversian canal systems connect with the surface and the marrow cavity by fine canals which transmit small blood vessels. Thus, the means for fluids for maintenance of the bone cells and organic material is provided (Bradley, 1960).

Histologically both bone types are not dissimilar, though differences exist during their developmental periods. The entire bone structure of each type of bone is not totally developed by intramembranous or intracartilaginous ossification. The terms, membrane and cartilage bone, are not based on the microstructure at maturity, but rather on the embryonic developmental process. Paradoxes exist in which bones initially classified as membrane later are comprised to a large degree of bone developed in cartilage and vice versa. The mandible is an example of the former, and the long bone shafts an example of the latter. Thus, it is the environment in which ossification occurs—not the process of ossification—that differentiates membrane from cartilage bone.

Intramembranous Ossification. Intramembranous ossification is preceded by the development of a primary fibrocellular mass center during the embryonic stage. Coincident with enzyme phosphatase synthesis by the differentiating osteoblast cells, the formation of a calcified organic matrix, which cements or affixes the fibers together, proceeds. The matrix or ground substance contains complex compounds, including protein polysaccharides. Formation and calcification of the bone fibrous matrix framework usually proceed simultaneously. The bone matrix which has not been calcified is known as *osteoid.*

The first vestige of bone is a small needle-shaped body or spicule formed on very small supporting fibers or trabeculae surrounded by osteoblasts, the bone-forming cells. Some of the osteoblasts become encased in matrix lamellae and become osteocytes. The dividing osteoblasts continue to add bone to the trabeculae. New trabeculae are formed and the centers of ossification expand. Primary

trabeculae connect with secondary trabeculae as growth proceeds. The trabeculae increase in thickness as new bone is formed on their surfaces. The bone continues to develop and expand with the new trabeculae formed peripherally together with free islands and nodules of bony tissue.

When the bones have practically occupied their prefixed definitive regions and are in the proximity of other bones, a peripheral bony border develops as the trabeculae interconnect. Thus, throughout the ossification period there is a gradual reduction in the open reticulum and marginal areas of the membrane bones.

Membranous Ossification. Membranous ossification involves appositional development of new bone on a preexisting surface such as the shaft region of a growing bone. In this light, it could be referred to as the final stage in endochondral ossification. Successive cycles of bone deposition and bone resorption take place on all growing bone surfaces, whether they be *periosteal, endosteal,* or in the *Haversian canals.* These cycles of intermittent tissue activity are at times distinguishable by lines which demarcate the cyclic growth periods.

Endochondral or Cartilage Bone. To trace the growth and development of endochondral bone in an orderly fashion, one logically starts with the cartilage precursor which differentiates from the mesenchymal blastema early in embryonic development. The precartilage state of chondrification (cartilage formation) first develops in the center of the mesenchymal mass and spreads peripherally as an intercellular cartilage-like matrix. Later, individual mesenchymal cells become oriented to form a perichondrium—which, in later stages of development, is known as the periosteum. The perichondrium develops during embryonic growth into a cartilaginous miniature of the individual skeletal parts. Eventually cartilage is replaced with bone as endochondral ossification progresses.

Endochondral Ossification. Bone endochondral ossification consists of a long sequence of cellular activities. These activities include multiplication, growth, degeneration, vascularization, formation of bone trabeculae, and remodeling of the bone tissue. Normally all these activities proceed in equilibrium. Lengthening of the bony shafts causes the metaphysial bone trabeculae to be extended continuously into the receding area of degenerating eroding cartilage. Cartilage cell division takes place in the epiphyseal plate. Cellular multiplication along the longitudinal columns of cartilage is reflected by an increase in size and formation of thin intercellular partitions. Thus, *growth of the epiphyseal cartilage plate is a process of interstitial—not appositional— growth such as is observed in bone.* Growth rate and thickness of the epiphyseal plate tend to parallel one another. Maturation of the cartilage cells is associated with the calcification of the cartilage matrix. The calcified cartilage is soon vascularized from the adjacent metaphysis. *Chondrolysis* (cartilage breakdown) produced by the osteoclast cells and vascular endothelium destroys the calcified cartilage, the transverse, and most of the longitudinal cartilage partitions (Pritchard, 1952). The vascularized, calcified cartilage (*primary spongiosa*) develops into active sites of bone deposition by osteoblastic activity that originates from the connective tissue cells associated with the proliferating vascular system. In some bones the periphery of epiphyseal plates and adjacent *metaphysial spongiosa* (Plate 16) develops an encircling collar of membrane bone; *i.e.*, a *perichondrial ring* on which bone apposition at one edge and bone resorption at the other advances the epiphyseal plate relative to the bone shaft. Longitudinal growth and ossification depend on the continued proliferation of this epiphyseal cartilage tissue. Endochondral ossification in birds (Sissons, 1956, in *The Biochemistry and Physiology of Bone*) is similar to that described in mammals, but the cartilage cells

do not appear as mature prior to vascularization.

Marrow and Marrow Cavity. Another important phase of bone growth and development involves the bone marrow and its cavity. During embryonic life, *hyaline* (as opposed to fibro- and fibroelastic cartilage) cartilage models are formed for many of the bones at a certain developmental stage. Each model is specific for each bone and each species. The specialized cartilage cells enlarge. Within these *hypertrophic* cartilage cells the appearance of glycogen, glycolytic enzymes, and alkaline phosphatase first becomes evident (Kyes & Potter, 1934). Concurrently, growth of the periosteum extends through the vascular mesenchyma in the areas of hypertrophic cartilage cells, which are replaced with primitive bone marrow. The mesenchymal cells—precursors of the periosteum—have the capacity to form hemopoietic and bone marrow cells. The erosion (breakdown) of the cartilage cells continues until the enlarged marrow cavity contains but fragments of the cartilage matrix and is relatively free of cartilage cells. This cartilage erosion process declines as it approaches the portion of the cartilage bone models that subsequently become the epiphysis and epiphyseal cartilage. Hereafter, the marrow and cavity increase as a continuation of bone size (Kyes & Potter, 1934). Aside from its hemopoietic functions, bone marrow participates in *osteogenesis—* bone formation. The endosteum, a condensation of stroma of bone marrow, consists of reticular (endosteal) cells which undergo transformation into active cells if they are required, and later revert again to reticular cells of the stroma (Plates 15 & 16).

III. COMPOSITION OF BONE

Bone is composed of inorganic and organic materials. These two fractions vary in proportions from one part of a skeleton or a bone to another. Water-free bone tissue contains about two thirds inorganic material (which gives bone its characteristic hardness) and one third organic material. Ninety to 96 percent of the organic fraction consists of collagen, insoluble scleroprotein, and ground substance. The ground substance is composed of various inadequately characterized protein polysaccharides. Water, on the average, accounts for about 8 percent of the volume and 20 percent of the weight of compact bone. However, water content of bone has been reported to be as high as 39 percent for man and 50 percent for domestic animals (Morrison, 1956). The inorganic material comprises about one fourth and one half, respectively, of the volume and weight of bone; the organic matrix accounts for the remainder. In bones of growing animals, the inorganic content is normally lower and the organic and water content greater.

Biophysical studies indicate that the mineral fraction of bone corresponds closely to that of apatite. Its composition is conveniently expressed empirically as $[Ca_3(PO_4)-2]n \cdot CaCO_3$. A portion of the carbonate present is replaced with hydroxyl, citrate, chloride, fluoride, and other anions. Some of the cations are replaced by magnesium, sodium, strontium, lead, and possibly other cations. Thus, an empirical formula cannot adequately express the complexity of combinations that can be present in bone at different sites under different conditions. The total organic matter content of different bones varies widely within a species and within similar bones between species. Compact bone invariably contains less organic matter than does cancellous bone. These differences are further compounded by age and level of nutrition.

An approximate composition of bone is: water, 45 percent; ash, 25 percent; protein, 20 percent; and fat, 10 percent. Bone ash of mammals is composed of approximately 36 percent calcium, 17 percent phosphorus, and 0.8 percent magnesium. Other minerals exist in much smaller amounts (Maynard & Loosli, 1956). Similarly, the composition of

dry, fat-free, marrow-free, bovine cortical bone contains about 25 percent calcium; 0.4 percent magnesium; 0.7 percent sodium; 0.05 percent potassium; 0.04 percent strontium; 12.5 percent phosphorus; 4 percent calcium carbonate; 0.8 percent citric acid; 0.08 percent chlorine; 0.07 percent fluorine; and 4.9 percent nitrogen (Armstrong & Singer, 1965).

A chemical analysis of bone to satisfy all disciplines is difficult to perform because of the possible oversight of existing minor constituents and also the fact that several of the known constituents are difficult to characterize. The following results[*] represent the chemical analysis of a section of diaphysis from a bovine femur, air-dried at 105° C:

	Percent by Weight
Inorganic matter (possibly up to 1 percent citrate) insoluble in hot water	70
Inorganic matter soluble in water	1
Collagen	18
Mucopolysaccharide-complex	0.2
Resistant protein material more resistant to solution in boiling water than collagen	1
Water loss	8
TOTAL	98.2

The total inorganic matter is 71 percent. The nitrogenous organic matter, or total weight of collagen, protein, and mucopolysaccharide, represents about 20 percent of the bone weight.

There is evidence that bone mineral exists in two distinct phases: a crystalline (apatite) phase, and an amorphous (noncrystalline) calcium phosphate phase (Eanes & Posner, 1965). There is apparently no direct transformation of the amorphous phase to the crystalline mineral phase. The minerals in the amorphous phase first must be dissolved and the resultant ions redeposited in crystalline form. Noncrystalline calcium phosphate in young rats 8 to 38 days of age may equal 36 to 69 percent of the total inorganic mineral of bone.

[*] Adapted from Eastoe & Eastoe (1954). Biochem. J. 57:453.

A. Water

About 50 percent of the body weight consists of intracellular fluid. A small portion of this is in the skeleton. By means of the watery fluid, minerals are transported to deposition sites during bone formation. When they are released by osteoclastic action, or during exchange, minerals are carried in solution to maintain the physiologic constancy required by the body.

Water content of bone varies with species, the age of the animal, nutritional status, and type of bone. The compact bone of an adult dog contains, by weight, 3.7 percent water, 72.2 percent mineral, and 24.1 percent organic matter. Representative percentages for the volume of bone occupied by water, mineral, and organic matter are, respectively: 8.2 percent, 53.6 percent, and 13.2 percent. Approximately 12.5 percent of the total water volume of compact dog bone is in canaliculi, Haversian canals, and osteocytes; 87.5 percent is in the organic matter (Robinson & Elliott, 1957). Neither normal drying at 100° C nor centrifugal forces can remove the water bound to the mineral crystals of compact bone. The water content of bone organic matter is relatively constant; however, the volume of organic matter is subject to change.

As bone ossification proceeds, the crystals of bone mineral continue to replace water. The water space between the crystals decreases in size until a condition of dense mineralization is attained. At maximal mineralization of bone, diffusion of new ions is restricted (McLean & Urist, 1968). Thus, a slower rate of ion exchange is expected in the relatively stable, highly mineralized, dense, compact bones.

B. Calcium and Phosphorus

Bone has a very active metabolism in regard to replacement of its mineral phase. About 29 percent of the inorganic phosphate in epiphyses of rat skeletons is renewed every 50 days, while 20 percent of the bone calcium is rapidly and reversibly exchanged in *in vitro* experiments (Falkenheim *et al.*, 1951). The very large bone mineral microcrystalline surface area, estimated to be 100 to 200 m²/gm of bone, may help to emphasize the rapid exchange of calcium and phosphate. Calcium and phosphorus account for about 27 and 12 percent respectively of dry, fat-free, compact bones in bovines, while potassium represents about 0.06 percent. Potassium is primarily an intracellular cation and has no special affinity for bone. Percentages for calcium and phosphorus are slightly larger and the percentage of potassium is lower than reported for the skeleton of man. Some calculations indicate that approximately 99 percent of the calcium and 80 percent of the phosphorus of the body is present mostly in the bones. A small proportion of this is in the teeth. The ratio of calcium to phosphorus in bone ash is about 2 to 1.

C. Sodium

Isotope dilution methods indicate that bone contains about 7 gm of sodium per kg of dry bone, but other techniques indicate 0.7 percent of dry, fat-free, compact bone is sodium. Thirty to 45 percent of the sodium and about 8 percent of the total body water of man is in bone (McLean & Urist, 1968). About 42 percent of the sodium in bone of canines and man is rapidly exchangeable.

D. Fluoride

The fluoride content of the skeleton is normally 0.02 to 0.05 percent. Fluoride is one of the most avid of all bone-seeking substances. In some species, a gradual increase in fluoride content of bone with age has been demonstrated. Twenty µg/100 gm of bone in elderly men is not unheard of. Fluoride, once it has been incorporated into the mineral crystal lattice, is comparatively stable—not readily mobilized. Excessive quantities of fluoride result in bones that lose their normal color, become thick, soft, and weak (Comar *et al.*, 1953), and have a reduced rate of growth.

E. Magnesium

Magnesium appears to be *chemisorbed* on the surface of crystals of bone mineral (McLean & Urist, 1968), and is associated with the deposition of bone salt. Some evidence indicates that magnesium is also an activator of alkaline phosphatase. When magnesium is deficient, bones tend to be brittle. Magnesium content of dry, fat-free, compact bone is about 0.41 percent, which represents 70 percent of the body content. It is estimated that one third of the bone magnesium is subject to mobilization to other tissues when the intake is inadequate.

F. Lipid

The presence of a small quantity of lipid in bone has been observed as sudanophil granules of fat associated with the osteoblast cells. The sudanophil granules apparently contain phospholipids. The principal lipid depot of bone is in the marrow, and the quantity varies with the state of nutrition. Composites of bone and bone marrow may contain 10 to 25 percent of ether-extractable materials. A substantial quantity of iron is also found in bone marrow.

G. Collagen

The genesis of collagen is closely associated with the appearance of fibroblast cells. Whether the collagen fibers arise with*in* the fibroblast cells or in the *inter*cellular spaces is a debatable question.

Cartilage and membranous connective tissues—the precursors of bone—contain collagen and polysaccharides. Together they constitute the organic matrix, which, when ossified, is bone. Aging is associated with a change in size, density, and, possibly, number of the collagen fibers, as well as a decrease in protein polysaccharide "mucopolysaccharide" content and solubility. Collagens from different sources exhibit wide variations in solubility. Bone collagen from various mammalian and avian species differs very little morphologically, *i.e.*, in fibril structure and amino acid or chemical composition. However, collagens are deficient in several of the amino acids essential to protein metabolism. Collagen of mammals and birds has a higher hydroxyproline and proline content and less methionine, serine, and threonine than collagen of fish. Collagen is the most abundant protein (second most abundant constituent) in bone. Some elastin and reticulin are present in much smaller quantities. More than 90 percent of the dry, fat-free weight of the organic matter of bone is collagen. The amino acid composition of bone collagen appears to be almost a replica of that in skin. Its physical properties are similar to those of collagen found in tendons and cartilage. The physiologic turnover of bone and tendon collagen is very slow.

The collagenous fibers of bone are more densely packed than are those in any other tissue of the body. All collagens are made up of fibrils with double crossbanding at regular intervals. The ground substance associated with the collagen fibers is characterized by the presence of protein polysaccharides containing hexosamines. The principal mucopolysaccharides that result from degradation, such as by hydrolyzation of protein bound to polysaccharides, are types of chondroitin sulfate components consisting of glucuronic and sulfuric acid. Additionally, hyaluronic acid and keratosulfate mucopolysaccharides are found in smaller quantities. A mucopolysaccharide as such is not present in live animals. Ground substance is the organized extracellular and interfibrillar ultrastructure component of all connective tissues.

Mast cells have also been related to collagen formation. These cells are numerous in bone marrow and loose connective tissue throughout the body. In rats, stressed by a deficiency of calcium or vitamin D, mass cells tend to accumulate in the bones. These cells migrate to, in, and under the endosteum when the rats stop growing (McLean & Urist, 1968).

H. Phosphatases

One of the first observable signs of alkaline phosphatase accumulation and activity in embryos of rodents, swine, fowls, and man appears in the developing perichondrium of hypertrophic cartilage cells of the vertebrae and ribs (Gomori, 1943). Activity of phosphatase is likewise seen in the ossifying cartilage, bones, and teeth of young animals. It is also present in most formation sites of ectopic bone—bone that develops in abnormal, out of place locations. In mature bone, most phosphatase activity is restricted to the cytoplasm of osteoblasts of the periosteum, the Haversian canals, the endosteum, and peripheral osteocytes. Osteoblastic cell enzymatic activity decreases as the cell becomes more completely enclosed and more deeply embedded in bone substance. Intense enzymatic activity is present in the periosteum osteogenetic layer surrounding a bone within hours after injury (Bourne, 1948). Phosphatase is associated with osteolysis, i.e., bone resorption, the precipitation of calcium as bone salt, the formation of organic matrix of bone, and, also, the continued growth of bone crystals. Deficiencies in matrix formation coincident with a diminished alkaline phosphatase activity in the epiphyseal plates are correlated with small stature and short midphalanges of the limbs. There is evidence (McLean & Urist, 1968) that alkaline phosphatase is involved in the synthesis of mucoprotein and fibrous protein, in both noncalcifiable and calcifiable tissues. Apparently, once a calcifiable matrix tissue is formed, it remains calcifiable whether alkaline phosphatase is present or not. Therefore, phosphatase does not appear to be essential, i.e., it need not be present, for calcification to occur. Alkaline phosphatase is not detectable in the glycolytic pathways.

Bone contains more phosphatase enzyme than does any other organ in the body; it has twice that found in the kidney and twenty times that found in the liver. The body apparently contains a spectrum of phosphatases with overlapping substrate preferences that maintain a certain degree of specificity. Growth and development of bones, bone components, and the exchange of mineral ions involve hormonal enzyme complexes in which alkaline phosphatase has an important role. The precise role(s) at present is rather obscure; however, the associated occurrences observed with and, in some instances, without the presence of alkaline phosphatase are documented.

I. Citrate

Most of the citric acid in bone is in the ossified regions and tends to parallel the ash content; very little is in the bone marrow. The citrate ion may be combined in a complex with calcium. In some species, 90 percent of the citric acid found in the body is in the skeleton. The citric acid content of bone declines slowly with age. Thus, the citrate ion appears to be associated with the vitamin D-parathyroid interplay involved with the solubility and mobility of bone salt (Sherman, 1947). A similar gradual decline is observed for lactic acid content of bone, which ranges from 0.1 to 0.3 percent. Bone citric acid values differ widely from one species to another. The citrate ion comprises from 0.2 to 2 percent, an average of 0.8 percent in bones of mammals; 1 to 5 percent in bones from fish; and 0.6 to 3 percent in bones from birds.

J. Glycogen

Glycogen is present in hypertrophic cartilage cells and preosseous mesenchymal cells, but apparently not in active osteoblasts. In the developing membrane of porcine bones, glycogen has been found in the osteogenic *mesenchyme preceding* centers of ossification and also in preosteoblast cells of the periosteum.

However, glycogen tends to disappear as the bone-forming cells and the first bony spicules appear. It reappears in the osteoblasts in maximal amounts during mineralization. Glycogen is abundant in many embryonic tissues that do not show any glycogen after maturation (Gutman & Yu, 1950).

Similarly, glycogen is apparently absent from differentiating osteoblasts in a fractured callus. The synthesis and breakdown of glycogen in bone tissue for the release of energy are similar to the synthesis and breakdown of anaerobic glycolysis in muscle and other soft tissues. The contribution of the energy released to formation or calcification of bone is still unresolved.

IV. FACTORS THAT INFLUENCE THE DEVELOPMENT OF BONE

The general shape or form of a bone is determined by factors which are not as yet completely clarified. Some embryonic bones placed in culture media isolated from other tissues, circulating blood, hormones, nervous influences, and muscle tension self-differentiate into the same general shape as expected *in vivo*. Heredity, hormones, and nutrition undoubtedly affect the rate of growth, shape, quality, and ultimate size of bone.

A. Genetic

Familial bone traits exist in many animal species. A recognized characteristic *shape* of the head in families and races of people and breeds and species of livestock is transmitted by genetic factors. *Dwarfism* and the associated stubby skeleton are other examples of genetic influence. Dwarfism is not uncommon in cattle, mice, man, and other species. It is sometimes referred to as hereditary pituitary dwarfism because the genetic effects may involve the development of and the secretions of the anterior lobe of the pituitary. On the other hand, some forms of *cretinism* appear to be due to a genetic block in the organification of iodine. Conversely, *gigantism* or *giantism* in man is associated with an unusual overall large size from time of birth. This phenomenon is genetically influenced. Still another genetic malady is *exostosis*, the abnormal bony outgrowth from the surface of bone. Similarly, *achondroplasia*, the reduction in growth of all cartilaginous bone-forming tissue that results in a shortening of the long bones, and *brachydactylia*, the abnormally short phalanges ob-

served in man and other animal species (March, 1941), are genetically endowed characteristics.

The mutant gene in the mouse referred to as *gray lethal* interferes with osteoclastic resorption of bone and results in the accumulation of bone where erosion would normally occur. In affected mice, the metaphysial trabeculae of the long bones persist; thus, excessive bone formations occur and the normal bone marrow cavities do not develop. Hereditary factors appear to be involved in *fibrous dysplasia* of bone in pigs. Principally, the nasal bones are involved, but some evidence indicates that the long bones are also concerned (Kowalczyk *et al.*, 1958).

Epiphyseal dysplasia, the ossification failure in the epiphyseal center, represents a serious deficiency in skeletal formation that is observed occasionally in man and swine. *Loose-jointedness* (Marfan syndrome), the results of defects in ligaments and tendons as well as excessive length of the round bones of the extremities, is attributed to an autosomal dominant gene. An autosomal recessive gene is responsible also for *hypophosphatasia*—a rare metabolic malady in which an alkaline phosphate deficiency results in the formation of bone lesions due to the inability of the osteoblasts to elaborate calcifiable organic matrix (Milch, 1960). Control of the complex genes and other factors responsible for the many "hereditary" bone maladies in man and beast must await further elucidation by geneticists and husbandrymen.

B. Hormones

The hormones of the *pituitary*, the *thyroid*, the *adrenals*, and the *gonads* affect bone growth and development. *Cretinism* (due to a thyroid deficiency) and *acromegaly* (associated with the excessive secretion of growth hormones in man) are hormonal manifestations which affect bone growth. Skeletal growth and maturation in rats are retarded following hypophysectomy or thyroidectomy. Growth retardation of the skeleton is dependent upon the age at the time of operation, but bone maturation apparently is not. Differentiation of the epiphyseal ossification centers can be restored in many thyroidectomized animals by administration of thyroid hormone. Similarly, growth hormone administered to immature hypophysectomized animals promptly results in growth of the epiphyseal cartilage, the adjacent spongia, and a reinvasion of the cartilage. Growth hormone of the anterior pituitary, estradiol, and parathyroid hormone administered to young animals tend to produce bones that are bigger and heavier than normal (Wilkins, 1953). Overdoses of thyroxine may cause premature closure of the epiphyses and cessation of growth— effects that sometimes result from hyperthyroidism. Normally, thyroxine not only is conducive to bone growth, but also to bone maturation; however, growth hormone stimulates growth without advancing maturation of the skeleton. The effect of growth hormone on long bones can be illustrated until closure of the epiphysis occurs (McLean & Urist, 1968). Growth in bone length is dependent on the invasion of the cartilage by capillaries and osteogenic cells and replacement of the cartilage by diaphyseal bone. Species specificity for growth hormone is an important consideration. Deficiencies of growth hormone, thyroxine, and vitamins A and C are known to impair chondroitin sulfate synthesis and thus the synthesis of collagen of bone. Testosterone and growth hormone have a synergistic stimulating influence on epiphyseal growth.

Excessive administration of adrenocorticotropic hormone (ACTH) or glucocorticoids has an inhibiting effect on *osteo-* and *chondrogenesis*, *i.e.*, bone and connective tissue development (Baker & Ingle, 1948). Long bones particularly tend to be severely affected. A general retardation of growth has been observed in some animals, particularly the young rat and chick, while in other animals an associated interference with bone resorption can be illustrated by the unresorbed spongy bone in the proximity of the epiphyseal plate. These effects tend to be temporary and differ from species to species. Estrogen administered in excess tends to produce similar effects in rats and mice. The relationship of estrogens to *osteogenesis* is best exhibited in birds by the growth of the spongy endosteum bone into the marrow cavity prior to the start of the egg-laying cycle (Kyes & Potter, 1934). Estrogens also tend to enhance development of collagen fibers. Both adrenal and sex steroids affect collagen formation but do not effect a change in established collagen. Some evidence indicates that estrogen opposes the stimulating effects of growth hormones.

The parathyroid gland secretion (PTH) and vitamin D are important synergistic, homeostatic, regulatory constituents involved in the metabolism of calcium and phosphate, and, thus, bone development. PTH has a direct influence on the mobilization of calcium from the skeleton, the absorption of calcium from the intestinal tract, and the excretion of phosphate by the kidneys. Vitamin D has a major influence on absorption of calcium during bone mineralization and also complements the parathyroid hormone in calcium mobilization from bone. The presences of vitamin D and PTH also enhance citrate formation in bone. In circulation, PTH is rapidly destroyed. Its deficiency leads to lowered blood calcium ion concentration; an excess has the opposite effect. Either condition, if prolonged, can produce important changes in bone. The circulating level of the calcium ion in the blood apparently acts directly on the parathyroid glands to regulate the secretion of PTH. High calcium levels inhibit PTH secretion so that calcium is deposited in bones. Conversely, low calcium levels stimulate PTH secretion so that calcium is mobilized from the bones. Evidence is accumulating that strongly suggests important influences of PTH at the subcellular and membrane levels throughout the body other than its accepted *calcemic* and *phosphaturic* roles. The mobilization of chondroitin sulfate—an important constituent of ground substance of bone connective tissue—may also occur under the influence of PTH. PTH also apparently has an important role in the release of lysosomal acid hydrolases from osteoclast cells. These enzymes aid in eroding bone during resorption. During these events, the release of lactic and citric acids lowers the pH, enabling the bone mineral to become soluble.

The complex interrelationships pertaining to endocrine abnormalities and their effects on bone growth and maturation are by no means fully understood. Much interest currently exists in these perplexing hormonal interrelationships.

C. Nutrition

Skeletal growth continues even under conditions of general nutritional restrictions during which total body weight is stationary or even declining. Adequate amounts of protein and minerals must be available for the maintenance of normal bone structure.

Skeletal growth and body growth parallel each other in a variety of nutritional disturbances; on the other hand, nutrient intake affects the mineral relationships in bone without an appreciable change in bone dimensional measurements, ash content, or rate of gain in body weight (Smith, 1931). Bone nutrition does not end with the normal cessation of body growth, since physiological demands are continually made on the mineral reserve, especially for calcium and phosphorus during reproduction and lactation. Continued bone nutritive deficiencies may result in serious bone problems later in life without any perceptible early symptoms. This is particularly true because bone size is an inherited trait and large bones can be deceptive, since size does not necessarily indicate density and strength of bones.

The effects of severe undernutrition and subsequent *rehabilitation* on bone growth and development of pigs have been described (Pratt & McCance, 1965). Perhaps it is sufficient to state that, in the case of swine, severe undernutrition for a year results in a slower rate of bone growth followed by a cessation of development in the length and

width of the long shaft bones. These phenomena are associated with a loss of fully hypertrophied cartilage cells. Intermittent growth activity of the cartilage and periosteal tissue can be observed respectively in the peripheral *chondrocytes* and layers of the *cortical* bone. Endosteal resorption of much of the bone present at the start of the undernutrition period produces a distinctive "rarity" in bone structure. Rehabilitation as much as one year later immediately results in an increase in body weight, but normal bone growth, with all of its manifestations, does not resume until about the twenty-eighth day of rehabilitation. Roughly, 80 to 90 percent of the normal bone length is attained during the recovery rehabilitation. Definite deformities, resulting from the tendency of bone diameters to be greater with abnormal curvations of the long bones, become apparent when compared to similar bones of control swine. Other species of animals can be expected to react not too differently under similar treatment.

Vitamin A. Vitamin A has a definite role in the control of bone osteoblastic and osteoclastic activities. Divergences from the normal activity of osteoblasts have been manifest by the development of bone in soft tissue where it is normally not found; abnormal activity of osteoclasts has been shown by their failure to absorb previously formed bone. Thus, a prolonged deficiency of vitamin A (hypovitaminosis A) leads to a reduction in chondroitin sulfate, abnormalities in the shapes of bones, and, also, the failure of certain bony *foramina* to enlarge. These abnormalities have been observed in the bones of dogs, rats, guinea pigs, rabbits, and young cattle. Coincident with the abnormal bone development, various nerves are subjected to undue pressures. Avitaminosis can result in a degeneration of bone cartilage and bone, a manifestation of nerve lesions, and, also, the constriction of nerves in the restricted bone nerve canals which is associated with, but not necessarily the cause of, blindness in calves, deafness in dogs, and muscle incoordination in swine, sheep, and cattle (Mellanby, 1947). Fragile long bones are also associated with excessive amounts of vitamin A. This fragility is due to the *proteolytic* activity on the bone matrix constituents by acid hydrolases released from lysosomes under the influence of large quantities of vitamin A (Fell, 1964).

Vitamin B. Bone abnormalities in newly hatched chicks and newborn rats are related to vitamin B_{12} deficiencies in the rations of the hen and pregnant mouse (Grainger *et al.*, 1954).

Vitamin C. Prolonged vitamin C deficiency results in weak *scorbutic bones* (due to scurvy, *i.e.*, scorbutus). "Scurvy lines" on the tibia and femur and fractures of the epiphyses are common results of lasting vitamin C deficiencies. Other manifestations include the malformation of the complex bone matrix so that osteoblasts take the form of reticular cells and the bone phosphatase activity decreases. Concurrently, a reduction occurs in the production of bone, because the abnormal tissue formed during the vitamin deficiency is not calcifiable, *i.e.*, the intercellular materials of the supporting tissues—bones and other collagenous tissues—are not produced or are of defective development (McLean & Urist, 1968). The synthesis of fibers and the structureless matrix of bone are subsequently inhibited by the effects of the deficiency of vitamin C. Vitamin C is essential to the normal development and function of all subcellular structures, *e.g.*, ribosomes and mitochondria, of all higher animals. These manifestations reflect the basic role of vitamin C in tissue metabolism.

Vitamin D. The most remarkable function of vitamin D is its ability to stimulate calcification of newly formed bone in the presence of calcium and phosphorus. This ability accounts for the application of the term "bone-forming nutrients" to these

three bone essentials (Maynard & Loosli, 1956). A deficiency of vitamin D is associated with the failure of developing bone to calcify even though osteoblastic activity continues. This condition is known as as rickets. Several sterols — dihydrotachysterol; ergosterol, D_2; and cholecalciferol, D_3—that possess antirachitic potencies and *calcemic* properties are known, but to a lesser degree than vitamin D. A suitable ratio between calcium and phosphorus in the presence of vitamin D is a prerequisite for optimum bone calcification. With an abundance of vitamin D present, the calcium to phosphorus ratio becomes less critical; however, no amount of vitamin D will compensate for severe deficiencies of either element. The physiological mechanism by which the vitamin activates calcification is unknown. Its influence on calcium and phosphorus metabolism is apparently intertwined with the activity of the parathyroid hormone and alkaline phosphatase. Hypercalcemia, the result of massive doses of vitamin D, is conducive to the deposition of calcium salts at uncommon sites of soft tissues and can lead to death. The condition is also referred to as hypervitaminosis D. Fortunately, a wide range exists between adequate and harmful amounts.

Minerals. The minerals in the body serve many indispensable functions. The mineral elements in bones provide the principal reservoir for mobilization as soluble salts when minerals from other sources are inadequate to meet tissue requirements. The osmotic pressure relations required for nutrient and waste transfer through cell walls are dependent on mineral salts. A vital portion of proteins in the nuclei of all body cells is phosphorus. The phospholipid components of all living cells and the iron of hemoglobin and also myoglobin serve vital functions. The delicate relationship of minerals is the principal factor in the maintenance of tissue acid-base equilibrium. Mineral elements serve a role in various enzyme systems, have specific metabolic functions, and provide an unexcelled rigidity and strength to the skeletal structure.

Catabolism and anabolism of carbohydrates, fats, and proteins are not independent of phosphate-containing adenosine triphosphate and nucleotides. Phosphorus compounds play an essential role in transformations of energy in the body. Cell division, reproduction, and transmission of hereditary characteristics are, in part, also dependent on the phosphorus in nucleoproteins.

In bones of rats, an adequate calcium intake is reflected by relatively short metaphyses indicative of a short supply of stored calcium and very active osteoblastic and osteoclastic cells which demonstrate the rapid calcium turnover for dietary purposes. The bones of these rats are porous, similar to those of very young animals (Sherman, 1947). Conversely, rats on an optimum calcium intake have long and dense metaphyses indicative of maximum calcium storage and diminished osteoblastic and osteoclastic activity which shows minimum calcium turnover and maximum storage of bone mineral. The bones of these rats are very dense, similar to those of adult animals. Calcium retention is favored when intake is low. This is probably due to an increase in the exchangeable fraction when calcium is deficient. Exchangeable calcium represents about 5 percent of the total amount in bones of animals. The exchangeable fraction decreases with age. An appreciable loss or gain by the body of either calcium or phosphorus can be expected to reflect a similar change in the other element.

Macro amounts of seven mineral elements are generally considered as necessities for animals: calcium, chlorine, magnesium, potassium, phosphorus, sodium, and sulfur. At least seven other elements are required in trace quantities: cobalt, copper, iodine, iron, manganese, selenium, and zinc. Additionally, chromium, fluorine, molybdenum, and possibly others appear to have an important metabolic role, especially in man. All of these elements plus aluminum, lead,

strontium, and others have been found in bone or bone marrow. The daily requirement of many of the elements is not known.

Sodium is insoluble and rather inert in bone. The ability of bone to release sodium ions to protect the H^+ ion concentration of the extracellular fluids appears dependent on parathyroid secretions. Sodium apparently is incorporated in a rather stable crystal fraction of bone from which it is not removed until released by bone resorption.

Zinc serves as a component of many metallo-enzymes, *i.e.*, alkalines, phosphatases, and dehydrogenases. Zinc deficiency in young birds is characterized by a shortening and thickening of the long bones of the legs and an enlargement of the hock joint. Severe zinc deficiencies in hens result in embryo malformations, primarily skeletal, *e.g.*, small limbs, absence of limbs, or absence of vertebral development. In Eygpt, dwarfism and hypogonadism in man is associated with a zinc deficiency. Vitamin D appears to increase zinc deposition in bone through its stimulatory effect on bone growth and calcification. However, vitamin D has no detectable influence on the zinc deficiency syndrome in pigs or rats.

Little or no difference in zinc concentration is observable in tissue of most organs of zinc-deficient animals; however, bone, hair, liver, and testes are exceptions. The precise role of zinc metabolism is not completely elucidated when deficient defects are apparent in protein metabolism, cell division, and growth.

Trace amounts of aluminum normally exist in bone; excessive amounts reduce phosphorus assimilation and are conducive to a decrease in bone ash and to ricket-like symptoms. Although manganese and copper constitute only a very small content of bone, manganese and copper deficiencies have been implicated in abnormalities of bone development in chicks, sheep, and cattle (Wolbach & Hegsted, 1953). Manifestations of experimentally developed manganese deficiency in animals include a retardation of size as well as structural and chemical anomalies of bone. These are associated with a suppression in the proliferation of epiphyseal cartilage cells, a decline in bone matrix formation, a failure of growth, and a general decrease in strength of bone. These observations are most apparent in rats and birds.

Excessive quantities of fluorine, lead, and strontium induce definite bone malsyndromes (Pantelouris, 1967). Prolonged, excessive intakes of fluorine contribute to softened bones of reduced strength. Young animals deposit and retain greater proportions of daily F^- intake than do mature animals. Animals, such as laying hens, that have a large rapid turnover of bone minerals, are able to shunt large amounts of F^- into the mineral phase and thus elude the toxic effects of the fluoride ions on bone cells (Johnson, 1964). The capacity of bone to accept and incorporate F^- is tremendous. An abnormal pattern of the bone matrix and bone cells is associated with continual excessive intakes of fluorine. This condition is known as fluorosis and is characterized by *hyperostosis*, an increased thickening of cortical bone and periosteum, and increased bone density, as well as the formation of bone spurs in the insertion of muscles and attachments of ligaments near bone articulations. Fluorine intake during bone mineralization not only increases its content in bone, but also increases the size of the bone mineral crystals which tends to reduce their surface area and solubility.

Molybdenum and copper apparently compete in the same metabolic pathways. Excess molybdenum intake in cattle induces bone abnormalities characterized by cartilaginous hyperplasia and reduced osteoblastic activity.

Lead is taken up and released by bone in a manner similar to the up-take and release of calcium and strontium. Lead and strontium are extraneous contaminates. Lead intoxication symptoms can occur if acidotic conditions are present. Radioactive strontium in sufficient amounts has a destructive effect on bone marrow hemocyto-blasts of the hemopoietic tissue. The effect is a rapid depletion of red blood cells and antibody-producing cells (Pantelouris, 1967).

Thus, vitamin, nutrient, hormonal, or mineral deficiencies can result in atypical bone development, especially if they are of prolonged duration during the time of active growth. Conversely, excesses or imbalances due to intake or metabolic disorders can likewise result in abnormal structural development of bone. The many ramifications attributed to genetic influence on bone development are of great moment, despite the fact that the mediating gene pool sources are most difficult to measure or document.

REFERENCES

Armstrong, W. D., & Singer, L. (1965). Composition and constitution of the mineral phase of bone. Clin. Orthop. *38*:170–90.

Baker, B. L., & Ingle, D. J. (1948). Growth inhibition in bone and bone marrow following treatment with adrenocorticotropin (ACTH). Endocrinology *43*: 422.

Bell, G. H., Cutherbertson, D. P., & Orr, J. (1941). Strength and size of bone in relation to calcium intake. J. Physiol. (London) *100*:299.

Bourne, G. H. (1948). Alkaline phosphatase and vitamin C deficiency in regeneration of skull bones. J. Anat. *82*:81.

Bradley, O. C. (1960). The Skeleton. Chap. 2. In: *The Structure of the Fowl*. Rev. T. Grahame (Ed.), London, Oliver & Boyd.

Brody, S. (1945). *Bioenergetics and Growth*. pp. 712–740, New York, Reinhold Publishing Corporation.

Comar, C. L., Visek, W. J., Lotz, W. E., & Rust, J. H. (1953). Effects of fluorine on calcium metabolism and bone growth in pigs. Amer. J. Anat. *92*:361–389.

Eanes, E. D., & Posner, A. S. (1965). Kinetics and mechanism of conversion of noncrystalline calcium phosphate to crystalline hydroxyapatite. Trans. N. Y. Acad. Sci. *28*:233–241.

Falkenheim, M., Underwood, E. E., & Hodge, H. E. (1951). Calcium exchange; the mechanism of adsorption by bone of CA^{45}. J. Biol. Chem. *188*: 805.

Fell, H. B. (1964). Some factors in the regulation of cell physiology in skeletal tissues. In: *Bone Biodynamics*. H. M. Frost (Ed.), Boston, Little, Brown, & Co.

Gomori, G. (1943). Calcification and phosphatase. Amer. J. Pathol. *19*:197.

Grainger, R. B., O'Dell, B. L., & Hogan, A. G. (1954). Congenital malformations as related to deficiencies of riboflavin and vitamin B_{12}, source of protein, calcium to phosphorus ratio and skeletal phosphorus metabolism. J. Nutr. *54*:33–48.

Gutman, A. B., & Yu, T. F. (1950). A concept of the role of enzymes in endochondral ossification. In: *Metabolic Interrelations*, Trans. 2nd Conf., p. 167–190, New York, Josiah Macy, Jr. Foundation.

Haines, R. W. (1947). The development of joints. J. Anat. *81*:33–55.

Hanson, F. B. (1919). The development of the sternum in sus scorofa. Anat. Rec. *17*:1–23.

Johnson, L. C. (1964). Morphologic analysis in pathology: The kinetics of disease and general biology of bone. In: *Bone Biodynamics*. H. M. Frost (Ed.), Boston, Little, Brown, & Co.

Keith, A. (1919). Chaps. 14–16. *Menders of the Maimed*. London, Oxford.

Kowalczyk, T., Anderson, R. A., Simon, J., & Baker, F. D. (1958). Fibrous dysplasia of the bone in swine. J. Amer. Vet. Med. Ass. *133*:601–604.

Kyes, P., & Potter, T. S. (1934). Physiological marrow ossification in female pigeons. Anat. Rec. *60*: 377–379.

Maynard, L. A., & Loosli, J. K. (1956). *Animal Nutrition*. R. A. Brink (Ed.), New York, McGraw-Hill Book Co.

McLean, F. C., & Urist, M. R. (1968). *Bone, Fundamentals of the Physiology of Skeletal Tissue*. Chicago, University of Chicago Press.

Mellanby, E. (1947). Vitamin A and bone growth: The reversibility of vitamin A-deficiency changes. J. Physiol (London) *105*:382–399.

Milch, R. A. (1960). Studies of alcaptonuria: Inheritance of 47 cases in eight highly interrelated dominican kindreds. Amer. J. Human Genet. *12*: 76–85.

Mørch, E. T. (1941). *Chondrodystrophic Dwarfs in Denmark*. Copenhagen, Munksgaard.

Morrison, F. B. (1956). Minerals in livestock feeding Chap. 5. In: *Feeds and Feeding*. 22nd ed., Ithaca, N. Y., Morrison.

Pantelouris, E. M. (1967). *Introduction to Animal Physiology and Physiological Genetics*. New York, Pergamon Press.

Pratt, C. W. M., & McCance, R. A. (1965). *Atrophic-osteoporotic Changes in the Long Bones of Severely Undernourished Pigs*, Proceeding of the Second European Symposim on Calcified Tissue. L. Richelle and M. J. Dallermagne (Eds.), University of Liege.

Pratt, C. W. M., & McCance, R. A. (1966). The structure of the long bones in pigs stunted by severe undernutrition. In: *Swine in Biomedical Research*. L. K. Bustad & R. O. McClellan (Eds.), Seattle, Frayn Printing Co.

Pritchard, J. J. (1952). Cytological and histochemical study of bone and cartilage formation in the rat. J. Anat. *86*:259.

Robinson, R. A. (1951). *Metabolic Interrelations*, Trans. 3rd Conf., p. 271, New York, Josiah Macy, Jr., Foundation.

Robinson, R. A., & Elliott, S. R. (1957). The water content of bone. I. The mass of water, inorganic crystals, organic matrix, and "CO_2 space" components in a unit volume of dog bone. J. Bone Joint Surg. *39-A*:167–68.

Ruth, E. B. (1932). A study of the development of the mammalian pelvis. Anat. Rec. *53*:207–225.

Sherman, H. C. (1947). *Calcium and Phosphorus in Foods and Nutrition*. New York, Columbia University Press.

Sisson, S. (1945). Arthrology. Chap. 2. In: *The Anatomy of the Domestic Animals*. J. D. Grossman (Ed.), Philadelphia, Saunders.

Smith, A. H. (1931). Phenomena of retarded growth. J. Nutr. *4*:427.

Wilkins, L. (1953). Disturbances in growth. Bull. N. Y. Acad. Med. *29*:280.

Wolbach, S. B., & Hegsted, D. M. (1953). Perosis: epiphyseal cartilage in choline and manganese deficiencies in the chick. Arch. Path. *56*:437–453.

Chapter 13

Lipids and Adipose Tissue

By R. S. Emery

LIPIDS combined with structural protein form the membranes essential to life in all cells. Adipose tissue is vital in supplying these structural lipids and in providing caloric homeostasis. Mobilization of body fat during prolonged underfeeding is a well known phenomenon but the equal importance of adipose tissue in daily energy homeostasis has not been widely appreciated. Lipid metabolism changes with, and partially controls, the various productive functions such as lactation, fattening, work, and growth. These more dynamic aspects of fat metabolism do receive special emphasis in this chapter, but without any intent to slight the structural and storage importance of lipids.

I. MORPHOLOGY

A. Development and Description

Adipose tissue differentiates from mesenchymal tissue. Embryonically related bone marrow tends to develop adipose tissue with age. This embryonic kinship to connective tissue explains the gross similarity of connective tissue to adipose tissue after fasting. Adipose tissue always distributes itself perivascularly and each cell touches one or more capillaries. The apparent lack of vascularity in adipose tissue is due to the swelling of cells by triglycerides, which form the depot fat in the central vacuole. When the central vacuole is collapsed by fasting, a fat pad looks like an exceptionally vascular piece of cartilage.

Fetal adipose tissue looks much like brown adipose tissue or liver, but the central vacuoles start to form after birth. Growth is largely due to hypertrophy since adipocytes (individual fat cells) can enlarge by ten diameters as the central vacuole fills. However, new cells do form, always along blood vessels. In extreme obesity, adipose tissue is seen in striated and nonstriated muscle due to newly formed cells along blood vessels. During this fattening, the number of cells in fat pads will double but the weight of the fat pad will increase many fold due to hypertrophy (Fig. 13–1).

When a fat pad regresses during restricted caloric intake, the central vacuole is depleted in a few weeks but months are required for the additional adipose cells to regress. Thus, the animal retains additional ability to make fatty acids and store fat for

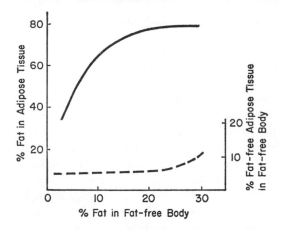

FIG. 13–1. The relationship between the fat content of the guinea pig body, the fat concentration in its adipose tissue, and the body content of fat-free adipose tissue. (*Adapted from Pitts, 1963. Ann. N.Y. Acad. Sci. 110*:11.)

some time (Hausberger in Renold & Cahill, 1965).

Adipose tissue is usually richly innervated and the adrenergic nerves are required for normal fat mobilization. If the nerve supply to a portion of the body is disturbed, fat accumulates due to a failure to mobilize fat (Barrnett in Kinsell, 1962).

B. Distribution and Function

Mammals developed an adipose organ as they evolved from amphibia. This organ is vital in the same sense as liver or kidney is, although it is diffused through the body. Adipose cells have the ability to store fat in a central vacuole surrounded by cytoplasm; other cells store only small amounts of fat as inclusions in their cytoplasm. Fat in adipose tissue generally should be considered a pool of calories for maintenance of homeostasis, but evolution has withdrawn some adipose tissue from this pool for mechanical purposes. Foot pads and orbital fat around the eyes are examples. Such mechanical tissue contributes little to metabolism and is excluded from this discussion. Lipid in nonadipose organs is largely structural and plays a noncaloric role.

1. STRUCTURAL LIPIDS

Cell membranes, endoplasmic reticulum, and mitochondria are a complex mixture of phospholipids, protein, and sterols (Van Deenen, 1965). These lipids, which equal about 2 percent of body weight, are vitally involved with absorption across membranes and with electron transport. Although the phospholipoprotein matrix of membranes is formed and destroyed as a unit, some fatty acid exchange occurs within the phospholipids (Shapiro, 1967). This fatty acid exchange permits modification of membrane properties and adaptation of the animal to its nutritional environment (Table 13–1).

Phospholipids contain most of the essential fatty acids in the body. These essential fatty acids must contain double bonds with *cis*-configuration at carbons 6 and 9 counting from the methyl end. Kidney damage

Table 13-1. Structure and Nomenclature of Lipids

I. Phosphatidic phospholipids

$$
\begin{array}{ll}
 & \quad\; H \\
(1\ \text{or}\ \alpha) & HC - O - R_1 \\
 & \quad\; | \\
(2\ \text{or}\ \beta) & HC - O - R_2 \\
 & \quad\; | \qquad\qquad O \\
 & \quad\qquad\qquad\quad \parallel \\
(3\ \text{or}\ \alpha) & HC - O - P - O - R_3 \\
 & \quad\; H \quad - \;\backslash \\
 & \qquad\qquad\qquad\quad O
\end{array}
$$

R_1 = ester-linked fatty acid; usually saturated.

R_2 = ester-linked fatty acid; usually unsaturated.

If either the α- or β-carbon is occupied by an hydroxyl rather than a fatty acid group, the compound is a lysophosphatide.

R_3 = phosphate-linked polar group, usually choline, to yield phosphatidylcholine (lecithin); sometimes phosphatidylethanolamine, (cephalin); and occasionally serine or another amino acid.

II. Nonphosphatidic phospholipids

These contain phosphate and a basic amine group but not glycerol. Most common is sphingomyelin of nerve tissues.

III. Triglycerides

R_1 and R_2 are similar to above except that R_2 usually equals palmitate in the fat of swine and milk.

R_3 = ester-linked fatty acid which has wide range of types.

and eczema result from a lack of essential fatty acids in the diet. Thus, linoleic acid is the C18:2 essential fatty acid technically named *cis, cis* 9,12-octadecadienoic acid. The bonds between these particular carbons cannot be desaturated by animals although they require the spatial properties given by such unsaturation. Where the position or configuration of the double bonds is not known, fatty acids are described by their number of carbons and unsaturated bonds. C18:2 means any eighteen carbon fatty acid with two sites of unsaturation.

Fatty acid specificity is only partial in phospholipids since, in case of deficiency, substitutions are made but with an attempt to keep similar molecular dimensions. The enzymatic machinery for building cell walls prefers certain building blocks but certain substitutions are permitted as long as the final wall meets the specifications for life. For instance, in ruminants, erythrocytes have a smaller lecithin and greater cholesterol content than those in nonruminants, suggesting an adaptation to available nutrients (Van Deenen, 1965).

The exciting roles of phospholipids in life processes are rapidly being discovered. These roles, however, are entirely different from the function of neutral lipids and adipose tissue in caloric homeostasis.

2. BROWN and WHITE ADIPOSE TISSUE AS AN ORGAN

Adipose tissue forms 10 to 20 percent of body weight, but 70 to 80 percent of this is fat, mostly triglycerides (Fig. 13–1). The nonfat portion of adipose tissue represents from 2 to 4 percent of body weight, which is comparable to the weight of liver and other major organs. White adipose tissue is easily recognized only when its cells are engorged with fat. Fat is first deposited subcutaneously and then in the peritoneal cavity throughout the mesentery and in well-formed pads about the kidney (perinephric) and ovaries (perimetrial). As obesity increases, adipose tissue appears in the fascicular connective tissue between muscle fibers,

giving the "marbling" observed in "prime" beef.

Fat can be mobilized, resulting in a decrease in size of that adipose tissue although most of the cells remain. Such adipose tissue blends into the connective tissue which formerly surrounded that fat pad. However, the former vascularity remains and thus identifies the fat-free adipose tissue.

Brown adipose tissue, sometimes referred to as hibernating glands, is a much different tissue than white adipose in the adult animal. Epinephrine or norepinephrine released into blood by fright or anxiety causes a rapid release of fatty acid from white adipose tissue but has little effect on brown adipose function. Yet brown adipose tissue can store and release fat as animals hibernate or arouse. Brown adipose tissue is extensively distributed in hibernating species but occurs only in the interscapular region of most animals (Joel in Renold & Cahill, 1965).

The dark color and rich vascularity give brown adipose tissue the appearance of liver. Like other nonadipose tissue cells, fat inclusions are dispersed throughout the cytoplasm and the central vacuole of depot fat cells is absent. This tissue may have evolved separately from adipose tissue to perform a special function akin to endocrine organs. To the extent that we all contain a little brown adipose tissue we may all be hibernators. Fetal adipose tissue has much the appearance of brown adipose tissue but differentiates after birth.

3. LIPID CONTENT OF THE BODY

Young animals or lean adults contain more lipid than can be accounted for as adipose tissue (Table 13–2). In man, structural lipids, phospholipids, and sterols account for the discrepancy between the adipose tissue content of some 10 percent and the lipid content of 18 percent. Structural lipids are only partially extractable by most procedures because of their close association with protein. Together with protein they form a rather constant part of the body mass. Since body water declines with age

Table 13-2. Relationship between Fat and
Body Composition

(*Adapted from Pearson, 1963. Ann. N.Y. Acad. Sci. 110:
291; Reid et al., 1963. Ann. N.Y. Acad. Sci. 110:327.*)

Species	Age (in months)	Water	Fat	Protein	Mineral
		(Percent of Body Weight)			
Cattle	17	61	16	18	4.5
Cattle	22	63	14	18	4.5
Pig	—	58	24	15	2.8
Sheep	8	67	13	16	3.4
Sheep	20	58	22	16	3.4
Horse	—	60	17	17	4.5
Man	—	59	18	18	4.3

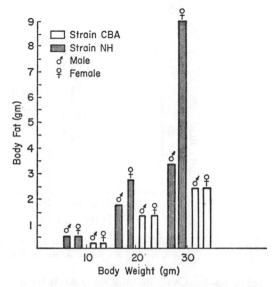

FIG. 13-2. The difference in development of body fat by two inbred strains of male and female mice which were fed the same diet to maturity. (*Adapted from Liebelt, 1963. Ann. N.Y. Acad. Sci. 110:723.*)

independently from its reciprocal relation with storage fat in adipose tissue, the percentage of structural lipids increases with age. The storage fat (triglyceride) in the central vacuole of adipose tissue is the variable component of body lipids and it varies in proportion to the water content. Triglyceride contains no water and dilutes the water content of the lean body mass. A change of 1 percent in the concentration of body fat in the pig changes the body water about 1.3 percent while a 1 percent change in the fat of cattle changes body water only 1.1 percent. Failure to lose weight during the first few weeks on a reducing diet is partially due to replacement of fat with water. The principle demonstrated here is universal but the quantitative degree varies with age, species, and physiological state.

Caloric balance affects the amount of body fat much more than species, age, sex, climate, or composition of the diet. When caloric intake exceeds the potential for growth of muscle and bone, fat is deposited. Young animals grow rapidly and are more difficult to fatten but the final distribution and amount of fat are similar whether fattening occurs throughout growth or near maturity (McMeeken, 1943). Females tend to have more fat, particularly in subcutaneous pads, than males. Studies with inbred mice demonstrate these trends, exceptions to these trends, and the feasibility of selecting for widely different patterns of fat deposition

within a species (Fig. 13-2). Generalizations about sex, species, and age effects on either the amount of fat or its distribution are apt to be erroneous.

II. COMPOSITION

A. Fatty Acids

Organ specificity for fatty acid composition tends to obscure the species differences. Since phospholipids prefer polyunsaturated fatty acids containing twenty or more carbons, these acids will form a higher proportion of the structural lipids in lean tissue than of the fatty acids present in the triglycerides of adipose tissue or even in those of liver, which sometimes contain substantial quantities of triglycerides as small cytoplasmic droplets. This may be noted by comparing the columns headed "phosphatides" with the other columns in Table 13-3. The more saturated depot fat obscures the unsaturated fat as animals are fattened for market or as they become obese. The subcutaneous fat tends to be more unsaturated, particularly in colder climates (Fig. 13-3). This tendency is attributed to the lower melting point of unsaturated fat and the

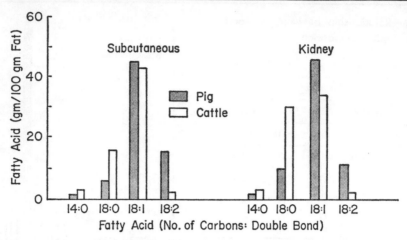

F<small>IG</small>. 13–3. Comparison of fat composition between species and location. (*Data from Church et al., 1967. J. Anim. Sci. 26:1296; Koch et al., 1968. J. Anim. Sci. 27:360.*)

Table 13-3. Comparison of Fatty Acids from Different Tissues in Cattle and Pig

(*Data from Hilditch & Williams, 1964. The Chemical Constitution of Natural Fats. 4th ed., Wiley, N.Y.; Koch et al., 1968. J. Anim. Sci. 27:360.*)

		Liver	
	Kidney Fat	Triglycerides	Phosphatides
Species	(gm saturated fatty acid/100 gm fat)		
Cattle	53	42	40
Pig	38	32	27

attempt of organisms to maintain a constant degree of firmness regardless of temperature.

Ruminants have larger proportions of saturated fatty acids than nonruminants due to extensive hydrogenation of dietary fat by the bacteria in the rumen (Tove, 1965). These rumen bacteria, working in the absence of oxygen, attempt to reduce all feed constituents in order to obtain energy by hydrogen oxidation. Microbial and chemical hydrogenation of fatty acids results in both displacement of double bonds and changes from the natural *cis*- to the unnatural *trans*-configuration. Neither rumen bacteria nor the animal can utilize the *trans*-acid as well as the *cis*- and hence, *trans*-C18:1 tends to accumulate with the saturated acids. Ruminant fat also contains more branched-chain, odd-carbon number,

and hydroxy fatty acids than nonruminant fat due to the metabolism of the rumen bacteria. The relatively large amount of short-chain fatty acids with fourteen or less carbons in ruminant fats has also been attributed to the ruminal bacteria. However, these short-chain acids tend to soften the fat and may represent adaptation to excess stearic acid (C18:0), which makes fat hard.

The fat of nonruminants can be markedly softened by feeding unsaturated fatty acids but the composition returns to normal a few weeks after the unsaturated fat is removed from the diet. Ruminants are resistant to this softening effect, but C18:1 displaces other fatty acids as the length of their feeding period is extended or as they become fatter (Bensadoun & Reid, 1965).

B. Nonlipid Components

Adipose tissue contains 1.2 to 2.6 mg N/gm fresh weight. This amounts to about 1.25 percent protein, which has been further characterized as 16 percent collagen, 7 percent nucleoprotein, and 77 percent cytoplasmic protein (Herrera & Renold in Renold & Cahill, 1965). Since approximately four fifths (80 percent) of adipose tissue is triglyceride, the protein would all be in the one fifth of the cell encasing the fat vacuole. This adipose tissue, when freed of deposited fat, should contain five times 1.25 percent, which equals 6.25 percent of protein. Thus, even the active portion of adipose tissue contains less protein than do other vital organs. This small quantity can be explained by inclusions of small fat vacuoles and some glycogen in the cytoplasm.

A protein turnover of 15 percent per day has been measured in adipose tissue; the value again is comparable to that in other vital organs. The activities of several enzymes involved in carbohydrate metabolism are about one half as much per unit of nitrogen in adipose tissue as in liver. A notable exception is the activity of glucose-6-phosphate dehydrogenase, which provides much of the hydrogen for fatty acid synthesis. Thus, carbohydrate metabolism in adipose tissue is specifically directed toward fat synthesis.

C. Fat Absorption and Blood Lipids

Bile salts, formed in the liver and secreted into the intestine, emulsify the dietary and endogenous fat, making it available to the intestinal and pancreatic lipases. As digestion proceeds, mono- and diglycerides are formed which further aid emulsification of the water-insoluble fat into micelles, exposing a large surface to the water-soluble fat-digesting enzymes called lipases. The relative slowness of this process delays gastric emptying and food transit which, in moderation, is beneficial to the animal. Glycerol, fatty acids, and monoglycerides move into the intestinal mucosa to start the fatty acid cycle (Fig. 13–4). All fat passes through this cycle and some is extensively recycled. Here is the control point for lipid metabolism.

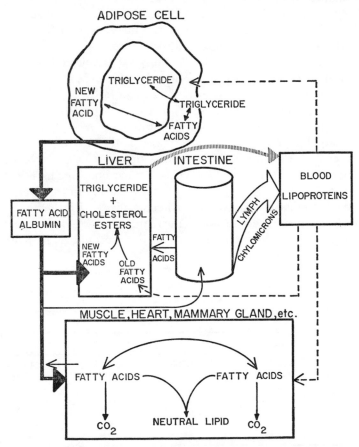

Fɪɢ. 13–4.   The fatty acid cycle.   Fatty acids enter the body through the intestine and mix with fat from the other organs.   The adipose tissue and liver control the availability of fat by controlling the blood fatty acids and lipoporteins.

1. LIPOPROTEINS

Blood fat is about 90 percent particulate. These lipoprotein particles are made only in the intestine and liver and those made by the intestine are called chylomicrons. Lipoproteins have been traditionally divided into density classes by centrifugation. Particles of different chemical identity are found within each lipoprotein class (Table 13–4). Thus, low-density lipoproteins generally migrate electrophoretically with the beta globulins and high-density lipoproteins migrate with the alpha globulins. However, more rigorous electrophoretic techniques will separate five lipoproteins from human blood and precipitation with organic solvents, salts, and organic polymers such as polyvinylpyrrolidone, dextran sulfate, or heparin produces additional fractions.

Tissues selectively remove fat from certain lipoproteins, with preference for the lower-density classes. However, selection must involve chemical characteristics other than density. The lipid in high-density lipoproteins is relatively stable and should be considered a storage fat. However, cholesterol moves between high- and low-density lipoproteins. High-density lipoproteins form a repository for free cholesterol, which shifts to low-density lipoproteins to transport free fatty acids as the need arises.

The intestinal mucosa mixes dietary fatty acids with those of endogenous origin and re-forms chylomicron triglycerides. Or, endogenous lipids can form all of the chylomicrons (Bierman in Renold & Cahill, 1965). Cholesterol esters are formed similarly. The energy and glycerol for these esterifications come primarily from glucose but the glycerol from intestinal lipolysis or that absorbed as monoglyceride can also be utilized. These esterification reactions occur in the endoplasmic reticulum, which is the same cellular structure responsible for protein synthesis. Electron radioautographs have clearly shown that both protein and triglyceride syntheses are required to make lipoprotein (Jersild, 1966).

Chylomicrons are secreted into the lymph and move through the thoracic duct into the general blood circulation. These primary lipoproteins are rapidly transformed in the blood and liver to secondary particles, and, finally, to the low-density lipoproteins. After a fatty meal the large primary particles disperse light, giving the lymph and serum a milky appearance referred to as alimentary lipemia. Lipemia is normally cleared in one or two hours by tissue removal and transformation of primary particles to smaller lipoproteins.

Liver is the only organ known both to remove lipoproteins from blood and to discharge new lipoproteins into blood. Fatty acids are carried to the liver by serum albumin and mixed with those synthesized in the liver and those removed by lipolysis from other lipoproteins. The liver then re-form

Table 13-4. Types of Particulate Blood Fat

(From Trams & Brown, 1966. J. Theoret. Biol. 12:311; Lees, 1967. J. Lipid Res. 8:396; Levy et al., 1967. J. Lipid Res. 8:463; Scanu, 1966. J. Lipid Res. 7;295.)

Characteristics	Type of Particle				
	Primary[1]	Secondary[1]	VLDL[2]	LDL[2]	HDL[2]
Density (g/cc³)	<1.006	<1.006	<1.006	<1.063	<1.063
Diameter (mμ)	200	70	20–70	20–30	7–20
Cholesterol and esters (%)	3	3	17	41	11
Triglycerides (%)	80–90	80–90	55	14	9
Phospholipid (%)	6	6	20	22	22
Protein (%)	2–4	2–4	8	23	58
Arginine/Isoleucine	1.8	1.8	0.8	0.6	2.2

[1] Types of chylomicrons.
[2] Very low-density lipoprotein (VLDL), low-density lipoprotein (LDL), high-density lipoprotein (HDL).

triglycerides by using both glycerol and glucose and adds specific proteins to form new lipoproteins (Trams *et al.*, 1966, Jones *et al.*, 1967). Poisons, dietary protein deficiency, choline deficiency, or other conditions which interfere with hepatic lipoprotein formation result in fat accumulation and eventual necrosis of the liver.

The fatty acids contained in blood glycerides are more subject to dietary variation than is depot fat. The values given for blood in Table 13–5 are comparable to those

Table 13-5. Composition of Blood Plasma Lipids

(*Data from Long, 1961. Biochemist's Handbook. Spoon, London, 1193 p.; Davis & Sachan, 1966. J. Dairy Sci. 49:1567; Evans et al., 1961. J. Dairy Sci. 44:475.*)

	Species			
	Man (Female)	Cat	Rabbit	Cattle (Lactating)
Lipid		(mg/100 ml plasma)		
Phospholipid	195	132	78	101
Glyceride	154	108	105	<37
TOTAL	617	376	243	396
Glyceride fatty acids: (% of glyceride)				
C16:0	20–28			17–27
C16:1	—			0.5–3
C18:0	4–8			25–40
C18:1	32–46			12–22
C18:2	11–38			8–14

given for tissue in Table 13–3 except for a somewhat greater degree of unsaturation. The ruminant is again notable for its high stearic acid content, and, also, for a low plasma triglyceride but a much greater cholesterol ester concentration than that found in most species. The cholesterol esters of plasma show a strong affinity for unsaturated fatty acids and contain as much as 75 percent C18:2. The human being has a relatively high total blood fat due to an increased concentration of all classes of lipids. Diet will not account for these species differences since man has twice the cholesterol ester concentration of the cat fed a similar ration and the rabbit has 2 to 10 times as much triglyceride as the cow, another herbivore. The hen has a peculiarly high plasma lipid when laying eggs, but is comparable to other species at other times.

2. FREE FATTY ACIDS

Blood plasma contains less than 1 mM (about 20 mg/100 ml) of free or nonesterified fatty acids, which are bound to serum albumin and would be very toxic if not so bound. These acids have a rapid flux and are replaced every few minutes. Their concentration can vary more than fivefold within an animal and within a few hours, depending on the nutritional state, energy demand, and anxiety. Oleic and palmitic acids predominate among the free fatty acids except in the ruminant, in which stearic acid can be of equal magnitude. Unsaturated acids and oleic acid, especially, show the most rapid fluctuation (Annison in Dawson & Rhodes, 1964).

Adipose tissue is the primary source of free fatty acids the release of which is under hormonal, neural, and nutritional control. Increasing blood glucose reduces blood fatty acid concentration by stimulating hepatic and adiposal re-esterification. Hepatic esterification results in release of a new lipoprotein which may go back to adipose tissue to release more fatty acid to start the cycle again (Fig. 13–4). Thus, the type and amount of fat deposited are controlled by the fatty acid cycle. Fasting, growth hormone, epinephrine, and sympathetic nerve stimulation augment release and concentration in blood of free fatty acids. Since most tissues utilize fatty acids for energy in direct proportion to the blood concentration, this cycle performs an important homeostatic mechanism by permitting the animal to draw on its fat stores to supply either most or as little as 10 percent of its energy (Brodie *et al.*, Carlson *et al.*, in Renold & Cahill, 1965).

III. PHYSIOLOGY

The role of lipids in the energy economy of an animal can be appreciated after studying Figure 13–5. Although practicality dictated that rats be used for the experiment, the

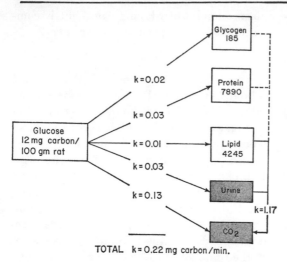

TOTAL k = 0.22 mg carbon/min.

FIG. 13–5. Energy flow through body lipids. Transfer rates between pools (k) are given in mg carbon per minute. Pool sizes are given in mg carbon per 100 gm rat. Uniformly labeled glucose was injected into postabsorptive rats, and the transfer rates measured before sacrificing the rats to determine pool sizes. (*Data from Shipley et al., 1967. Amer. J. Physiol. 213:1149.*)

glycerides is about equal to that through glucose. Thus, some fatty acids are re-esterified within the tissue without forming lipoproteins and some others are esterified and recycled as cholesterol esters. The fatty acid cycle is costly. About one fourth of the carbohydrate calories is lost in fatty acid formation and another 6 percent is lost in re-esterification.

A. Sites and Methods of Lipid Synthesis

Formation of fatty acids and other lipids by the animal insures a constancy of body composition independent of dietary supply. Fattening pigs or lactating cows seldom eat enough fatty acids to supply their needs for production. Moreover, the lipid intake from plants is highly unsaturated and makes the animal very flabby unless the fatty acids are rebuilt during metabolism.

1. ORGANS IN WHICH FAT IS SYNTHESIZED AND UTILIZED

Rapid transport of lipid between tissues and the influx of dietary fat make it surprisingly difficult to determine which organs are most important for synthesis. Most of the data have been collected from rats, but limited data from other species justifies extrapolation to man and farm animals. Thus, in the nonlactating animal the liver is responsible for 3 to 50 percent of fatty acid synthesis, the intestine for 5 to 15 percent, and adipose tissue for 50 percent or more. In addition, many other tissues have a limited ability to synthesize fatty acids, and during lactation, the mammary gland becomes a major site of this activity. Brown adipose tissue has a rate of fatty acid synthesis and turnover which is comparable to that of liver (Favarger in Renold & Cahill, 1965).

Adipose tissue is of secondary importance for synthesis of cholesterol and phospholipids. The liver and other tissues are the predominant site for synthesis of these compounds. Hepatic cholesterol synthesis is regulated by supply so that cholesterol con-

data should apply also to other animals. The rate of glucose carbon transformation to fat was only about one third of that to protein (k = 0.01 versus k = 0.03) or one half of that to glycogen, but all of the glycogen and about 60 percent of the protein will eventually come back into the glucose pool.

Remembering that fat has 2.25 times the energy concentration of carbohydrate, the energy produced by the glucose flux through lipid to carbon dioxide is exceeded only by the direct oxidation of glucose. These rats had been fasted and were mobilizing fat about 100 times faster than they were making it. Fat was supplying more than 80 percent of their energy requirement. Under conditions of feeding, the rate of lipid formation would be greater and its oxidation slower. Although these rats contained almost twice as much protein as fat in their bodies, the caloric value of the fat was about equal to all of the nonfat portion of the body.

The caloric flux through free fatty acids is estimated at 20 times the maintenance energy of the animal. Obviously, most of these fatty acids are re-esterified and not oxidized. The energy flux through tri-

centration in the body resists change with altered dietary supply.

Refeeding after a fast causes a large increase in the rate of fatty acid synthesis, particularly in the liver. This increase explains the wide range, 3 to 50 percent, for the hepatic contribution to total fatty acid synthesis in the rat. The effects of refeeding are more dramatic in rats than in most other species; moderation is required for extension of this trend to other species.

2. Species and Nutrient Comparisons

Adipose tissue has not evolved as a specialized organ in fish and fowl to the extent that it has in mammals. Therefore, in fowl, synthesis of fatty acids in the liver exceeds that in adipose tissue.

Ruminants, because of nutrient conversion to acetic acid by rumen bacteria, have adapted to utilize acetate in preference to glucose for fatty acid synthesis. Rats clearly prefer glucose for fatty acid synthesis. Incorporation of glucose into lipid is 1 to 10 times more in rats than in cows. Species, organ, and substrate differences for oxidation (CO_2 production) are not nearly as marked as those for fat synthesis. Finally, hepatic fat synthesis in the cow is negligible, but in the rat it is appreciable even considering that the rat has about five times as much adipose as hepatic tissue.

3. Methods of Fatty Acid Synthesis

The malonyl-CoA pathway clearly dominates fatty acid synthesis (Martin & Vagelos in Renold & Cahill, 1965; Tove, 1965). In this pathway, acetate is first activated to acetyl-CoA, which is then carboxylated in the presence of ATP and a biotin-containing enzyme to form malonyl-CoA (Fig. 13–6). Increasing magnesium and bicarbonate concentrations within the physiological range stimulate this reaction (Smith & Dils, 1966). In the third step, malonyl-CoA condenses with a fatty acyl-CoA, and, after decarboxylation plus reduction by two molecules of NADPH, a fatty acid two carbons longer is produced. These reactions occur on a soluble complex of enzymes in which the growing fatty acid is bound to a pantetheine-containing protein termed "acyl-carrier protein" (Rao & Johnston, 1967). The complex tends to dissociate from the carrier protein during fasting, when fatty acid synthesis is reduced. Synthesis is also inhibited by fatty acids or acyl-CoA and it is stimulated by citrate. More glucose, such as is provided by better feeding, will increase the cellular levels of citrate and remove the fatty acids by stimulating triglyceride formation.

Fatty acid synthesis requires a large supply of hydrogen from NADPH (Fig. 13–6). The dehydrogenation of glucose-6-phosphate with subsequent decarboxylation to pentose-phosphate supplies most of the NADPH. Metabolism of glucose by way of pentose-phosphate is generally increased in tissues making fatty acids.

Most fatty acids are further elongated, desaturated, or both by an enzyme complex in the endoplasmic reticulum of the cell (Mohrhauer et al., 1967). Like the malonyl-CoA pathway, ATP and coenzyme A are required but NADH can partially replace NADPH and acetyl-CoA can replace malonyl-CoA under at least some circumstances.

$$\text{Acetate} + \text{ATP} + \text{CoA} \xrightarrow[\text{Mg}^{++}]{(1)} \text{Acetyl CoA} + \text{AMP} + \text{PPi}$$

$$\text{Acetyl CoA} + \text{ATP} + \text{HCO}_3^- \xrightarrow[\text{Mg}^{++}]{(2)} \text{Malonyl CoA} + \text{ADP} + \text{Pi}$$

$$\text{Acetyl CoA} + 7\,\text{Malonyl CoA} + 14\,\text{NADPH} \xrightarrow{(3)} \text{Palmitate} + 14\,\text{NADP} + \text{CO}_2$$

Fig. 13–6. Fatty acid synthesis via malonyl-coenzyme A (CoA). PPi = inorganic pyrophosphate. Pi = inorganic orthophosphate. NADP = nicotinamide-adenine dinucleotide phosphate, formerly triphosphopyridine nucleotide. NADPH = reduced form.

This system is inhibited by fatty acids and by lack of insulin.

B. Mobilization and Utilization

The concept of fat mobilization and transport to another tissue for utilization was presented with the fatty acid cycle in section II-C and in Figure 13–4. This section deals with the control of the fatty acid cycle through nutritional and hormonal control of the molecular processes in individual organs.

1. Coordination Between Organs

The fatty acids of lipoproteins are removed from blood by the liver, adipose tissue, lactating mammary gland, and, to a lesser extent, by other tissues. The classical lipoprotein lipase, or clearing factor lipase, is released from tissues and activated by heparin. This enzyme is inhibited by bile salts. Liver has a different lipoprotein lipase, which is not activated by heparin

(Higgins & Green, 1967). It is now apparent that lipoprotein lipase attacks the 1,3-position of triglycerides while a monoglyceride lipase attacks the 2-position. Specificity is not absolute, but certain triglycerides and the 1,3-position are preferentially attacked (Payza *et al.*, 1967). Other esterases liberate fatty acids from cholesterol esters.

Free fatty acids are liberated from triglycerides in tissues by still other lipases. Adipose tissue has a hormone-sensitive lipase in the endoplasmic reticulum which is strongly activated by epinephrine and norepinephrine. Fatty acids are removed from phospholipids by a series of phospholipases, each of which has positional specificity. Thus, animals can control the organ from which lipid is mobilized, the type of lipid withdrawn, and even which fatty acid is withdrawn.

Free fatty acids in blood are transferred

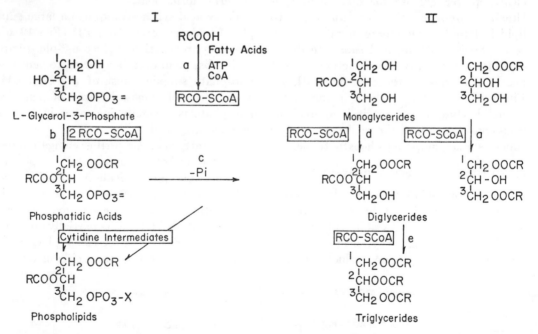

Fig. 13–7. The alternative pathways of diglyceride synthesis in the intestinal mucosa: I, The glycerol phosphate-phosphatidic acid pathway. II, The direct monoglyceride pathway. Enzymes involved are (a) (fatty) acid: CoA lipase (AMP) (6.2.1.3); (b) acyl CoA:L-glycerol-3-phosphate acyl transferase (2.3.1.15); (c) L-α-phosphatidate phosphohydrolase (3.1.3.4); (d) acyl CoA:monoglyceride acyl transferase; (e) acyl CoA:diglyceride acyl transferase. (Enzymes d and e have been neither isolated, nor purified, nor officially named as yet.) The glyceride numbering system used here follows the well-established system in general use for the triosephosphates. (*From Senior, 1964, J. Lipid Res. 5:495.*)

in about equal portions to liver, striated muscle, and splanchnic area, including the heart. In short, uptake is equivalent to blood flow times fatty acid concentration. The percentage of extraction and mobilization decreases with increasing chain length and degree of saturation. The mammary gland is an exception to this generalization since it does not have a net uptake of free fatty acids.

Synthesis of glycerides and phospholipids occurs in the endoplasmic reticulum, where the concentration of cytidine compounds controls the extent of phospholipid synthesis (Fig. 13–7). The importance of the phosphatidic acid pathway is emphasized by the high requirement for glycerol-3-phosphate (about 8 mM) for optimum triglyceride synthesis in most tissues. A mixture of palmitate and unsaturated fatty acid gives better synthesis than does any single fatty acid, but the exact fatty acid mixture which is optimum for triglyceride synthesis is probably determined by the species and tissue (Brindley et al., 1967). Since fatty acids are bound to a compound similar to coenzyme A when synthesized, it would be a definite energetic advantage for the animal to utilize this fatty acyl compound directly for glyceride synthesis. Such a reaction has been identified (Rao & Johnston, 1967).

2. Hormonal and Nutritional Effects

Fatty acid and glyceride synthesis is very sensitive to both the energy and insulin supply. This sensitivity can be partially explained by an increased flux of glucose through the phosphogluconate-pentose pathway, which yields more NADPH, and partially by an increased supply of glycerol phosphate for glyceride synthesis. However, a full explanation requires other factors which are well-defined.

Lipoprotein lipase activity generally parallels the need and ability of a tissue to utilize fatty acid. Fasting one day causes muscle lipoprotein lipase activity to double producing a decline of 50 percent lipoprotein lipase in muscle. This diverts fat from adipose tissue to muscle and further adaption occurs with longer fasting. Since exchange of glyceride fatty acid between plasma and liver is faster than between plasma and adipose tissue, hepatic regulation of fat metabolism also occurs. Final control of the fatty acid cycle is centered in the liver since hepatectomy abolishes the cycle (Carlson et al., in Renold & Cahill, 1965).

Free fatty acid release from adipose tissue is suppressed 80 percent by insulin in the fed animal but insulin has little effect in the nonfed animal. Insulin also stimulates protein synthesis in adipose tissue. Among the many effects of growth hormone, fatty acid release is one of the most sensitive. Thyroxine and cortical steroids are required in a permissive sense for norepinephrine or epinephrine stimulation of the lipase, which releases fatty acids. Fasting stimulates this same lipase, but the mechanism is not known (Brodie et al., in Renold & Cahill, 1965). Hormone-sensitive lipase loses its activity in about an hour in the absence of hormone, but can be reactivated by incubation with ATP and epinephrine. This epinephrine effect is clearly a case of hormonal control of enzyme activity rather than hormone induction, but the sustained effect of corticoids and growth hormone probably involves formation of new enzymatic protein.

Species differences in hormone sensitivity of adipose tissue have been reported without appropriate regard to other variables. Thus, older adipose tissue, due either to age of the animal or due to earlier formation of a particular fat pad, will be less active and less responsive to hormones than will younger tissue. The epididymal fat pads of rats is one of the more reactive adipose tissues and should not be compared to a subcutaneous storage tissue such as the hump of a camel or the tail of fat-tail sheep, which are relatively unreactive within the species (Khachadurian et al., 1967). Also, such comparisons are usually made on a fresh weight basis, ignoring the fact that only the nonfat portion of adipose cells is metabolically active. On a fresh weight basis but not on a fat-free basis, adipose tissue from a fat animal will usually be less reactive than that from a lean animal. Nevertheless, the rat seems to be unusually responsive to both insulin and the catecholamines, while ruminants are relatively unresponsive. The pig is notably resistant to hormonal stimulation of fatty acid release, a condition also noted in mice with hereditary obesity.

Subcutaneous fat is most resistant to free acid release; perirenal, genital, and mesenteric adipose tissues are less resistant in the order given. A similar relationship exists for fat incorporation from lipoproteins in which mesenteric tissue is several times more active than subcutaneous fat tissue. These tissue activities are inversely proportional to the tissue age as determined by the stage of animal development at which they are matured.

Tissue utilization of lipids is strongly influenced by the concentration of lipid which reaches them (Table 13–6). The amount of free fatty acid oxidized to carbon dioxide for production of energy in liver is constant over the entire physiological range of about 0.1 to 1.0 mM, indicating saturation of the mitochondrial enzyme system at a low concentration. The mitochondrial system for partial oxidation of fatty acid to aceto-

acetic and β-hydroxybutyric acids is not saturated until the concentration reaches about 0.5 mM. Ketone body formation yields only about 17 percent as much energy as complete oxidation, so that returns are diminished and feed efficiency is reduced. Although percentage of fatty acid which is metabolized declines as the concentration increases, the percentage oxidized declines most dramatically. The concentration of lipoprotein triglycerides reaching the cell exceeds that of free fatty acids, but the metabolic utilization increases only slowly and to the maximal blood concentration. With both sources of fat, the highest concentrations are toxic and fatty liver followed by necrosis would result from continued exposure. In muscle, in which oxidation is the primary fate of fatty acid, oxidative utilization increases with concentration. Nevertheless, as the concentration of fat in blood increases, the animal will oxidize less of it and divert more to production of ketone bodies and storage fat, with some consequent loss of energetic efficiency.

C. Fattening and Lactation

Obesity in man is analogous to fattening in animals. Considerations in this section are drawn from both human and animal sources

Table 13–6. Hepatic Lipid Metabolism as a Function of Lipid Concentration

(*Recalculated from Ontko, 1967. Biochim. Biophys. Acta 13:13; Ontko & Jackson, 1964. J. Biol. Chem. 239:3674.*)

Substrate	Esterified Fatty Acid		Carbon Dioxide		Acetoacetic and $\beta(OH)$butyric Acid	
	(μg)	(% of Substrate)	(μg)	(% of) Substrate)	(μg)	(% of Substrate)
Free Fatty Acid:[1]						
5.1 (0.2 mM)	29	28	17	17	47	46
12.8 (0.5 mM)	77	30	22	8	110	43
20.6 (0.8 mM)	82	20	19	4	96	23
25.6 (1.0 mM)	72	14	17	3	59	12
Triglycerides:[2]						
20	—	—	9	2.2	18	4.5
100	—	—	24	1.2	49	2.5
390	—	—	47	0.6	95	1.2
780	—	—	56	0.4	109	0.7

[1] Palmitic acid, ^{14}C-labeled.
[2] Chylomicrons labeled with C[14]-palmitic acid. Due to a change of incubation time and use of different rats, the rates of metabolism of chylomicrons and free fatty acids are not directly comparable.

and applications are appropriate to either man or animal. Weight gain in mature animals is about 80 percent glyceride fat. Thus, 1.2 kg gain is equivalent to 1 kg of fat deposited. In the growing animal, weight gain consists of the formation of protein and structural lipid until the caloric intake exceeds the limits of growth rate, when the excess calories are deposited as fat. Lactation can be considered a special process of fattening in a growing animal in which milk is a growth tissue with about equal proportions of fat and protein. A cow producing about 30 kg of milk would be excreting about 1 kg of fat per day, the amount which is equivalent to the best gains in fattening cattle.

1. Physiological Control of Fattening

The deficiency of adipose tissue and hepatic fat synthesizing enzymes in newborn animals is reflected by a reduction in blood fat. The data for calves (Table 13–7) are applicable to most species, with appropriate corrections for physiological age. Concentrations of free fatty acids are high at birth, decline rapidly when nursing starts, and then gradually attain adult values. Total plasma lipids and all esterified fatty acid concentrations increase until weaning and then decline to adult values. The relationship between age and body fat has been discussed previously (Fig. 13–2).

As adipose tissue ages, its metabolic activity declines, fatty acid release slows, and fattening commences. Subcutaneous adipose tissue is the first to differentiate and form central fat vacuoles. Later this process moves to adipose tissue in the abdominal cavity. The exact pattern varies with species and strains within a species. Intramuscular adipose tissue develops only with maturity of the animal and does not accumulate fat unless there is both caloric excess and lack of exercise. Thus, the condition we strive to attain in market grade cattle would approach pathological obesity in humans. The art of animal feeding is to attain the obesity without the pathology as evidenced by fatty degeneration of hepatic and other tissues and failure of the circulatory system.

Insulin causes a general hyperplasia and hypertrophy of adipose tissue at concentrations which have only small effects on other tissues. The fatty acid desaturase system is particularly sensitive. Insulin administration and the feeding of a low-fat diet reduce the effect of aging on adipose tissue. The relationship between insulin and hereditary obesity has not been established, but the implications are obvious.

Some forms of hereditary obesity are associated with increased glycerokinase activity in adipose tissue, which has the same effect as an increase in the glucose supply to the tissue. A decrease in fatty acid release has also been associated with hereditary obesity and fattening in pigs. However, the dominant factor in fat accumulation is caloric intake, which offsets the small differences between both type of ration and species of animal (Bensadoun & Reid, 1965). Dietary fat causes obesity largely because it increases the caloric intake.

Table 13-7. Plasma Lipids in Calves

(*Data from Shannon & Lascelles, 1966. Austr. J. Biol. Sci. 19:831.*)

Age	Total Fat	Phospholipids	Triglyceride (mg/100 ml)	Cholesterol Ester	Free Fatty Acids
Birth	140	51	7	37	21
6 days	295	129	25	103	8
1 month	392	169	13	161	8
2 months	359	156	21	139	6
6 months	314	142	21	118	5

2. Physiological Control of Milk Fat Production

Secretion of milk fat requires lactation. Under estrus cycle, hormones, pregnancy, and parturition, the functional mammary cells develop from undifferentiated epithelial tissue. With lactation, the mammary gland forms a lipoprotein lipase complex to transfer blood fat to the secretory cells, a malonyl-CoA pathway to synthesize 14-carbon and 16-carbon fatty acids, a special enoyl-hydrase to reduce β-hydroxybutyric acid, and the microsomal system for chain elongation, desaturation, and glyceride formation.

The fatty acids of milk with less than 16 carbons are formed in the gland while those with more than 16 carbons are derived from blood. Palmitic acid is derived from both sources. Most of the carbon for fatty acid synthesis in ruminants comes from blood acetate and β-hydroxybutyrate. The glucose and amino acid supply is only adequate to form the protein, lactose, and energy. In nonruminants, glucose from blood would largely replace acetate and β-hydroxybutyrate, with a consequent reduction in production of very short-chain fatty acids. Nonruminant milks have a higher content of polyunsaturated fatty acids and a lower content of stearic acid than do ruminant milks, but this difference reflects the supply of fatty acids from blood rather than synthesis in the mammary gland (see Table 13–5).

Acetate is the major precursor for fatty acid synthesis in both ruminants and nonruminants, but in nonruminants the acetate is obtained by mitochondrial oxidation of glucose to citrate and then cleavage of the citrate to acetate after it is transferred into the cytoplasm, where the fatty acid synthesizing enzymes are found. In ruminants, glucose is committed to complete oxidation once it enters the mitochondria and is found only in the glycerol portion of milk fat. Acetate from the rumen is the major precursor for fatty acids with up to 16 carbons.

The longer chain fatty acids of milk come from blood lipoproteins (Tove, 1965) and, particularly, the lipoproteins precipitable with dextran sulfate or heparin. Triglyceride synthetase is the rate-limiting reaction for milk fat formation; the kind of fatty acid available affects this enzyme (Kuhn, 1967). Blood fat is transferred to the mammary gland and secreted from the endoplasmic reticulum without passing through the Golgi apparatus (Stein & Stein, 1967).

Dietary control of milk fat secretion is superimposed upon the on-off control of lactation. In a diet of high caloric density, replacing hay with grain will depress milk fat synthesis in ruminants. It was previously hypothesized that decreased ruminal acetate production causes this phenomenon, but the hypothesis was discarded with the realization that fatty acids changed in milk are derived from blood fat and there is little change in the acetate (Fig. 13–8). Glucose infusion causes a similar depression in milk fat and alters composition. These same rations affect the fatty acid synthesizing enzymes of adipose more than those of mammary tissue (Fig. 13–9). Rations which stimulate ruminal propionic acid also stimulate retention of fat in adipose tissue and depress milk fat secretion (Weiss et al., 1967). In other words, the cows fatten instead of secreting milk fat. Addition of sodium bicarbonate or magnesium salts to the diet stimulates transfer of blood fat to the mammary gland and, thus, milk fat secretion.

D. Abnormal Lipid Metabolism

The most common abnormality of lipid metabolism is obesity from excess caloric intake. As discussed previously, there are genetic changes which increase triglyceride deposition and suppress fatty acid release from adipose tissue. To some extent, selecting farm animals for increased fatness may be selecting for some such abnormality.

Lipomas or tumors of adipose tissue are usually nonmalignant increases in adipose tissue due to a combination of reduced fatty acid mobilization and increased synthesis of both fatty acids and triglycerides. These growths occur as localized lumps in both subcutaneous and mesenteric fat. Occasionally,

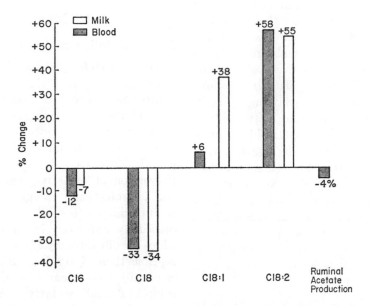

FIG. 13–8.   Percent change in ruminal acetate production and fatty acids of milk and blood triglyceride caused by feeding a milk-fat depressing ration. (*Data from Davis & Sachan, 1966. J. Dairy Sci. 49:1567; Davis, 1967. J. Dairy Sci. 50:1621.*)

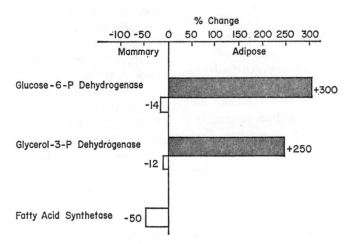

FIG. 13–9.   Percent change of several enzymes of adipose tissue and mammary gland caused by feeding a milk-fat depressing ration. (*Data from Opstvedt et al., 1967. J. Dairy Sci. 50:108.*)

they become malignant (Gellhorn & Benjamin in Renold & Cahill, 1965).

Abnormal fatty acids can interfere with the fatty acid cycle or find their way into structural fats where they cause a number of problems such as reduced absorption from the intestinal tract, fatty livers, ataxia, and dystrophy. Vitamin E deficiency shows how lack of a natural antioxidant permits peroxidation of unsaturated fatty acids in the animal. Since phospholipids contain most of the unsaturated fatty acids, they are the primary site of the lesion. Oxidized fatty acids of dietary origin are poorly absorbed and cause intestinal disorders (Schultz et al., 1962). In a similar abnormality, phytanic acid accumulates in the serum because it can be substituted for fatty acids in triglyceride synthesis but is not available to lipases (Laurell, 1968).

Fatty liver, which leads to necrosis in its extreme form, is usually caused by interference with lipoprotein formation and, thus, export of fat. Increased hepatic fat synthesis may be a factor in some cases, but increased mobilization of fat from adipose tissue is more common. All hepatoxins are capable of producing fatty liver, probably by interfering with the protein synthesis required for lipoprotein formation. Alcohol increases hepatic uptake of fatty acids and prevents their export as lipoprotein, but does not block triglyceride formation. Fasting and diabetes have a similar effect. Essential dietary agents such as choline, vitamin B_{12}, or methionine which prevent fatty liver also seem to be involved in lipoprotein synthesis (Shapiro, 1967).

Lipodystrophy, a hereditary deficiency or absence of adipose tissue, results in an insulin-resistant hyperglycemia and hyperlipemia. Both of these symptoms are the kind that would be predicted from the failure of adipose tissue to remove the influx of fat and carbohydrate from blood after a meal. Kidney and cardiac disorders usually accompany lipodystrophy. A more common and less serious disorder is a hereditary absence

of one or more of the lipoprotein classes from blood (Senior in Renold & Cahill, 1965).

IV. NECROBIOLOGY

A. Peroxidation and Rancidity

Unsaturated fatty acids react with oxygen to form lipoperoxides, aldehydes, keto-acids, hydroxyacids, epoxy compounds, and polymers of these compounds. This nonenzymatic reaction is self-catalyzing but is accelerated by metals and hematin compounds. There are also lipoxidase enzymes. Prevention of oxidative deterioration consists of protection from oxygen and inclusion of antioxidants to block the chain reaction caused by self-catalysis. Free fatty acids which are liberated by lipases are more susceptible than glycerides. Rancid flavor is caused by a combination of these oxidation compounds and free fatty acids.

The fat deposited in adipose tissue or secreted in milk is well protected by the lipoprotein membrane which surrounds the central fat vacuole or the milk fat globule. However, these membranes contain lipases which are activated when the membrane is damaged. When this membrane is ruptured by processes such as rendering, churning, or homogenization, conditions must be adequate to destroy lipase activity in order to preserve the flavor and nutritive value of fat. The liberated fat must be protected from oxygen, light, elevated temperatures, copper, iron, hemoglobin, lipases, and lipoxidases.

Phospholipids are both unsaturated and dispersed throughout the carcass. For this reason, muscle and liver are both more susceptible to rancidity than fat. Tissues of nonruminants, because of their higher polyenoic acid content, are more susceptible than tissues of ruminants. For example, beef muscle contains about 0.5 percent of polyenoic acid while pork muscle contains about 1 percent. Refrigeration, good nutrition, and management of market livestock are the best means of controlling rancid flavor (Schultz et al., 1962).

B. Influence of Nutrition and Management

Oxidized fats are not very digestible and will cause abdominal pain and a fatty diarrhea (steatorrhea) if consumed. Those oxidized products which are absorbed cause toxicity symptoms ranging from reduced growth to enlarged liver. Free radicals produced during peroxidation destroy oxidation-susceptible vitamins such as vitamin E, carotene, and essential fatty acids. Unsaturated feed-grade fats must be stabilized with synthetic or natural antioxidants to prevent these problems. Fats for household cooking are usually wholesome when they reach the consumer, but repeated exposure to high temperature causes problems. Frying fats are frequently hydrogenated to increase stability and add texture. Unfortunately, this process also causes migration of the unsaturated bonds, which destroys essential fatty acids and reduces the food value.

To insure against rancidity in meat and protect the animal against deficiency of essential fatty acids and vitamins, it is important to feed only well-stabilized fats and to insure an adequate intake of the natural antioxidants, vitamin E and selenium. Pale, exudative muscle in pigs is similar to the syndromes of vitamin E and selenium deficiency in young calves and lambs except that early mortality in ruminants may prevent the marketing problem. The free fatty acid release from adipose tissue caused by nutritional or other stress just prior to marketing will aggravate the quality control problems.

V. RESEARCH TECHNIQUES

A. Synthesis

Study of fatty acid synthesis is essentially an investigation of the soluble enzymes involved (Martin & Vagelos in Renold & Cahill, 1965). The effect of variation of cellular metabolite levels on the composition and amount of end product deserves more attention (Smith & Dils, 1966). The microsomal desaturase and elongation mechanism has received only enough attention to show that it is sensitive to hormonal and nutritional status (Mohrhauer et al., 1967). The nature of the enzyme complex which synthesizes fatty acids has been described much better than it has been interpreted. The current problem is to relate fatty acid synthesis and its control to metabolism of organs in various physiological states and species (Tove, 1965).

Incorporation of labeled substrate into the lipids of a tissue, either in vivo or in vitro, is a good way to study relationships between species and organs (Steinberg & Vaughan in Renold & Cahill, 1965). However, such studies do not yield information on biochemical control mechanisms unless they are accompanied by sufficient data on enzyme and substrate levels within tissues. For example, glyceride formation could control fatty acid synthesis (Kuhn, 1967). Whenever isotope data is used to estimate the rate of synthesis of a product, it is important to consider dilution of the label by endogenous stores. For example, the radioactivity per gram of adipose tissue after feeding C^{14}-acetate to a lean animal will almost certainly be greater than after feeding it to a fat animal simply because the newly formed acids will be diluted by more preformed lipid in the fat animal. Many people have interpreted such data to mean that a lean animal synthesizes fat more rapidly than a fat animal; readers and researchers are warned against such errors.

B. Composition

Gas chromatography has greatly facilitated identification of the individual fatty acids. There are a great many variations of this procedure and the choice of methods depends upon the tissue, the mixture of fatty acids involved, and whether free fatty acids are to be included (Katz & Keeney, 1966; Shapiro, 1967). Thin-layer chromatography is the method of choice for separating the various lipid classes such as triglycerides and phospholipids. Although this procedure was used in a number of references cited in this chapter, standard reference books should be consulted. If a limited number of separations are to be made, commercially prepared kits are reasonably priced and convenient.

Chromatography is a continuous extraction sequence in which the sample is partitioned between a stationary phase and a moving phase. In gas chromatography, the stationary phase is chosen for its ability to dissolve the constituents to be separated, while solubility in the mobile phase is largely controlled by temperature. A temperature of 10 to 50° C above the boiling point of the test substance is usually most satisfactory. With thin-layer or liquid-column chromatography, the water of hydration in the supporting material usually forms the stationary phase. A moving phase is chosen for its ability to dissolve the test constituent, *i.e.*, similar materials dissolve each other. For example, if a polar sample is not moving fast enough in relation to the solvent (moving phase) a more polar solvent should be chosen. Chromatography is largely an empirical art; the user should not hesitate to try modifications of published techniques if they serve his purpose.

Lipoproteins have traditionally been identified by flotation in the ultracentrifuge (Bierman in Renold & Cahill, 1965). This mild separation technique removes contaminating serum proteins but also requires expensive instrumentation and a great deal of labor and lacks sensitivity. Precipitation with dextran sulfate and specific antisera is an economical preparative procedure that reproduces the biological system more closely than does the ultracentrifuge. Electrophoretic separations on paper, cellulose acetate, or gels can be employed for both quantitation and characterization, but only small amounts of material can be used conveniently.

Lipids are subject to hydrolysis and oxidation. Degradation can be minimized by low temperature, exclusion of oxygen, and minimal contact with polyvalent cations. Since all methods entail some risk of degradation, the experimenter must examine the integrity of his system by comparison of several techniques.

C. Metabolism

Perfusion of organs with either lipids or lipid precursors is a sophisticated means of studying their metabolism. The blood supply of a particular organ can be isolated within the animal or the organ can be removed. Blood flow and oxygenation are usually deficient with either procedure (Scow in Renold & Cahill, 1965). However, the relationship between blood vessels and tissue cells is at least partially maintained. Since these preparations have a very limited life, it is difficult to study the effect of diet or management on lipid metabolism within an animal.

If the blood supply to an organ is known, measurement of the arterial and venous concentrations of metabolites will provide an estimate of metabolism. This technique has been extensively used with the mammary gland but collateral blood supply leaves some uncertainty about both concentration and blood flow. It has also been used with liver by surgical installation of catheters, but the diffuse vascularity of adipose tissue makes permanent installation of a catheter difficult. The arteriovenous difference approach is very useful for studying the effect of diet and management on lipid metabolism within an animal. The ability to study the relationship between several organs by this technique has not yet been exploited.

The metabolic activity of individual organs or cellular fractions can be studied with tissue obtained by biopsy or slaughter (Hirsch & Goldbrick in Renold & Cahill, 1965). A wide range of procedures are available, each with its advantages and limitations. The technique of incubation of tissue slices with various substrates provides a convenient form of organ perfusion but has a disadvantage in that the normal relationship between blood vessels and cells has been lost. To identify which individual steps control a metabolic process within the cell and to remove interference by endogenous substrates, tissues are homogenized and enzymes or enzyme complexes isolated by some combination of centrifugal and chemical procedures.

The best research technique is the one which most simply tests the hypothesis or solves the problem. There are many unanswered questions concerning lipid metabolism and any research project should start with a definition of the question, and then proceed to a general approach to its answer. Good techniques produce good research only when they are applied to worthwhile questions.

REFERENCES

Bensadoun, A., & Reid, J. T. (1965). Effect of physical form, composition and level of intake of diet on the fatty acid composition of the sheep carcass. J. Nutr. 87:239–244.

Brindley, D. N., Smith, M. E., Sedgwick, B., & Hubscher, G. (1967). The effect of unsaturated fatty acids and the particle-free supernatant on the incorporation of palmitate into glycerides. Biochim. Biophys. Acta 144:285–295.

Dawson, R. M. C., & Rhodes, D. N. (Eds.) (1964). Metabolism and Physiological Significance of Lipids. London, Wiley.

Higgins, J. A., & Green, C. (1967). Properties of a lipase of rat-liver parenchymal cells. Biochem. Biophys. Acta 144:211–220.

Jersild, A., Jr. (1966). A radioautographic study of glyceride synthesis in vivo during intestinal absorption of fats and labeled glucose. J. Cell Biol. 31:413–427.

Jones, A. L., Ruderman, N. B., & Herrera, M. G. (1967). Electron microscopic and biochemical study of lipoprotein synthesis in isolated perfused rat liver. J. Lipid Res. 8:429–446.

Katz, I., & Keeney, M. (1966). Characterization of the octadecenoic acids in rumen digesta and rumen bacteria. J. Dairy Sci. 49:962–966.

Khachadurian, A. K., Kamelian, M., & Adrouni, B. (1967). Metabolism of sheep adipose tissue in vitro. Amer. J. Physiol. 213:1385–1390.

Kinsell, L. W. (Ed.) (1962). Adipose Tissue as a Organ. Springfield, Charles C Thomas.

Kuhn, N. J. (1967). Regulation of triglyceride synthe-

sis in the parturient guinea-pig mammary gland Biochem. J. 105:225–231.

Laurell, S. (1968). The action of lipoprotein lipase on glyceryltriphytanate. Biochim. Biophys. Acta 152:80–83.

McMeeken, C. P. (1943). Principals of Animal Production. Christchurch, New Zealand, Whitcombe & Tombs.

Mohrhauer, H., Christiansen, K., Gan, M. V., Deubig, M., & Holman, R. T. (1967). Chain elongation of linoleic acid and its inhibition by other fatty acids in vitro. J. Biol. Chem. 242:4507–4514.

Payza, A. N, Eiber, H., & Tchernoff, A. (1967). Studies with clearing factor. IV. Fatty acid exchange reaction catalyzed by clearing factor. Proc. Soc. Exp. Biol. Med. 124:771–774.

Rao, G. A., & Johnston, J. M. (1967). Studies of the formation and utilization of bound CoA in glyceride biosynthesis. Biochim. Biophys. Acta 144:25–33.

Renold, A. E., & Cahill, G. F., Jr. (Eds.) (1965). Adipose tissue. Sec. 5. In: Handbook of Physiology. Washington, D.C., Amer. Physiol. Soc.

Schultz, H. W., Day, E. A., & Sinnhuber, R. O. (Eds.) (1962). Symposium on Foods; Lipids and Their Oxidation. p. 442, Westport, Conn., AVI Publishing.

Shapiro, B. (1967). Lipid metabolism. Ann. Rev. Biochem. 36:247.

Smith, S., & Dils, R. (1966). Factors affecting the chain length of fatty acids synthesized by lactating rabbit mammary glands. Biochim. Biophys. Acta 116:23–40.

Stein, O., & Stein, Y. (1967). Lipid synthesis, intracellular transport and secretion. II. Electron microscopic radioautographic study of the mouse lactating mammary gland. J. Cell Biol. 34:251–263.

Tove, S. B. (1965). Fat metabolism in ruminants. In: Physiology of Digestion in the Ruminant. R. W. Dougherty (Ed.), p. 399, Washington, Butterworth.

Trams, E. G., Brown, E. A., & Lauter, C. J. (1966). Lipoprotein synthesis. I. Rat plasma lipoprotein composition and synthesis from radioactive precursors. Lipids 1:309–315.

Van Deenen, L. L. M. (1965). Phospholipids and biomembranes. Prog. Chem. Fats Lipids 8:1–127.

Weiss, R. L., Baumgardt, B. R., Barr, G. R., & Brungardt, V. H. (1967). Some influence of rumen volatile fatty acids upon carcass composition and performance in growing and fattening steers. J. Anim. Sci. 26:389–393.

Chapter 14

Skin, Wool, and Hair

By E. H. Dolnick

THE purpose of this chapter is to review the growth, development, and replacement of animal fibers and to introduce the reader to the complex array of factors affecting growth of the skin and its appendages. A knowledge of the active and quiescent periods of the hair or wool growth is necessary in animal husbandry practices as well as in research endeavors so that correct evaluations may be made of the influences exerted by nutrition, climatic variation, and the inherent physiological state of the animal.

I. MORPHOLOGY

A. Skin

Epidermis. The skin is made up of two distinct components: epidermis and dermis (Plate 17). The epidermis develops as a single layer in the early fetus. This soon becomes two-layered and then many-layered. The outermost protective fetal layer, the periderm, is a transitory structure and is lifted off by the developing hairs as they erupt. The inner layer of basal germinative cells of the epidermis, the *stratum germinativum*, provides the cells from which the intermediate layers of the epidermis are

formed. The definitive surface layer, the *stratum corneum*, is made of thin, scale-like, keratinized cells.

Epidermal-Dermal Junction. The boundary between the epidermis and the dermis is straight and uniform in appearance except in animals that are relatively hairless such as man and the pig. In these, rete pegs—or epidermal projections—provide definite areas for the blood capillaries and for the terminations of tactile nerves. These pegs are also called ridges or dermal papillae and may be likened to patterns of hills and valleys.

Dermis. The dermis contains various elements: hair follicles, sebaceous glands, sweat glands, nerves, blood vessels, and lymphatics for draining away material in the tissue fluid. Just below the dermis is a loose type of subcutaneous tissue filled with tissue fluid and fat. When present, the *panniculus carnosus* muscle forms a sharp histologic boundary between the dermis and the subcutaneous layer. This muscle is prominent in many animals, including the rabbit, cat, and mink. In sheep, it is found over the thorax and a large part of the flank. It terminates dorsolaterally short of the mid dorsal line and is absent in the lumbar and hip regions (Ryder, 1955).

B. Structure of Hair and Wool

Hairs consist of a central core of cells, the medulla, which is surrounded by the cortex. The cortex in turn is covered with an outer layer of cells, the cuticle (Plate 18). The medulla is absent in lanugo (fetal hairs of man) and in fine wool of sheep.

The medullary portion of the hair is made up of a ladder-like arrangement of one or more rows of cells. It may be absent or discontinuous, as in man, or it may be continuous along the entire length of the hair, as in the mouse. The cells of the cortex are spindle-shaped and longitudinally oriented. Pigment may be present in both the medulla and the cortex, but is usually absent from the cuticle. The cuticle consists of thin, transparent, cornified, epithelial cells which are sometimes arranged like overlapping shingles. The pattern of the cuticular scales varies from region to region in the individual hair.

The crimp in wool is attributed to the bilateral pattern of the cortex, designated as ortho- and paracortex (Fig. 14–1). Fine

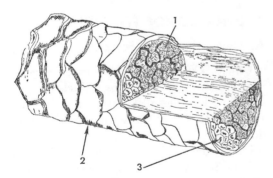

Fig. 14–1. Diagrammatic representation of the bilateral arrangement of the cortex of a wool fiber: (1) orthocortex, (2) cuticle, and (3) paracortex.

crimped alpaca fibers of the huacayo variety (von Bergen, 1963) have the bilateral pattern of fine wools. In fibers that form waves rather than crimp, such as mohair fibers of the adult Angora goat and the coarser types of wool (*e.g.*, Lincoln), a ring of paracortex

surrounds the orthocortex, producing a radial arrangement instead of a bilateral one.

The hair within the follicle is surrounded by three sheaths: the inner root sheath, the outer root sheath, and the connective tissue sheath. The inner root sheath extends from the sebaceous gland to the bulb or root of the hair. The outer root sheath, which is essentially an invagination of the epidermis, is a continuation of the basal cell layer and extends to the bulb of the hair. The connective tissue sheath is found fully developed only around the lower half of the root. It is absent from the lanugo hairs of man and may be absent in very fine hair and wool. If a growing hair is plucked, the inner root sheath comes away with the plucked hair while the outer root sheath and the connective tissue sheath remain within the skin.

C. Fiber Types and Patterns

While animal hairs have many forms, they may be simply classified into two broad categories: guard hairs and underfur. The guard hairs are laid down first in fetal development and are the outer hairs of the coat. They are associated with three structures, namely, the sweat gland, the sebaceous gland, and the *arrector pili* muscle. This muscle, when it contracts, not only changes the position of the hair so that it stands vertically, but also helps to constrict the superficial blood vessels of the skin. The hairs of the underfur make up the undercoat and appear slightly later in development. They are finer, more uniform in diameter, and generally crimped or wavy. Except for an occasional sebaceous gland lobe, there are no accessory structures for the follicles of the underfur.

Guard hairs may be (a) the common variety, which are straight in form, (b) tylotrichs, which are also straight in form but slightly longer, and (c) heterotypes. Heterotypes contain features of both guard hair and underfur since they are coarse at the proximal end and fine and wavy at the distal end. Approximately 10 percent of the guard hairs of the animal coat are tylotrichs. They are distinguished histologically by a thick pad of epidermal cells (Straile, 1960). This form of guard hair has a sensory function.

The arrangement of follicles in groups is best seen in histological sections of the skin oriented to show hairs in cross-section at the level of the sebaceous gland. A follicle group consists of one guard hair (tylotrich or common variety) flanked by two heterotypes. Underfur hairs of varying number surround each guard hair in satellite fashion. The primary and secondary fibers of sheep correspond to the guard hair and the underfur of other animals. In fine-woolled sheep, the primary follicle that produces a hairy type of fiber in the birthcoat will subsequently form a wool fiber.

D. Glands of the Skin

Sweat Glands. There are two kinds of sweat glands: apocrine and eccrine. Apocrine glands are usually associated with hair or wool follicles, while eccrine glands are not. The apocrine gland is an accessory structure of each primary wool fiber and each guard hair (Plate 18), and is a prominent feature in the skin of most mammals. It is absent, however, in many rodents. In man, the rudiments of apocrine glands are present over most of the body in the fetus, but they disappear shortly after birth and become restricted principally to the region of the axilla, the inguinal region, and the perianal skin.

The eccrine glands are the true secretory glands and produce the clear aqueous sweat which cools the body by evaporation. The distribution of eccrine glands over the whole body surface is found only in man. In most other mammals these glands are found primarily in the hairless regions such as the foot pads. While there are eccrine glands in the skin of monkeys, they are not very active and the skin remains dry in hot weather (Montagna, 1965). Sweat, when it collects on the skin, is termed "sensible perspiration" as opposed to insensible vapor or perspiration, which is evaporated as fast as it is formed. Cattle sweat visibly only from the muzzle, while dogs and cats sweat freely from the foot pads as well as from the muzzle. Animals, except for the horse and man, do not sweat from the general body surface.

Sebaceous Glands. The fully developed sebaceous gland secretes sebum *in utero*. The gland first appears as a small swelling in the upper third of the follicular structure from cells of the presumptive outer root sheath. Secretion is of the holocrine type since it is formed by the complete disintegration of the gland cells. The sebaceous gland duct empties into the neck of the follicle. The secretion of the gland gives gloss to the coat and helps to maintain a proper skin texture.

Yolk (A Skin Gland Secretion of Sheep). The secretion of the sweat glands, suint, and the secretion of the sebaceous glands, sebum, combine to form yolk in the fleece of sheep. An outstanding property of suint is its hygroscopic nature; suint may absorb as much as 5 times its weight of water. A good supply of yolk is associated with good condition for wool growth and may run as high as 50 percent or more of the fleece weight.

II. DYNAMICS OF FOLLICLE GROWTH

A. Development of Follicles

General Pattern of Development. Factors inherent in the skin are principally responsible for the initiation of hair follicles. Even in the anomaly of hairlessness produced by aberrant developmental processes, primitive hair follicles are formed. The first follicles to develop in all mammalian fetuses are those of the tactile hairs, which are followed rapidly by follicles of the general body surface. The initial formation of an early hair follicle consists of a solid mass of epidermal cells that invades the underlying dermis. This mass soon becomes shaped like a plug and the thickened lower part of the downgrowth is invaginated by a small group of mesodermal cells (Fig. 14–2). The mesodermal cells then form the papilla,

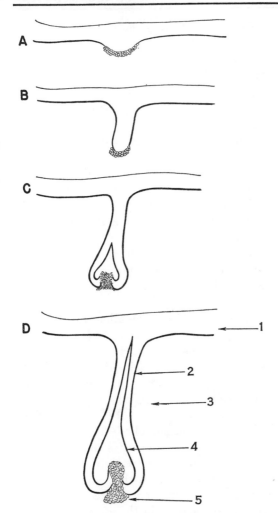

A

B

C

D —— 1

—— 2

—— 3

—— 4

—— 5

FIG. 14–2. Diagrammatic sketch of the stages of hair follicle development: A, appearance of anlage; B, follicle in the shape of a plug; C, formation of hair bulb; and D, fully formed hair in early follicle. The epidermis is designated as (1); the hair follicle (2); the dermis (3); the developing hair (4); and the papilla (5).

which lies directly beneath the presumptive germinal layer, or matrix plate, of the developing hair.

The outer root sheath of the follicle, the matrix cells of the hair bulb, and the basal cell layer of the epidermis are all germinative in nature. The basal cell layer replaces the cells that are being sloughed off at the skin surface; the outer root sheath gives rise to new hair germs; and the matrix cells evolve to make up the developing hair once the hair germ has formed. The connective tissue

sheath serves for the insertion of the *arrector pili* muscle, a bundle of smooth muscle fibers which are formed into a band. This muscle is generally attached to the central guard hair or to the central primary wool fiber of the follicle group.

Follicle Development in Sheep. In merino sheep the follicles of the tactile hairs around the eyes and lips develop on or about the thirty-eighth day of fetal life. Follicles on the crown of the head appear shortly before the fiftieth day. On approximately the sixtieth day primary follicles are present over the whole body surface and these follicles continue to develop until about the eighty-fifth day (Carter, 1955). The secondary follicles produce the fine fleece fibers of wool and make their appearance after the eighty-fifth day. In mature merino sheep, the coat is made up entirely of wool; the coarse hairs are restricted mainly to the head, legs, and tail. Before birth, the primary follicles of the fleece produce the hairy fibers of the birth-coat of the lamb; after birth, these same follicles produce fine wool in the wool breeds.

B. Hair Growth and Replacement

Growth Cycle. Hair growth occurs in cycles in which periods of active growth alternate with periods of rest. Dry (1926) divided the cycle into three separate phases: *anagen* (period of growth), *catagen* (involutional stage), and *telogen* (period of rest). Hairs in the catagen phase are found infrequently, since this stage is of extremely short duration (Van Scott, 1958). There is wide variation in the length of the growing and resting phases of the hair cycle in the different species. For example, the resting phase in merino and rambouillet sheep is very brief and only a very small percentage of the fleece is in the telogen phase at any one time. The growth phase, anagen, is of several years' duration. Although the majority of fibers grow continuously from season to season, a small proportion of these are shed each season.

In contradistinction, the anagen phase is

brief in young mice; it lasts only about 17 days. The catagen phase occurs between the seventeenth and the nineteenth day of the cycle, while the telogen phase covers a period of 10 days or less. As the animal matures, the growth sequences are spaced progressively farther apart, and the period of rest may last several months.

At the completion of growth, all follicles cease to form medullary substance, the papilla shows signs of atrophy, and the bulb degenerates to form a club- or brush-shaped structure (Plate 18). The lower half of the hair follicle, including the bulb, is a transient structure. The base of the hair with its inner root sheath ascends to the level of the sebaceous gland. The outer root sheath produces a cord of cells, the epithelial strand, and as the hair follicle retreats from the lower portion of the dermis, the epithelial strand draws the papilla along with it. The papilla then regresses to the primordial state and becomes a free cluster of connective tissue cells. Within a given period, a new epithelial hair germ is set aside by proliferation of the cells of the outer root sheath and the hair growth cycle is renewed. The old hair is retained in the hair follicle by its closely investing sheath and remains there during the resting period alongside the newly developing hair.

Shedding. There is a tendency for certain breeds of sheep to shed seasonally. Wild sheep, Icelandic sheep, the blackhead Persian sheep in South Africa, the karakul sheep, the Wiltshire sheep, and many other breeds shed their coat annually. In the British Mountain breeds, the coat is only partly shed. Some Scottish sheep undergo a partial molt in the summer. Cattle shed their coat twice a year, but the autumn molt is not so well defined as the spring molt. In cattle, molting occurs twice a year, even in a tropical environment (Dowling & Nay, 1960).

Many fur-bearing animals, such as mink, weasel and muskrat, also lose their coat twice a year while the fox molts only once

a year (Bassett & Llewellyn, 1948). The summer coat of the adult mink is reddish in color and is retained from mid-July to early August. Hairs of the new winter coat make their first appearance during the early part of September and the pelage is ready for pelting by late November.

There is some uniformity in the manner in which the molt occurs in most mammals. The fall molt begins on the abdomen and proceeds from the rump toward the head, and at the same time spreads progressively over the lateral surface to the back. The pattern of the spring molt constitutes the reverse process of the fall molt. It begins on the head and back regions and spreads ventrally and posteriorly.

The Prime Skin. The skins of fur animals are taken when they are fully furred and in their best condition. The hairs of prime skins have grown to their full length and are in the telogen phase of the growth cycle. When viewed from the leather side, a prime pelt will be creamy white in color, since hairs in the telogen phase are colorless at their root ends. The skins of fur animals that are not prime are generally taken out of season and have hairs that are still in the anagen stage. Pigmented areas of varying size which show up in the untanned leather side of the pelt are indicative of hairs in the anagen phase. The size of the pigmented area will be governed by the number of hairs involved.

C. Coat and Fleece Character

Sheep. Most mammals have two coats— an outer one of long, coarse hairs which serves as protection, and an inner one of short, soft, fine hairs which provides warmth. All sheep originally had an outer coat of coarse, medullated fibers in addition to the fine fibers of the undercoat. The elimination of the outer coat through selection left only the fine inner coat of wool. The birthcoat of the newborn lamb still shows the two coats: the longer, coarse, primary hair and the short, fine fibers of the definitive fleece. The coarse wool hairs of the general body surface are gradually shed and are replaced by the fine wool fibers. The head, legs, and tail (if retained), however, will continue to grow fairly coarse, medullated fibers throughout the life of the sheep.

Angora Goat. The coat of the adult Angora goat is made up mainly of mohair fibers which are long, white, and lustrous in appearance. Uniform locks of spiral ringlets form a fleece. While coarse fibers are found intermingled with mohair, their numbers are small. Most of the coarse fibers of the fetus and early kid are replaced by mohair in the adult. The number of medullated fibers in the well-bred mohair is normally below 1 percent (von Bergen, 1963).

Cattle. Hair types have not been studied as extensively in cattle as in sheep. Cattle of European origin have a thick type of winter coat and a shorter, glossy coat in summer. Although there are no underfur hairs, the guard hairs of the heterotype follicles are slightly smaller in size than those of the central primary follicles.

III. CONNECTIVE TISSUE

A. Epidermal-Dermal Relationship

The epithelial basal cell layer of the epidermis rests on a connective tissue sheet known as the basement membrane. This membrane serves as a boundary between the dermis and the epidermis and is an important adjunct. The underlying dermis in the intact animal provides support and protection and serves as a medium for exchange of materials.

B. Connective Tissue Fibers and Cell Types

The fibers of the dermis are classified as collagenous, reticular, and elastic, and occur interwoven in the dermal tissue. The fibroblast is the most common cell type. It retains the capacity to regenerate and is involved in the formation of collagenous fibers.

The fat cell is another predominant cell type, and is found in the loose connective tissue of the dermis, generally in the vicinity of blood vessels. Filling the space between the fibrous and cellular elements is a complex, gel-like matrix.

Collagenous fibers are massed into heavy bundles which are flexible and have great tensile strength. Elastic fibers are thinner and anastomose freely, forming a delicate network in the spaces between the collagenous bundles. Reticular fibers, like elastic tissue fibers, form a delicate network; however, reticulin is basically collagen. The principal difference between collagenous and reticular fibers is one of size rather than of physical or chemical makeup.

C. The Active Fat Cell

For a long while, adipose tissue was considered to be an inert storage organ and in no way involved in the energy metabolism of the body. It is becoming more and more evident that the adipose cell takes a very active part in the synthesis and breakdown of fat. The fat cell consists of two compartments: a large one, serving as a relatively inert storage site in which the exchange or turnover is very slow, and a smaller one, consisting of lipids in a very rapid state of turnover.

Adipose tissue is one of the most vascular tissues in the body. In the adult, when fat is deposited, it tends to be laid down first in the neighborhood of blood vessels. In the fetus this vascular arrangement is defined before fat is deposited.

The fetal fat cell is characterized by its small size and granular cytoplasm. Fat cells in late fetal life form well-defined masses of tissue arranged in lobular fashion. The collagenous fibers in the loose connective tissue of the dermis provide a matrix for the aggregations of grapelike clusters of mulberry-shaped cells found between the hair follicles.

After birth, the cells have the form of the usual mature signet-ring type of fat cell. The cytoplasm in these is displaced to the periphery and the nucleus is eccentrically located. Small islands of mulberry-shaped cells containing large nuclei and granular cytoplasm are still evident, however, in the early postnatal animal.

Fetal fat cells have been seen to divide (Bell, 1909). Cell division, however, does not occur in the mature adipose cell since such cells lose this capacity the moment lipid appears. When fat disappears from the cell, the cell reverts to an embryonic form which is mesenchymal in nature. These remain quiescent until they again become filled with a store of fat. Islands of adipogenic tissue, identical to developing fetal fat tissue, can be seen in serial sections made of adipose tissue in adult man (Simon, 1965).

It is an interesting phenomenon that the adipogenic type of fat cell appears coincident with new hair growth. In the early anagen phase the adipose cells are small in size and number. As hairs reach the telogen phase, however, they are of the large storage type and completely fill the dermis from the base of the shortened telogen follicle to the lower limits of the dermal boundary.

There is a selective disposition of fat, not only in the loose connective tissue of the skin but also in the subcutaneous tissues, the omentum, and the mesenteries, indicating that fat accumulates in specific cells which have a definite regional distribution. Selective disposition of fat is also found in the hump of the camel and in the tail of fat-tailed sheep. The specificity of fat cells is further suggested by the occasional occurrence of fat tumors, which develop as circumscribed benign growths irrespective of the development of adipose tissue in the body.

IV. PHYSIOLOGY

A. Nutrition

Effects of Inadequate Nutrition. Nutrition influences the quantity and quality of animal fibers. This influence, however, is observed only during the growth phase of the fiber growth cycle. Since the wool of sheep has an extremely long growth cycle, the results of nutritional influences are expressed more radically than they are in other mammals. The duration of the growth cycle for wool is six years or more in the wool breeds as contrasted with several weeks or several months for the hair growth cycle in other mammals (Ryder & Stephenson, 1968).

Fibers in the telogen phase of the growth cycle (wool or hair) remain in the skin for a time before they are shed. They are not vulnerable to intrinsic or extrinsic factors during this period, since the papilla in the telogen phase has atrophied and is no longer a source of blood supply. Of equal importance is the fact that fibers in the telogen phase are completely keratinized at their root ends and therefore inert.

If the level of nutrition is adequate, the fibers of the sheep's fleece grow at a fairly constant rate. In general, however, wool grows fastest in the summer and early fall, when sheep on pasture have a high level of nutrition. A low plane of nutrition can cause a decrease in rate of wool growth and in diameter and length of the fiber. The affected fiber then shows a constriction. If this constriction, or thinning, is severe, a complete break across the wool staple results. A depression in wool growth may also occur with the end of pregnancy and the beginning of lactation (Yeates, 1965). Underfeeding of ewes during the critical stages of pregnancy produces poor quality wool.

Effects of Protein Quality and Protein Deficiencies. Loss of hair or wool is often associated with a deficiency of one or more nutritional elements. The body requirements of animals kept on inadequate protein diets must be satisfied before fiber growth can make substantial progress. In mink that are fed a diet poor in protein quality, hair germs in the skin fail to develop to their fullest capacity, resulting in thinly furred pelts. Diets used to test the ability of sheep to utilize poor quality roughage can cause pronounced shedding of the secondary wool fibers. In the series studied by Lyne (1964), the poorest diet contained only 2.6 percent crude protein *ad libitum*. Schinckel & Short

PLATE 17

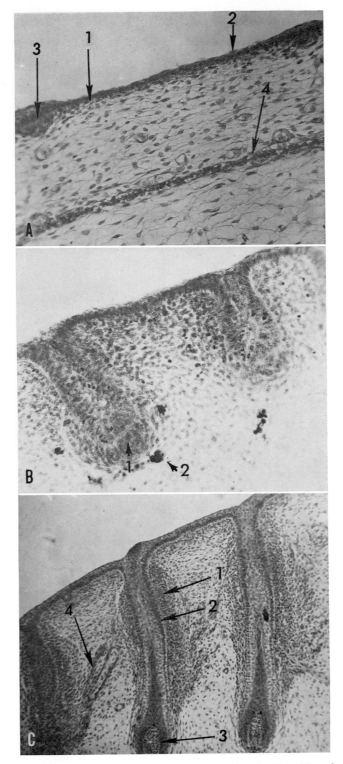

A, Skin of a 17-day-old rabbit fetus: (1) basal cell layer of the epidermis; (2) peridermal cell of outer protective fetal layer; (3) hair follicle anlage; and (4) *panniculus carnosus* muscle.   (200 ×)

B, Skin of a 19-day-old rabbit fetus: (1) papilla of a tactile hair follicle; and (2) blood capillaries.   (400 ×)

C, Skin of a 21-day-old rabbit fetus.   Note strong, dense, dermal sheath of the tactile follicle.   When present, in the pelage follicle, the dermal, or connective tissue sheath is restricted to the root end: (1) connective tissue sheath of tactile follicle; (2) outer root sheath; (3) papilla; and (4) forerunner of *arrector pili* muscle.   (400 ×)

PLATE 18

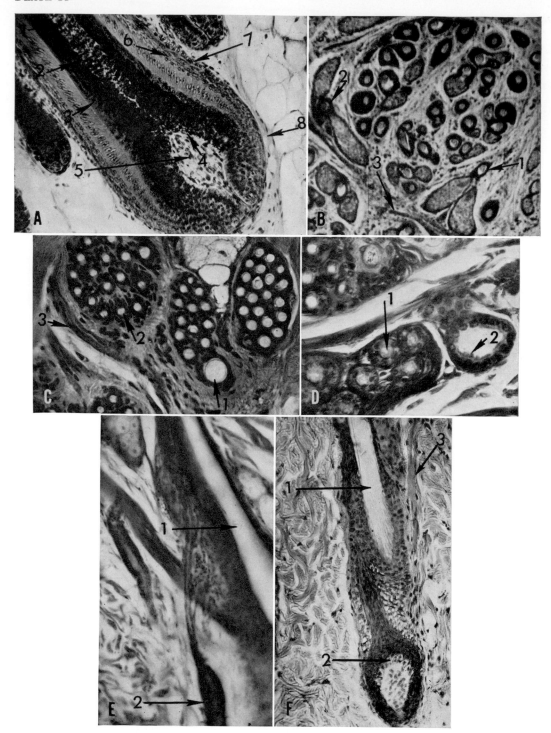

A, Photomicrograph of root end of a growing guard hair follicle: (1) medulla; (2) cortex; (3) cuticle of the hair; (4) matrix plate; (5) papilla; (6) inner root sheath; (7) outer root sheath; and (8) connective tissue sheath. (200 ×).

B, Photomicrograph showing the trio arrangement of follicles in the skin of rambouillet sheep. (100 ×). All satellite follicles are secondary ones: (1) Primary follicle; (2) heterotype follicle; and (3) *arrector pili* muscle.

C, Photomicrograph showing the trio arrangement of follicles in the skin of mink: (1) guard hair follicle; (2) heterotype follicle; and (3) *arrector pili* muscle. (400 ×).

D, Hair follicle (1) and sweat gland (2) in the skin of mink shown in cross section (400 ×).

E, Hair follicle (1) and sweat gland (2) in the skin of mink shown in longitudinal section (200 ×).

F, Anagen and telogen phase hairs of the rat lying side by side in the same follicle: (1) root end of shedding club hair; (2) matrix plate of growing hairs; and (3) *arrector pili* muscle (200 ×).

(1961) found that the more adverse the nutrition, the greater the amount of shedding of the secondary fibers. The number of primary fibers shed under such conditions is small.

Effects of Mineral Deficiencies. Information concerning the importance of metals in the growth of hair and wool is meager. It is known that all animals require small amounts of trace minerals. Hair and wool contain small quantities of lead, arsenic, and magnesium in addition to zinc, iron, copper, and selenium. Nonessential materials such as lead and arsenic may be taken up from environmental sources (fencing and cages); the dietary essential minerals such as zinc, copper, iron, and selenium are required in the formation of fiber.

The enzyme tyrosinase, a copper protein complex, is needed in the production of pigment (Frieden, 1968). Diets deficient in copper will produce graying of hair in black rabbits; in cattle they produce a dull, discolored coat. Black-woolled sheep with a copper deficiency show almost complete disappearance of pigment, although some pigmentation may persist in the form of gray banding. The follicles of these sheep that are fed deficient diets lose their normal ability to impart the characteristic crimp. The wool of such sheep is weak and lustrous in appearance.

Molybdenum poisoning in cattle and sheep will cause a loss of pigment which can be restored by adding cooper to the diet. It is known that molybdenum affects the utilization of copper. The alternate feeding and withholding of molybdenum produces a novel pattern of black and white banding in the fleece.

Zinc is always present in hair in relatively high concentrations. Zinc deficiency in young rats causes atrophy of the hair follicles and loss of already developed hair shafts (Follis, 1958). It is not clear whether zinc is essential for the development of the earliest fetal hair buds, and whether it is actually biologically essential for the growth of hair in the adult animal. Zinc deficiency in man, and also in swine, sheep, and calves, leads to lesions of the skin.

Effects of Vitamin Deficiencies. Biotin deficiency in the mink alters the underfur to a grayish-white color. Deficiencies in biotin, choline, para-aminobenzoic acid, and pantothenic acid have been implicated in the graying of hair.

Deficiencies in vitamins A and E have been observed to cause impaired hair growth in the rat. Pantothenic acid deficiency causes a thinning of hair in swine and monkeys and a loss of hair in mice. In the riboflavin-deficient rat, the fur becomes ragged and encrusted with a dark brown material (Lorincz, 1954).

B. Influence of Physical Environment

New fibers arise periodically. The hair growth cycle is governed directly by heredity, hormones, the photoperiod, and only indirectly by temperature. A sudden drop in temperature can accelerate growth of hair only if growth is already in progress.

The rhythm of wool growth is seasonal and under the influence of length of the day. A gradual reversal of wool growth rhythm is obtained by reversing the seasonal rhythm of the length of the day, in which case maximum wool growth is produced in winter rather than in summer.

The molting process is also cyclical. It precedes, or is concomitant with, new fiber growth. It, too, cannot be initiated through the direct effect of temperature change; however, it can be inhibited by moving the animals to an environment of another temperature. The change of coat in weasels from brown to white in early winter is delayed when these animals are moved indoors.

Individuals and breeds vary greatly in their response to the shedding stimuli. This is evident in the ratio of guard hair to underfur; this ratio varies in the same animal from year to year. The stimulus to molt is activated through the anterior pituitary gland, but is dependent upon the amount of light received through the eyes. It is known that the seasonal change in the length of day is the most important environmental variable affecting molting and new hair growth in most mammals.

Cattle can be made to grow long, coarse coats in the summer and short, glossy ones in the winter by reversing the pattern of the length of the day. Out-of-season molting can also be induced in an animal such as the mink. When subjected to varying photoperiods, starting with a short day and then increasing the length, minks will molt and come into their winter coats early.

C. Hormonal Influences

While the thyroid plays an important role in growth of the animal fiber, it is effective only insofar as it is capable of regulating the general body metabolism. The removal of the thyroid in young lambs leads to a reduction of wool growth, probably because of metabolic changes which, in turn, lead to a diminished consumption of feed. Excess thyroxine that is administered to untreated sheep will increase wool output. This increase, however, is offset by an increase in feed consumption. In thyroidectomized mink, the normal cyclic activity of the hair follicles is inhibited . Follicles in the telogen phase remain static in the skin.

In most mammals hair growth spreads gradually, in the form of a wave, from one region of the body to another, in synchronous fashion. It may start on the head and back and from there gradually spread ventrally and posteriorly over the body. After adrenalectomy, hair growth has the appearance of being more rapid if contrasted with the poor physical state of the animal. Dieke (1948) found that at 16 weeks of age adrenalectomized black rats had already attained the sixth cycle of hair growth, whereas the controls were only beginning the fifth cycle. Adrenelectomy, however, has no effect on the rate of growth of the individual follicle; it can accelerate the initiation and spread of the hair growth wave (Montagna, 1958).

Adrenocortical carcinomas induce already present, dormant hair follicles to become active, thus producing hirsutism or excessive growth of hair in women. Cortisone compounds used in the treatment of baldness in man (alopecia areata) can initiate hair growth, but beneficial results of such treatment are transitory at best.

Removal of the pituitary gland will delay the completion of the fiber growth cycle and will also cause a reduction in the number of fibers produced. In mammals other than sheep, the hair is soft, infantile, and somewhat longer than in the normal animal.

D. Genetic Factors

Hairlessness. Anomalies of the hair coat are well documented in the literature. Complete hairlessness has never been observed inasmuch as hair follicles do develop and hairs are formed within these follicles. Hairs break off at or below the skin surface for several reasons: (a) they may be unable to escape through the follicular pore, where sebaceous glands give rise to cysts, (b) imperfect keratinization will produce hairs of poor tensile strength and hairs that break easily, and (c) club hairs in some instances do not form following the maturation of the first hair and the follicle soon degenerates.

Ordinary baldness in man may be aggravated by overproduction of androgen (Behrman, 1952). On the whole, however, baldness is influenced by genes and the normal aging process. Reduction of the blood supply to hair follicles in man results in loss of hair.

Lethal hairlessness in cattle is characterized by arrest of the development of hair follicles at an early stage (Hutt & Saunders, 1953). In nonlethal hypotrichosis the hair follicles are formed, but they possess no dermal papillae. In semihairless cattle, the early coat of fine, curly hair is shed to form a pattern of patchy, coarse hair.

In some hairless mice, the anomaly is not evident until after the shedding of the first coat (Thigpen, 1930). Primitive hair follicles are in evidence, but aberrant developmental processes prevent the full development of the hair itself.

Another type of hairlessness phenotypically different from the naked type occurs in mink and is characterized by an almost complete lack of underfur (Shackelford, 1950). The guard hairs that are present are curled and fine in texture; the skin is wrinkled and tends to blister easily.

Hirsutism in Man. Some types of hirsutism such as hypertrichosis universalis and idiopathic hirsutism are genetic in origin. Idiopathic hirsutism is possibly produced by a defect in the hair follicle or in the intermediary metabolism of sexual steroids. Hypertrichosis universalis, as exhibited in the "dog man" of the circus, involves most of the body surface. Scalp hair grades into the lanugo-like hair on the forehead and face. Pubic and axillary hairs are also of the lanugo type. This anomaly is evidence of arrested development, since lanugo hairs are produced rather than the coarser ones.

V. RESEARCH METHODS

There are many avenues of approach available for the study of the development and growth of fibers. The techniques described below have been used with good success.

Tissue Culture. The growth, direction, and organization of cells into definite units or patterns which subsequently make up the appendages of the skin can be directly observed through the medium of tissue culture (Lasnitski, 1964). Hardy (1949) was one of the first to succeed in cultivating *in vitro* mouse skin from explants. She observed that the time of formation of the follicles in an explant was related to the age of the animal from which the tissue was obtained, not to the period of cultivation. Once differentiation was begun, the hair developed within the follicle although no tissues other than epidermis and underlying dermis were present. While these results indicate that growth is possible in surroundings in which there is neither blood supply, nervous stimulation, nor influence of other tissues, they do not preclude the possibility that other tissues do have an effect on the skin before morphological differentiation takes place.

Hair Root Studies. An assessment of the dynamics of hair or wool growth can be made with Van Scott's method (1957) for determining anagen and telogen counts of plucked fiber root ends (Fig. 14–3). This is a simple, effective method for obtaining the percentage of growing hairs. It is also useful for detecting abnormalities which may occur in the root ends of fibers after experimental treatment. Hairs in the telogen phase are distinguishable from those in the anagen phase in that they have an epithelial sac. Hairs in the anagen phase have a keratogenous zone above the distinctive bulb end as well as internal and external root sheaths. although these may be lost in plucking,

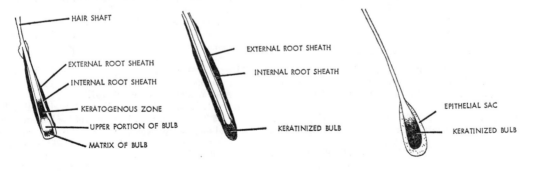

ANAGEN   CATAGEN   TELOGEN

FIG. 14–3. Distinctive characteristics of plucked hair roots. (*Adapted from Van Scott et al., 1957. J. Invest Dermat. 29:197–204.*)

Hairs in both the catagen and telogen phases have a keratinized bulb, but the epithelial sac is not in evidence in the catagen stage.

Radioactive Markers. Radioactive markers may be used for measuring the rate of wool growth. Downes and Lyne (1959) injected radioactive cystine into sheep at recorded intervals, prepared autoradiographs of the fibers after several days' growth, and measured the growth rate directly.

Plucking, Shearing, or Dye-banding. Other methods for studying fiber growth include the shearing of one side of an animal or plucking or shearing a small area of the coat to observe the new hairs as they make their appearance. Plucking permits a more accurate determination of the time required for complete regrowth of hair under different experimental conditions (Mohn, 1964). Dyebanding is practiced when large numbers of animals are involved in field experiments. In wool sheep, the dye is applied in narrow bands approximately 10 cm long at the skin surface after the fleece has been opened or spread apart on the animal (Chapman & Wheeler, 1963).

REFERENCES

Bassett, C. F., & Llewellyn, L. M. (1948). The molting and fur growth pattern in the adult silver fox. Amer. Midl. Naturalist *39*:597–601.

Behrman, H. T. (1952). *The Scalp in Health and Disease.* St. Louis, C. V. Mosby Co.

Bell, E. T. (1909). On the histogenesis of the adipose tissue of the ox. Amer. J. Anat. *9*:413–435.

Carter, H. B. (1955). The hair follicle group in sheep. Anim. Breed. Abstr. *23*:101–116.

Chapman, R. E., & Wheeler, J. L. (1963). Dye-banding: a technique for fleece growth studies. Aust. J. Sci. *26*:53–54.

Dieke, S. H. (1948). The effect of removing various endocrine glands on the hair cycles of black rats. Endocrinology *42*:315–319.

Dowling, D. F., & Nay, T. (1960). Cyclic changes in the follicles and hair coat in cattle. Aust. J. Agri. Res. *11*:1064–1071.

Downes, A. M., & Lyne, A. G. (1959). Measurement of the rate of growth of wool using cystine labelled with sulphur 35. Nature (Lond.) *184*:1884–1885.

Dry, F. W. (1926). The coat of the mouse (*Mus musculus*). J. Genet. *16*:287–340.

Follis, R. H., Jr. (1958). *Deficiency Disease.* Springfield, Ill., Charles C Thomas.

Frieden, E. (1968). The biochemistry of copper. Sci. Amer. *218*:102–114.

Hardy, M. H. (1949). The development of mouse hair *in vitro* with some observations on pigmentation J. Anat. (Lond.) *83*:364–384.

Hutt, F. B., & Saunders, L. Z. (1953). Viable genetic hypotrichosis in Guernsey cattle. J. Hered. *44*: 97–103.

Lasnitski, I. (1964). Tissue culture of skin. In: *Progress in the Biological Sciences in Relation to Dermatology.* A. Rook & R. H. Champion, (Eds.) Vol. II, pp. 483–493, England, Cambridge University Press.

Lorincz, A. L. (1954). Nutritional influences. Chapter 27. In: S. Rothman, *Physiology and Biochemistry of the Skin.* Chicago, University of Chicago Press.

Lyne, A. G. (1964). Effect of adverse nutrition on the skin and wool follicles in merino sheep. Aust. J. Agri. Res. *15*:788–801.

Mohn, M. P. (1964). Hair growth. In: *Laboratory Technique in Biology and Medicine.* V. M. Emmel & E. V. Cowdry (Eds.), pp. 194–195, Baltimore, Williams and Wilkins.

Montagna, W. (1958). Chapter 21. In: *Biology of Hair Growth.* W. Montagna & R. A. Ellis (Eds.), New York, Academic Press.

Montagna, W. (1965). The skin. Sci. Amer. *212*: 56–66.

Ryder, M. L. (1955). Studies on the nutrition of wool follicles in sheep: anatomy of the general blood supply to the skin. J. Agri. Sci. (Camb.), *45*: 311–326.

Ryder, M. L., & Stephenson, S. K. (1968). *Wool Growth.* New York, Academic Press.

Schinckel, P. G., & Short, B. F. (1961). The influence of nutritional level during pre-natal and early postnatal life on adult fleece and body characters. Aust. J. Agri. Res. *12*:176–202.

Shackelford, R. M. (1950). *Genetics of the Ranch Mink.* New York, Pilsbury Publishers Inc.

Simon, G. (1965). Histogenesis. Chapter 11. In: *Handbook of Physiology, Section 5, Adipose Tissue.* A. E. Renold & G. F. Cahill, Jr. (Eds.), Washington, D. C., Amer. Physiol. Soc.

Straile, W. E. (1960). Sensory hair follicles in mammalian skin: the tylotrich follicle. Amer. J. Anat. *106*:133–148.

Thigpen, L. W. (1930). Comparative histological studies on hairless mammals. University of Pittsburgh Abstracts of Theses *6*:191–197.

Van Scott, E. J., Reinertson, R. P., & Steinmuller, R. (1957). The growing hair roots of the human scalp and morphologic changes therein following amethopterin therapy. J. Invest. Derm. *29*:197–204.

Van Scott, E. J. (1958). Response of hair roots to chemical and physical influence. Chapter 18. In: *Biology of Hair Growth.* W. Montagna & R. A. Ellis (Eds.), New York, Academic Press.

von Bergen, W. (1963). *Wool Handbook.* Vol. I, New York, Interscience Publishers.

Yeates, N. T. M. (1965). *Modern Aspects of Animal Production.* Washington, D. C., Butterworth.

IV. Nutritional Requirements for Growth

Chapter 15

Energy Requirements

By W. P. Flatt and P. W. Moe

I. ENERGY USES

ENERGY is required for practically all life processes. In animals, these processes include maintenance of blood pressure and muscle tone, heart action, transmission of nerve impulses, ion transport across membranes, reabsorption in the kidney, protein and fat synthesis, the secretion of milk, and the production of wool and eggs. The utilization and transformation of energy by living organisms are of fundamental importance to the whole field of biology. Therefore, it is desirable to know the amounts of energy required by an animal to perform various functions so that the most efficient utilization of feedstuffs and other resources can be planned. Among the variables that have been shown to influence the energy requirements of animals are species, body size, age, sex, type and level of growth and production, activity, and environmental conditions.

The economic production of animal products is dependent upon meeting the total nutritional requirements of the animal. To do this, it is necessary to know the nutritional value of the feeds as well as the nutritional requirements of the animal. Protein,

minerals, and vitamins are discussed in other chapters, so only the role of energy will be stressed in this chapter. It is well known, however, that without the other essential nutrients, energy utilization would be impaired and animal growth (even life itself) would cease. A deficiency of energy is manifested primarily as a lack of growth, body tissue losses, or reduced production rather than by specific signs such as those which characterize the deficiency of some specific nutrients. Energy deficits have sometimes gone undetected for extended periods, resulting in lowered production of meat, milk, and fiber. Numerous guides have been formulated to assist the nutritionist in planning rations that will minimize the possibilities of nutritional deficiencies. Some of these guides are discussed in this chapter.

II. ENERGY TERMINOLOGY AND DEFINITIONS

Energy is usually defined as the capacity to do work. Work is defined as the product of a given force acting through a given distance (Lehninger, 1965). Some of the terms that are used to express energy are ergs, joules, calories, kilocalories, megacalories, therms, and BTU's (British thermal units). The gross energy, digestible energy, metabolizable energy, "available" energy, productive energy, net energy, and physiological fuel values of diets are usually expressed as calories, kilocalories (Calories), or megacalories (or therms) per unit of food or feed. Other expressions of energy values of feeds include kilograms of total digestible nutrients (TDN), starch equivalents (SE), and fodder units. National Research Council Publication 1411 (1966) contains a discussion of energy terminology, particularly as related to expressing the energy requirement and nutritional values of diets for domestic animals. A clear understanding of these terms is essential in any discussion of energy metabolism.

A. Calories and Other Units Expressing Energy

A *calorie* (*cal*) is the amount of heat required to raise the temperature of 1 gm of water from 14.5 to 15.5° C. This is equivalent to 4.184 joules. In nutritional usage, the *kilocalorie* (kcal) (1 kcal = 1,000 cal) is most often used, and in human nutrition it is referred to as a kilogram calorie or as a "large calorie" and always spelled "Calorie," with a capital C to distinguish it from the "small calorie." In animal feeding, the term most frequently used is the megacalorie (Mcal) (1 Mcal = 1,000 kcal = 1,000,000 cal), but *therm* (1 therm = 1 Mcal = 1,000 kcal) is also used to express the net energy or productive energy of rations. The preferred term, however, is the megacalorie because the word "therm" is ambiguous; it has been defined in different contexts as 1 kilocalorie, 1 megacalorie, or 100,000 BTU's.

Any form of energy may be converted to another form, so that electric energy, radiant energy, and chemical energy can be converted to heat. Chemical energy in particular is measured as heat and is expressed in calories. Electrical measurements can be standardized more accurately than heat measurements, however, so that a calorie is now officially defined in electrical terms (Kleiber, 1961).

1 calorie   = 4.1868 international joules
1 joule   = 1 voltcoulomb
1 volt   = 1/1.0183 of the electromotive force of a standard Weston cell at 20° C.
1 coulomb = 1/96494 of the amount of electricity carried by 1 gm equivalent (107.88 gm silver, for example).

Although animal nutritionists rarely use British thermal units to express the heat production of animals or the energy value of feeds, engineers commonly use this term. One BTU is equivalent to 0.252 kcal, and is measured as the amount of heat required to raise the temperature of 1 pound of water 1° F.

B. Gross Energy

Gross energy (*GE*) is the amount of heat, measured in calories, that is released when a substance is completely oxidized. This measurement is made in a bomb calorimeter containing 25 to 30 atmospheres of oxygen. The heats of combustion (gross energy) of most feeds for ruminants are approximately 4.4 to 4.5 kcal/gm. The gross energy of most nutrients varies from 3.8 kcal/g for glucose to about 9.4 kcal/g for fat. The average gross energy values of the three principal feed constituents are estimated to be (Brody, 1945):

	kcal/gm
Carbohydrates	4.10
Protein	5.65
Fats	9.45

The gross energy of some materials that are most likely to be of interest to animal nutritionists are listed in Appendix I.

C. Digestible Energy

1. APPARENT DIGESTIBLE ENERGY (DE).

The food intake gross energy (GE_i) minus fecal energy (FE), including the undigested food and the metabolic body and bacterial residues fraction of the feces, is called *apparent digestible energy*.

$$DE = GE_i - FE$$

2. TRUE DIGESTIBLE ENERGY (TDE).

True digestible energy can be estimated by accounting for the metabolic fecal energy (MFE), but this is not usually done in practice.

$$TDE = GE_i - (FE - MFE)$$

Unless specified otherwise, the term "digestible energy" refers to apparent digestible energy.

3. TOTAL DIGESTIBLE NUTRIENTS (TDN).

Digestible energy is comparable to *total digestible nutrients* (TDN) in that in rations usually fed to ruminants there are approximately 2000 kcal DE/lb TDN (Swift, 1957). This figure can vary, depending upon the species of animal and the composition of the ration. TDN is defined as digestible crude protein plus digestible carbohydrates plus 2.25 times the digestible crude fat. The crude fat is multiplied by 2.25 because of the higher energy content of fat than carbohydrates.

$$TDN = DP + DNFE + DCF + 2.25 \ (DEE)$$

in which DP is digestible crude protein. DNFE is digestible nitrogen free extract. DCF is digestible crude fiber. DEE is digestible ether extract (fat). TDN is a measure of digestible energy, although it is expressed in units of weight or percent rather than energy.

D. Metabolizable Energy

Metabolizable energy (*ME*) is the gross energy of the feed consumed minus fecal energy minus energy in the gaseous products of digestion (GPD), and minus urinary energy (UE). In human nutrition, in which combustible gas losses are negligible, the term which is comparable to metabolizable energy is *physiological fuel value* (PFV). Various "corrections" are made by some investigators for nitrogen retained or lost from

the body (ME_n) (Rubner, 1901) or for heat of fermentation (ME_c) (Blaxter, 1962).

$$ME = GE_i - FE - GPD - UE$$
$$ME_n = ME - (\text{Nitrogen balance} \times 7.45 \text{ kcal})$$
$$ME_c = ME - 0.8 \text{ GPD (GPD is predominantly } CH_4)$$

The factor often used to calculate ME for ruminant rations from DE is 0.82 (N.R.C. Publ. 1232, 1964; 1349, 1966). This is equivalent to about 1640 kcal/lb TDN, but estimates have ranged from 1616 to 1814 kcal/ME/lb TDN (3563 to 4000 kcal ME/kg TDN) (Brody, 1945). Although the ME value of the diet for ruminants averages about 82 percent of the DE value, this is only an approximation. Values for single diets may be 80 to 90 percent of the DE value. The fact that ME values which appear in tables of feed composition are usually calculated as 82 percent of the DE value rather than being measured reduces their value for comparing feeds. A measured value of ME is more precise than DE as an expression of energy value, but true measured values are less common since measurement of the energy losses as urine, methane, and feces is required. Measurements of urine energy are fairly routine, but the measurement of gaseous methane losses requires more specialized equipment.

The ME/DE ratio is affected by the nature of the diet and the level of feed intake. Low quality diets result in larger proportions of energy lost as methane. The loss is increased at higher levels of feed intake. The proportion of GE_i lost as methane ranges from about 3 to 10 percent, depending on these conditions. Associated with the methane loss is the loss of the heat of fermentation. The heat of fermentation is usually considered to be about 80 percent of the amount lost as methane. This is the basis for computing ME_c by subtracting $0.8 \times$ methane energy from ME. The actual heat of fermentation, however, is not easily measured. In measurements of heat production by respiration calorimetry, the carbon dioxide produced in the rumen fermentation is partially absorbed by the blood stream and becomes mixed with the carbon dioxide expired from the lungs. Diets consumed by monogastric animals are normally of more constant composition and digestibility than those ingested by ruminants. In monogastric animals the fecal energy loss more adequately represents the total digestive loss. In swine, the ME value of diets may be computed according to the following relationship (N.R.C. Publ. 1599, 1968):

$$\text{ME (kcal/kg)}$$
$$= \text{DE}_{\text{(kcal/kg)}} \times 0.96 - \frac{(0.202 \times \text{protein \%})}{100}$$

E. Net Energy

Net energy (*NE*) is the difference between metabolizable energy and heat increment (HI):

$$\text{NE} = \text{ME} - \text{HI}$$

Heat increment is the increase in heat production (HP) following consumption of food when the animal is in a thermoneutral environment, and it includes the heat of fermentation (HF) and the heat of nutrient metabolism (HNM). Net energy includes the amount of energy used either for maintenance only (NE_m) or for maintenance plus

production ($\text{NE}_m + \text{NE}_p$). The net energy for production may be defined more specifically as net energy for fattening (NE_f), for milk production (NE_l), for growth (NE_g), for pregnancy (NE_{preg}), or for work (NE_{work}).

$$\text{NE} = \text{GE}_i - \text{FE} - \text{GPD} - \text{UE} - \text{HI}$$
$$\text{NE} = \text{NE}_m + \text{NE}_f + \text{NE}_l + \text{NE}_g + \text{NE}_{preg} \ldots$$
$$+ \text{NE}_{work}$$

Figure 15–1 illustrates the relative magnitude of the losses of energy in the form of feces, urine, methane, and heat production by lactating dairy cows fed a hay and concentrate ration. It also clarifies the definitions of energy discussed above.

The net energy value of a feed is dependent not only on the digestive losses discussed above, but on the nature of the diet, the species of animal involved, and the physiological function for which it is used. Net energy is a measure of how adequately the nutrients which are absorbed meet the metabolic needs of the animal.

III. METHODS OF EXPRESSING ENERGY VALUES OF FEEDS AND REQUIREMENTS OF ANIMALS

A. Digestible Nutrients

The expressions of energy value used most widely in the United States are TDN and

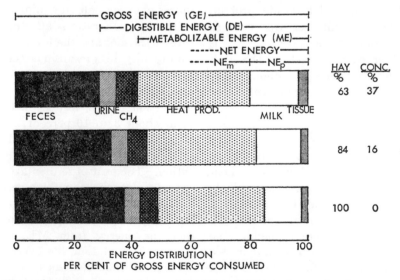

Fɪɢ. 15–1. Illustration of energy terminology and the different systems of expressing the energy value of feeds. The bar chart shows relative energy losses when a mixed ration is fed to a lactating dairy cow.

DE, which are based on digestibility measurements. This preference has been due to the relative ease of making these measurements and the accumulation of vast amounts of such information on a great variety of feedstuffs. These expressions are useful when comparing feedstuffs within a given class in which the differences among feeds are not great. Expressions of energy requirements given in terms of TDN or DE are adequate for animals for whom the requirements were actually determined on feedstuffs of similar quality to those being fed. These systems are not adequate, however, when comparing feeds of widely varying composition such as forages and concentrates. The TDN and DE methods systematically overestimate the energy value of poor quality forages in relation to concentrate feeds. This deficiency has been known for many years, and several attempts have been made to correct it by adjusting either the feed value or the listing of energy requirements.

B. Estimated Net Energy

The ENE values listed by Morrison (1956) were derived by using the available energy balance data (obtained from respiration trials and calorimetric studies) and some empirical adjustments based on a subjective evaluation of related data. In addition to these values, many more have been calculated according to the equation of Moore *et al.* (1953), which is as follows:

$$\text{ENE} = 1.393 \text{ TDN} - 34.63$$

in which ENE is expressed in megacalories per 100 pounds of dry matter and TDN is expressed as percent or pounds of TDN per 100 pounds of dry matter.

The ENE value, regardless of the method of computation or estimation, is actually a net energy value for *fattening*, and represents a prediction of the megacalories of tissue that should be deposited when a given quantity of feed is consumed. Although the system was not designed for use specifically for dairy cattle, it has been adapted and

used for this purpose. The energy requirements for maintenance, reproduction, and milk production, which are expressed as megacalories of estimated net energy, are summarized by Morrison (1956). The requirements for maintenance of a 1100 lb cow (499 kg) are estimated to be 5.6 to 6.3 Mcal ENE/24 hr, which corresponds to 27.3 to 30.9 kcal $\text{ENE}/\text{kg}^{0.87}$. For milk production, 0.29 to 0.30 Mcal ENE are required per pound of 4 percent fat corrected milk (which has 0.34 Mcal). The reason for the apparently low requirements for maintenance and milk production is that the efficiency of utilization of metabolizable energy is greater for these functions than for fattening. If values are expressed in terms of net energy for fattening, then less appears to be required for maintenance and milk production. This difference leads to the anomaly that only 300 kcal ENE is required to form 340 kcal milk.

C. California Net Energy System

Another net energy system that is currently being used in the United States is the one developed by the California workers for growing and finishing beef cattle (Lofgreen & Garrett, 1968). The basic principle is the same as for other net energy systems, but the method of deriving values for feeds and for computing rations differs. The California workers obtained net energy values for feeds by carrying out comparative slaughter trials and measuring energy retention by means of the specific gravity of the carcass. On the basis of their studies and those reported by earlier workers, they proposed that rather than a single net energy value, two values should be used. One value, called NE_m, would apply to feeds which are fed at or below maintenance level while another value, called NE_g, would be the net energy for feeds which are fed above maintenance level and represent the part used for gain of body tissue. The reason for using two separate values for each feed is that the partial efficiency of energy utilization is higher for maintenance (k_m) than it is for fattening (k_f). This difference has been demonstrated by many workers. Many of the difficulties that have been associated with using a single net energy value to evaluate rations, especially when rations differing widely in composition are fed at different planes of nutrition, can be attributed to this difference in efficiency. One of the assumptions made in this system is that the net energy for maintenance is equal to the fasting metabolism of the animal. This was found to be 77 ± 5 kcal/kg/24 hr, for both steers and heifers. Another assumption is that there is a linear relationship between metabolizable energy intake and energy retention, provided that all the measurements are made above maintenance. Since the caloric value of body weight gain differs between steers and heifers and is also influenced by rate of gain, separate equations were derived. They were as follows:

Steers:

$$NE_g = 52.72\ g + 6.84\ g^2$$

Heifers:

$$NE_g = 56.03\ g + 12.65\ g^2$$

NE_g is kcal of energy deposited per unit of body weight expressed as $kg^{3/4}$ and g is the daily empty weight gain in kg. Some of the net energy values of feeds are given by Lofgreen & Garrett (1968). For a given ration, the NE_m and NE_g values may be calculated. The amount used for maintenance can be estimated by using a table or calculated by using 77 kcal/$kg^{3/4}$. The remainder of the feed consumed is available for productive purposes. By knowing the NE value of the ration and the caloric value of body tissue deposited, it is possible to calculate the rate of gain. Conversely, it is possible to calculate the amount of a given ration that must be consumed in order to obtain the desired rate of gain.

The above system has been used extensively by commercial feedlot operators, nutrition consultants working with the feeding industry, and by others throughout the western United States. In spite of the apparent complexity of using two values for feed and calculating separately the requirements for maintenance and production, the method has been employed successfully in a highly competitive industry and has been considered to be worth the extra effort by those who have used it.

D. British "Net Energy" System (Based on Metabolizable Energy)

Although the new British system of evaluating feeds and expressing the energy requirements of animals appears to be different from the net energy systems discussed above, in practice it is almost identical to the California system. The Agricultural Research Council of Great Britain (1965) adopted the system proposed by Blaxter (1962) which is based on using metabolizable energy and a series of equations for calculating the energy requirements of animals. Many reasons are given for adopting this system instead of the conventional net energy system, but the main one is that metabolizable energy is utilized with a different efficiency for different physiological functions. This same reason is given by the California workers for using NE_m and NE_g rather than ENE. An additional factor taken into consideration by the British system is the influence of the plane of nutrition on the metabolizable energy value of the ration. The effect of level of intake is corrected as follows: deduct $9.6 - 0.11\ Q_m$ from the observed ME value; Q_m is the ME value of the total diet measured at the maintenance level of intake.

The maintenance needs are expressed as metabolizable energy, with the amounts based on the fasting metabolism of the animal and the nutritional value of the ration.

The equation which takes into consideration ration effects is as follows:

$$k_m = 54.6 + 0.30\ Q_m$$

in which k_m is the efficiency of utilization of metabolizable energy for maintenance and

Q_m is the metabolizable energy of a diet as a percentage of its gross energy determined at the maintenance level of feeding. Fasting metabolism estimates for each species of livestock are tabulated.

For fattening, the equation is:

$$k_f = 0.81\ Q_m + 3.0$$

in which k_f is the efficiency of utilization of metabolizable energy for fattening (Fig. 15–2). No equations were given for lactation, but the values listed indicate that it was assumed that efficiency of utilization of ME for milk production was 65 to 70 percent and that the relationship was curvilinear rather than linear.

The British system has much to recommend it from the scientific standpoint because it makes possible a study of the component parts of the problem and avoids the necessity of multiple listing of net energy values for different functions. On the other hand, it is complex and difficult to understand. Once the information is available for use with this system, it will be a simple matter to calculate "net energy" values for the *rations* (not the individual ingredients).

E. Rostock Net Energy for Fattening System (NE_f)

Another major system, which is based on calorimetric studies carried out primarily with fattening animals, is the NE_f system proposed by the workers in Rostock, East Germany. For several years they have been carrying out energy balance studies and checking the results obtained by a pioneer in this field, Oskar Kellner. Kellner's work led to the development of the starch equivalent system that has been used so widely throughout the world. This form of net energy, in which the energy retention in the body (primarily as fat) is expressed in terms of the amount of starch required to form the body fat, has been widely accepted and easily understood. As in the case with ENE, NE_f and starch equivalent values may be calculated from digestible nutrients. The equation for cattle reported by the Rostock workers was as follows:

$$NE_f = 1.71\ DP + 7.32\ DEE + 2.01\ DCF + 2.01\ DNFE$$

in which NE_f is Mcal net energy for fattening and DP, DEE, DCF and DNFE are diges-

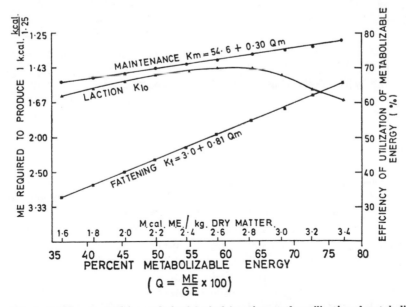

FIG. 15–2. Effect of ration composition and physiological function on the utilization of metabolizable energy by ruminants based on British feeding standards. The amount of metabolizable energy required to produce one kilocalorie of body gain (fat) and milk, and the efficiency of use of ME below maintenance are shown (A.R.C., London, 1965).

tible crude protein, digestible ether extract, digestible crude fiber, and digestible nitrogen free extract expressed in kilograms.

Correction factors are used to adjust the NE_f value of mixed rations for cattle for the effect of variation in concentration of digestible energy in the total diet on the efficiency of conversion to body tissue. These adjustment factors for cattle are as follows:

Digestibility of the Energy in the Ration (%)	Correction Factor
67 — 80.0	1.00
65 — 66.9	0.97
63 — 64.9	0.96
61 — 62.9	0.95
59 — 60.9	0.93
57 — 58.9	0.91
55 — 56.9	0.89
53 — 54.9	0.87
51 — 52.9	0.84
50 — 50.9	0.82

The Rostock system is a means of predicting animal response in terms of Mcal of body tissue deposited. Except for the method of deriving the values, it is very similar to the others that have been discussed. The widespread acceptance of the starch equivalent and related systems based on this same principle is an indication that it is not too complex for use in common practice.

F. Net Energy for Lactation (NE_l)

In the previous systems, the energy requirements for milk production were expressed in terms of TDN, ME, or NE_f (or some other expression of net energy for fattening.) This practice has led to some difficulties, for the influence of ration composition is not so great for lactation and lipogenesis, which occurs in a lactating cow, goat, or ewe, as it is for a fattening animal. During the past decade, a considerable number of energy balance trials have been conducted with lactating dairy cows. The results indicate that k_l and k_m are influenced by concentration of energy in the ration (percent ME) in a very similar manner, and in a manner different from k_f. This finding could serve as the basis for a much simpler system for use with dairy cows. Instead of using NE_m and NE_f and NE_l to calculate rations for lactating cows that are also gaining body tissue, it should be possible to express the energy requirements and the feed values in terms of NE_l. In this system, the energy requirements will be a direct expression of the energy in milk. The amount of energy required to produce 1 kg of milk containing 750 kcal of energy will be .750 Mcal NE_l.

IV. METABOLISM

A. Energy Transformations in Animals

It is outside the scope of this chapter to discuss thermodynamics and bioenergetics in detail, but a general knowledge of these subjects is helpful in understanding the factors that influence efficient utilization of energy by animals. Several excellent articles and chapters in textbooks have been written on these subjects, including those by Krebs and Kornberg (1957), Lehninger (1965), Kleiber (1961), Brody (1945), and Blaxter (1962). Baldwin (1968) has written a concise review of the theoretical calorific relationships that are applicable to nutrition and bioenergetics. This section was adapted from his article.

Many of the energy transformations and expenditures associated with food utilization for maintenance, growth, and production can be explained in physiological and biochemical terms. Release of energy from feedstuffs occurs during digestion, absorption, storage, mobilization, the synthesis of the pyrophosphate bond(s) of adenosine triphosphate (ATP) at the expense of the free energy released upon oxidation of foodstuffs, and, finally, during hydrolysis of pyrophosphate bonds of ATP during the performance of physiological work functions. Examples of these physiological processes will be discussed and related to specific nutritional energetic terms, which are only arbitrary divisions because the processes cannot always be separated and may be interdependent. The energy expenditures are considered in the order in which they might be expected to occur when a feed is consumed by an animal.

The following abbreviations are commonly used in describing intermediary energy metabolism. Some have been used in the models presented in Table 15–1.

ATP = adenosine triphosphate
ADP = adenosine diphosphate
UTP = uridine triphosphate
UDP = uridine diphosphate
P_i = inorganic phosphate
NAD = nicotinamide adenine dinucleotide or coenzyme I
$NADH_2$ = reduced nicotinamide adenine dinucleotide
$NADPH_2$ = reduced nicotinamide adenine dinucleotide phosphate
α-GP = α-glycerophosphate
PP_i = pyrophosphate
AMP = adenosine 5'-phosphate
Ac-CoA = acetyl coenzyme A
CoASH = reduced coenzyme A
RNA = ribonucleic acid
DNA = deoxyribonucleic acid
DPN = diphosphopyridine nucleotide (synonym for NAD)
$DPNH_2$ = reduced diphosphopyridine nucleotide
TPN = triphosphopyridine nucleotide (synonym for NADP)
$TPNH_2$ = reduced triphosphopyridine nucleotide
TPP = thiamine pyrophosphate
IDP = inosine diphosphate
ITP = inosine triphosphate
GDP = guanosine diphosphate
GTP = guanosine triphosphate

1. DIGESTION

Energy release due to bond breakage and the energy loss associated with the synthesis of pancreatic protein are two examples of processes that occur during digestion.

breakage of the peptide bond during digestion.

Similarly, the free energy of glucosidic-bond hydrolysis is approximately 4.3 kcal/M and the heat of combustion of 100 gm of starch is 420 kcal/M. Starch hydrolysis during digestion releases 2.5 kcal of heat per 100 gm starch hydrolyzed or 0.6 percent (2.5 per 420) of the heat of combustion of starch.

The conversion of triglycerides to glycerol and fatty acids releases about 0.1 percent of the energy content of the triglyceride. Losses due to bond breakage similar to those that occur during digestion occur during the mobilization of glycogen, protein, and fat from body tissues during fasting. Emphasis is placed upon the fact that these losses are relatively insignificant.

Synthesis of Digestive Secretions. The loss of energy associated with the synthesis of digestive secretions can be appreciable. The energy cost of synthesizing protein is on the order of 5 M of ATP/M of amino acid incorporated. If a 200 kg bovine which consumes 7500 kcal of digestible energy per day secretes 344 gm of pancreatic protein per day (Keller *et al.*, 1959), the estimated energy loss due to synthesis of the pancreatic protein is 310 kcal, or 4.1 percent of the daily energy intake.

$$\left[5\text{M ATP} \times \frac{344 \text{ g pancreatic protein}}{100 \text{ gm/M}} \times 18 \text{ kcal/M ATP} = 310 \text{ kcal} \right]$$

Bond Breakage. The free energy of peptide bond hydrolysis is approximately 3 kcal/M, and the average molecular weight of the amino acids of most proteins can be assumed to be 100 gm/M. Thus, the energy release per 100 gm of protein or per mole of peptide bond hydrolyzed during digestion would be 3 kcal. The heat of combustion (gross energy) of 100 gm of protein is about 570 kcal, so about 0.5 percent (3 per 570) of the energy content of protein is released due to

This example does not account for any energy cost if the amino acids secreted as pancreatic protein are not completely reabsorbed as amino acids and transported back to the pancreas at no net cost.

2. ABSORPTION

The absorption of glucose and amino acids is an active (ATP-requiring) process. If the absorption of 1 M of glucose from the digestive tract requires the use of 1 M of ATP,

Table 15–1. Energy Costs of Storage and Tissue Turnover

(From Baldwin, 1968. J. Dairy Sci. 51:104.)

A. Model for Estimating the Energy Cost of Glycogen Storage

glucose + ATP $\longrightarrow$ glucose-6-phosphate + ADP

glucose-6-phosphate $\longrightarrow$ glucose-1-phosphate

glucose-1-phosphate + UTP + H_2O $\longrightarrow$ UDP glucose + 2 P_i

UDP glucose + glycogen(n) $\longrightarrow$ glycogen(n+1) + UDP

UDP + ATP $\longrightarrow$ UTP + ADP

glycogen(n+1) + P_i $\longrightarrow$ glucose-1-phosphate + glycogen(n)

glucose-1-phosphate + H_2O $\longrightarrow$ glucose + P_i

Sum: 2 ATP + 2 H_2O $\longrightarrow$ 2 ADP + 2 P_i

$$\text{Energy cost} = \frac{2 \times 18 \text{ kcal/M ATP*}}{686 \text{ kcal/M glucose}} \times 100 = 5.2\%$$

* Calculated on the basis of 686 kcal/M of glucose, which yields 38 ATP/M when totally oxidized according to accepted pathways, $\frac{686}{38} = 18$ kcal/M ATP lost as heat. The often quoted 8 kcal/M ATP represents only the heat loss upon hydrolysis of the terminal pyrophosphate bond of ATP. In addition, 10 kcal/M of ATP are lost during the formation of the terminal pyrophosphate bond during glucose oxidation.

B. Model for Estimating Energy Losses Associated with Storage of Glucose as Fat.

12 glucose + 24 ADP + 48 NAD $\longrightarrow$ 24 Ac-CoA + 24 ATP + 48 $NADH_2$

24 Ac-CoA + 21 ATP + 42 $NADPH_2$ $\longrightarrow$ Palmityl-CoA + 21 ADP + 42 NADP

0.5 glucose + 1 ATP + 1 $NADH_2$ $\longrightarrow$ α-GP + 1 ADP + 1 NAD

3 Palmityl-CoA + α-GP $\longrightarrow$ Triglyceride

Sum: 12.5 glucose + 2 ADP + 5 NAD $\longrightarrow$ Triglyceride + 2 ATP + 5 $NADH_2$

Energy costs: 12.5 glucose at 686 kcal/M = 8,575 kcal

Tripalmitate (7,597 kcal) + others (273 kcal) = 7,870 kcal

$$\frac{705 \text{ kcal heat (i.e. } 8,575-7,870)}{8,575 \text{ kcal}} \times 100 = 8.2\%$$

C. Model to Estimate Energy Cost Associated with Turnover of the Cori Cycle.

glucose + 2 ADP + 2 P_i $\longrightarrow$ 2 lactate + 2 ATP (muscle)

2 muscle lactate $\longrightarrow$ 2 liver lactate (via blood)

2 lactate + NAD^+ $\longrightarrow$ 2 pyruvate + 2 $NADH_2$ (liver)

2 pyruvate + 2 CO_2 + 2 ATP $\longrightarrow$ 2 oxalacetate + 2 ADP + 2 P_i (liver)

2 oxalacetate + 2 ATP (equivalents) $\longrightarrow$ 2 phosphoenolpyruvate + 2 ADP + 2 CO_2 (liver)

2 phosphoenolpyruvate + 2 $NADH_2$ + 2 ATP $\longrightarrow$ glucose + 2 NAD + 2 ADP + 4 P_i (liver)

liver glucose $\longrightarrow$ muscle glucose (via blood)

Sum: 4 ATP + 4 H_2O $\longrightarrow$ 4 ADP + 4 P_i

Energy cost = 4M ATP $\times$ 18 kcal/M ATP = 72 kcal/turn of Cori cycle.

D. Model to Estimate Energy Costs Associated with the Turnover of Triglycerides in Adipose Tissue.

triglyceride + 3 H_2O $\longrightarrow$ 3 fatty acids + glycerol (adipose)

glycerol (adipose) $\longrightarrow$ glycerol (liver)

1 glycerol (liver) + 1 ATP $\longrightarrow$ $\frac{1}{2}$ glucose (liver) + 1 ADP + 1 NAD + 1 H_2O + 1 P_i + 1 $NADH_2$

$\frac{1}{2}$ glucose (liver) $\longrightarrow$ $\frac{1}{2}$ glucose (adipose)

$\frac{1}{2}$ glucose (adipose) + 1 ATP + 1 $NADH_2$ $\longrightarrow$ α-glycerol-phosphate (adipose) + 1 ADP + 1 P_i + 1 NAD

3 fatty acids + 3 ATP + 3 CoASH $\longrightarrow$ 3 fatty acyl-CoA + 3 AMP + 3 PP_i

3 PP_i + 3 H_2O $\longrightarrow$ 6 P_i

3 AMP + 3 ATP $\longrightarrow$ 6 ADP

3 fatty acyl-CoA + α-glycerol-P $\longrightarrow$ triglyceride + 3 CoASH

Sum: 8 ATP + 8 H_2O $\longrightarrow$ 8 ADP + 8 P_i

Energy cost = 8M ATP $\times$ 18 kcal/M ATP = 144 kcal/M triglyceride then (18 kcal/M ATP/686 kcal/M glucose) 2.6 percent of the potential energy value of glucose would be expended during absorption.

3. STORAGE

Storage of Glucose as Glycogen. Animals are continuous metabolizers, but they eat only periodically. Therefore, a portion of the dietary metabolites must be stored for later use as sources of energy. Glucose can be stored as either glycogen or fat. Only about 30 percent of the dietary carbohydrates may be stored as glycogen (West and Todd, 1962) and the remainder is either utilized directly or stored as fat. The cost of glycogen storage estimated from the model in Table 15–1 is approximately 5 percent.

A portion of the heat loss associated with glycogen formation occurs soon after the consumption of food whereas the portion of the heat loss associated with the mobilization of glucose from glycogen (4.1 kcal/M) occurs later.

Storage of Glucose as Fat. The energy costs associated with the storage of glucose as fat may be estimated by the model shown in Table 15–1 to be about 8 percent. The general model involves the conversion of glucose to acetyl-CoA, which is converted to fat. The model is constructed so that only energy release occurring soon after consumption of feed is considered. The net energy loss associated with the intermediate storage of glucose as fat is also about 8 percent. In calculating net energy loss, the energy relationships associated with the conversion of glucose to acetyl-CoA were not considered because these would occur with or without intermediate storage of glucose as fat. Instead, energy losses associated with the mobilization of fat and degradation of fatty acids to acetyl-CoA are considered.

In summary, the estimated costs of absorption of glucose from the intestine (2.6 percent) plus the costs of pancreatic secretion (4.1 percent) and storage of glucose as glycogen (5 percent) or fat (8 percent)

yield estimates of heat increments of about 11 to 14 percent associated with the feeding of carbohydrates to animals under maintenance conditions. The estimated energy costs of storage appear to be the major component of the heat increment of carbohydrate utilization, and their contribution would be expected to vary, depending upon the amount and form of glucose storage.

4. PROTEIN AND AMINO ACID UTILIZATION FOR ENERGY

When protein and amino acids are used as sources of energy, the heat increment is estimated to be about 20 to 40 percent. Krebs (1963) has calculated heat increments of about 20 percent for protein utilization for energy, not including costs of digestion and absorption. The energy cost for urea synthesis is 2 ATP molecules per atom of protein nitrogen (except for arginine). The extent of coupling of oxidative processes with phosphorylation and the energy losses due to partial oxidation and excretion (12 percent) are also discussed by Krebs (1963).

B. Metabolic Pathways

1. FORMATION OF ATP

Two major pathways of carbohydrate utilization operate in animals. One of these involves glycolysis via the Embden-Meyerhof pathway and terminal oxidation via the tricarboxylic acid pathway. The other involves a combination of hexose monophosphate cycle and hexose monophosphate glycolysis coupled with terminal oxidation via the tricarboxylic acid pathway. The relative contributions of these alternate pathways vary with species, diets, tissues, and physiological status. The efficiencies of ATP formation via these alternate pathways are approximately equivalent, so the intermediate aspects of the two pathways will not be considered. Total oxidation of glucose via the Embden-Meyerhof glycolysis and TCA (tricarboxylic acid) cycle oxidation yields 38 M of ATP. Current estimates of the energy content of the terminal pyrophosphate bonds of ATP are about 8 kcal per M (Long, 1961). This means that during the oxidation of 1 mole of glucose, 304 kcal of energy are effectively trapped in the form of ATP. The heat of combustion of glucose is 686 kcal per mole, so the efficiency of ATP formation during the oxidation of glucose is 44 percent (304/686). Thus, 56 percent of the potential energy available from glucose is lost as heat during ATP synthesis. On the basis of similar reasoning, the heat losses associated with the oxidation of palmitate, propionate, and acetate for the formation of ATP are 57, 61, and 62 percent, respectively. These energy losses are components of the basal energy expenditure of animals.

2. Utilization of ATP for Maintenance Processes.

The energy (ATP) utilized in connection with maintenance of muscle tone, heart action, kidney reabsorption, transmission of nerve impulses, protein resynthesis, and ion transport across membranes can only be estimated. The operation of the Cori cycle, the turnover of triglyceride in the adipose tissue, and the replacement of intestinal epithelium illustrate a portion of the total energy costs of maintenance.

Table 15–1 illustrates the energy cost estimates associated with turnover of the Cori cycle. In resting human subjects, about 0.25 M of glucose per day is recycled via the Cori cycle so that approximately 1 percent of the daily energy expenditure is associated with Cori cycle activity.

The energy costs associated with the turnover of 1 M of triglyceride (Table 15–1) result in an energy loss of 144 kcal. The energy expenditure associated with turnover of triglycerides in rats amounts to approximately 15 percent of the digestible energy intake at maintenance (Ball, 1965).

About 500 gm of protein must be synthesized per day by an adult ruminant to support the rapid replacement of the cells of the intestinal epithelium. Approximately 450 kcal per day must be expended in the re-

placement of the intestinal epithelium (5 ATP/100 gm protein synthesized $\times$ 18 kcal/M ATP $\times$ 500 gm protein). This replacement represents approximately 6.0 percent of the daily energy intake of a 200 kg bovine.

3. Energy Utilization for Productive Processes.

Productive processes represent another type of function requiring energy. Calculations based on biochemical pathways can be compared to estimates of the efficiency of utilization of metabolizable energy (ME). The efficiency of ME utilization for maintenance is related to the efficiency with which foodstuffs are utilized as substitutes or replacements for body stores. The heat increment below maintenance reflects energy expenditures associated with the utilization of foodstuffs in excess of those associated with the utilization of body stores such as glycogen and fat. For production, the efficiency of ME utilization reflects the efficiency with which foodstuffs are incorporated into the product, while heat increment estimates reflect the energy expenditures required for the incorporation of foodstuffs into the product.

Milk Synthesis. Milk synthesis is one example of a productive function. Some assumptions were made in calculating energy costs for lactose, milk fat, and protein synthesis. These were as follows: acetate to propionate absorbed from the rumen was 2.5 to 1; the balance of amino acids absorbed was ideal for milk synthesis; and, only energy costs occurring after absorption need be considered in relating metabolizable energy to net energy.

Lactose synthesis was considered to occur totally at the expense of propionate, which, after absorption, was converted to glucose in the liver and then transported to the mammary gland for conversion to lactose. The estimated metabolic efficiency of lactose synthesis from propionate is 78 percent.

Milk protein synthesis was assumed to

occur from an almost perfect balance of amino acids, with propionate serving as the source of energy. The costs associated with uptake, transport, and rearrangement of amino acids were assumed to be nil. Based on present concepts of protein synthesis, 2 ATP equivalents are required for the activation of 1 M of amino acid to form acyl-s-RNA; 1 ATP equivalent (GTP) is expended during peptide bond formation; and about 2 ATP equivalents are required for regeneration of transfer and other RNA after the synthesis of 100 gm of protein. The estimated ATP requirement for the synthesis of 100 gm of milk protein is thus 5 M of ATP. To supply this ATP, 0.295 M of propionate must be oxidized. The input-output relationship is as follows:

> Input: 567 kcal/M of average amino acid + 108 kcal/0.295M of propionate = 675
> Output: 570 kcal/M of protein = 570
> Efficiency = 570/675 = 84%

The estimated metabolic efficiency for milk protein synthesis from amino acids and propionate is thus about 84 percent.

To simplify estimation of the metabolic efficiency of fat synthesis, it was assumed that the efficiency of palmitate incorporation into milk fat is representative of overall efficiency and that acetate is the sole precursor of palmitate.

The overall metabolic efficiency of milk synthesis is 76 percent. The estimate compares favorably with observed efficiencies of utilization of metabolizable energy for milk synthesis of 70 percent (Blaxter, 1962). In the formulation, the estimates of metabolic costs associated with the synthesis of the major components of milk were considered. Energy costs associated with the secretion of minor milk components and maintenance of the gland might be relatively large. The rate of turnover of cytoplasmic protein in the mammary glands of rats may be as much as 3 percent of the cytoplasmic protein turned over per hour (Emery & Baldwin, 1967). If a high rate applies to cows, the efficiency estimate above would be reduced.

In view of an estimated efficiency of milk synthesis of 76 percent as compared to real efficiency estimates as high as 70 percent, it seems reasonable to suggest that milk secretion can occur at close to maximal theoretical efficiency and with a minimal heat increment.

Growth and Fattening. The theoretical efficiency estimates obtained for protein and fat synthesis in the above formulation for milk synthesis can also be employed to estimate theoretical efficiencies for growth and fattening. The theoretical efficiency of growth in rapidly growing lambs is estimated in Table 15–2 to be 78 percent, a value

Table 15–2. Estimation of Efficiency of Growth in Young Lambs

(From Baldwin, 1968, J. Dairy Sci. 51:104.)

Composition of Gain	Gain (g)	Gain (kcal)	Estimated ME Required above Maintenance (kcal)	Efficiency
Protein	926	5,278	6,280	0.84
Fat	414	3,880	5,400	0.72
Total	1,340	9,158	11,680	0.78

which compares favorably with estimates of the efficiency of utilization of ME supplied above maintenance for growth of young animals of greater than 70 percent (Blaxter, 1962). An alternate means of expressing the comparison of observed and theoretical efficiencies can be devised based upon the fact that when productive performance is considered, the difference between input above maintenance expressed as metabolizable energy (ME) and output expressed as energy in product or net energy (NE) represents the heat increment (HI = ME − NE). According to this relationship, the estimated, theoretical heat increment of growth is 22 percent (100 − 78) and the observed heat increment is 30 percent (100 − 70), indicating that almost 75 percent (22/30) of the energy costs (heat loss) associated with growth can be accounted for as biochemical costs. This means of expression emphasizes the functional nature of heat increment and helps dispel notions that the term "heat increment" implies waste.

It is acknowledged that the simplifying assumptions that were made in this section may be questionable. They were made in order that physiological processes and biochemical reactions could be related to classical nutritional terms. It is obvious that a great deal of additional biochemical and physiological data must be accumulated befor meaningful models can be formulated.

V. ENERGY REQUIREMENTS OF ANIMALS

A. Feeding Standards

1. GENERAL COMMENTS.

It is apparent from the preceding discussion that the metabolism of animals involves a large number of metabolites. The measurement of the amount of each metabolite required for a specific physiological function is a complex matter. There is also the question of how much of each of these nutrients is supplied by a given feedstuff. The adequacy of the nutrients supplied by the digestion of a feed in meeting the metabolic needs of the animal determines how much of that specific feed is required for a given animal producing a known amount of end product. This brings us into the whole area of expressions of energy value of feeds and the energy requirements of animals. The complexity of the problem is exemplified by the number and variety of feeding standards currently in use throughout the world.

2. TABLES OF ENERGY REQUIREMENTS.

The tables of requirements used most frequently in the United States, based on the National Research Council recommended nutrient allowances, are given in the appendix. The energy required by each animal species is listed separately for each physiological function including maintenance, growth, and production. This is done for convenience of computation and simplification of use. However, there is a *single total requirement* for the animal. Animals are not fed by giving one

feed for maintenance and another for production. Several feeds may be given, but the metabolites supplied by each are used for all physiological functions.

Some major differences are apparent in the requirements listed for animals which are kept for extended periods of time for breeding and milk production and those which are grown rapidly for meat production. Meat animals are usually allowed *ad libitum* consumption of food because of the necessity for rapid and economical growth. The nutrient requirements of growing swine and poultry are therefore best described by minimum concentrations of energy, protein, vitamins, and minerals in the diet. The minimum energy concentration suggested for growing and finishing swine is 3100 to 3500 kcal DE/kg of diet and for growing and laying hens it is 2450 to 2600 kcal ME/kg of diet.

When *ad libitum* levels of feed intake are required for the desired animal response, the relative importance of listings of absolute nutrient requirements is decreased, but the need for the exact energy value of feeds is greatly increased. In this case, the major management decisions concern not how much to feed the animal, but the relative monetary cost of different feeds that could be given to obtain the desired gain.

With ruminants, the questions of energy value and energy requirements present additional considerations because materials that vary widely in chemical and physical properties are commonly used. Microbial fermentation in the rumen makes it possible for poor quality roughages to be utilized as energy sources. This is not the case with nonruminants. In the case of the dairy cow, the feed is limited most of the time rather than given *ad libitum*. Thus, it becomes even more important that the total nutrient requirements as well as the nutritional value of various diets be known.

Because of the tremendous variation in milk production, milk fat percentage, and body size among dairy animals and because many processes such as maintenance, milk production, gestation, and fattening may occur simultaneously, it is more convenient to present additive requirements. The requirements for each activity may be computed separately for a given animal; the sum of these provides the total energy requirement. In doing this, however, it must be remembered that the processes and the requirements are not completely independent.

Maintenance. When an animal is neither lactating nor pregnant and is neither gaining nor losing body tissue, it is said to be at "maintenance." The maintenance requirement is a measure of the amount of feed energy necessary to maintain this animal for long periods of time without gain or loss in body weight. The use of energy for maintenance is probably the least understood of all physiological functions because the energy is not used for a product which can be directly measured. The end product is the maintaining of life processes of the animal: motion such as walking, the movement of internal organs such as the heart, diaphragm, and digestive tract, and the maintenance of proper body temperature.

Because of the need to maintain body temperature, it is less clear how much of the dietary energy is required for energy compounds such as ATP and how much is needed simply as heat. In the thermoneutral range of temperatures, the metabolism of energy compounds to produce ATP and other intermediate stores of energy yields in sufficient amounts of heat to maintain the body temperature of the animal without metabolism of additional metabolites for this purpose alone. At environmental temperatures below the critical temperature, however, the maintenance requirement increases. This reflects the more rapid heat loss from the animal and the need to replace this heat with a higher rate of metabolism and greater heat production. In this particular instance, the heat increment of poor quality diets could be used to good advantage.

For convenience, it is desirable to separate the requirement of the producing animal into the separate requirements for production (milk or growth) and maintenance. The maintenance requirement is that portion of the total energy requirement which is not proportional to milk production or growth. In this fashion, all of the increases in body function such as increased metabolic rate, increased activity of digestive tract associated with higher levels of feed intake, and other functions which accompany the productive processes are attributed to the increased production. The remainder of the requirement is attributed to maintenance. This portion is usually considered to be a function of body size or metabolic body size (kg body weight raised to the 3/4 power).

Growth. Nature has accorded a very high priority to the process of growth. Undernutrition of young growing animals results in continued growth, even if the undernutrition is severe enough to cause a decrease in body weight. Some protein deposition and bone growth continue, and the energy is provided from the reserves of fat in the body of the animal. This type of growth occurs at the expense of subcutaneous, intramuscular, or intra-abdominal fat (Blaxter, 1962).

For normal growth, however, the energy intake must exceed the animal's maintenance needs and the calorific value of the body tissue formed. The calorific value of the body tissue increases with age because juvenile growth contains more water, protein, and bone mineral and less fat than does later growth. Increased levels of feed intake result in accelerated growth rates, but may also increase the ratio of fat to protein, which is deposited as body gain. For example, dairy calves which are consuming a diet with adequate protein respond to increasing levels of feed intake by depositing a progressively higher proportion of fat than protein.

The distinction between growth and fattening is not clear, for both occur simultaneously. The factors that affect growth include heredity, diet, and environmental conditions. There is a genetic limit to the daily synthesis of body protein in growth; even on the same ration, some animals will fatten while others continue to deposit protein.

The net efficiency of the utilization of energy for growth is affected by the partial efficiencies as well as the total amounts of fat and protein deposited. In studies with young calves which are fed only milk, the efficiency of utilization of metabolizable energy is about 85 percent, and the concentration $\left(\frac{ME}{GE}\right)$ of ME in the milk diet is approximately 95 percent. These efficiencies are considerably greater than those noted for mature animals.

3. Effect of Age and Breed

Young cattle, sheep, and goats are essentially nonruminants, so that part of the effect of age on energy utilization is the transition to diets higher in fiber. This transition as well as the energy losses due to fermentation and the increasing proportion of fat in the body tissue, tends to increase the energy requirements per unit of weight increment.

Ritzman and Colovos (1943) noted that the efficiency of utilization of a ration by growing heifers was greater for growth than for fattening. Although this observation is commonly made, further research is needed before definite conclusions of the effect of age on energy utilization can be drawn.

The fasting metabolism of cattle and sheep tends to fall markedly with age. This decrease should lead to a lower maintenance requirement for older animals, and thus leave a greater proportion of the ration for productive purposes. The limited information available about other species also indicates a slight decline in rate of fasting metabolism with age.

Breed and sex have an influence on energy requirements also, but the effects appear to be relatively small (A.R.C., 1965). In cattle, however, Japanese workers have noted relatively large breed differences (Hashizume *et al.*, 1962). Their native cattle (Japanese Black breed) had appreci-

ably lower rates of fasting metabolism than did the European breed (Holstein). This lowered maintenance requirement could affect the apparent efficiency of energy utilization in the same manner as noted above for age effects.

4. Effect of Muscular Work and Activity.

One of the factors that affects the energy requirements of an animal is the energy expenditure due to muscular work. Approximately 2.0 kcal/kg/24 hr is required for standing. For walking, about 95 kcal/kg/mile is required for sheep and 79 kcal/100 kg/mile for cattle. The work of ascent is about 196 kcal/100 kg/1000 ft for sheep and 207 for cattle. The work of descent and the energy costs of grazing have not been determined with sufficient accuracy to warrant the inclusion of an estimate.

5. Effect of Environmental Temperature.

Most of the values obtained in calorimetric studies were made with animals in a thermo-neutral zone. The effects of temperature on energy requirements vary with the relative humidity, wind velocity, and length of hair coat (insulation) of the animal. In addition, the type of ration and level of feed intake influence the effects of temperature on energy utilization of animals. These effects are discussed in Chapter 4 as well as by Blaxter (1962). In general, however, energy requirements are increased if an animal has to use extra energy to maintain body temperature.

6. Practical Considerations.

The actual requirements for growth depend to a very great extent on the use which the animal serves. For example, for meat animals the most desirable situation is the maximum possible growth rate to obtain the greatest economic efficiency. The faster the growth rate the more efficient the animal because of the smaller amount of feed which will be required in order to reach a given live weight. For this reason, the major concern is usually the selection of feeds which will give maximum growth rates when the animal is fed *ad libitum*. The amounts of energy required to achieve certain growth rates are given in the appendix tables. These figures may also be used to indicate the growth rate to be expected when a certain amount of feed is actually consumed.

In animals that are kept for breeding and milk production, the optimum growth rate is often somewhat less than the maximum possible. This is especially true for dairy cattle. Experiments have shown that dairy heifers grown at the maximum growth rate suffer more reproductive problems than heifers grown at a moderate rate (Reid *et al.*, 1957). Although these animals produced more milk during the first period of lactation, the milk production during succeeding lactation periods was uniformly inferior.

VI. METHODS OF STUDYING ENERGY METABOLISM OF ANIMALS

A. Comparative Slaughter Techniques

An accurate measurement of energy value of feeds requires the measurement of the actual amount of energy retained by the animal or produced as a useful product, or the measurement of all forms of energy loss. The former method is better adapted to studies of growing and fattening animals, in which the amount of energy stored can be measured by slaughter techniques. This method was employed by Lofgreen (1964) for the determination of net energy value of feeds for fattening beef cattle. The body energy retention was measured as the difference in total body energy of groups of animals slaughtered before and after a feeding period. The body energy was estimated from measurements of the specific gravity of the carcass, which is influenced by the amount of body fat. The net energy value of the feed is the additional energy stored as body tissue per unit increase of food consumed. The amount of heat produced is the difference between the measured energy balance and the metabolizable energy intake.

B. Energy Balance Studies

In dairy animals, however, the slaughter technique is not appropriate. Energy losses are more adequately measured in complete energy balance trials which involve exact measurement of all forms of energy intake and energy loss. The most difficult form of energy loss to measure is that of heat production. Measurements of heat production began with Lavoisier and LaPlace nearly 200 years ago when they enclosed an animal in a chamber which contained ice. The heat produced by the animal melted the ice, and the amount of ice melted multiplied by the latent heat of ice indicated the total heat production.

Since that time, elaborate techniques have been developed for measuring heat production although all are adaptations of two main types. The first is a direct measurement of the heat liberated by the animal and the second is the calculation of heat production from measurement of the respiratory exchange of the animal (Plates 19 & 20).

1. DIRECT CALORIMETRY.

Calorimeters in which the heat given off by an animal is measured directly have been used in this country and in Europe. With direct calorimetry the heat is measured either by the increase in temperature of a known volume of water or by the electrical current generated as heat passes across thermocouples. The adiabatic calorimeter was used in the large animal calorimeter constructed by Armsby at Pennsylvania State University at the beginning of this century. In this calorimeter the temperature of an outer water jacket was continuously monitored and adjusted so as to prevent the loss of heat from the calorimeter. The heat was removed from the calorimeter by means of water circulating in pipes inside the chamber. The increase in temperature of this water as well as the amount of water circulated provided the measure of heat production by the animal.

More recently, direct calorimeters have employed the gradient layer principle (Ben-

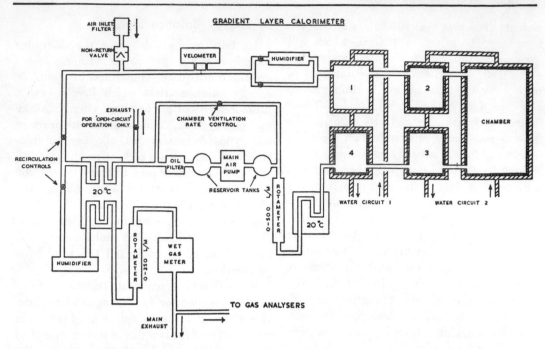

GRADIENT LAYER CALORIMETER

FIG. 15–3. Schematic diagram of a gradient layer calorimeter for measuring the heat production of sheep, wine, and other animals weighing up to 100 kg. (*Courtesy of Dr. J. Pullar & Dr. K. L. Blaxter.*)

zinger & Kitzinger, 1949; Benzinger *et al.*, 1958; Pullar, 1958). When heat is conducted across the wall of the chamber, a temperature gradient is produced. This temperature gradient is detected very precisely by copper-constantan junctions which produce an electrical potential proportional to the rate of heat flow.

Comparative measurements of direct and indirect calorimetry have shown that the two methods give excellent agreement. For this reason and because the direct calorimeters are necessarily more complex and expensive to operate, most nutritional studies are now carried out by means of indirect calorimetry.

2. INDIRECT CALORIMETRY.

It may be noted that in the chemical reaction converting glucose to carbon dioxide and water, 6 moles of oxygen are consumed and 6 moles of carbon dioxide are produced. The ratio of carbon dioxide produced to oxygen consumed is 1:1 or 1.00. This ratio, termed the respiratory quotient (RQ), serves to indi-

cate the type of nutrient being metabolized. The RQ is near 0.7 when fats of average composition are metabolized. The RQ is lower with fats because the relatively larger amounts of hydrogen in this material require more oxygen for complete combustion. The RQ with proteins is less constant because of the many uses which protein serves in the body; however, metabolism of protein normally results in an RQ near 0.8. RQ values in excess of 1.00 may be obtained when dietary carbohydrates are used for the production of body fat.

The measurement of RQ along with the total oxygen may be used to compute the total heat production of the animal. With this method, the amount of heat produced per liter of oxygen consumed for various values of RQ may be obtained from tables. An alternate method, and one which is used more widely at the present time, is the use of a single equation relating heat production to the respiratory exchange. The following equation was proposed by Brouwer and adopted by the Third Symposium on Energy

PLATE 19

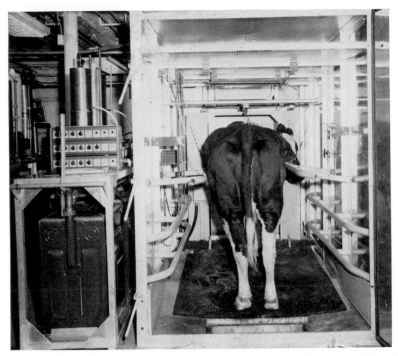

A, A cow in an open-circuit respiration chamber at the Energy Metabolism Laboratory, Beltsville, Maryland. The gas metering and sample collection units are on the left, outside the chamber. (*USDA Energy Metabolism Laboratory, Beltsville, Maryland.*)

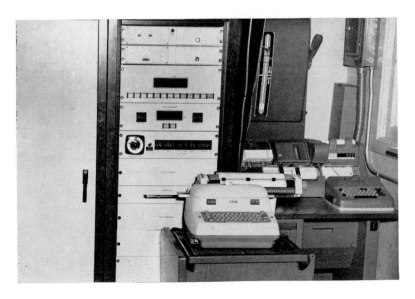

B, Automatic recording equipment, including a data logger, automatic remote typewriter, and automatic cardpunch. Data on gas volume, gas composition, environmental conditions, animal activity, and other information related to measurements of the energy metabolism of animals are automatically recorded at regular, pre-selected intervals. (*USDA Energy Metabolism Laboratory, Beltsville, Maryland.*)

PLATE 20

A, A bomb calorimeter being used by a technician to measure the heat of combustion of samples collected in an energy metabolism experiment. (*USDA Energy Metabolism Laboratory, Beltsville, Maryland.*)

B, An induction furnace and gasometric analyzer being used by a technician to measure the carbon content of feed and other biological materials. (*USDA Energy Metabolism Laboratory, Beltsville, Maryland.*)

C, Gas analysis by a technician using infrared gas analyzers for carbon dioxide and methane and a paramagnetic analyzer for oxygen. The technician is introducing a sample of gas into the analyzer and is recording the identification on a strip-chart recording potentiometer. (*USDA Energy Metabolism Laboratory, Beltsville, Maryland.*)

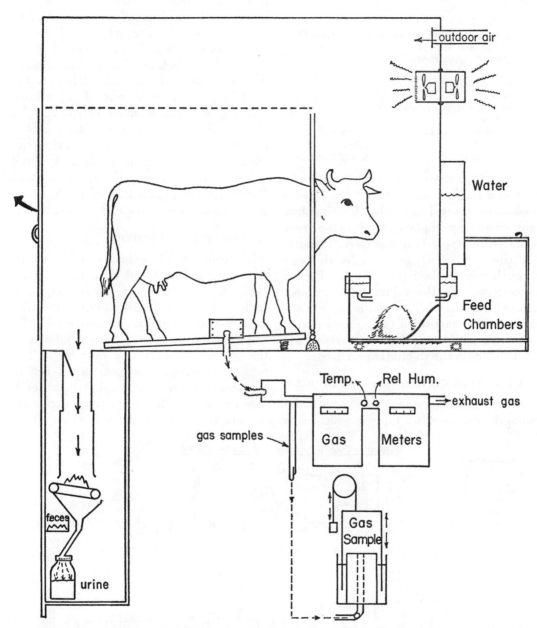

Fɪɢ. 15–4. A cow in an open-circuit respiration chamber at the Energy Metabolism Laboratory, Beltsville, Maryland. The gas metering and sample collection units are on the left, outside the chamber. Such a chamber is used to measure heat production and gaseous exchange in cattle and other large animals. (*USDA Energy Metabolism Laboratory, Beltsville, Maryland.*)

Metabolism at Troon, Scotland (E.A.A.P., 1965):

Heat (kcal) =
3,866 O_2 + 1.200 CO_2 — 0.518 CH_4 — 1.431 N
in which O_2 =
liters of oxygen consumed
CO_2 =
liters of carbon dioxide produced
CH_4 =
liters of methane produced
and N =
grams of nitrogen excreted in the urine

3. OPEN-CIRCUIT RESPIRATION CHAMBERS.

The actual measurement of the respiration exchange of animals may be accomplished by several methods. The method most commonly employed involves the use of open-circuit respiration chambers. Animals are enclosed inside an air-tight chamber through which air is circulated. Precise measurement of the volume of air drawn through the chamber as well as determination of the composition of the air entering and being withdrawn from the chamber are used to obtain the respiration exchange.

4. CLOSED-CIRCUIT RESPIRATION CHAMBERS.

Other types of respiration chambers include the closed-circuit type, in which air is recirculated. The composition of the air inside the chamber is maintained at normal levels by the addition of oxygen to replace that consumed by the animal and the removal of the carbon dioxide by trapping it in an absorbent such as potassium hydroxide. The advantage of this method is that the need for exact measurement of gas volume and for determination of composition is eliminated. The measurements of the respiratory exchange are made gravimetrically. This method is particularly well suited to work with small laboratory animals. With large farm animals, however, the cost of absorbents becomes an important factor.

5. CONFINEMENT METHOD.

The confinement method is used for short-term measurements of heat production. With this system the animal is completely enclosed in a chamber which has no facilities for the exchange of air. The carbon dioxide and methane are allowed to accumulate and the oxygen concentration decreases. The rate of change in gas composition during the short experimental period provides a measurement of the respiratory exchange. An adaptation of this method which allows

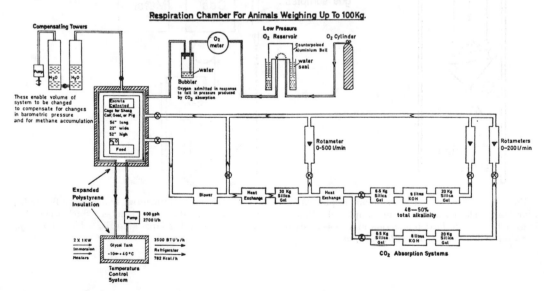

FIG. 15–5. Diagram of a closed-circuit respiration chamber used to measure the heat production of sheep, goats, calves, pigs, or other animals weighing up to 100 kg. (*Courtesy of Dr. K. L. Blaxter, Rowett Research Institute, Bucksburn, Aberdeen, Scotland.*)

continuous measurement has been described by Pullar and his associates (1967). A method of replacing the air inside the chamber with fresh outdoor air very rapidly is coupled with continuous analysis of air inside the chamber. Thus, the chamber is alternately closed, flushed rapidly, then closed again. Integration of the gas composition curves between flushing periods provides a measure of the total respiratory exchange.

6. OTHER TECHNIQUES.

Other methods of measuring the energy metabolism of animals include using masks or tracheotomized animals to facilitate the collection of respiratory gases. Insensible weight loss is another technique that has been used. For detailed information about the methods, Blaxter's book (1962) is highly recommended.

REFERENCES

Agricultural Research Council (1965). The nutrient requirements of farm animals, No. 2, Ruminants; Technical Reviews and Summaries. Agricultural Research Council, London.

Baldwin, R. L. (1968). Estimation of theoretical calorific relationships as a teaching technique. A review. J. Dairy Sci. *51*:104–111.

Ball, E. G. (1965). Some energy relationships in adipose tissue. Ann. N.Y. Acad. Sci. *131*:225.

Benzinger, T., & Kitzinger, C. (1949). Direct calorimetry by means of the gradient principle. Rev. Sci. Instr. *20*:849–860.

Benzinger, T. H., Minard, R. G., & Kitzinger, C. (1958). Human calorimetry by means of the gradient principle. J. Appl. Physiol. *12*:2.

Blaxter, K. L. (1962). *The Energy Metabolism of Ruminants.* Springfield, Illinois, Charles C Thomas.

Brody, S. (1945). *Bioenergetics and Growth.* New York, Reinhold Publishing Corp.

Brouwer, E. (1965). *Report of Subcommittee on Constants and Factors.* In: *Energy Metabolism.* E.A.A.P. Publ. No. 11. K. L. Blaxter (Ed.), p. 441, London, Academic Press.

Emery, R. S., & Baldwin, R. L. (1967). Turnover of several mammary enzymes during lactation. *Biochim. Biophys. Acta 136*:223.

Hashizume, T., Kaishio, Y., Ambo, S., Morimoto, H., Masubuchi, T., Abe, M., Horii, S., Tanaka, K., Hamada, T, & Takahashi, S. (1962). Metabolism of matter and energy in cattle. III. On the maintenance requirement of energy and protein for Japanese Black breed and Holstein breed cows in Japan. Bull. Natl. Inst. Agric. Sci. (Japan) Series G, No. 21:213–311.

Keller, P. J., Cohen, E., & Neurath, H. (1959). The proteins of bovine pancreatic juice. J. Biol. Chem. *234*:311.

Kleiber, M. (1961). *The Fire of Life—An Introduction to Animal Energetics.* New York, John Wiley & Sons, Inc.

Krebs, H. A. (1963). The Metabolic Fate of Amino Acids. In: *Mammalian Protein Metabolism.* H. N. Munroe & J. B. Allison (Eds.), New York, pp. 165–170, Academic Press.

Krebs, H. A., & Kornberg, H. L. (1957). Energy transformations in living matter. Ergeb. Physiol. *49*:212.

Lehninger, A. L. (1965). *Bioenergetics. The Molecular Basis of Biological Energy Transformations.* New York, W. A. Benjamin, Inc.

Lofgreen, G. P., & Garrett, W. N. (1968). A system for expressing net energy requirements and feed values for growing and finishing beef cattle. J. Anim. Sci. *27*:793–806.

Long, C. (1961). *Biochemists' Handbook.* Princeton, New Jersey, D. Van Nostrand Co., Inc.

Moore, L. A., Irvin, H. M., & Shaw, J. C. (1953). Relationships between TDN and energy values of feeds. J. Dairy Sci. *36*:93–97.

Morrison, F. B. (1956). *Feeds and Feeding.* 22nd ed. Ithaca, New York, Morrison Publishing Co.

National Research Council, Committee on Animal Nutrition (1964). *Joint United States-Canadian Tables of Feed Composition.* National Academy of of Sciences, N.R.C. Publ. 1232, Washington, D.C.

National Research Council, Committee on Animal Nutrition (1966). *Nutrient Requirements of Domestic Animals.* III. Nutrient Requirements of Dairy Cattle. National Academy of Sciences, N.R.C. Publ. 1349, Washington, D. C.

National Research Council, Committee on Animal Nutrition (1966). *Glossary of Energy Terms and Their Biological Interrelationships.* National Academy of Sciences, N.R.C. Publ. 1411, Washington (1968).

National Research Council, Committee on Animal Nutrition (1968). *Nutrient Requirements of Domestic Animals.* II. Nutrient Requirements of Swine. 6th Revised ed. National Academy of Sciences, N.R.C. Publ. 1599, Washington, D. C.

Nehring, K., Schiemann, R., & Hoffman, L. (1969). *The Energy Metabolism of Farm Animals.* Proposal for A New System of Energetic Evaluation of Food on the Basis of Net-Energy-Fat Worked Out in March 1966. K. L. Blaxter, G. Thorbek, & J. Kielanowski (Eds.), Proceedings of the Fourth Conference on Energy Metabolism of the European Association of Animal Production. Newcastle Upon Tyne, Oriole Press Ltd.

Pullar, J. D., Brockway, J. M., & McDonald, J. D. (1969). *Energy Metabolism of Farm Animals.* A Comparison of Direct and Indirect Calorimetry. K. L. Blaxter, G. Thorbeck, & J. Kielanowski (Eds.), Proceedings of the 4th Conference on Energy Metabolism of the European Association of Animal Production. Newcastle Upon Tyne, Oriole Press Ltd.

Pullar, J. D. (1958). *Direct Calorimetry of Animals by the Gradient Layer Principle.* Proceedings of the

First Symposium on Energy Metabolism of the European Association of Animal Production. Publ. No. 8. Copenhagen.

Reid, J. T., Loosli, J. K., Turk, K. L., Trimberger, G. W., Asdell, S. A., & Smith, S. E. (1957). Effect of nutrition during early life upon the performance of dairy cows. Proc. Cornell Nutr. Conf. Buffalo, New York.

Ritzman, E. G., & Colovos, N. F. (1943). Physiological requirements and utilization of protein and energy by growing dairy cattle. Univ. N. H. Agric. Exp. Sta. Tech. Bull. 80.

Rubner, M. (1901). Der energiewert der kost des menschen. Z. Biol. *42*:261–308.

Schneider, B. H. (1947). *Feeds of the World, Their Digestibility and Composition*. W. Va. Agric. Exp. Sta. Morgantown, W. Va.

Swift, R. W. (1957). The calorie value of TDN. J. Anim. Sci. *16*:753–756.

West, E. S., & Todd, W. R. (1962). *Textbook of Biochemistry*. 3rd ed., New York, The Macmillan Co.

REFERENCES ON BELTSVILLE STUDIES
(Not referred to in the text)

Moe, P. W., & Flatt, W. P. (1969). Automatic Data Logging Equipment for Energy Metabolism Studies with Large Animals. Proceedings of the Fourth Symposium on Energy Metabolism of the European Association of Animal Production. Newcastle Upon Tyne, England, Oriole Press Ltd. (In Press.)

Flatt, W. P., Moe, P. W., Moore, L. A., Hooven, N. W., & Lehmann, R. P. (1969). Energy Utilization by High Producing Dairy Cows. I. Experimental Design, Ration Composition, Digestibility Data, and Animal Performance During Energy Balance Trials. Proceedings of the Fourth Symposium on Energy Metabolism of the European Association of Animal Production. Newcastle Upon Tyne, England, Oriole Press Ltd. (In Press.)

Flatt, W. P., Moe, P. W., Munson, A. W., & Cooper, T. (1969). Energy Utilization by High Producing Dairy Cows. II. Summary of Energy Balance Experiments With Lactating Holstein Cows. Proceedings of the Fourth Symposium on Energy Metabolism of the European Association of Animal Production. Newcastle Upon Tyne, England, Oriole Press Ltd. (In Press.)

Flatt, W. P., Van Soest, P. J., Sykes, J. F., & Moore, L. A. (1958). A Description of the Energy Metabolism Laboratory at the U. S. Department of Agriculture, Agricultural Research Center in Beltsville, Maryland: Principles, Methods, and General Aspects. Proceedings of the First Symposium on Energy Metabolism of the European Association of Animal Production. Publ. *8*:101–109. Copenhagen.

Flatt, W. P., & Coppock, C. E. (1965). Physiological Factors Influencing the Energy Metabolism of Ruminants. In: *Physiology of Digestion in the Ruminant*. R. W. Dougherty, R. S. Allen, W. Burroughs, N. L. Jacobson, and A. D. McGilliard (Eds.). Butterworth. Washington.

Chapter 16

Protein Requirements

By W. H. Pfander

I. PROTEIN

PROTEINS are large molecules consisting of multiple combinations of amino acids joined by peptide linkages between the amino acids and further shaped by linkages formed by S-S and hydrogen bonding. They occur as essential constituents of all living cells. Each protein present in the different types of cells of each species has a specific number of amino acid residues joined in a genetically determined sequence. Only a few complete structures have actually been determined. Since structural details determine reactivity and especially solubility and areas available for enzyme attack, detailed knowledge concerning proteins can be valuable in the study of nutrition. Current data provide limited information about the amino acid sequence in common proteins which, due to the crude isolation measures, may actually be contaminated. In a feed, the conditions are more complex. A grain, oil seed, or animal by-product will contain a number of specific proteins and many other nitrogenous materials in a heterogenous mixture. It is difficult and time-consuming to partition a feed into its specific proteins; therefore, several

methods for expressing the nitrogen or protein composition of feedstuffs and the requirements of animals have been used.

A. Crude Protein (CP)

After several specific proteins were obtained and analyzed, they were found to contain approximately 16 percent nitrogen. This became the basis for converting nitrogen to protein. It was recognized that all of the material containing nitrogen was not true protein so that it became convenient to refer to it as "crude protein." In the usual method of analysis, the chemist determines the nitrogen content of the sample and multiplies his result by 6.25 to obtain the percent of protein. Proteins vary in their nitrogen content, ranging from 18 percent in certain grains to 15.2 percent in milk. In most cases, specific nitrogen-containing compounds are expressed relative to 16 gm nitrogen, again inferring that all proteins are similar. Many other nitrogenous substances which are not true proteins are present in nature. This heterogenous group is often called nonprotein nitrogen (NPN). Examples of NPN compounds include:

Allantoin
Amides
Amines, *e.g.*, histamine
Amino acids
Ammonia
Choline
Creatine → creatinine
Glutathione
Hippuric acid
Nitrate, nitrite

Oxytocin
Peptids
Pigments
Purines
Pyrimidines
Urea
Uric acid
Vitamins, *e.g.*, nicotinic acids

These substances include materials that are vitamins as well as building blocks for nucleic acids and various catabolic end products. NPN usually is more abundant in young, very wet tissues than in mature tissue.

B. Macromolecules

Most of the structural material containing nitrogen is, in fact, protein (P). The bones, muscles, vital organs, and skin of animals, and the leaves, seeds, and germs of plants all contain identifiable proteins of molecular weights ranging between 13,000 and over 300 million. The genetic material of cells, DNA, also contains nitrogen and is classed as a macromolecule. Many of the proteins are combined with other materials. Lipoproteins are common constituents of cell membranes and glycoproteins are found in reproductive fluids and nervous tissue. Proteins were traditionally classified according to solubilities. More modern analytical techniques indicate that many of these "proteins" are, in fact, mixtures, but many

of the properties associated with *in vitro* solubility may also be significant in nutrition. The more soluble proteins are rapidly attacked by rumen microorganisms and by digestive enzymes (Table 16–1).

C. Building Blocks and Residues

When protein molecules are exposed to certain enzymes or are autoclaved in an enclosed vial with 6N acid, peptide bonds are hydrolyzed and amino acids are released. Those present in most normal tissue are shown in Figure 16–1. In addition to the common amino acids shown, many others which are derived from them are present in small amounts in biological fluids or specific tissues. It is generally accepted that certain of the common amino acids are not formed in mammalian tissues and must be provided in the diet. However, others can be produced by the body if a suitable nitrogen donor and a carbon chain are present. Many bacteria can form amino acids from ammonia so they may be independent of external sources of amino acids. The fluids and tissues include many other nitrogenous compounds. These are produced from specific dietary amino acids.

D. Example Composition of Plant and Animal Tissues

The partial amino acid composition of certain plant fractions and animal tissues or products is shown in Appendix II.

Table 16–1. Classification of Proteins

Types		*Main Characteristic*	*Examples*
Simple proteins (yield only amino acids)	Albumins	Water-soluble	Egg albumin
	Globulins	Ppt by $\frac{1}{2}$ saturated $(NH_4)_2SO_4$	Beta globulin
	Glutelins	Soluble in dilute acid or alkali	Wheat glutelin
	Prolamines	Soluble in 70% ethanol	Zein
	Keratins	Insoluble	Hoof and hair
Conjugated proteins (combined with materials other than amino acids)	Phosphoproteins		Casein
	Lipoproteins		
	Glycoproteins		
	Nucleoproteins		
	Metalloproteins		
	Porphyrinoproteins		Hemaglobin
Derived proteins—produced by breakdown or alteration of other groups			Denatured proteins

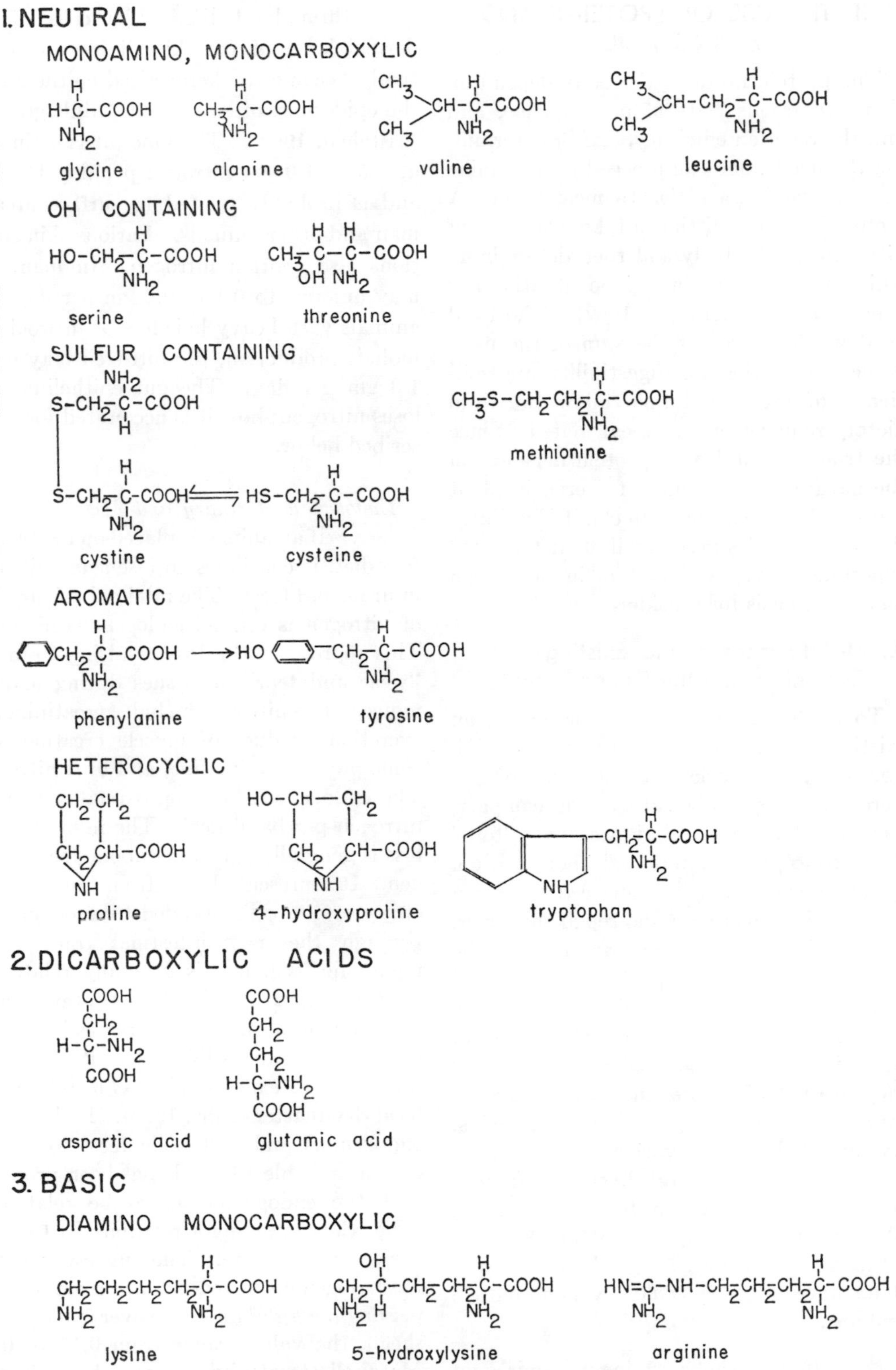

I. NEUTRAL

MONOAMINO, MONOCARBOXYLIC

glycine alanine valine leucine

OH CONTAINING

serine threonine

SULFUR CONTAINING

methionine

cystine cysteine

AROMATIC

phenylanine tyrosine

HETEROCYCLIC

proline 4-hydroxyproline tryptophan

2. DICARBOXYLIC ACIDS

aspartic acid glutamic acid

3. BASIC

DIAMINO MONOCARBOXYLIC

lysine 5-hydroxylysine arginine

FIG. 16–1. Amino acids occurring in natural proteins.
Amino acids: $R–CH_2–CH(NH_2)-COOH$

II. THE USE OF PROTEINS AND AMINO ACIDS

The protein needs of the body depend on tissue functions, rate of performance, and and the reaction efficiency. After determining the needs, one can proceed to determine how to provide a ration to meet them. A factorial system listing all known uses of nitrogen by the body and then determining which are operating in a given situation has been proposed (Mitchell, 1962). The total need would represent the sum of the uses. Some assumption for digestibility and efficiency of use would then allow the total dietary requirement to be calculated. Since the true digestibility of protein depends on the nature of the binding material in plant walls and on treatment effects, it is unlikely that all variables needed will be determined experimentally. The factors involved can serve as a basis for considering needs.

A. Maintenance of the Existing Organism and the Dynamic State

To have a living, growing organism, an existing body structure must be maintained. Formerly, the various tissues and organs were considered to be rather rigid compartments, but isotope techniques have altered these concepts and replaced them with a dynamic concept. All tissues can be shown to be subject to erosion and repair; however, certain tissues and organs are more labile than others. Serum proteins, liver, and intestinal wall "turn over" rapidly. The keratins are more stable, and while those in the epidermis may be shed to the exterior, they are not often broken down enough to contribute to the body pool of amino acids. Example half-lives of various proteins have been calculated as: rat liver, certain enzymes, 3 hours; rumen bacterial protein, 20 hours; rumen protozoal protein, 38 hours; rat liver, plasma, gut, 7 days; human serum, 10 days; red blood cells, 50 days; and muscle proteins, average, 180 days.

Epithelium. The proteins of hair and skin are largely keratins. Hair continues to grow throughout life. The cells of the dermal layer of the skin divide and eventually become the keratinized epithelium of the epidermis, often termed "adult growth" (Mitchell, 1962). The amount is estimated at 0.5 to 1.0 gm nitrogen per day for man and is probably related to surface area in man and other animals. Various skin secretions also contain nitrogen. In man, this may amount to 0.1 to 0.2 gm per day. In animals with heavy hair loss or in wool and mohair production, the nitrogen may equal 1.0 gm per day. The gut epithelium also loses nitrogen; how it is accounted for is described below.

Endogenous Urinary and Metabolic Fecal Losses. If an animal is placed on a nitrogen-free diet, it continues to lose some nitrogen in urine and feces. The minimal urinary loss of nitrogen is termed endogenous nitrogen. This represents the loss resulting from the "wear and tear" of tissues during maintenance. The nitrogen includes creatinine, an excretion product of muscle creatine, and some purine, pyrimidine, and urea nitrogen. The endogenous loss approximates 2 mg of nitrogen per basal kcal. The minimal fecal loss is generally called metabolic fecal nitrogen. It represents losses from the digestive enzymes and cells abraded by food passing through the gastrointestinal tract. The metabolic fecal loss is generally related to the total dry matter intake. However, since metabolic fecal nitrogen increases if the diet is high in undigestible fiber, it would be preferable to express this loss in relation to fecal dry matter (Ellis, 1956). Estimates of endogenous and metabolic fecal losses are shown in Table 16–2. It has been proposed that the endogenous losses be related to $W^{0.73}$, thus implying a relationship to basal energy expenditure. The suggested values for endogenous nitrogen range from 0.20 gm per day per $kg^{0.73}$ for cattle over 200 kg. For sheep, the values range from 0.17 to 0.09. Metabolic fecal nitrogen may be estimated as 5 gm per kg DM intake. Young animals

Table 16–2. Endogenous and Metabolic Fecal Nitrogen
(*From the Literature.*)

	Weight (kg)	Endogenous Nitrogen (mg/kg BW)	Gm Metabolic Nitrogen per	
			kg Food DM	kg Fecal DM
Sheep	33	32	2.4	7.2
Calf	40		2.7	
Cattle	400	22	5.3	
Cattle, African			3.7	
Buffalo		19	3.4	6.3
Nonruminant average . .			2.2	
Adult ruminant average .			4.5	

have greater losses and "native breeds" have lower losses than the average mature animal.

Regeneration. Since the body proteins are constantly undergoing breakdown, the tissues must be constantly regenerated. If the body is not 100 percent efficient in recapturing the amino acids lost from tissues, additional dietary amino acids will be needed for replacement. The destruction of red blood cells and the recapture of iron are well established, but the fate of the amino acids in heme is less well documented. The requirement for regeneration is most intense following accidents or illness. It is established that extra dietary protein is desirable during recovery periods and, presumably, the nature of regeneration will affect the needs.

B. Reproduction

The total protein used in reproduction is generally small in comparison to that needed for maintenance. A notable exception is that required by birds with high laying rates, or in females such as the guinea pig, who produce large litters in relatively short times.

Male. The semen contains from 0.5 to 1.5 gm nitrogen per 100 ml. Eighty-five to 95 percent of this is protein-bound. Normal ejaculates for several species indicate that the usual breeding rates might increase the protein needs of bulls by 1 gm per day. Boar ejaculates, however, may contain 9 gm of protein so that lot breeding could cause a significant increase in protein required.

Female. The main increase in protein use during the reproductive cycle of females is for protein deposition in the fetus. An example of this in swine is shown in Appendix III. During the last few weeks, the amount of protein deposited in the gravid uterus is large enough to require special provision to meet these needs. This might be 7 gm nitrogen per day in the ewe during the last month of pregnancy and 30 gm per day in the cow during the ninth month. An egg contains between 6 and 7 gm protein. Therefore, a hen laying at a 70 percent rate will need an average of 4 to 5 gm available protein, an amount equal to 4 percent of the feed. During the reproductive cycle, some nitrogenous compounds may be lost in uterine and vaginal secretions. Except for the menstrual losses of the human female, these are not great. Other physiological changes take place which tend to increase the total nitrogen requirement. These include an increase in serum proteins associated with a 10 percent increase in blood volume during early pregnancy and some increase in liver and intestine proteins.

C. Growth of Tissues and Organs

Tissues and organs vary in composition of amino acids. Typical examples are shown in Table 16–3. Since organs develop at different rates, one would expect the amino acid needs to reflect the intensity and amount of growth of the tissues. Muscle may contain 50 percent and connective tissue 25 percent of the total proteins, so they are major factors in determining the overall requirement.

Fetal. In most domestic animals, the young are well developed at birth. The protein requirements for the developing rat are low and can be omitted if growth hormone is injected. Apparently the female can obtain protein from her body to provide that needed in the young.

Postnatal. At birth, some of the enzyme systems are not well developed; in ruminants, the forestomachs are very small. The order of postnatal growth of the major tissues is bone, muscle, fat. This implies a period of maximum deposition of protein. This period of growth comes at about 450 pounds live weight in cattle (Pfander, 1968). Organs develop at different rates and vary in composition and in total mass. Organs supplying an important fraction as a percent of body weight are: gastrointestinal tract, 6; blood, 5; liver, 2; lung, 1; and kidney and brain, over 0.5. At puberty, there is a marked increase in rate of growth of gonads, so that some increase in protein may be implied.

Defensive Mechanisms. Since the young pig is born with no immune proteins and may not start to form antibodies for several weeks it is dependent on those absorbed in early life. Antibodies are absorbed from colostrum during the first 36 hours of life while a strong trypsin inhibitor is present and before hydrochloric acid secretion has become established. Protein absorption at this stage is relatively nonspecific. Immune globulins are high in threonine and tryptophan (gamma globulin, 8.4 and 2.9 percent versus 7.0 and 1.0 percent in albumin); these amino acids might be critical in 4 to 8 week old pigs.

D. Milk, Egg, and Fiber Production

The protein and amino acid composition of milk, eggs, and wool is shown in Appendix II. In addition to the amino acids in the product, it would be essential to provide an additional supply for the growth of tissues which form the product and to cover the inefficiencies associated with the diversion of amino acids to uses other than protein formation. Obviously, rate of production will be a major determinant of total requirement.

III. CONVERTING FEED TO ANIMAL NEEDS

Except for a few hours after birth or during certain disturbances, proteins do not pass directly from food to the body. A series of events, not yet completely understood, must proceed systematically before the proteins in feeds are absorbed and transported to the tissues.

A. Digestion

Digestion may be defined as those events preparing the food for absorption. The mechanical aspects of breaking the hull or cell wall and reducing feed to particle size can be done by the teeth or by mechanical devices and can be considered the first step in the digestive process.

Enzymes. Enzymes are usually called biological catalysts. They are special proteins which perform difficult tasks at body temperature. The digestive enzymes hydrolyze other proteins to amino acids. The enzymes are added in sequence as the food passes down the gastrointestinal tract. The organ, enzymes, and the products formed are shown in Table 16–3. The digestive enzymes attack soluble proteins, but not keratins. The rate of hydrolysis is very rapid for milk and meat products and slower for zein and other insoluble proteins. Milk and meat products can be made less soluble by heat treatment. Some plants, *e.g.*, soybeans, contain an antitrypsin or other enzyme inhibitors which are destroyed by heat. The digestibility of such products is improved by proper heat treatment.

Microbial Fermentations. Anaerobic bacteria in the forestomachs of functional

Table 16–3. Organs and Enzymes Involved in Protein Digestion

Organ	Enzyme[1]	Substrate	End Product
Stomach	Pepsin	Proteins	Proteoses
	Parapepsins	Endopeptids of tyrosine and phenylalanine	Peptids
	Rennin	Casein	Paracasein (curd)
Pancreas	Trypsin	Arginine and lysine, endopeptids	Polypeptids
	Chymotypsin	Tyrosine endopeptids, phenylalanine endopeptids, tryptophan endopeptids, methionine endopeptids	Peptids
	Pancreatic peptidase	Peptid bonds adjacent to neutral amino acids	
	Carboxypeptidase A	Exopeptids of phenylalanine, tyrosine, tryptophan, leucine	Free amino acids
	Carboxypeptidase B	Exopeptids of basic amino acids	Free amino acids
Intestinal mucosa	Leucine amino peptidase, amino peptidase, di- and tri-peptidases	Peptid with terminal amino group	Free amino acids

[1] Many of these are secreted as inactive precursors which are activated under the influence of other enzymes, pH, or activators.

ruminants attack proteins. The proteolytic activity is high and deamination of small molecular weight nitrogenous compounds is rapid. The rate of attack has been correlated to ease of solubility in the classical sense. Casein and gelatin are rapidly attacked; certain naturally occurring proteins such as fibrin and zein are attacked slowly (Ellis *et al.*, 1956). Although casein and egg proteins are attacked readily, heat treatment of the protein decreases the rate of proteolysis. Rumen ammonia concentration peaks rapidly after a ruminant eats a soluble protein. The NPN components are also rapidly converted to ammonia. These steps require the presence of urease, nitrate reductases, and many other specific and nonspecific enzymes provided by the mixed cultures of the rumen. Many rumen bacteria require and most use ammonia as a nitrogen source. Certain organisms have been shown to use heme and certain dipeptides. Amino acids can serve as a nitrogen source for some bacteria, but cellulolytic organisms require ammonia. Cysteine is usually included in the media as a reducing compound in *in vitro* experiments. Numerous organisms respond favorably to supplements of branched

chain volatile fatty acids, which are most often supplied by the residual carbon chain of certain amino acids. It is difficult to show a favorable effect from adding these materials to the diet although they are essential for the formation of leucine, isoleucine, and valine. Some organisms require carbon dioxide in the initial stages of growth. The carbon dioxide becomes incorporated into threonine, aspartic acid, and glutamine. The protozoa obtain protein by engulfing bacteria and food particles. Others may use soluble nitrogen. The bacterial and protozoan cells are passed from the rumen-reticulum to the omasum. Considerable fragmentation may occur, but the bulk of the organisms disintegrate in the abomasum and are digested in a manner similar to that of digestion of feedstuffs in the stomach of monogastric animals. In fact, many of the studies on digestive enzymes have been made with bovine tissues. The composition of typical rumen bacteria is shown in Appendix II. Some bacteria contain complex carbohydrates in the cell wall which may protect them from enzymatic attack. Protozoan proteins are more readily digested than those of bacteria and they may make up 20 percent of the microbial mixture. The high content of RNA and DNA in microbial cells may limit the utilization of microbial nitrogen in ruminants to 85 per cent of that present (Ellis & Pfander, 1965).

Consideration of Other Species. Numerous bacteria live in the intestinal tracts of all animals. Their role in nitrogen metabolism has been largely ignored. In those species practicing coprophagy, microbial protein could contribute to the total dietary nitrogen. The horse can tolerate urea, but its exact use of nitrogen has not been documented.

B. Absorption of Products of Digestion

Mechanisms. Amino acids are actively transported across the intestinal wall. The L-forms are generally more readily transported than the corresponding D-forms. It appears that certain amino acids are more readily absorbed than others, but their relationships change when the mucosa is supplied with different ratios of amino acids. At equimolar concentrations, methionine, isoleucine, and valine were absorbed most rapidly, and were followed in order by the neutral amino acids and the dicarboxylic acids. Within any group, the essential amino acids were absorbed more rapidly than the dispensable ones (Adibi *et al.*, 1967). Other techniques have indicated that methionine and alanine are rapidly absorbed. It may be that the acids compete for the available absorption sites. Leucine and isoleucine and/or valine have been shown to be mutually inhibitory (Szmelcman & Guggenheim, 1966).

Amounts. The capacity of the absorption system is limited, but the evidence available suggests that the long length of intestines provides a reserve capacity to compensate for overloading. Animals digest and absorb proteins well even after part of the intestinal tract has been removed. It is generally found that active transport is impaired by an overload of amino acids, but few free amino acids are excreted in the feces, possibly because those escaping absorption are metabolized by bacteria in the large intestine. Various amines are thought to be formed by such bacteria. A temporary overload might be produced either by rapid eating of high protein diets, by a liquid diet that is not coagulated in the stomach, or by excess supplements of free amino acids. The latter possibility is of concern in supplementing the relatively insoluble proteins of cereal diets since, if the supplement loads the absorption sites before the proteins are digested, an effective imbalance of amino acids may be produced.

C. Transport and Cellular Uptake

During absorption, the amino acid concentration of the portal blood rises, and investigators have sought to use this change as a measure of the effectiveness of absorption. The portal system should reflect absorption, but transamination may occur in the gut wall, changing the expected concentration of alanine, glutamate and aspartate. The amino acids are generally believed to be transported in the free state, but the high concentration of glutamine in blood indicates that there are alternate systems. Studies of the uptake by cells are complicated by the dynamic nature of tissue proteins, which may result in release and uptake at the same time. Amino acids are transferred to intercellular space and subjected to metabolic alterations. The concentration in circulating plasma may be 4 mM and the concentration within cells 30 mM. The concept of an amino acid pool as a reservoir for holding the amino acids in transit has been challenged. A true pool should be homogenous, and what is usually called the pool has many components. Amino acids are present in higher concentrations in tissue cells than in plasma, and the relative concentration is not always related to blood concentration in a simple ratio. Cell uptake can be influenced by hormones, but all tissues do not respond in the same way. Liver and muscle provide good examples of the types of responses obtained. The liver is a catabolic tissue and increases its uptake of amino acids under the influence of adrenal hormones and destroys protein under the influence of insulin. Muscle increases amino acid uptake under the influence of growth hormone and insulin, but loses protein when adrenal corticoids are given. Certain sex hormones are organ-specific. Estradiol causes the rat uterus to increase its uptake of tracer amino acids, but does not influence the uptake of other tissues (Noall *et al.*, 1957). Since nutrition can influence hormone production, it appears that a primary basis for nutritional-endocrine interaction will be found at the level of cellular uptake of amino aicds. Vitamin B_6 may also be a factor in the interaction since it is involved in both transport and metabolism. In some species, the time delay can be much less and low protein diets accentuate the need for simultaneous supplementation.

D. Metabolism of Amino Acids

Several distinct pathways for manipulating amino acids are available. Some are specific for only one stereoisomer of a given acid, but others show broad specificity.

Deamination. The liver and kidney contain an enzyme active against D- but not L-amino acids. D-amino acid oxidase catalyzes the following reaction:

$$\underset{\underset{H}{|}}{\overset{\overset{NH_2}{|}}{RC}}-COOH + FAD + \tfrac{1}{2}O_2 \rightarrow RC\overset{\overset{O}{\parallel}}{-}COOH + NH_3$$

A weak L-amino acid oxidase is present in tissues, but is relatively unimportant. The only powerful L-amino acid oxidase is glutamic dehydrogenase, which is specific for glutamic acid. In rumen bacteria and some other anaerobic bacteria, mutual oxidation-reduction occurs between pairs of amino acids. A response of this type is called the

Stickland Reaction, after the discoverer, (Stickland, 1934). A typical example is:

$$\underset{\underset{NH_2}{|}}{\overset{\overset{H}{|}}{CH^3-C}\;COOH}$$

$$+ \qquad\qquad + 2H_2O \rightarrow 3CH_3COOH$$

$$\underset{\underset{NH_2}{|}}{2CH_2COOH}$$

$$+$$
$$3NH_3$$
$$+$$
$$CO_2$$

Alanine most often plays a role in this reaction, but proline, isoleucine, leucine, and valine are known to be involved in rumen reactions and to produce branch-chained volatile fatty acids (El-Shazly, 1952).

Transamination. Transamination is a major device for transferring amino groups. The two major systems involve aspartic and glutamic acids. Typical reactions include:

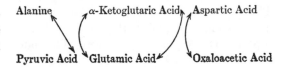

The system involving aspartic acid is most widespread in animal tissues, but reactions involving other α-oxo acids are a major basis for the formation of the dispensable amino acids. The key role of glutamic acid is shown by the fact that following administration of a given radioactive amino acid and subsequent isolation of the amino acids from protein, the administered radioactive acid is present in highest amount followed by glutamic acid. Other acids show some activity.

Decarboxylation. Biological amines are formed by the decarboxylation of amino acids. The enzymes for the reaction are most common and specific in bacteria, but are also found in plant and animal tissues.

While the amines are biologically important, the pathway is not quantitatively important as a means of altering amino acids. Typical examples are the formation of histamine and tryptamine from histadine and tryptophan. The source of amines appearing in urine is disputed by several investigators. When antibiotics were given to control bacteria (not proven by direct counts), the amount of certain amines was reduced (Perry *et al.*, 1966).

Synthesis of Macromolecules. The synthesis of proteins is a function of the cell. Most proteins remain in the cell or daughter cell which produced them, but the liver and the endocrine and exocrine glands release special proteins into the blood or ducts. A simplified version of the elaborate synthetic mechanism is outlined below. Deoxyribonucleic acid (DNA) in the nucleus is a double-stranded spiral of purines, pyrimidines, deoxyribose, and phosphoric acid held together by physical-chemical forces such as hydrogen bonding. The bases involved are adenine, guanine, cytosine, and thymine. These structures contain the basic genetic information concerning the nature of the protein to be produced by the cell. The code for each amino acid to be added is provided by a combination of three of the bases in sequence. Sixty-four "code words" are possible; two different words may produce the same amino acid and some possible words have not been matched to any amino acid. Before protein synthesis can occur, one of the DNA strands is activated to produce a corresponding strand of ribonucleic acid (RNA). This compound differs from the DNA by having ribose instead of deoxyribose and uracil instead of thymine. The RNA is able to migrate from the nucleus and is called "messenger RNA" (mRNA). The mRNA molecules then combine with ribosomal particles to form a template for synthesis. The amino acids to be used for synthesis are activated and combined with soluble RNA molecules (sRNA) specific for the amino acid by aminoacyl-sRNA synthe-

tases. The sRNA molecules line up with the template, and a series of events which have not yet been documented, take place which completes the bonding needed to form the protein. A complete protein molecule can be formed in a few minutes, indicating that amino acids are incorporated at the rate of several per second. Certain amino acids such as hydroxyproline may be produced by alteration of the preformed protein chain. Certain antibodies, *e.g.*, actinomycin D, can block the reaction and amino acid analogs can replace a natural amino acid in the sequence. In order for these synthetic reactions to be completed, all of the amino acids requested by mRNA must be present. This implies that the supply of amino acids in the cell must be maintained at a concentration permitting synthesis. If an amino acid supplement for a deficient protein is fed 12 hours after the protein, synthesis is reduced.

E. Excretion

Nitrogen is excreted by several routes. Some pathways appear to be incidental to other activities, but major routes and substances excreted are of hereditary significance. A number of abnormal excretory materials have been identified.

Urea Formation and Detoxifying Mechanisms. Ammonia, the major product formed by deamination of amino acids, is toxic to mammalian tissue. Some animals, especially fish, excrete it unchanged into surrounding fluids; birds and lizards convert it to uric acid. The mammal uses two mechanisms for removing it. Small amounts are rapidly converted to glutamine by a system which is the first line of defense against ammonia toxicity. The major pathway is via urea, which is formed in the liver.

Kidney. The kidney has a major role in controlling nitrogen elimination. Since water flow or filtration rate may have a marked influence on nitrogen loss, it is essential that water intake be controlled or recorded in studies involving nitrogen excretion (Arroyave *et al.*, 1966).

The kidney filtrate at Bowman's capsule is similar to the NPN content of blood plasma. The amount of filtrate will depend on pressure effects, flow rates, and osmolar load.

The tubules reabsorb many of the substances filtered by the glomerulus. Most of the L-amino acids are filtered and completely reabsorbed. The reabsorption of histidine and glycine is 97.6 and 98.4 percent. However, B-alanine and methylhistidine are reabsorbed less completely (Wright & Nicholson, 1966). Certain neutral L-amino acids have been reported to interfere with phosphate reabsorption (Michael & Drummond, 1967). Urea is filtered and not actively reabsorbed. About 40 percent may be reabsorbed by passive diffusion, but the percentage is increased by low flow rate or high concentration and reduced by diuresis. Under some conditions, nearly all of the creatinine and about 95 percent of the uric acid are reabsorbed.

Ammonia formation by the kidney is part of a homeostatic control mechanism for cation preservation. $NH_3 + H^+ \rightarrow NH_4^+$, which can spare the fixed cations such as Ca^{++} and K^+. Glutamine is the major source of ammonia and it is secreted by the tubules into the urine.

The concentration of nitrogenous components in urine during fast and in periods of low nitrogen and high nitrogen intakes shows that urea varies with the protein and energy level in the diet. It is less on a nitrogen-free diet than during fast, when protein stores are being depleted. The ammonia level will often be higher on a high-protein intake because the phosphorus and sulfur from the protein must be excreted as acids. The kidney uses ammonia and hydrogen to neutralize them. However, another condition may contribute to high urinary ammonia nitrogen. Improper preservation of urine will allow bacteria to grow and urease will convert urea to ammonia.

The constant feature of urine excretion is the level of creatinine. This factor is more stable in males than females.

Skin. Nitrogen is constantly lost through the skin. This process was discussed under maintenance needs.

Digestive Tract. The nitrogenous losses in feces are normally low in animals on low-fiber diets containing proteins that have not been excessively heated. The Atwater calorie factors of 9.4.4 kcal/gm used in human nutrition assume apparent digestibilities of proteins of 97, 85, and 75 percent for animal, fruit, and cereal proteins (Watt & Merrill, 1950).

Undigested nitrogen may represent more than half the fecal nitrogen in animals if feedstuffs which are high in keratin or whose cell walls are undigestible are included in the diet. The components of metabolic fecal nitrogen are the unabsorbed residues of enzymes and other proteins secreted into, and cells abraded from, the digestive tract. It has been estimated that bacteria contribute 40 percent of fecal nitrogen.

IV. DETERMINING AND MEETING REQUIREMENTS

A. The Concept of "Essential" Amino Acids

Essential amino acids are those which cannot be synthesized by the body at a rate sufficient to allow normal growth or to maintain nitrogen equilibrium in the adult (Rose, 1938). They must be supplied in the ration. The term is convenient from the standpoint of ration formulation, but leads to confusion in practice. The nonessential amino acids are also essential to the body machinery, but mechanisms are available for their formation in the tissues. The term "normal growth" may be confusing. In general, as knowledge of quantitative requirements and genetic selection has increased, growth rates also increase. What was "normal growth" to Rose may be only 50 to 75 percent normal in today's well-organized rat colony.

Specific Requirements. The estimated requirements for several species are summarized in Table 16–4.

Table 16–4. The Estimated Amino Acid Requirements of Several Species
(*From the literature.*)

	Infant mg/kg	Man gm/day	Chick % Diet	Dog mg/kg BW	Rat % Diet	Swine % Diet
Arginine		0	1.2	48	0.2	0.2
Histidine		0	0.4	23	0.3	0.2
Isoleucine	90	0.7	0.75	70	0.5	0.5
Leucine	425	1.1	1.4	105	0.8	0.6
Lysine	170	0.8	1.1	68	0.9	0.70
Methionine + cystine	85	1.1	0.75	53	0.6	0.5
Threonine	87	0.5	0.7	44	0.5	0.5
Phenylalanine + tyrosine	169	1.10	1.3	86	0.9	0.45
Tryptophan	30	0.25	0.2	12	0.15	0.13
Valine	161	0.8	0.85	65	0.7	0.50
Glycine	0	0	1.0	0	0	—

Nonspecific Nitrogen and Sulfur. Tissue Needs: In addition to the essential amino acids, some nonspecific nitrogen must be supplied. Most proteins contain only 50 per cent or less of their nitrogen in the essential amino acids. Albumin has 65 percent and collagen 25 percent. The synthesis of proteins obviously will require extra nitrogen if only essential amino acids are given. In purified diets, extra nitrogen can be added as nonessential amino acids or by combinations of amino acids, diammonium citrate, and/or urea. When all of the nitrogen comes from amino acids, extra energy may be needed to maintain nitrogen balance or growth. Some evidence for a specific requirement for proline or glutamic acid is available. This probably reflects the extra demand for these acids at high growth rates (Breuer *et al.*, 1964).

Role of Microorganisms: Rumen microorganisms can reduce inorganic sulfur and incorporate it into their protoplasm as amino acids. Sulfur-35 given as sulfur or sulfate has been recovered in the proteins of blood, milk, wool, and tissue. If large amounts of hydrogen sulfide are formed, there would be a risk of precipitating some of the essential minerals.

B. Estimation of Requirements of Animals

The investigator who is establishing re-quirements has many criteria from which to select the basis on which he will form his estimates. Usually experiments are done on groups of animals selected to reduce experimental variation rather than to include population extremes. Therefore, the estimated requirement may be misleading. The user of requirement tables will be aided by having available full experimental details on animal age, sex, source, and experimental environment. Some measure of animal variability should also be included.

1. Growth or Production Trials

Growth trials, sometimes referred to as "feed and weigh" trials, have been most frequently used. They formed the basis of the original attempts by Osborne and Mendel to evaluate proteins and to establish the qualitative and quantitative amino acid and protein requirements for growth. Since growth requires appreciable protein deposition, the method is basically valid. It can be improved, however, by obtaining information on carcass composition. Trials with swine often show equal growth rates at the lower of two amino acid or protein levels, but the pig which is receiving the higher level is more efficient, has larger muscles and less fat than his counterpart. With the availability of whole body counters, these measurements can be obtained without sacrificing the animal and would be valuable adjuncts to any growth trial conducted on animals to be retained for reproduction or at periods terminating before slaughter.

2. NITROGEN BALANCE

The limitations of the growth trial cited above encouraged the search for other criteria and the nitrogen balance trial came into favor. This technique is most effective when conducted under highly standardized conditions. It involves a preliminary period for the animal to become adjusted to the metabolism cage and the ration, and requires considerable care in accounting for all of the nitrogen. The magnitude of the losses are 1 percent for gaseous loss, variable loss from ammonium from collecting containers not containing acid, and losses of uncollected urine or feces. Nitrogen intake is usually overestimated since the animal will usually waste some feed, the amount depending on the physical form, the nutritional balance, and the level of feeding (Martin, 1966). Least absolute waste, but not always the least percentage waste, is associated with liquid diets fed at low intake levels. The short-term nitrogen balance technique usually overestimates the value of a protein and is limited by not being able to assess the relative usefulness of a protein source or amino acid mixture throughout the growth period. In many trials reported in the literature, the balances reported do not seem reasonable when compared to the gains reported. Part of this disparity is probably due to changes in weight of the contents of the digestive tract, but the stress associated with the cage life may actually alter normal protein deposition patterns. The biological value (BV) procedures are refinements of the nitrogen balance methods. In addition to the usual balance trial, a second trial is run with the animal fed a nitrogen-free diet to determine the endogenous urinary and metabolic fecal losses. It is assumed that these losses are similar to those occurring on the test protein. The Thomas Mitchell equation for BV may be developed as follows:

N_{UE} = urinary nitrogen on nitrogen-free diet
N_{MF} = fecal nitrogen on nitrogen-free diet
N_I = nitrogen intake on test protein
N_U = urinary nitrogen from test protein
N_F = fecal nitrogne from test protein

Truly digestible (or absorbed) N, $N_A = N_I - (N_F - N_{MF}) = N_I - N_F + N_{MF}$

Retained N, $N_R = N_A - (N_U - N_{UE}) = N_A - N_U + N_{UE}$

or $N_R = N_I - N_F + N_{MF} - N_U + N_{UE}$

$$BV = \frac{N_R}{N_A} \times 100 = \frac{N_I - N_F + N_{MF} - N_U + N_{UE}}{N_I - N_F + N_{MF}} \times 100$$

The BV method is subject to the same errors which exist in any balance technique. Furthermore, the values of a given protein will vary with the level fed, the low levels producing higher values. The main advantage of this method is to produce an objective evaluation of a series of proteins under defined conditions. From this standpoint, it allows comparisons to be made between laboratories of methods that may be more precise than growth methods. Many other methods for arriving at a biological evaluation have been proposed.

3. BLOOD AMINO ACID LEVELS

Blood is the major transport system for the nutrients en route to the cells. Physiologists have long tried to analyze blood to determine if the nutrients are in adequate supply or if end products of digestion are being absorbed. The portal blood should be preferred for these measures, but technical difficulties usually cause investigators to use systemic blood. Interest in these techniques for evaluating amino acid status was revived when chromatographic procedures became available (Moore & Stein, 1951). Since the blood amino acid levels represent a balance between intake from absorption, release from tissue proteins, metabolic alterations, and cellular uptake, it seems logical that the best results would be obtained when growth is maximal. The amino acids in the blood are sampled at a time when they most closely reflect the difference between the mixture coming from the intestine and that required for anabolism.

Furthermore, adequate energy is needed to reduce the use of amino acids for purposes other than synthetic activities. Since adrenal corticoids favor liver catabolic activity, stress conditions should be avoided. These conditions have been most closely obtained for chicks in Scotts' laboratory, (Zimmerman & Scott, 1967a, 1967b), where a very effective amino acid diet based on blood amino acid concentration has been formulated. The system has been used with mixed swine rations to indicate differences in dietary protein, but their results were difficult to interpret because values for tryptophan were not determined.

This procedure has great potential for determining availability of amino acids from different dietary proteins, evaluating the effect of processing on availability, and determining the limiting amino acids at different stages of the growth cycle. Additional data are needed on normal values, the effect of energy intake, and the relative effects of recycling of nitrogen through the gut wall. The experiments must be planned to minimize disturbances in regular feeding patterns. More data on the storage time of amino acids within cells are needed to make the technique more meaningful.

4. Other Blood or Urine Components

Since urea is the major route for excretion of excess nitrogen, blood urea nitrogen (BUN) values rise under conditions that limit urea excretion or increase formation. In physiological conditions, BUN correlates well with protein intake and the BUN value can be used as an index of protein adequacy. BUN levels are higher when animals are fed equal amounts of poor quality protein than when they are fed complete protein. BUN has been suggested as a measure for evaluating the adequacy of protein nutrition in ruminants (Preston et al., 1965). Since there is some diurnal variation of BUN and levels fluctuate more on NPN than on protein diets (Schnakenberg, 1968), care is

needed in selecting sampling times for BUN measures. A high concentration of urea in urine may be taken as an index of protein excess or imbalance. In general, the greater the percent of urinary nitrogen present as creatinine, the higher the BV of the protein. The blood proteins are depleted when animals are fed a diet with inadequate amounts of protein or one having low levels of essential amino acids. In preliminary nutritional status surveys, blood proteins can be used to get a general idea of the adequacy of the diet with respect to protein; but, since proteins of other organs may be depleted before serum protein falls, the procedure lacks sensitivity.

5. The Factorial Method

Mitchell (1962) described and gave examples of the factorial method. The method involves a number of assumptions:

Metabolic rate predicted from weight
Endogenous nitrogen at 2 mg nitrogen per basal kcal
Rate of gain
Composition of gain
Efficiency of utilization of absorbed protein for maintenance and growth
Metabolic fecal nitrogen
True digestibility of protein
Losses through the skin
Additional nitrogen at certain seasons of the year or stages of growth as required for hair growth and udder development.
Deposition in products of conception and economic products, e.g., milk and wool.

Calculations such as these are interesting in that they show the limitations in our present knowledge, but it is doubtful if they can serve as more than first approximations of the requirement for a given animal. They would probably be more useful for herd averages and as a negative control for evaluating given rations. The calculations do not consider whether the protein requirement of rumen organisms is relatively greater than that of animal tissues, and they make no allowances for recycling of protein. Therefore, the British scientists hedged a bit by setting a lower limit on protein intake for ruminant rations.

6. COMMITTEE EVALUATION

National and international bodies appoint select committees to review the existing data and prepare tables of estimates of requirements. Data on the requirement for protein in the presence of adequate supplies of other nutrients are limited. The task of these groups is usually one of trying to extrapolate from minimum data to generalized situations although they may have to rationalize conflicting evidence from two or more groups of investigators. Quite often, values of investigators are averaged without regard to the precision of determination on the number of animals used by the several investigators who give conflicting reports. The amino acid requirements of the growing pig, as outlined by the National Research Council and a British committee, are summarized in Appendix IV. The higher requirements of the British committee may reflect a greater interest in the production of lean pork than that of the Americans, but it is probably also a reflection of the higher protein and more limiting feeding practiced in Britain.

C. Expressing Requirements

Since animals may be rationed on an individual basis by giving them a certain mass of feed or by allowing them free choice of a prepared mixture, two common systems of expressing requirements exist. One system is based on amount per day, the other on concentration in the ration. Most farm animal requirements have been on the concentration basis, but human requirements are stated on a weight basis. It is known that energy level influences protein requirements in ruminants and that both energy and protein alter the requirements of poultry and swine for essential amino acids. Therefore, some indication of energy level may become a part of a protein or amino acid requirement.

Nitrogenous Components. The first approximation of a requirement for an animal will be in terms of total nitrogen in the diet. Generally, this need is expressed as crude protein. This system ignores differences in digestibility and in amino acid composition, giving the same value to old shoe leather as it does to choice steak or oil meals. As long as similar feeds or feeds of known characteristics are compared, the system is very useful.

Digestible protein has been suggested as a more meaningful criterion for reporting protein value, and some requirement tables use this figure. Unfortunately, the digestibility of proteins is altered by processing and may be affected by other dietary constituents. Furthermore, apparent digestibilities of a given protein may vary, depending on the level at which it is fed. Tabular values of estimated digestible protein content may give false values for a given dietary situation, especially if there is an associative effect of feeds. *In vitro* methods have been developed to estimate the digestibility or availability of specific proteins or mixtures. The three most frequently used are the microbiological assay, the trypsin digest, and the colorimetric assay based on fluorodinitrobenzene. These procedures can be conducted within a few days and provide useful information. Some feed manufacturers may include one or more of these criteria as part of their specifications for protein ingredients. The artificial rumen has been used for evaluating dietary nitrogen in ruminant diets. If the coefficient of digestibility of crude protein is plotted against percent dietary protein, the relationship, $Y = 70 \log X\text{-}15$, is obtained (Glover *et al.*, 1957). This or a similar equation is probably the best estimate of available protein in ruminant diets. A correction for increased metabolic fecal loss based on estimated departures in DM digestibility from the usual ranges can be applied. Alternatively, DM digestibility can be estimated by *in vitro* techniques.

If protein nutrition in monogastric animals is believed to be essentially amino acid nutrition, it follows that any rationing program must be based on supplying the specific amino acids needed by the animal being considered. The British ARC Poultry Committee selected this procedure, but since essential amino acids make up only about one half of most proteins, an additional increment of nonessential amino acids is needed. It can usually be assumed that any combination supplying the essential amino acids will also supply the nonessentials. The problems encountered are greater than those discussed under protein. In addition to the uncertainties associated with estimating the protein digestibility of a given product, the vexing question of deciding if all amino acids will be digested to the same degree also requires an answer. The fluorodinitrobenzene test is actually an attempt to determine the available lysine. Some evidence suggests that the coefficient of digestibility of methionine in soy protein is lower than that of the total protein. Thus, a major problem in using amino acid composition will be to provide digestibility estimates for the amino acids in the proteins. The second problem is that of obtaining good estimates of the amino acids in a given product. Analytical techniques are based on hydrolyzed proteins. Thus, the methods used influence the loss of amino acids during hydrolysis. Hydrolysis may convert some compounds not normally available to the animal to amino acids. If tabular values are used for ration formulation the ration may be found to vary from the expected value. Plants may deposit several proteins of markedly different amino acid composition at different rates without changing the total nitrogen. Furthermore, when protein of corn has been increased by added nitrogen fertilizer, the zein fraction has increased more rapidly than the other proteins. An attempt to evaluate the possible variation in mixed diets was made. If the mean composition of a given feed is determined and the standard deviation is known, use of a value of the mean -1σ for formulating rations might be desirable (Pfander & Tribble, 1957).

Relation to Other Dietary Components. In addition to deciding which nitrogenous factor to use in expressing requirements, the question of how to relate the requirement to the total ration must be considered.

The moisture content of grains may vary between 5 and 15 percent, depending on storage conditions. If a 16 percent crude protein ration is formulated from corn containing 15 percent water and 9 percent protein, it is possible for the same corn to produce a 17 percent crude protein ration if the moisture is 5 percent. Currently, most requirements are expressed on an "as fed" basis with the implied assumption of 90 percent dry matter.

It is probable that the animal actually requires a finite amount of protein. This need can be met by eating more of a low-protein or less of a high-protein diet. In the former situation, there will be excess energy and fat will accumulate.

D. Factors Altering Requirements

Genetic Variation. Genetic research has shown that certain animals have a deficiency or absence of enzymes which are involved in some phase of protein metabolism. Any time such a block occurs, inefficiencies can be expected to be present. Certain strains of chickens have different arginine requirements (Nesheim & Hutt, 1962), and Rose and Wixom (1955) found variations in the amount of amino acids needed to maintain nitrogen equilibrium in men. The coefficient of variation ranges from 6 percent for sulfur amino acids to 40 percent for tryptophan. No two individuals have exactly the same requirements for nutrients. Since organs and tissues grow at different rates, such variation should be considered. The differences between liver and muscle are quite large for some amino acids, so that if conditions favor development of one tissue over another, the dietary requirements for amino acids should change accordingly. Different genetic strains can therefore be expected to show differences in requirements.

Environment. The external environment can influence protein requirements. High environmental temperatures cause lower food intake and may require ration adjustment. Low temperatures increase intake and could lower percentage protein needs. Alternatively, controlled environment could be a stabilizing factor on nutrient requirements. Low-protein diets are synergistic with bacterial infection and roundworm parasitism. Good protein nutrition should accompany efforts to control these factors.

Some of the interactions between hormones and protein deposition were discussed in Section II. Both types of hormones and concentrations are important variables in determining the effect of protein needs. Generally, hormones increasing anabolic activity will increase the need for dietary protein. The effects of catabolic hormones can be made less severe by proper protein nutrition. The best examples are found in the use of high protein diets for patients recovering from wounds such as burns. Large amounts of thyroxine cause toxicity, but the effect varies with protein source.

E. Evaluating Dietary Sources

A major task for nutritionists is to develop simple, reliable tests for evaluating dietary sources. The same considerations discussed under requirement determination underlie the choice of method. Fewer problems are associated with purified proteins than with mixed proteins.

Assay Procedures for Amino Acids. Before the amino acid content is determined, the protein or mixture must be hydrolyzed. The usual method is to place the protein in an evacuated tube or in nitrogen with 6N hydrochloric acid and heat for at least 24 hours at boiling temperatures. Acid hydrolysis destroys tryptophan, cystine, and some other amino acids. It may combine amino acids with carbohydrate and change availability. Samples with high fat content may need to be extracted before hydrolysis. Many feed additives can interfere with subsequent procedures, and care may be needed to remove them from the feed.

Biological: Standardized techniques with bacteria are available. Most of the values in the current literature were obtained by this procedure. Under good conditions, the assay can give values within ±10 percent of the true mean. Lyman and his associates (1956) list details of procedures for various amino acids. Sometimes an enzymatic digest is used.

Chemical: Many specific tests for individual amino acids were developed and used to identify and measure amino acids. These were reviewed and evaluated by Block & Bolling (1951). The current procedures involve separation by chromatography (paper, ion exchange, gels) and colorimetric detection of ninhydrin derivatives. The amino acids can be converted to derivatives and separated by gas-liquid chromatography (Gehrke & Stalling, 1967). Hydrolysis by acid or enzymatic treatment is a necessary first step for proteins. These methods are especially suitable for free amino acids in biological fluids and for the hydrolysates of purified proteins.

Assay Procedures of Mixed Protein. For over 50 years, investigators have tried to assign comparative values to the proteins of feedstuffs. Of necessity, these required biological evaluation.

Growth Trials: Osborne & Mendel utilized a protein efficiency ratio (PER) arrived at by dividing gain by protein consumed. The most balanced proteins resulted in higher values. Modifications of this system have been used to measure partial responses of body parts. Liver and blood regeneration has been used. Other modifications include NPU and slope ratio analysis of dose response curves. The last named procedure incorporates desirable features of modern biological assay procedures and should provide a sound basis for evaluating the relative efficiency of proteins.

Biological Value: The basic determinations needed for the calculation of biological value were given previously. The technique gives very precise but often unrealistic values because it is determined at low protein levels. Mitchell (1962) defends this system with vigor because it does rank proteins in order of growth-promoting qualities. However, it is often possible to increase growth from a given protein by feeding more of it, thus meeting the required amino acid levels and letting the body convert excesses to carbon chains used for energy and urea. Thus, either 12 percent whole egg protein or 16 percent crude protein from soybean meal and casein has promoted equivalent growth and feed efficiency. The protein efficiency was better on the egg protein (Pfander & Tribble, 1957). The inability of the BV procedure to reflect the ability of higher levels of protein to meet growth requirements is a serious limitation. Another limitation is that the results cannot be used to predict the value of combining two or more proteins of known values into a ration. A BV method based on the percent of creatinine nitrogen excretion was used for ruminants (Blair et al., 1959). With refinement, this procedure could be valuable and deserves more testing than it has received.

Index: After the biological value of whole egg was determined to be 100, the concentration of each essential amino acid was taken as being present in ideal amounts. The concentration of each amino acid in test proteins was expressed as a percent of that in egg. The lowest value was then assigned to the protein. This was called the chemical score. Oser (1951) used an index method expressing the amino acids as a percentage of those in egg and then making two adjustments. Values over 100 are taken as 100 and negative values taken as 1. The logs of each ration are then determined, the total determined, and the antilog of the mean taken. This method gives the geometric means of the ratios or the essential amino acid index. The method can be modified by including cystine and tyrosine to the extent that they can replace methionine and phenylalanine for a given species. Supplemental D-L mix-

tures could be included, or amino acids could be omitted if they are not required. The essential amino acid index correlates well with biological value, but it might need to be adjusted for availability. In fact, an essential amino acid index based on enzymatic digests rather than feedstuff has been proposed (Sheffner et al., 1956). The basic assumption of these methods, that egg protein represents an ideal standard, has been challenged.

F. Selecting Supplements

"Good Quality" Protein. Regardless of the criteria used for evaluation, certain proteins consistently receive superior ratings. These have been referred to as high-quality proteins. Generally, they are high in those amino acids most needed for tissue synthesis and have a balance of essential amino acids similar to that of the tissues they will form. Generally, these proteins occur at levels of 35 to 70 percent of the feed ingredients and form the basis for most protein supplements. The general distribution pattern in cereal grains, the major food and feedstuffs, is one of high leucine and low levels of lysine and varying but reduced amounts of one or more of the following: methionine, tryptophan, threonine, and histidine. Supplements that are highly regarded contain amino acids that are limiting in grains but may provide excess amounts of others. Current practice is generally centered around the use of dairy, fish, meat, and poultry products to supply lysine; soybeans to provide lysine and tryptophan; and a wide selection of products to supply sulfur amino acids. With poultry rations, close attention must also be given to arginine and glycine sources.

Synthetic Amino Acids and/or Analogs. If the established essential amino acid requirements of any species are totalled and the figure is doubled to provide nonspecific nitrogen, the protein equivalent of the mixture is less than the established protein requirement. Rats grow well on purified casein diets containing 10 percent protein and 0.15 percent methionine. Similar studies on other species suggest that if the needed amino acids could be added to the ration to correct the specific deficiency, total protein needs could be reduced. Synthetic amino acids are available, but, in many cases, at prices exceeding those of providing them in natural foodstuffs. Methionine or its hydroxyanalog is now widely used in poultry feeds, and lysine has been used to supplement cereal products. Certain microbes produce and excrete amino acids into the media. By selecting strains that excel in this characteristic, microbiologists can obtain purified amino acids of L-configuration. Such products are now commercially available. Since chemical synthesis produces D-L amino acids, the nutritive value of the D is of interest. Most studies indicate that D-methionine is available and D-tryptophan is partially available. D-lysine is not available. The D-amino acids, if present in large amounts, interfere with the utilization of L-forms. Thus, a small percentage of D-amino acids might not influence performance, but larger amounts would cause toxicity or reduced gains.

The Problem of Unbalanced Proteins. An excess of an amino acid can be as disruptive to an organism as a deficiency. In fact, if one amino acid is present in great excess, it may impose a relative deficiency on some of the others. This effect of a great excess of one amino acid has been termed "amino acid imbalance." The effect is much greater on low-protein diets. Great care is needed in adding supplements of a single amino acid to a protein. In fact, if the added amino acid is not the first limiting or if it is added in excess of the amount needed to balance the absorbed amino acids, the truly limiting amino acid in the first instance and the second limiting in the second will be deficient and performance will be reduced.

4. NPN SOURCES AND POTENTIAL

The utilization of nonprotein nitrogen has been studied in greater detail in ruminants than in monogastric animals. In ruminants, the problem is essentially one of providing cultural conditions favoring a maximum proliferation of microorganisms without producing substances toxic to the host. In order to accomplish this, the rate of growth of organisms needs to be synchronized with the passage of ingesta so that a rapidly growing culture can be continued. Substrate must not be limiting, and all needed materials must be present at the microenvironment surrounding the dividing cell. This condition is best achieved by multiple feedings of a well-fortified diet high in soluble carbohydrate. Diethylstilbestrol may improve NPN utilization.

Currently, urea is the major NPN source. It can be regarded as a physiological component of the medium since it is introduced into the rumen via saliva and by diffusion through the rumen wall, especially when animals are on low-protein diets. Organisms with urease activity are normal rumen inhabitants. Synthetic urea can provide essentially all the nitrogen required for maintenance and some growth, and, after a proper adjustment period and supplementation, maintain milk production in dairy cows (Virtanen, 1963). Calculations based on ^{15}N urea indicate that overall efficiency of urea and ammonium carbonate is similar, and slightly less than that of food proteins. This work needs confirmation under optimum conditions for NPN use. Diammonium phosphate can provide both nitrogen and phosphorus. Other NPN sources known to be used if proper supplements are provided include: biuret, nitrate, ammonium chloride, amides, amines, and intermediates. In the future, most supplemental nitrogen for ruminants will probably come from NPN sources; however, there is a possibility of recycling some nitrogen from by-products not used by humans as a direct food source. Animals use nonspecific nitrogen to form the nondietary essential amino acids; but since most proteins contain an excess of these amino acids in relation to requirement, little interest exists in providing nonspecific nitrogen for practical rations. If the essential amino acids become available in large amounts or if genetic selection permits the growth of feedstuffs with large concentrations of essential amino acids, NPN could become an increasingly important factor in diets of many animals. NPN and lysine have been shown to improve a corn diet (Kies *et al.*, 1967). That diammonium citrate and glutamate have often been used with more success as nonspecific nitrogen than have other amino acids from cecal synthesis remains to be established. Animals with a functional cecum which practice coprography can probably use some NPN to advantage in synthesizing essential amino acids.

Exotic Sources. Man's search for extra protein (and energy) leads him to explore exotic sources. Seas contain a wide array of biological materials which are not harvested and the species now used are not managed efficiently. Fossil fuels and woods can be altered and used as food for microorganisms, or perhaps altered and used as feed for animals. Nitrogen excreted by animals is mostly wasted today. Research on the possibility of recycling this material is largely related to man in space, but basic principles developed there may eventually find application in routine living. Perhaps the Oriental system of using pigs to utilize human waste and chickens to utilize waste from pigs will become a common feature of future living. Current concern about such systems is that pathogenic organisms or their products and feed additives may also be transferred with the nutrients and that waste products will accumulate and become toxic. Certain flavors and odors will need to be removed.

REFERENCES

Adibi, S. A., Gray, S. J., & Menden, E. (1967). The kinetics of amino acid absorption and alteration of plasma composition of free amino acids after intestinal perfusion of amino acid mixtures. Amer. J. Clin. Nutr. *20*:24–33.

Arroyave, G., Jansen, A. A., & Torrico, M. (1966). Razón nitrógeno ureico/creatinina como indicador del nivel de ingesta proteica. Arch. Nutrición latinoamer. *16*:203–212.

Blair, J. W., Page, H. M., & Erwin, E. S. (1959). Proposed method for determining biological value of protein in ruminants. Proc. Soc. Exp. Biol. Med. *100*:459–461.

Block, R. J., & Bolling, D. (1951). *The Amino Acid Composition of Proteins and Foods.* 2nd ed. Springfield, Ill., Charles C Thomas.

Breuer, L. H., Pond, W. G., Warner, R. G., & Loosli, J. K. (1964). The role of dispensable amino acids in the nutrition of the rat. J. Nutr. *82*:499–506.

Ellis, W. C. (1956). The Utilization of Nitrogen by Lambs. M. S. Thesis, University of Missouri, Columbia.

Ellis, W. C., Garner, G. B., Muhrer, M. E., & Pfander, W. H. (1956). Nitrogen utilization by lambs fed purified rations containing urea, gelatin, casein, blood fibrin, and soybean protein. J. Nutr. *60*:413.

Ellis, W. C., & Pfander, W. H. (1965). Rumen microbial polynucleotide synthesis and its possible role in ruminant nitrogen utilization. Nature (London) *205*:974.

El-Shazly, K. (1952). Degradation of protein in the rumen of the sheep. 2. The action of rumen microorganisms on amino acids. Biochem. J. *51*:647–653.

Gehrke, C. W., & Stalling, D. L. (1967). Quantitative analysis of the twenty natural protein amino acids by gas-liquid chromatography. Separation Sci. *2*:101–138.

Glover, J., Duthie, D. W., & French, M. H. (1957). The apparent digestibility of crude protein by the ruminant. 1. A synthesis of the results of digestibility trials with herbage and mixed feeds. J. Agri. Sci. *48*:373–378.

Kies, C., Fox, H. M., & Williams, E. R. (1967). Effect of non-specific nitrogen supplementation on minimum corn protein requirement and first-limiting amino acid for adult man. J. Nutr. *92*:377–383.

Lyman, C. M., Kuiken, K. A., & Hale, F. (1956). Essential amino acid content of farm feeds. Agri. & Food Chem. *4*:1008–1013.

Martin, A. K. (1966). Some errors in the determination of nitrogen retention of sheep by nitrogen balance studies. Brit. J. Nutr. *20*:325–337.

Michael, A. F., & Drummond, K. N. (1967). Inhibitory effect of certain amino acids on renal tubular absorption of phosphate. Canad. J. Physiol. Pharm. *45*:103–114.

Mitchell, H. H. (1962). *Comparative Nutrition of Man and Domestic Animals.* Vols. I & II, New York, Academic Press.

Mitchell, H. H., Carroll, W. E., Hamilton, T. S. & Hunt, G. E. (1931). Food requirements of pregnancy in swine. Ill. Agri. Exp. Sta. Res. Bull. No. 375.

Moore, S., & Stein, W. H. (1951). Chromatography of amino acids on sulfonated polystyrene resins. J. Biol. Chem. *192*:663–681.

Nesheim, M. C., & Hutt, F. B. (1962). Genetic differences among white leghorn chicks in requirements of arginine. Science *137*:691.

Noall, M. W., Riggs, T. R., Walker, T. R., & Christensen, H. N. (1957). Endocrine control of amino acid transfer. Science *126*:1002–1005.

Oser, B. L. (1951). Methods for integrating essential amino acid content in the nutritive evaluation of protein. J. Amer. Diet. Ass. 27:396–402.

Perry, T. L., Hestrin, M., Macdougall, L., & Hansen, S. (1966). Urinary amines of intestinal bacterial origin. Clin. Chim. Acta 14:116–123.

Pfander, W. H. (1968). Integration of the protein deposition rates in cattle. Unpublished calculations from Missouri growth experiments.

Pfander, W. H., & Tribble, L. F. (1957). Amino acid composition of swine rations and amino requirements of weanling pigs. Mo. Agri. Exp. Sta. Res. Bull. No. 626.

Preston, R. L., Schnakenberg, D. D., & Pfander, W. H. (1965). Protein utilization in ruminants. I. Blood urea nitrogen as affected by protein intake. J. Nutr. 86:281.

Rose, W. C. (1938). The nutritive significance of the amino acids. Physiol. Rev. 18:109–136.

Rose, W. C., & Wixom, R. L. (1955). The amino acid requirements of man. J. Biol. Chem. 217:997–1004.

Schnakenberg, D. (1968). Diurnal Variation in Blood Urea and Rumen Ammonia Nitrogen. M. S. Thesis, University of Missouri, Columbia.

Sheffner, A. L., Eckfeldt, G. A., & Spector, H. (1956). The pepsin digest residue (PDR) amino acid index net protein utilization. J. Nutr. 60:105–120.

Stickland, L. H. (1934). Studies in the metabolism of the strict anaerobes (Genus clostridium). Biochem. J. 28:1746–1759.

Szmelcman, S., & Guggenheim, K. (1966). Interference between leucine, isoleucine, and valine during intestinal absorption. Biochem. J. 100:7–11.

Virtanen, A. I. (1963). *Proc. XII Cong. Assoc. Scand. Agric. Sci.*, Helsinki.

Watt, B. K., & Merrill, A. L. (1950). Composition of foods. U. S. D. A. Handbook 8.

Wright, L. A., & Nicholson, T. F. (1966). The plasma levels, filtered loads, excretion rates, and clearances of a number of ninhydrin-positive substances by normal fasting dogs. Canad. J. Physiol. Pharmacol. 44:195–201.

Zimmerman, R. A., & Scott, H. M. (1967a). Plasma amino acid pattern in chicks in relation to length of feeding period. J. Nutr. 91:503–506.

Zimmerman, R. A., & Scott, H. M. (1967b). Effect of fasting and of feeding a non-protein diet on plasma amino acid levels in the chick. J. Nutr. 91:507–508·

Chapter 17

Mineral Requirements

By I. A. Dyer

ARISTOTLE believed that there were only four elements: earth, air, fire, and water. Since his time these "elements" have been partitioned into many sub-units. Among the inorganic sub-units, at least 17 and probably 19 minerals are known to be required by animals. The knowledge that certain elements are required is not new, but the number now known to be required would have been unbelievable, even at the beginning of the twentieth century. Scientists have long known that traces of most elements exist in animal tissue but thought they were contaminants rather than functional entities. Many of the elements now believed to be contaminants will be shown to be essential as more "unnatural" feeding regimens are used and as analytical procedures improve.

Minerals have many functions relating directly or indirectly to animal growth. They contribute to the rigidity of the bones and teeth, and are an important part of protein and lipid fractions of the animal body. In addition, they preserve cellular integrity by osmotic pressures and are a component of many enzyme systems which catalyze metabolic reactions in biological systems. As an example of diversity of function, calcium is used in large amounts by the body for synthesis of osseous tissue. Conversely, cobalt has a regulatory role, through vitamin B_{12}, rather than contributing quantitatively to tissue synthesis, yet its effect on growth is as dramatic as that of calcium. Phosphorus and certain other elements contribute structurally and have many regulatory functions. Minerals involved in several metabolic processes are more likely to be interrelated with other minerals than those involved in a single function. Knowledge of the minerals which are known to be interrelated is increasing rapidly (Fig. 17–1). As knowledge expands in the field of mineral metabolism, many additional interrelationships undoubtedly will be shown.

Chelation (a process by which a metal atom is sequestered) of minerals is employed in some feeding regimens. Chelation has many medical functions such as removal of certain isotopes or poisonous metals from the body and the deactivation of bacteria and/or viruses by depriving them of needed metals. Nutritionally, chelators are used to deliver trace metals to the site where they are needed. The binding (chelation) must

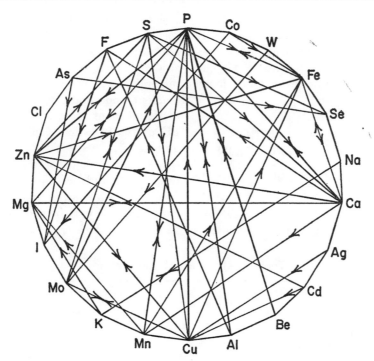

FIG. 17–1. Mineral interrelationships in animal metabolism. The arrows indicate antagonism between elements. For example, calcium is antagonistic to manganese. Magnesium and calcium are mutually antagonistic. (*Adapted from drawing by A. D. Tillman, Oklahoma University.*)

be weak, that is, the affinity of the chelator for the metal must be less than the affinity of the tissue for the same element. The element is then released to tissue.

There are many natural and synthetic chelating substances. Among the well-known ones occurring in nature are hemoglobin (containing iron), vitamin B_{12} (containing cobalt), and cytachrome oxidase (containing copper and iron). Many compounds act as synthetic chelators. One of the more prominent ones is ethylenediamine-tetra acetic acid (EDTA), which sequesters divalent cations. Even aspirin and citric acid serve as chelating compounds.

The science of chelation as it relates to the nutrition of domestic animals is new. This phenomenon, much more than techniques heretofore used, offers possibilities of regulating the amount of a given metal ion at the cellular level.

I. SOURCE

Animals have two primary sources of the

inorganic elements, through natural feeds and supplementation. Even though a large portion of the required minerals may be provided by plants, mineral supplementation is a necessary practice in properly nourishing the animal. The mineral content of the plant depends primarily upon the plant species, abundance of the element in the soil, and the conditions (pH, moisture, and others) affecting the plant growth, and, consequently, the mineral uptake by the plant. Because of the many factors affecting composition, the mineral content of plants will vary both within and among species. The latter results in gross differences (Appendices V & VI). Because of mineral composition variability, average values should be used only as guides in calculating the mineral intake of the animal from plants.

Minerals from animal tissue compared with those from plants often are more available to other animals. These elements have already been digested and absorbed, whereas those in plant tissue may be bound in indigestible cellulosic compounds. Further, the supply from animal tissue is more nearly proportional to the requirement. These factors do not within themselves assure a higher mineral availability, however, since many exceptions can be found.

II. ABSORPTION, FUNCTION, DEFICIENCY, AND EXCRETION

The absorption of minerals was poorly understood until isotopic techniques were developed for tracing the elements through the body. The primary reason for the con-

fusion was the fact that, because of the absorption and excretion into the gut, and the reabsorption cycle, the conventional balance studies did not indicate the net utilization of minerals. The absorption of minerals is dependent upon many factors, including level of the element ingested, age of the animal, pH of the intestinal contents, state of the animal with respect to deficiency or adequacy of the element, and the presence of other antagonistic minerals or nutrients (Table 17–1). Earlier investigators did not consider these variables; consequently, their reports often lack agreement.

Table 17–1. Mineral Absorption

Element	Primary Absorption Site	Mechanism	Form in Which Absorbed	Conditions Favoring Absorption
A. MACROELEMENTS				
Calcium	Duodenum and jejunum (most species)	Active transport dependent upon or stimulated by vitamin D; simple diffusion in presence of vitamin D	Ca^{++}	Low calcium and phosphorus diets; low phytate, oxylate, and phosphate level in the diet; presence of vitamin D; acid condition in gut; presence of lactose, citric acid, and certain amino acids; youth
Chloride	Small intestine	Active transport	Cl^-	
Magnesium	Small intestine	Active transport	Soluble Mg^{++} salts	Low level of ammonium
Phosphorus	Duodenum, primarily in distal portion	Active transport	Free phosphates, primarily	Presence of potassium; relatively low calcium, iron, strontium, magnesium and beryllium; acid condition
Potassium	Small intestine, primarily	Active transport	K^+	
Sodium	Small intestine, primarily with some being absorbed from the stomach	Active hexose transport	Na^+	
Sulfur	Small intestine		Sulfates, cystine, methionine, some as hydrogen sulfide	
B. MICROELEMENTS				
Arsenic			Organic forms and as arsenic trioxide	
Barium	Very low absorption			
Chromium	Small intestine		Hexavalent	Absence of gastric juice; achlorhydria

Table 17–1. Mineral Absorption (Continued)

Element	Primary Absorption Site	Mechanism	Form in Which Absorbed	Conditions Favoring Absorption
Cobalt (vitamin B_{12})	Small intestine, primarily; some throughout the tract; species-dependent		Co^+ and as B_{12}	B_{12} intrinsic factor; sorbital; adequate iron and pyridoxine; deficiency of folic acid; low temperature; prefeeding
Copper	Throughout digestive tract, depending upon species and diet. In rats, stomach, duodenum, jejunum, and ilium in the order listed	Active transport	Cu^{++} and in carbonates or H_2O soluble compounds, e.g., cupric carbonate, cuprous oxide, and cupric sulfate	Increased gut acidity, e.g., low level of calcium; low molybdenum, sulfate, and zinc
Fluorine	Small intestine		Soluble compounds, as NaF	Low calcium and aluminum
Iodine	Throughout intestinal tract		Inorganic iodides	
Iron	Proximal duodenum but some throughout the tract	Active transport	Fe^{++}	Acid condition favored by mannose, fructose, and glucose; adequate dietary protein; body's need for iron
Manganese	Small intestine, primarily	Active transport	Mn^{++}	Relatively low concentration of calcium, phosphorus, and iron.
Molybdenum	Throughout the tract		Hexavalent, water-soluble forms	Low dietary sulfate
Nickel				
Selenium	Small intestine, distal portion		Organic and inorganic; soluble compounds more readily absorbed	Relatively low levels of sulfur, tellurium, and arsenic; acid conditions
Strontium				
Tin				
Titanium				
Vanadium			Oxytartrate	
Zinc	Duodenum, primarily; some throughout the tract		Zn^{++} or as carbonate, oxide, or sulfate	Low dietary content of phytic acid

The elements have a myriad of functions. Each of those known to be required is briefly discussed individually and as it relates functionally to other elements. A more complete coverage is presented of the trace elements (Underwood, 1962), copper (Peisach et al., 1966), iodine (Pitt-Rivers & Trotter, 1964), selenium (Rosenfeld & Beath, 1964), zinc (Prasad, 1966). A general discussion of macro- and microelements (Sandstead, 1968) and an advanced treatise on the elements have been published (Comar & Bronner, 1960, 1961, 1964, 1965).

All animals are subject to mineral deficiencies. These may be caused by (a) a suboptimal amount of a given element in the feed; (b) imbalance of another mineral or nutrient which decreases absorption; (c) any condition which increases the rate of passage of the element through the gut or body; and

(d) a metabolic antagonist which causes the animal to require a greater quantity of the element. The terms "imbalance" and "deficiencies" are not synonymous, but either condition may lead to the other.

The pathway and rate of excretion of the inorganic elements vary. Some are excreted almost entirely in the feces, others almost entirely in the urine, while still others are excreted through both routes. Additionally, certain elements, particularly sodium, are lost in sweat. Iron may be lost in sizable quantities during menstrual cycles. Integument loss from the body also accounts for minor losses of many elements.

A. Macroelements

1. CALCIUM

Calcium is required in large quantities by growing animals, largely for skeletal development (Appendix VII). The skeleton ash content is approximately 26 per cent of the weight of the functional bone. Calcium phosphate, calcium carbonate, and magnesium phosphate or carbonate make up approximately 85, 14, and 1 percent respectively of the bone ash. Bone contains approximately 99.5 percent of the calcium in the body. Although osteogenesis (see Chapter 12) constitutes the major demand for this element for growing animals, calcium also has several regulatory functions. It participates in metabolic functions including osmoregulation, muscle contraction, blood coagulation, ATPase activation, decrease in cell permeability, and reduction in nerve irritability (Table 17–2). Calcium also is required in large amounts during lactation; bovine milk contains approximately 0.12 percent of this element.

Calcium is absorbed primarily in the proximal portion of the gut. An acid condition, adequate vitamin D, low phytate, oxalate, fat, and phosphate are conducive to calcium absorption. Absorption rate decreases with the age of the animal.

Growth (gross weight increase) of the bone is completed early in life. Calcium metab-olized after growth serves to replenish the calcium loss and also in a regulatory role. The metabolic functions requiring calcium depend upon a constant minimal amount. Conversely an excess of calcium may result in mineral imbalances having profound physiological effects. Parakeratosis in pigs occurs when excess calcium is fed. The calcium increases the chelation of zinc by phytic acid (Byrd & Matrone, 1965). Even though excess zinc may prevent parakeratosis induced by calcium, a reduction of the latter is the preferred method of preventing this anomaly.

Excess calcium has many other undesirable effects. Stott (1968) indicated that excess calcium may increase the incidence of parturient paresis in mature and aged dairy cattle. High dietary levels of this element may also impair absorption of calcium, phosphorus, copper, and manganese.

The effects of calcium deficiency would vary, depending upon the age of the organism. In the young, rickets may be a result of calcium deficiency. In the adult, a deficiency in calcium is known as osteomalacia. Rickets is characterized by a decreased concentration of hydroxyapatite in the organic matrix of the bone and prebone cartilage. In this condition the organic matrix synthesis is more rapid than the inorganic deposition in this tissue. Osteomalacia is characterized by a decreased content of hydroxyapatite in the bone matrix. In the latter anomaly, the calcium had been deposited during osteogenesis and later withdrawn. Calcium deficiency may also result in an enlargement of the parathyroid, widespread hemorrhage, lesions in the digestive tract, and vision impairment from cataract formation (Follis, 1958).

The calcium which is not absorbed and that returned from the body to the gut is excreted in the feces. The majority of calcium is lost through the feces. Normally the urinary loss is small, but it can be increased by vitamin D, which causes an increase in calcium absorption (Capen *et al.*, 1968). In the study of Capen and his associates,

Table 17–2.   Summary of Primary Mineral Functions

Element	*Some Functions*
A. MACROELEMENTS	
Calcium	Osteogenesis; decreases nerve irritability; necessary for blood clotting; decreases cell membrane permeability; ATPase activator; needed for osmoregulation
Chloride	Maintenance of proper osmotic concentrations; carbon dioxide transport; solubility of proteins; activates salivary amylase
Magnesium	Osteogenesis; enzyme activator (kinases, mutases, enolase, ATPases, cholinesterase, alkaline phosphatases, isocytricdehydrogenase and arginase); helps decrease tissue irritability
Phosphorus	Osteogenesis; necessary for fat and carbohydrate metabolism (components of ATP, DNP +, glucose-1-phosphate, phosphoproteins, phospholipids, and nucleic acids)
Potassium	Electrolyte and water balance; intracellular osmotic concentration; activates enzyme systems such as pyruvic kinase and those having to do with phosphorylation of creatine; increases heart beat; increases tissue irritability
Sodium	Electrolyte and water balance (neutrality regulator); exchange with potassium ions during nerve and muscle action; increases tissue irritability; osmotic pressure regulation
Sulfur	Component of cysteine methionine, biotin, sulfolipids, cystine, sulfonated polysaccharides and many other metabolites
B. MICROELEMENTS	
Aluminum; arsenic; barium; boron; bromine	No known function in animals
Chromium	Has an insulin-like effect on glucose metabolism
Cobalt	Activator of certain peptidases; required in the synthesis of vitamin B_{12}
Copper	Prosthetic group of hemocyanins; increases iron absorption; necessary for formation of erythrocytes; component of flavoproteins; cofactor for tyrosinase, ascorbic acid oxidase cytochrome oxidase, uricase, plasma monoamine oxidase, and ceruloplasmin
Fluorine	Prevention of dental caries; no specific growth function known
Iodine	Needed for thyroxin synthesis (homeostatic regulator)
Iron	Component of hemoglobin and myoglobin; component of cytochromes, cytochrome oxidase and xanthine oxidase
Manganese	Needed for synthesis of cholesterol (mevalonic kinase); osteogenesis; normal functioning of the reproductive system; needed for bone phosphatase and may be a component of arginase; muscle ATPase and choline esterase may depend upon Mn^{++}
Molybdenum	Needed for zanthine oxidase, aldehyde oxidase, and iron flavoproteins. Molybdenum deficiency has not been produced in animals but has been in birds
Nickel	No definitely known function in animals; may be associated with pigmentation; bound to RNA
Selenium	Needed in prevention of muscular dystrophy in laboratory animals, exudative diathesis in chicks, and hepatic necrosis in rats
Strontium	No definitely known function in animals, although some suggest its possible need for bone calcification
Tin; titanium	No known function in animals
Vanadium	Depresses cholesterol and phospholipid biosynthesis; reduces ATP, CoA; stimulates MAO; may be necessary for normal bone and teeth formation
Zinc	Cofactor of carbonic anhydrase, and certain dehydrogenases, and phosphatases; required for growth; chicks require zinc for bone and feather development; may be required for RNA synthesis.

urinary calcium excretion increased approximately eightfold (from .045 to .36 percent) at 20 days after feeding 30 million units of vitamin D daily. Factors which increase parathyroid activity cause a withdrawal of calcium from osseous tissue and thereby result in more calcium being available for urinary excretion. Some calcium is lost during sweating in animals which use this mechanism for cooling. Man may lose up to 20 mg calcium per hour during profuse sweating.

2. Chlorine

Chlorine is found in animal tissue primarily as the chloride ion. This ion comprises approximately 0.15 percent of the total body weight. It is located primarily in extracellular fluids, but some intracellular chloride is present, particularly in the red blood cell. The chloride in the gastric juice is in approximately the same concentration as in the blood. In the gastric juice, however, it appears with the hydrogen ion rather than with sodium.

Chloride is absorbed primarily from the small intestine. Much of the chloride which is present as hydrochloric acid in the stomach is absorbed as it passes distally along the digestive tract. It is absorbed in large amounts from the distal digestive tract, perhaps to prevent a major loss of this ion in the feces.

Chloride functions in carbon dioxide transport through the chloride shift. In this shift, the bicarbonate content of plasma is increased by exchange with plasma chloride, which enters the red cell (Oser, 1965). Chloride helps regulate osmotic pressure. It functions also as an activator (for example, salivary amylase) and in the digestion of protein. It makes up about 63 percent of the total anions in extracellular fluid. This metabolic pool is subject to change, depending largely upon the level of intake.

Chloride deficiencies are not widespread. When a deficiency does exist, it generally is the result of an excess loss due to a digestive tract or metabolic malfunction.

Chloride is excreted primarily through the kidneys. Sweat may represent a sizable loss in certain species. Some chloride is lost through the feces, particularly when normal or greater than normal amounts are ingested. Chloride excretion via the urine or sweat is proportional to sodium excretion.

3. Magnesium

Quantitatively, magnesium content is low in animal tissue. About 40 percent of the body magnesium is in the soft tissue, and the remaining 60 percent is in the osseous tissue. Bone ash contains approximately 0.6 percent magnesium. Like phosphorus, magnesium is a constituent of all living cells. The content of magnesium, unlike that of calcium and phosphorus, remains almost constant in the tissue as the animal matures.

As much as 30 percent of the ingested magnesium is absorbed, primarily from the small intestine. It apparently is absorbed largely in the form of soluble magnesium salts. When ammonium and phosphates are abundant in the ration, an insoluble ammonium-phosphate-magnesium compound may be formed which is nonabsorbable or lowly absorbable. During a deficiency of magnesium, it is preferentially withdrawn from the bone to make up for the lack. The withdrawal rate, however, is not rapid enough to maintain a normal blood magnesium level.

Magnesium is a divalent cation which preferentially activates many enzyme systems. Among those activated include kinases, mutases, ATPases, cholinesterase, alkaline phosphatases, enolase, isocitric dehydrogenase, and arginase (Table 17–1). Other divalent cations such as manganese may substitute for magnesium in many enzyme catalyzed reactions. Thus, some enzymes are not specific for one element.

Another function of magnesium is to influence central nervous system (CNS) irritability by activating cholinesterase, which breaks down acetylcholine. Thus, CNS is in a hyperirritable state during a magnesium

PLATE 21

Mineral Deficiencies.

A, Magnesium deficiency. The pig on left received 413 ppm and the one on the right 70 ppm magnesium. Leg weakness and slow growth are evident on the low magnesium diet. (*Courtesy of Plumlee & Beeson, Purdue University.*)

B, Phosphorus deficiency in pig. Note rickets in rear legs of pig on the left. (*Courtesy of Plumlee & Beeson, Purdue University.*)

C, Copper deficiency. Note white zone in wool growth when copper was omitted from the diet.

PLATE 22

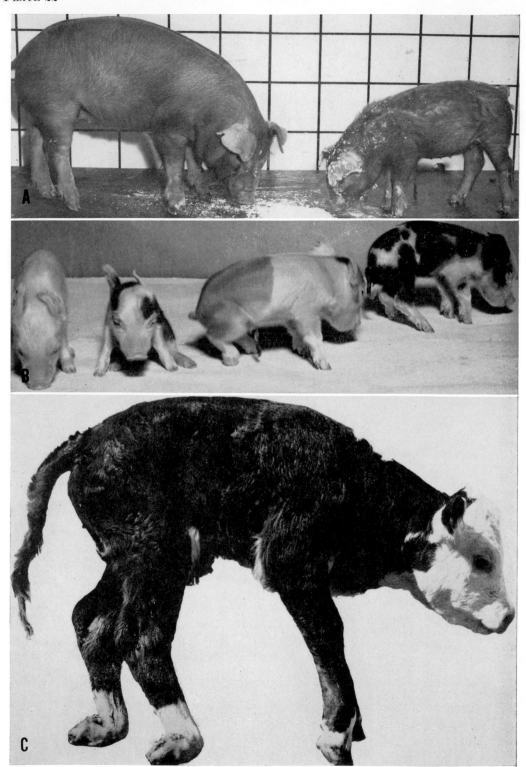

Iron and Manganese Deficiencies.

A, Effect of iron deficiency on growth and appearance of pigs. Pig on left received 115 ppm and one on the right 15 ppm iron. Note reduced growth and rough hair coat (right). (*Courtesy of Beeson & Plumlee, Purdue University.*)

B, Manganese deficiency in pigs. Note weakness of legs and poor sense of balance. (*Courtesy of Plumlee & Beeson, Purdue University.*)

C, Manganese deficiency in calf. A deficiency of this element may result in deformed limbs. Dam of this calf had received 16 ppm manganese. (*Rojas & Dyer, 1965. J. Anim. Sci. 24:664.*)

PLATE 23

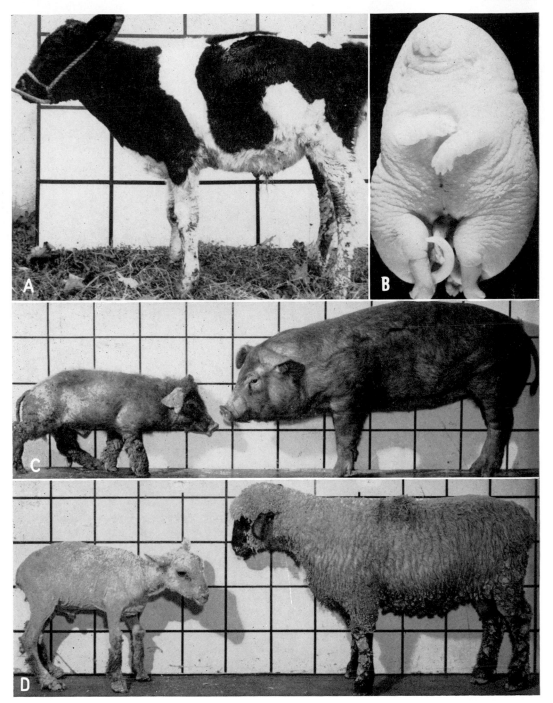

Zinc Deficiency in Animals.

A, Zinc deficiency of calf. Calf received low-zinc-purified diet. Note parakeratosis on rear legs. (*Miller, 1962. J. Nutr. 76.*)

B, Dam of these cojoined twins was given 9 ppm zinc during gestation. (*Hurley, 1968. Fed. Proc. 27:193.*)

C, Zinc deficiency in pig. Note dermatosis (parakeratosis) in pig. Pigs received 0 and 50 ppm added zinc, respectively. (*Courtesy of Smith & Beeson, Purdue University.*)

D, Zinc deficiency in lamb receiving 3 ppm zinc. Note loss of wool. (*Courtesy of Ott, Smith, & Beeson, Purdue University.*)

deficiency and conversely, in a depressed state when tissue magnesium is in excess. Further, less acetylcholine is released in the presence of excess magnesium.

Magnesium deficiency is not generally noted in animals unless severe diuresis (natural or induced), toxemia, hepatic cirhosis, or diarrhea is present. As noted above, when excess dietary ammonium ions and phosphate are present, magnesium may be poorly absorbed. The latter condition results in "grass tetany" in the bovine. This convulsive tetany resembles hypocalcemic tetany.

Magnesium is excreted primarily through the urine, but generally it is conserved by the body. Most magnesium filtered by the glomerulus is reabsorbed by the renal tubules.

4. Phosphorus

Phosphorus is required by all animals, although, quantitatively, less of it is required than of calcium. Approximately 80 percent of this element is contained in the osscous tissue, and makes up approximately 16.5 percent of the bone ash in the different species, including man. The remaining 20 percent is in the soft tissue.

Phosphorus is rapidly absorbed. It appears in general circulation within five minutes after entering the duodenum. Its absorption, primarily from the distal duodenum, is enhanced in acid media and in the presence of low concentrations of beryllium, calcium, magnesium, and strontium (Table 17–1). The efficiency of absorption depends upon several dietary factors including the form in which it is ingested, the pH of the intestinal fluids, the calcium-phosphorus ratio, and the amount of vitamin D present.

Both inorganic and organic phosphates are absorbed, but the former is more highly absorbed than the latter by many species. The reason is that, in many plants, phosphorus is in the form of phytic acid (inositol hexaphosphoric acid). Not only is phytic acid less well utilized in some species but it forms an insoluble compound, phytin (a calcium-magnesium salt of phytic acid) which reduces the absorbability of calcium and magnesium. If sufficient phytase is present in the duodenum, phytin can be degraded. Consequently, the phosphorus will be more highly absorbable.

Phosphorus found in the soft tissue has a myriad of functions. It is concerned with the intermediary formation of lecithins in fat metabolism. It is a constituent of phospholipids. Phosphorus plays a vital role in carbohydrate metabolism in the formation of hexosephosphates, creatine phosphate, and adenylic acid. Phosphorus also functions in protein metabolism by its presence in nucleoproteins and phosphoproteins (Oser, 1965).

The ratio of phosphorus to calcium is important in mineral metabolism. The amount of each element and the form in which it is ingested influences this ratio. Vitamin D and the parathyroid hormone help maintain the calcium and phosphorus within acceptable physiological limits. Vitamin D is necessary for the absorption of calcium from the gut and may be involved in calcium deposition. Conversely, when necessary, the parathyroid hormone causes a dissolution of calcium from the bone in order to maintain a constant blood level. This hormone also controls tubular phosphorus reabsorption.

Phosphorus deficiencies generally do not occur if the protein intake of the animal is moderate to high. A severe phosphorus deficiency results in gross skeletal deformities typified by rickets in the young. In this deformity, the thoracic cavity is greatly reduced. The growth of young is retarded and voluntary activity is restricted. Parathyroid hyperplasia has been observed in the rat, but it is not due to a phosphorus deficiency in all species. Many metabolic changes occur as a result of a phosphorus deficiency. These changes, however, may not be observed during a mild deficiency and, except by use of chemical or biological analyses, may be indistinguishable from other deficiencies, even during a gross deficiency.

Phosphorus is excreted both through the feces and urine. Since it is highly absorbable (about 70 percent of the amount in the human diet), a high percentage of the total is excreted in the urine. When dietary phosphorus is reduced its loss from the body is likewise reduced. Schaefer and his associates (1961) indicated that as much as 99 percent of renal tubular phosphorus may be reabsorbed during periods of phosphorus restriction. The loss through the feces is from that portion which is not absorbed and that which has been absorbed and excreted into the duodenum. This endogenous loss does not appear to be of major consequence since it normally would represent no more than 20 to 30 percent of the total fecal phosphorus. Further, endogenous phosphorus is reduced as age progresses.

5. POTASSIUM

Potassium, like sodium and chlorine, is well distributed in animal tissue. It is present as the potassium ion. It represents about 0.20 to 0.35 percent of the live weight, depending upon species and state of potassium nutrition. Potassium is the primary intracellular cation, thus constituting the "base" of body cells.

Potassium is absorbed as an ion primarily in the small gut. It has many functions in the body. Its presence, like that of sodium, increases nerve irritability. In this function it tends to counter the effects of calcium and magnesium. Potassium, which is primarily bound to protein, influences osmotic equilibrium and facilitates several enzyme reactions. These include pyruvic kinase and myosin ATPase. This element aids the uptake of neutral amino acids in higher animals. In certain unicellular organisms, a potassium deficiency results in slow growth, apparently by limiting the rate of transfer of amino acids from aminoacyl-sRNA to polypeptides (Lubin, 1964). Potassium also influences carbohydrate metabolism. This influence, at least partially, is mediated through the element's effect on the

adrenal gland. A deficiency of potassium causes a hypertrophy of the adrenals and postprandial hyperglycemia. Concurrently, the glycogen stores are depleted.

Potassium and magnesium are functionally related. A magnesium deficiency increases the elimination of potassium. As potassium is abnormally excreted to the point of cellular deficiency, sodium may replace some of it intracellularly.

Except under pathological conditions, deficiencies and excess of potassium are relatively rare although animals consuming high-grain diets often do not ingest sufficient amounts of it. This has been observed in ruminants (Telle *et al.*, 1964).

Potassium deficiency may be caused from consuming a diet low in this element or by diarrhea, diuresis, or acidosis. Animal products, nuts, and most green plants contain sufficient potassium to prevent a deficiency. Grains contain only about one half that required by the ruminant and probably other animals. A deficiency of potassium is manifested by muscular weakness, increased irritability, mental disorientation, cardiac irregularities, and reduced rate of growth.

The major excretory pathway is through the kidney. The homeostatic mechanism is sufficiently developed so that, except under pathological conditions, this cation is conserved. Excess urinary excretions of potassium may be induced by the administration of such compounds as acetazolamine and parathormone.

Factors governing the rate of potassium excretion include the ratio of sodium to potassium in the distal tubule, the capacity of the tubular cells to reabsorb sodium, and the amounts of sodium and hydrogen available for exchange with potassium. The amount of potassium excreted apparently is not dependent upon the filtered load of this ion.

6. SODIUM

Sodium comprises only about 0.2 percent of the total weight of the body. It is distributed throughout the body and is found primarily in the extracellular fluids, with the blood plasma and interstitial fluids having the greatest portion of sodium. It is the chief extracellular cation. Some sodium is also stored in the bone.

Carnivorous animals can secure enough sodium from their diet. Herbivore, however, must have their diet supplemented with sodium since plants generally have a small amount of this element. Plants not only contain low amounts of sodium, but they also have a high potassium content which causes an increase in sodium excretion.

Sodium is absorbed rapidly. The primary absorption site is the small intestine, although some is absorbed from the stomach. Most sodium is ingested as sodium salts such as chloride, bicarbonate, lactate, phosphate, and proteinate. It is then dissociated and absorbed as a sodium ion.

Sodium has fewer metabolic functions than phosphorus. Nevertheless, it is essential for animals. It is the primary alkaline plasma electrolyte, providing about 92 percent of the alkalinity of this fluid. In human plasma, the electrolyte distribution is approximately as follows (mEq/l) (Oser, 1965):

	Bases				*Acids*			
Na^+ +	K^+	Ca^{++} +	Mg^{++}	=	Cl^- +	HCO_3^- +	proteinate$^-$ +	others$^-$
154	5	5	3	=	106	28	17	16
			167	=	167			

Sodium is necessary for amino acid and glucose transport across the mucosa and cell membranes. One of the major functions of sodium appears to be connected with regulating acid-base balance and nerve irritability. The role of sodium in nerve irritability is as follows (Oser, 1965):

$$\frac{Na^+ + K^+ + OH^-}{Ca^{++} + Mg^{++} + H^+};$$ the greater the numerator, the greater the irritability and vice versa.

Dietary deficiencies of sodium are rare for man but occur regularly in domestic animals. Man does not require less sodium, but rather, his diet, which contains animal products, has a higher content of this element than does the plant diet of domestic animals. Sodium deficiency may become apparent following excess body fluid loss precipitated by excess sweating, diarrhea, vomiting, diuresis, and cortical insufficiency. An insufficiency of sodium results in muscular cramps, general weakness, and, finally, vascular collapse.

The excretion of sodium is primarily through the kidneys. Considerable sodium can be lost through sweat in certain species. Little sodium is lost in the feces. The homeostatic mechanism for controlling sodium metabolism is very efficient, since animals can survive for long periods on exceptionally low sodium intakes.

7. SULFUR

The body is composed of approximately 0.15 percent sulfur. This element occurs primarily in the organic form and is well distributed in the protein tissue of the body. It is an integral component of cystine, cysteine, methionine, biotin, and other compounds.

Animal foods contain many sulfur-bearing compounds such as thiamine, biotin, lipoic acid, glutathione, CoA, ergothionine, taurine, sulfalipids, and sulfated polysaccharides, which contribute to the animal's sulfur pool.

Sulfur absorption occurs primarily in the small intestine. In the ruminant, the sulfates—cystine, methionine, and hydrogen sulfide—supply the major dietary sources of sulfur. From among the different sources, inorganic sulfur is less well absorbed than organic sulfur in some, but not all, species.

As sulfur is metabolized, it becomes an integral part of many metabolites and end products in the animal. It is functionally related to calcium, copper, molybdenum, selenium, and zinc, and it is antagonistic to all of them except zinc. Sulfate, when ingested in excess amounts, reduces liver copper storage. When the amount of sulfate was increased from 1 to 6 gm per day in the diet of the sheep, hepatic copper storage was reduced 28 mg when the molybdenum content was held constant at 14.7 mg per head daily (Dick, 1956).

In the rat, inorganic sulfur is utilized in a way comparable to the utilization of organic sulfur (Michels & Smith, 1965). In this species, the inorganic sulfur appears to be well absorbed and purportedly is needed to prevent an increase in the amino acid requirement. The response of rats to avitaminoses E is influenced by the ratio of neutral to inorganic sulfur.

A sulfur deficiency should not occur under normal dietary regimens. The proteins contain considerable sulfur. In addition, sulfur is obtained from other sources, particularly sulfates. In ruminants a sulfur deficiency may occur if a nonprotein nitrogen is substituted for protein. Such deficiencies are seldom observed, however, since not all the protein is replaced by nitrogen and because sulfur may be obtained from other sources.

A deficiency of sulfur is most obvious in lambs, in which it is required for both body functions and wool growth. In this species, growth rate of the body and wool is reduced when dietary sulfur is quantitatively inadequate.

Sulfur is excreted both in the feces and the urine, depending upon the form in which it is administered and the amount given. There is also a hair and dermal loss, although the latter sources are comparatively small.

B. Microelements

1. COBALT

Cobalt is found in minute amounts in most animal tissue. The kidney has the highest concentration, averaging 0.20 to 0.25 ppm on a dry matter basis. The liver has the next highest concentration, with from 0.15 to 0.20 ppm. As with other elements,

tissue concentration is changed as dietary levels increase or decrease. The newborn has tissue levels of cobalt quite comparable to adults of the species; thus, cobalt is readily transported across placental membranes. In animals that synthesize vitamin B_{12} in the intestinal tract, tissue levels of cobalt indicate the animal's B_{12} status when cobalt is ingested orally. When cobalt is injected, tissue levels may be high but vitamin B_{12} level low because cobalt was not at the site of synthesis of vitamin B_{12}. In the latter case, vitamin B_{12} content of the blood or of the rumen would be a better indication of B_{12} adequacy in ruminants.

Cobalt, or more specifically, vitamin B_{12}, is absorbed in the proximal portion of the digestive tract in ruminants. In monogastrics in whom B_{12} is synthesized distally, it is also absorbed in that region. As is indicated in Table 17–1, the absorption of cobalt (vitamin B_{12}) is enhanced by several conditions: namely the intrinsic factor (probably a mucoprotein which is found in the gastric juice of "healthy" individuals); compounds such as sorbital; adequacy of iron and pyridoxine; deficiency of folic acid; and, low temperature.

The primary, and perhaps the only, function of cobalt is the synthesis of vitamin B_{12} (cobalamin). Cobalt represents about 4 percent of the vitamin B_{12} molecule. Carnivores can derive vitamin B_{12} from their diets. Since higher plants do not contain vitamin B_{12}, it must be synthesized in the digestive tract of herbivores, thus necessitating the dietary presence of cobalt.

Since most monogastric animals do not synthesize vitamin B_{12}, a cobalt deficiency, to date, has been limited to ruminants. When a cobalt deficiency exists, appetite is lost, and progressive wasting, polychromasia, and macrocytic anemia ensue.

Cobalt is excreted primarily via the bile and is lost in the feces. This route of excretion permits animals, particularly monogastrics, to synthesize B_{12}, absorb it, and thereby recycle cobalt.

2. COPPER

Copper is an integral part of many body processes. Biologists have long recognized that copper is a constituent of different biological materials, but it was not until the latter part of the nineteenth century (1897) that copper was reported to function as a respiratory compound. About thirty years later, copper was shown to be necessary for hemoglobin formation (Hart *et al.*, 1928). The exact role of copper in hemoglobin synthesis has not been completely explained.

Higher animals contain approximately 2 ppm copper in their tissue at maturity. The young of the same species contain about twice this level. Comparable tissue from different species often varies widely in its copper content. It is believed that this is more nearly related to their dietary regimen than to species differences *per se*.

Copper absorption occurs in different segments of the intestinal tract, depending upon species and diet. In the rat and perhaps other animals, considerable copper is absorbed from the stomach. In other species under different dietary regimens, copper may be absorbed from the colon. An acid condition is conducive to absorption of copper while molybdenum, sulfides, sulfates, and zinc apparently reduce its absorbability.

Plasma copper is loosely bound to protein while being transported throughout the body. This copper may be stored, primarily in the parenchymal cells of the liver, heart, and kidneys, or released for incorporation into copper-containing enzymes or compounds such as hemocuprein and ceruloplasmin. Copper is a cofactor for tyrosinase, ascorbic acid oxidase, cytochrome oxidase, uricase, plasma monoamine oxidase, and other enzymes. Additionally, copper is essential for erythropoiesis and it also increases iron absorption.

Through tyrosinase, copper is vitally concerned with melanin metabolism; thus, it is concentrated in such areas as the iris and hair follicles. Wool or hair looses its characteristic color when copper is deficient. If the deficiency is severe the line of color demarcation is visibly discernible and the characteristic crimp in wool is lacking.

Apparently molybdenum and copper compete for the same metabolic sites. Accordingly, when excess molybdenum is present copper is excreted to the extent of causing a deficiency. Deficiencies of copper lead to impaired reproduction through fetal death and resorption in some species. In other species, neonatal ataxia results. The latter may reflect a borderline deficiency while the former would result from a gross deficiency.

A copper deficiency also may result in extensive skeletal abnormalities. This is typified by bowing of the legs and an enlargement of joints. These limbs are more fragile than normal bones. Microcytic hypochromic anemia develops in most species, while in swine a normocytic normochromic anemia characterizes a copper deficiency.

Copper is excreted primarily through the bile into the intestinal tract. Copper, like many of the other elements, may be recycled many times before being lost from the body. A small amount is lost in urine and through dermal pathways.

3. FLUORINE

The essentiality of fluorine for growth has not been unequivocally established in either human or domestic animals. On the contrary, many data have accumulated since 1945 which show definite benefits to the human population in reducing dental caries and malocclusions. The mechanism by which fluorine exerts this action is not clear, but it has been shown to cause larger, more perfectly formed hydroxyapatite crystals when fluorine is present in adequate amounts. Use of fluorine to maintain dental health in most animals other than man seems not too relevant. This is particularly true of economic species kept only short periods for meat, milk, wool, and hair production.

Fluorine is present in many foods, particularly those from the sea. Animals readily absorb this element and accumulate it in the bone proportional to intake. Since fluorine is cumulative, bones of the older animals in a species have a higher concentration of this element.

Some loss of fluorine occurs, primarily through the urine. The loss, however, is seldom equal to the intake of fluorine.

4. IODINE

Iodine, like iron, has long been recognized in human medicine. More recently, its efficacy in preventing goiter and regulating metabolic rate in animals has been recognized.

Iodine is distributed throughout the body. Some occurs in every tissue, but the thyroid "traps" most of the iodine entering the body and, therefore, most of the iodine is in the thyroid gland rather than being equally distributed throughout the tissues. The thyroid represents approximately 0.2 percent of the body weight, yet approximately 75 percent of the iodine is held in this gland. Although the percentage of total body iodine remains relatively stable, the total amount

in any tissue varies with intake, age of the animal, and, to some extent, thyroid gland activity.

Iodine is not only well absorbed throughout the whole digestive tract but also appears to be well absorbed through the skin. Little iodine is found in the feces, indicating that almost all iodine administered orally is absorbed. Iodine in foods is almost entirely in the form of inorganic iodide. It apparently is absorbed best in this form.

Iodine in conjunction with tyrosine is needed for the synthesis of thyroxine. It is primary in the regulation of metabolic rate of the organism. Thyroxine is often regarded as the only compound having thyroidal efficacy, but other iodonated compounds and some with bromine substituted for iodine have efficacies higher than thyroxine. The thyronine nucleus is essential for activity while substitutions in the 3, 5, and 3′ positions are essential for maximum efficiency (Underwood, 1962). Figure 17–2 depicts the iodine circuit in the animal body.

The amount of iodine ingestion *per se* does

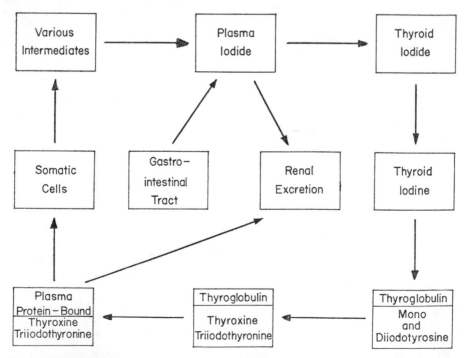

FIG. 17–2.   Schematic drawing of the metabolic pathway of absorbed iodide.

not insure against a deficiency. Some persons or animals in "goiter belts" have higher intakes of iodine than do those in "nongoiter belts," yet a higher percentage of the former group shows indications of a deficiency, *e.g.*, hyperplastic goiter, low metabolic rate, and cretinism. Thus, the synthesis and subsequent function of thyroxine are much more involved than the mere bringing together of tyrosine and iodine.

Iodine deficiency in domestic animals may result in the offspring either being born dead or dying soon after birth. Those that live have enlarged thyroids that are easily noticeable. In swine, not only is the thyroid enlarged, but the bodies are hairless.

Iodine, like other anions, is excreted almost entirely through the urine. Under normal conditions, urinary iodine is sufficiently constant to serve as an indicator of the thyroid function of the animal except in certain metabolic anomalies in which this constancy would not exist.

5. Iron

Iron has been known to be of value nutritionally for many centuries. It is found in amounts ranging from 60 to 90 ppm on a fat-free basis in most adult animals. The newborn of some species, the rabbit particularly, contains approximately twice the level found in the adult. The majority of iron is found in the hemoglobin; it makes up approximately 0.34 percent of this molecule. Considerable concentrations also are present in myoglobin and transferrin. Storage iron is largely in the form of ferritin and hemosiderin. In the liver, the former is preferentially stored until a level of 400 to 600 mg ferritin iron per 100 gm of liver is attained; then it remains relatively constant while hemosiderin stores increase.

Iron is poorly absorbed; the amount stored generally ranges from 3 to 10 percent of that in the diet. It is absorbed primarily in the proximal duodenum, although some is absorbed from other segments of the digestive tract, including the stomach. Absorbed iron is largely in the ferrous state. In some

species (rat and dog), ferric iron appears to be equally well absorbed. Since iron in most plants is in the ferric state, for optimum absorption, it must be released and reduced in most animal species. An acid condition favors the reduction of ferric iron and its subsequent absorption.

Although iron has many functions, one of the major ones is that of being a component of heme (ferroprotoporphyrin). Heme combines with globin to form hemoglobin. When saturated, this compound carries to the tissue one mole of oxygen per mole of hemoglobin or 1.36 ml of oxygen per gram of hemoglobin. In man, the hemoglobin cycle from synthesis to catabolism to synthesis requires approximately 125 days. Iron functions also as a component of cytochrome oxidase and xanthine oxidase.

A deficiency of iron results in hypochromic anemia. In this anemia, the circulating hemoglobin is reduced through either a decrease in erythrocytes or a lowered concentration of hemoglobin in the red cell. The reduced oxygen-carrying capacity results in weakness, fatigue and dyspnea on exertion, headache, and palpitation.

Several factors other than a low dietary concentration may result in a deficiency of iron. Animals that are heavily parasitized, have bleeding ulcers, or have impaired absorption may be deficient in this element. Further, to compensate for the increased requirement or loss, the fast-growing young, the pregnant dams, and females having menstrual cycles require more iron in their diet than do other animals.

The normal animal conserves its body iron by limiting excretion. This low excretion may not be a "metabolically planned" means of conserving iron. Rather, it most likely represents the animal's inability to eliminate this element since the evolutionary process has not provided the animal with a well-defined mechanism for eliminating it. The urine contains only small amounts of iron, about 0.2 mg daily in the human being. While the amount excreted through the bile is greater, a large proportion of the latter may be reabsorbed rather than lost in the feces.

6. MANGANESE

Manganese is distributed throughout the body in minute amounts but tends to concentrate preferentially in the mitochondrial tissue. Tissues having high concentrations of manganese include bone, liver, pancreas, pituitary gland, kidney, brain, and heart, the amounts in each decreasing in the order listed. The total amount stored ranges from 0.17 to 0.29 mg per kg body weight in man (Cotzias, 1958); presumably this amount would be comparable to that in other higher animals. The level of manganese in these tissues can be increased or decreased, depending upon dietary intake. This has been demonstrated in many species.

Manganese is absorbed primarily from the small intestine. Its absorption is decreased by excesses of calcium, phosphorus, or iron. It is poorly absorbed into general circulation unless a tissue deficiency of this element exists. In the absence of a deficiency, manganese may be absorbed and then passed to the liver, where it is returned to the gut by the bile.

Manganese activates bone phosphatase, deoxyribonuclease, and arginase, and may be essential for ATPase and cholinesterase activity. Some enzyme systems activated by manganese may also be activated by other divalent cations, particularly magnesium.

Manganese deficiencies have been shown in most higher animals other than man. A deficiency of manganese has been associated with bone deformity (Dyer & Rojas, 1965). During manganese deficiencies a reduction in the mucopolysaccharide content of preosseous tissue leading to osteogenesis retardation has been observed. The time at which manganese is made available has a profound effect on placental transfer by the dam and on the osteogenesis of her young. Rats supplemented on the fourteenth day of pregnancy produced normal young while those supplemented on the sixteenth day produced ataxic young with poor livability (Hurley &

Everson, 1963). Manganese uptake by the complete genitalia in gestating rats was highest from days 14 to 16 (Rojas et al., 1966). This uptake into the tissue was greater at 4 hours than at 85 minutes or at 8 hours after intravenous injection.

Reproduction is impaired when manganese is limited. In some species, females do not have a normal cycle, while in others the cycles are normal but the animals fail to conceive. Repeated matings fail to settle manganese-deficient cattle until human chorionic gonadotropin is administered. These data (Rojas et al., 1966) and the fact that testicular degeneration has been observed in males on low manganese intakes suggest that manganese also has a function in the reproductive process.

Manganese is eliminated from the body primarily through the bile into the intestinal tract. Very little manganese is excreted via the urine.

7. MOLYBDENUM

Molybdenum is distributed throughout the body, with the concentration varying among the different tissues. Hepatic tissue normally contains from 2 to 4 ppm on a dry matter basis. The kidney has the second highest concentration in most species except the chicken (Higgins et al., 1956). Tissue concentrations of molybdenum can be changed greatly since they are directly proportional to dietary intake. In one study, liver concentrations of molybdenum were increased from 1 to 11 ppm on a dry matter basis by increasing dietary molybdenum from 1 to 30 ppm (Davis, 1950).

Molybdenum is readily absorbed throughout the digestive tract. It is highly absorbed as a molybdate. Excess dietary sulfates reduce molybdenum absorbability. Marked reduction in total body molybdenum was demonstrated by varying both molybdenum and sulfate intakes (Dick, 1956). When molybdenum intake was 0.3 mg per head daily and inorganic sulfate was 0.9 gm, total body molybdenum was 92.9 mg. When the molybdenum was held constant and the inorganic sulfate increased to 6.3 gm, total body molybdenum content was reduced to 16.8 gm. This was due partially to decreased absorption and partially to increased excretion as a result of the inorganic sulfate.

Molybdenum is a cofactor for xanthine oxidase and aldehyde oxidase, both of which are important in purine metabolism. Deficiencies of molybdenum have not been noted in higher animals under normal dietary regimens even when molybdenum antagonists such as tungstate were administered. On the other hand, a molybdenum deficiency was induced in chickens by feeding tungstate (Leach & Norris, 1957). This was done with chicks from molybdenum-depleted hens. Excess molybdenum causes an increased loss and subsequent deficiency of copper.

Absorbed molybdenum, as is true for other anions, is excreted primarily through the kidneys. As noted above, any substance which reduces absorption of this element results in an increased fecal loss.

8. SELENIUM

Selenium, until about ten years ago, was considered a toxic element to be avoided rather than an essential dietary constituent. There was ample reason for this logic because the state of science was not such that selenium could be proved to be an essential element. It was sufficiently concentrated in certain regions to cause high morbidity and mortality in livestock during their sustenance from feeds grown only in seleniferous areas. Man living in seleniferous areas appears not to suffer, possibly because much of his food is imported from nonseleniferous areas or because the unstable organic selenium compounds are lost in the cooking process.

Selenium is readily absorbed, preferentially in the anterior duodenum (Hill, 1968), and transported in the plasma to all parts of the body. This element is stored primarily in the liver and kidneys.

The exact metabolic role of selenium has not been elucidated although it is now regarded as an essential element. Earlier work suggested a specific function separate from that of vitamin E. However, more recent studies indicate that its function is to insure efficient utilization of vitamin E or, more generally, to serve as a nonspecific antioxidant (Oldfield et al., 1963). Its presence in adequate quantity prevents an excess accumulation of calcium, phosphorus, magnesium, sodium, and potassium in the muscle of lambs (Schubert et al., 1961).

Selenium is excreted via several pathways, the greatest loss being through the urine, the next greatest through the feces, and the next through expired air. Animals containing high concentrations of selenium also lose some in hair, hoofs, and dermal decrement.

In the United States, it is not permissible to use a selenium supplement for feeding meat animals because of the element's low requirement and its toxicity when improperly used. Certain natural feeds contain high concentrations of selenium and these may be used in selenium-deficient regions.

9. ZINC

Zinc has been recognized as an element since 1509 (Prasad, 1966) and probably was used by earlier civilizations for making bronze or brass. The need for zinc in rat growth (Todd et al., 1934) and in the prevention of parakeratosis in swine (Tucker & Salmon, 1955) established its essentiality. Subsequent to these findings, zinc has been shown to be needed to support growth and well being in most other species.

Zinc is poorly absorbed; less than 10 percent of the ingested amount is absorbed in most species. Absorption varies, depending upon the form in which the element is ingested. Zinc carbonate, sulfate, and oxide, and metallic zinc are equally well absorbed. Zinc absorption is reduced in the presence of excess phytic acid. The soft tissues constitute a large pool of exchangeable zinc which is in equilibrium with plasma zinc. Tissues such as the liver and kidney cortex which have a high mitochondrial content are highest in zinc content.

Zinc was shown to be a constituent of

carbonic anhydrase long before it was shown to be essential for animals. Zinc constitutes 0.33 percent of this molecule and is essential for its activity. It is also a part of pancreatic carboxypeptidase, liver alcohol dehydrogenase, and glutamic dehydrogenase. Several other enzymes, including arginase, carnosinase, histidine deaminase, lecithinase, enolase, serum alkaline phosphatase, and oxalocetic decarboxylase, may be activated by zinc as well as certain other divalent cations.

Zinc deficiency results in decreased growth, reduced feed efficiency, alopecia, and dermatititic skin lesions. A deficiency of this element is accentuated by excess calcium in the diet.

Zinc is excreted primarily in the feces. The greater portion of that in the feces represents nonabsorbed zinc. Some of the absorbed zinc reenters the gut through the pancreatic duct, with a small amount being excreted through the bile. In one study, less than 5 percent of ingested zinc was excreted in the urine, indicating a limited ability of the kidney to void this element (Underwood, 1962).

Examples of various mineral deficiencies may be observed in Plates 21, 22, and 23.

III. FACTORS AFFECTING REQUIREMENTS

At the tissue level, domestic animals require essentially the same amount of minerals per comparable unit of function performed. This does not imply, however, that the dietary intake should be equal. On the contrary, differences in the percentage concentration of the diet should be expected. Factors affecting dietary mineral requirements and examples follow:

(a) The amount of food ingested per unit of weight. The young bovine will consume feed equal to approximately 3 percent of its body weight, while the rat will consume feed in quantities approximately 7 percent of its body weight. Assuming other factors affecting requirement and the dietary mineral concentration to be equal, the rat will be re-

ceiving 2.33 times more minerals per unit of weight than the bovine.

(b) Growth rate. A change in growth rate or any other function involving a mineral will change the amount of that mineral required for the particular function. Thus, animals growing rapidly require more minerals than those growing slowly or those fed for maintenance alone.

(c) Species. Some species are more productive than others, and the products of some are high in a given mineral. As an example, milk and the whole egg (including shell) are high in calcium. Such species-dependent functions require extra precautions in the mineral nutrition of the organism.

(d) Ambient temperature. Animals using evaporative cooling to lose body heat require more sodium in a hot climate than would be required by the same animals in a cooler climate. The chlorine loss would be proportional to the sodium loss.

(e) Kind of feed. Natural feeds are not digested completely. If the feed is not digested, the minerals contained therein are poorly digested; hence, a higher dietary intake will be required to meet the mineral needs. Examples of foods that are poorly digested are those containing a high percentage of lignin or cellulose.

(f) Form in which the element is ingested. Some forms of an element are much more available to the animal than others. Various examples are listed in Table 17–1.

(g) Dietary balance of other nutrients. The level of fat in the diet greatly affects the amount of calcium available for absorption. The formation of calcium soaps may affect not only the calcium absorption but also the absorption of other elements through a change in intestinal pH.

(h) Level of antagonists in the ration. The amount necessary to ingest is dependent not only upon the total requirement but also upon the amount of the antagonist which may reduce the percentage absorbed or metabolized.

(i) Age. Younger animals absorb a higher

Table 17–3. Signs and Symptoms of Deficiencies and Excesses of Minerals

Elements	Deficiencies	Excesses
	A. MACROELEMENTS	
Calcium	Reduced growth, particularly of bone; osteoporosis and osteomalacia; hyperirritability and tetany; hemorrhage; parathyroid enlargement	Idiopathic hypercalcemia; milk alkali syndrome; hypercalcuria; renal calculus
Chloride	Alkalosis; deficiency of potassium; renal lesions; achlorhydria; hyperexcitability	
Magnesium	Irritability of CNS; susceptibility to atherosclerosis in some species; vasodilation	Depression of CNS; depression of cardio-vascular system; reduces renal oxalate stone deposition
Phosphorus	Decreased body growth; decreased bone growth; rickets or osteomalacia; calcium citrate renal calculi; renal inanition	
Potassium	Adrenal hypertrophy; postprandial hyperglycemia; reduced tissue glycogen; increased renal ammonia production; increased susceptibility to infection; growth reduction	Hypertrophy of zona glomerulosa; hyper-kalemia
Sodium	Decreased growth	Hypertension; degenerative disease of arterioles and glomeruli
Sulfur	Reduced methionine, cysteine, thiamine, and biotin synthesis	
	B. MICROELEMENTS	
Cobalt	Loss of appetite; emaciation (condition indistinguishable from general inanition); listlessness; macrocytic anemia with polychromasia; incoordination	
Copper	Anemia (type, species dependent); depressed growth; hair or wool depigmentation; bone anomalies; neonatal ataxia; reduced reproductive efficiency	Hemolysis, jaundice, and death (species-dependent)
Fluorine	Enamel density reduction	Teeth change in shape, size, color, orientation, and structure; loss of appetite and death
Iodine	Goiter; reduced rate of cellular oxidation; reduced mental ability and physical stature; reduced integument integrity	
Iron	Hypochromic microcytic anemia; listlessness and fatigue; depressed growth; reduced resistance to infection	Excesses not absorbed in normal subjects
Manganese	Defective ovulation; testicular degeneration; congenital deformity; perosis in birds	
Molybdenum		Scouring; loss of weight; discolored hair coat
Selenium	Liver necrosis (some species)	Loss of hair, particularly from mane and switch; soreness of feet and sloughing of hoofs; incoordination (blind staggers); atrophy of heart; cirrhosis of liver; excess salivation
Zinc	Lesions of skin, mucocutaneous junctions, and epithelial structures; atrophic seminiferous tubules; retarded growth of testes and secondary sex organs in male; reduced growth in some species	

percentage of minerals than do older animals of the same species. Also, they require more minerals because of the growth process.

(j) Health of animal. Anomalies affecting metabolism influence the dietary requirement for minerals (Table 17–3).

Since there are so many factors affecting requirements for minerals, specific dietary needs are applicable for one and only one set of conditions. Hence, undue credence should not be placed on a published list of mineral requirements for any species. The "requirements" listed in Appendix VII are intended to serve only as a guide in formulating mineral mixtures for the maintenance and growth of animals.

IV. RESEARCH TECHNIQUES

Newer techniques for analyzing plant and animal tissue for their mineral content have resulted in rapid advances in understanding mineral metabolism in animals. Until recently, elements were determined in the laboratory by using time-consuming chemical separations. The time required was so great that a minimum number of samples were analyzed and generally only one or, at best, a few elements were determined on a given sample. This method retarded progress in general, particularly in the discovery of the myriad of mineral interrelationships.

Mass spectroscopy offers an excellent means of scanning samples for their mineral contents. This system of analysis is sometimes criticized in that it is more subjective than other methods and, therefore, less accurate and precise. Atomic absorption spectroscopy is less subjective. It too is rapid, but only one element can be determined at a time. One limitation to both of these procedures is that they are not highly sensitive; therefore, certain trace elements cannot be quantitated when they apppear in submicro amounts. Neutron activation analysis, especially the procedure for nondestructive analysis of the sample, appears to be the method of choice of most researchers studying mineral metabolism in animals. The nondestructive procedure shows promise

of being rapid, objective, very sensitive, and precise. When this procedure is fully developed, it will be an epoch in the history of mineral nutrition.

REFERENCES

Byrd, C. A., & Matrone, G. (1965). Investigation of chemical basis of zinc-calcium-phytate interaction in biological systems. Proc. Soc. Exp. Biol. Med. *119*:347–349.

Capen, C. C., Cole, C. R., & Hibbs, J. W. (1968). Influence of vitamin D on calcium metabolism and the parathyroid glands of cattle. Fed. Proc. *27*: 142–152.

Comar, C. L., & Bronner, F. (Eds.). *Mineral Metabolism.* Vol. I A(1960), B (1961), Vol. II A (1964), B (1965), New York, Academic Press.

Cotzias, G. C. (1958). Manganese in health and disease. Physiol. Rev. *38*:503–532.

Davis, G. K. (1950). The influence of copper on the metabolism of phosphorus and molybdenum. *Symposium on Copper Metabolism.* W. D. McElroy & B. Glass (Eds.), p. 216–229, Baltimore, Johns Hopkins Press.

Dick, A. T. (1956). Molybdenum in animal nutrition. Soil Sci. *81*:229–236.

Dyer, I. A., & Rojas, M. A. (1965). Manganese requirements and functions in cattle. J. Amer. Vet Med. Ass. *147*:1393–1396.

Follis, R. H., Jr. (1958). *Deficiency Disease.* Springfield, Ill., Charles C Thomas.

Hart, E. B., Steenbock, H., Waddell, J., & Elvehjem, C. A. (1928). Iron in nutrition. VII. Copper as a supplement to iron for hemoglobin building in the rat. J. Biol. Chem. *77*:797–812.

Higgins, E. S., Richert, D. A., & Westerfield, W. W. (1956). Molybdenum deficiency and tungstate inhibition studies. J. Nutr. *59*:539–559.

Hill, C. H. (1968). Studies on the absorption of [75]selenium in chicks. Fed. Proc. *27*:417.

Hurley, L. S., & Everson, G. J. (1963). Influence of timing on short-term supplementation during gestation on congenital abnormalities of manganese deficient rats. J. Nutr. *79*:23–27.

Leach, R. M., & Norris, L. C. (1957). Studies on factors affecting the response of chicks to molybdenum. Poultry Sci. *36*:1136.

Lubin, M. (1964). Intracellular potassium and control of protein synthesis. Fed. Proc. *23*:994–1001.

Michels, F. G., & Smith, J. T. (1965). A comparison of the utilization of organic and inorganic sulfur in the rat. J. Nutr. *87*:217–220.

Oldfield, J. E., Schubert, J. R., & Muth, O. H. (1963). Implication of selenium in large animal nutrition. J. Agri. Food Chem. *11*:388–390.

Oser, B. L. (1965). *Hawk's Physiological Chemistry.* 14th ed., New York, McGraw-Hill.

Peisach, J., Aisen, P., & Blumberg, W. E. (1966). *The Biochemistry of Copper.* New York, Academic Press.

Pitt-Rivers, R., & Trotter, W. R. (1964). *The Thyroid Gland.* Vols. I & II, London, Butterworth.

Prasad, A. S. (Ed.) (1966). *Zinc Metabolism.* Springfield, Ill., Charles C Thomas.

Rojas M. A., Dyer, I. A., & Cassett, W. A. (1966). Effect of stage of pregnancy on manganese transfer in the rat. Proceedings of the Eighth International Congress on Nutrition, pp. 265–266, Hamburg, Germany.

Rosenfeld, I. & Beath, O. A. (1964). *Selenium.* New York, Academic Press.

Sandstead, H. H. (1968). Present knowledge of the minerals. *Present Knowledge in Nutrition.* 3rd ed., pp. 117–125, New York, The Nutrition Foundation, Inc.

Schaefer, K. E., Hassen, M., & Niemoller, H. (1961). Effects of prolonged exposure to 15% CO_2 on calcium and phosphorus metabolism. Proc. Soc. Exp. Biol. Med. *107*:355–359.

Schubert, J. R., Muth, O. H., Oldfield, J. E., & Remmert, L. F. (1961). Experimental results with selenium in white muscle disease of lambs and calves. Fed. Proc. *20*:689–694.

Stott, G. H. (1968). Dietary influence on the incidence of parturient paresis. Fed. Proc. *27*:156–161.

Telle, P. O., Preston, R. L., Kintner, L. D., & Pfander, W. H. (1964). Definition of the ovine potassium requirement. J. Anim. Sci. *23*:59–66.

Todd, W. R., Elvehjem, C. A., & Hart, E. B. (1934). Zinc in the nutrition of the rat. Amer. J. Physiol. *107*:146–156.

Tucker, H. F., & Salmon, W. D. (1955). Parakeratosis or zinc deficiency disease in pigs. Proc. Soc. Exp. Biol. Med. *88*:613–616.

Underwood, E. J. (1962). *Trace Elements in Human and Animal Nutrition.* 2nd ed., New York, Acdemic Press.

Chapter 18

Vitamin Requirements

By L. S. Jensen

VITAMINS are a group of chemically unrelated organic compounds that are essential for life and a normal rate of growth in animals. The beginning of the vitamin story was in the 1890's, when Eijkman discovered that rice hulls cured a disease in fowl and related this to the disease beriberi in man. An embellished but interesting account of Eijkman's discoveries and those of other early investigators of vitamins is given by Silverman (1958). The first crystalline vitamin was isolated in 1926, and since that time some 15 or more different vitamins or vitamin groups have been chemically identified and shown to be necessary for growth. Prior to the isolation of the first vitamin, the names "fat-soluble A" and "water-soluble B" were used to distinguish between two distinct growth factors for rats. As new factors were discovered, different letters were used for designation or subscript numerals were applied to the letter B. Now that the chemical structure of each vitamin is known, organizations such as the International Union of Pure and Applied Chemistry have recommended the use of distinct names related to chemical structure. Common names such as vitamin A and B_{12} persist, however, not only among the lay public but among scientists as well.

I. SOURCE AND AVAILABILITY

Animals obtain the necessary vitamins for growth in four general ways. The varied concentrations of vitamins in the different feedstuffs comprising the rations of animals are a common source. A second major source is through microbial synthesis in the alimentary tract of animals. Maternal transfer, either through the uterus to the fetus in mammals or to the embryos via the yolk and albumin in birds, is a critical source of vitamins for prenatal growth and viability and growth during the first weeks after birth. Depending on species and appropriate conditions, some vitamins can be synthesized within the tissues of animals, so that no exogenous supply is required for these particular vitamins.

Diversity of importance of microbial synthesis of vitamins in the alimentary tract is well demonstrated in farm animals. On the one hand, monogastric animals such as swine and chickens have only a limited supply of vitamins available through intestinal synthesis, which is related in part, at least, to the rate of food passage and length of the digestive tract in these species. Birds are particularly susceptible to vitamin deficiencies; indeed, they have played a major role in the discovery of many of the vitamins. On the other hand, in ruminants, microbial synthesis of certain of the vitamins in the rumen provides a major source of these vitamins (Table 18–1). All B-complex

Table 18–1. Evidence for Synthesis of B-complex Vitamins in the Rumen of a Holstein Heifer

(*Adapted from Wegner et al., 1941. Proc. Soc. Exp. Biol. Med. 47:90.*)

Vitamin	*Vitamin Content (µg/gm)*	
	Ration	*Rumen Contents*
Thiamine . . .	3.3	4.5
Riboflavin . . .	5.7	24.6
Pyridoxine . . .	2.5	4.5
Pantothenic acid .	10.2	26.6
Niacin	19.0	50.0
Biotin	0.07	0.08

vitamins and one of the fat-soluble vitamins are adequately supplied by this synthesis. The young ruminant, however, must rely upon a dietary source of the B-complex vitamins before the rumen becomes functional and is adequately stocked with bacterial and protozoan organisms which synthesize the vitamins. A monogastric herbivore such as the horse, which has a large cecum where cellulolytic activity occurs, also relies upon microbial synthesis in the intestinal tract for a part of its vitamin supply.

A prime example of differences among species in tissue synthesis of a vitamin is ascorbic acid, which is required in the diet of man and a limited number of other animals, but is synthesized in the tissues by all domestic animals. Several other organic chemicals are required for growth of lower forms of animals but are not termed vitamins because they are not required by higher animals. An example of such a compound is lipoic acid, which is required for the growth of *Tetrahymena geleii* and other microorganisms, but

apparently can be synthesized in the tissues of all higher animals.

All vitamins are classified in two main categories—fat-soluble and water-soluble. Fat-soluble vitamins are soluble in organic solvents such as hexane or diethyl ether but are relatively insoluble in water while those that are classified as water-soluble are relatively insoluble in organic solvents. Although this division of vitamins is made on the basis of a single physical property, the two classes differ significantly in the way in which they function biochemically within the tissues of animals.

A. Fat-soluble Vitamins

Vitamins A, D, E, and K are the fat-soluble vitamins. Although essential fatty acids are not usually included with vitamins, they fit more logically with this class of nutrients and will therefore be discussed here.

A dietary source of vitamin A (retinol) is required by all species of higher animals, and this vitamin is most apt to be deficient in the feeds available for animals. The only source of true vitamin A in natural foods is the tissues of other animals. However, plants synthesize vitamin A precursors such as the carotenes, which can be converted to true vitamin A after being consumed by animals. The main and most efficient site of this conversion is the intestinal wall although there is some evidence that carotene can be converted to vitamin A in other tissues within the body. Vitamin A precursors are pigmented substances which are found in green forages such as alfalfa and grasses, but are generally absent from cereal grains and plant protein concentrates. An exception is yellow corn, which contains cryptoxanthin. Some higher animals such as man and cattle accumulate carotene in the liver and other tissues. There is a marked difference in carotene content of blood plasma and milk among breeds of dairy cattle, with Guernseys having the highest. A number of other animals such as goats, swine, rats, rabbits, and guinea pigs do not accumulate carotenoids in their body tissues, since they efficiently convert carotenes to colorless vitamin A in the intestinal wall. The yellow color observed in birds' eggs is primarily due to xanthophylls, which are not precursors of vitamin A.

Because of the nature of the chemical structures of retinol and its precursors, the vitamin activity of these compounds can be easily lost by oxidation or enzymatic and photochemical action. Livers of marine fish are very rich sources of vitamin A. The oil from these livers constituted a major supplement for animal feeds for two or three decades. Now vitamin A is chemically synthesized and dispersed in a protective matrix or homogenized into solid gelatin-sugar beadlets to retard its decomposition.

Vitamin D is obtained naturally by animals from irradiated feedstuffs or from synthesis within their own bodies under appropriate environmental conditions. Two closely related sterol compounds, 7-dehydrocholesterol in animals and ergosterol in plants, are the major precursors of vitamin D. Exposure of these compounds to ultraviolet irradiation breaks a carbon-to-carbon bond in one of the rings of sterol to form active vitamin D. Irradiated ergosterol is vitamin D_2 (ergocalciferol) and irradiated 7-dehydrocholesterol, vitamin D_3 (cholocalciferol). Either compound is active for most animals, but vitamin D_2 is relatively inactive for chicks, turkeys, and New World monkeys. Green leafy forages exposed to sunlight during the curing process are rich sources of vitamin D_2, but ergosterol is not activated in actively growing plants, so that pasture contains little of the active vitamin. However, 7-dehydrocholesterol in the skin of animals consuming pasture is exposed to solar irradiation and thus provides vitamin D_3 within the tissues.

Fish liver oils and some fish body oils contain relatively high quantities of vitamin D_3. The origin of the large quantity of this form of the vitamin in fish puzzled scientists for a long time because, even though animals can synthesize 7-dehydrocholesterol, most of the saltwater fish do not frequent the upper few meters of ocean water where solar irradiation is effective in activating the sterol. The puzzle was solved recently when it was discovered that fish contain enzymes which rupture the carbon-to-carbon bond providing an effect similar to that of ultraviolet irradiation (Blondin et al., 1964). Irradiated yeast is a good source of vitamin D_2 for animal feeds. Synthetic vitamin D_3 in a stabilized form is now readily available and is used extensively as a supplement in poultry rations.

Vitamin E occurs naturally in several different forms, but alpha tocopherol is the predominant one and the most active physiologically for animals. Cereal grains are fair natural sources of tocopherol but most plant oils are rich sources, with wheat germ oil being one of the richest. Green forages contain considerable vitamin E, but animal products are not particularly rich in the vitamin. Synthetic alpha tocopherol, esterified and treated to increase its stability, is now available for supplementing animal rations.

Vitamin K is the only fat-soluble vitamin synthesized by microorganisms in the alimentary tract, and this synthesis is a major source of the vitamin for most species of higher animals. Birds are the only animals in which a deficiency of vitamin K can be easily developed without using antimicrobial drugs in the alimentary tract or antimetabolites of vitamin K. In some animals such as rats and rabbits, vitamin K is obtained through consumption of feces (coprophagy). Placing these animals in wire-floored cages will not prevent coprophagy because they consume the feces directly from the anus. When cups are placed over the anal region to prevent coprophagy, a vitamin K deficiency is readily developed (Table 18–2). In addition to microorganisms, plants also synthesize vitamin K. Green leafy plant material is a good source of this vitamin. Animal products also contain substantial quantities of vitamin K.

Lineolic acid and arachidonic acid are the

Table 18–2. Development of Vitamin K Deficiency Rats by Prevention of Coprophagy

(*Adapted from Barnes and Fiala, 1959. J. Nutr. 68:603.*)

	Prothrombin Time (in seconds)	
Weeks	Coprophagy Allowed	Coprophagy Prevented
1	18	28
3	17	78
5	17	80
7	17	100

* Weanling rats were fed a purified diet with no added K and coprophagy was prevented by covering the anus with small plastic cups.

main essential fatty acids, but linolenic acid is also partially effective in preventing some aspects of an essential fatty acid deficiency. All of these compounds are long-chain fatty acids containing two or more double bonds. Plant oils such as corn, wheat germ, and safflower are rich sources of linoleic acid, and in general the cereal grains and other plant feedstuffs available for ration formulation adequately meet the needs of animals for these nutrients. Feedstuffs of animal origin are not generally good sources of these factors, but the fatty acid composition of carcass fat of monogastric animals such as the pig and chicken directly reflects the dietary composition of fatty acids. Pigs fed a diet with a high content of polyunsaturated fatty acids will have a "soft" fat that is liquid at room temperature.

B. Water-soluble Vitamins

The water-soluble vitamins include the B-complex vitamins, which are composed of thiamine, riboflavin, niacin, pyridoxine, pantothenic acid, cobalamin, folic acid, and biotin. Ascorbic acid (vitamin C), choline, and inositol are also included in this category. The B-complex vitamins are readily synthesized by microorganisms in the rumen of cattle, sheep, and other polygastric animals so that a dietary source of these vitamins is generally unnecessary. Cobalamin (vitamin B_{12}) and choline are the only B-complex vitamins that may be inadequately supplied by ruminal synthesis.

Cobalt is an integral part of the vitamin B_{12} molecule (Fig. 18–1), and if this mineral element is deficient in the ration, the microorganisms are unable to synthesize sufficient vitamin B_{12} to supply the host animal. Intestinal synthesis of B-complex vitamins also serves as an important source of these vitamins in nonruminants, either by direct absorption of the vitamins from the alimentary tract after synthesis or by coprophagy. However, a dietary source is necessary for most monogastric animals.

Niacin differs from other B-complex vitamins in that it can be synthesized in most species of higher animals from the amino acid, tryptophan, through a series of enzymatically catalyzed steps. The cat is a notable exception in that tryptophan is not converted to niacin by this species. A reli-

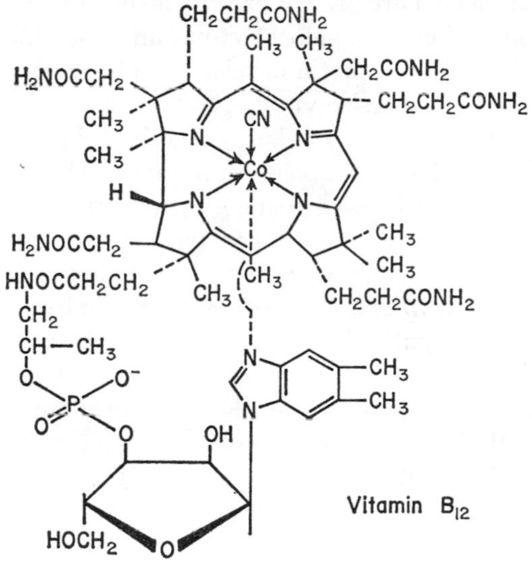

Vitamin B_{12}

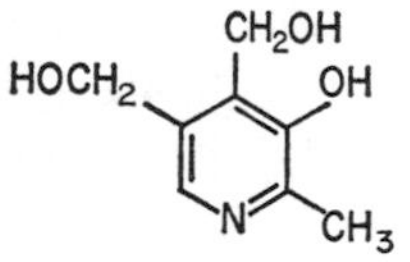

Pyridoxine

FIG. 18–1. A comparison of the chemical structures of two B-complex vitamins. Pyridoxine is relatively simple in relation to vitamin B_{12}. Seven years were required to establish the structure of the latter after its isolation in pure form in 1948.

ance on corn as a major part of the diet accounted for the high incidence of pellagra in people and black tongue disease in dogs in the southern part of the United States prior to the discovery of niacin. Corn is low in both niacin and tryptophan. Enzymes containing vitamin B_6 are required in the conversion of tryptophan to niacin, and evidence exists that other B-complex vitamins, including riboflavin and biotin, are required in this synthetic pathway.

Most higher animals do not require ascorbic acid in the diet since this vitamin is readily synthesized from glucose. Primates, including man and monkeys, guinea pigs, and certain fish species are unable to synthesize ascorbic acid, and thus require a dietary supply. In man, all enzymatic steps necessary for conversion of glucose to ascorbic acid are present except the one for the conversion of L-gulonolactone into ascorbic acid. With a biotin deficiency, rats are unable to synthesize vitamin C.

Choline also can be synthesized in the tissues of higher animals, but apparently the rate of synthesis in young growing animals is inadequate to meet their metabolic needs and, therefore, a dietary source of choline is necessary. This is particularly true in birds. *De novo* synthesis of choline proceeds by way of a phosphatidyl precursor from the amino acid serine and requires the transfer of labile methyl groups from methionine or other methyl donors. Adequate dietary methionine spares the requirement for dietary choline. Folic acid and vitamin B_{12} also are involved in the synthesis of choline since adequate levels of these vitamins in the diet also reduce the dietary requirement for choline. The necessity of inositol as a dietary essential for animals is somewhat questionable. Whether it is synthesized in the tissues of animals has not been clearly established, but it is known that it can be synthesized by microorganisms in the intestinal tract.

A brief summary of some of the best sources of the water-soluble vitamins is presented in Table 18–3. Certain materials are

Table 18—3. Some Rich Natural Sources of the Water-soluble Vitamins

Vitamin	Source
Thiamine	Yeast, cereal grains, plant protein concentrates
Riboflavin	Yeast, green leafy forages, milk products
Niacin	Yeast, distillers' solubles, rice and wheat bran
Pyridoxine	Yeast, cereal grain, animal tissues
Pantothenic acid	Yeast, alfalfa meal, liver
Cobalamin	Animal tissues
Pteroylmono-glutamic acid	Peanuts, liver, green leafy forages
Biotin	Yeast, distillers' solubles, liver
Ascorbic acid	Citrus fruits, potatoes, green leafy vegetables
Choline	Plant protein concentrates, animal tissues, wheat germ
Inositol	Cereal grains, liver, molasses

particularly rich in many of the B-complex vitamins—for example, dried brewers' yeast. Green, leafy plants are rich in many of the water-soluble as well as the fat-soluble vitamins. A unique vitamin is cobalamin, because it is not synthesized by plants. All of the other vitamins, including the fat-soluble ones, are synthesized either as the true vitamin or as a precursor in plants. Cobalamin can only be synthesized by microorganisms, but is found in animal tissues because animals obtain it from the diet or through microbial synthesis in the alimentary tract. All water-soluble vitamins are available in concentrated form through chemical or microbial synthesis from pharmaceutical companies. They are relatively inexpensive to use as supplements in animal and human diets.

II. BIOCHEMICAL FUNCTION

The fat-soluble vitamins, including the essential fatty acids, were among the first to be established as essential nutrients, but the biochemical role has not been completely established for any of these factors. On the other hand, the B-complex vitamins, which for the most part were discovered after the fat-soluble factors, have quite well established biochemical roles in animal metabolism. The fact that the B-complex vitamins are water-soluble, are components of enzyme systems, and are functional in the metabolism of microorganisms aided greatly in the establishment of their biochemical role. Biochemical research techniques with lipids have lagged behind those with enzymes and other water-soluble compounds. Furthermore, the fat-soluble vitamins, other than possibly vitamin K, are not required in the metabolism of microorganisms.

A. Fat-soluble Vitamins

The only biochemical role that has been clearly detailed for vitamin A in animal metabolism is its function in vision, which was largely evolved over several years by the investigations of Wald (1960). A deficiency of vitamin A in animals results in night blindness, an inability of the eyes to adapt to low intensity vision. The light-sensitive pigment in the rod photoreceptors in the retina of the eye necessary for scotopic vision is rhodopsin (visual purple), which is a combination of a *cis*-isomer of vitamin A aldehyde (retinine) and a protein, opsin. In the presence of light, rhodopsin is bleached through a series of chemical reactions to the protein and all *trans*-retinine (Fig. 18–2). Since 11-*cis*-retinal is synthesized in the tissues from retinol (vitamin A), an insufficient supply of retinol in the diet will cause a reduced level of photochemical receptor pigment in the eye.

Although the function of vitamin A in vision is an extremely important one, it is not essential for life. The discovery of the

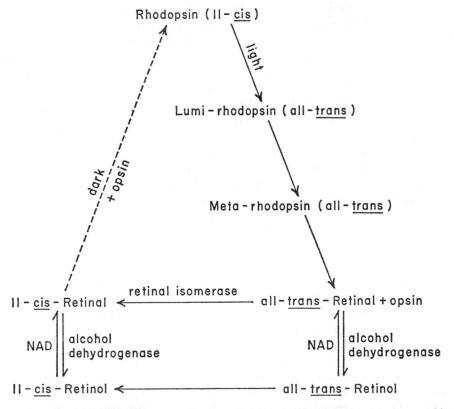

FIG. 18–2. Role of vitamin A in vision at low light intensities. A similar scheme involving the combination of 11-*cis* retinal with a different protein operates at high intensities in the cones of the eye. (*Adapted from Roels In: The Vitamins. Sebrell & Harris (Eds.), Vol. I, 1967.*)

biochemical role of vitamin A in vision has not led to the discovery of the exact biochemical mechanism involved in the growth function of vitamin A. Vitamin A acid (retinoic acid) fulfills most of the functions of vitamin A outside the reproductive system, but will not substitute for vitamin A in the synthesis of rhodopsin. The biochemical function of vitamin A in growth is thus entirely different from that in the vision cycle. Vitamin A is involved in the stability of cells and subcellular particles, in carbohydrate metabolism possibly due to hormone synthesis, and in the metabolism of several lipids such as coenzyme Q, squalene, and other sterols. It also participates in the synthesis of some proteins, but in an unknown manner, in the growth of mucus-secreting cells, and in mucopolysaccharide synthesis. In none of these relationships, however, is it known if vitamin A plays a direct role, and in no case has the exact biochemical function of vitamin A been established.

The biochemical role of vitamin D in animal tissues is involved in facilitating transfer of calcium ions from the mucosal to the serosal side of the intestinal tract. Vitamin D also facilitates the transport of divalent cations other than calcium, but increased absorption and retention of phosphate associated with vitamin D appears to be a secondary effect of the vitamin on calcium absorption. In addition to its effect on calcium absorption, vitamin D may also have some local action in bone and other tissue. The exact biochemical mechanism of vitamin D in calcium transport has not been established. A specific calcium-binding protein appears in the rachitic chick or rat duodenum some hours after administration of vitamin D (Taylor & Wasserman, 1967). Appearance of the same protein also is observed in the uterus of egg-laying, vitamin D-deficient hens following vitamin D administration. Other work indicates that vitamin D is converted to an active metabolite within the tissues. Recent findings support the hypothesis that a vitamin D

metabolite, in combination with a nuclear receptor molecule, initiates the synthesis of a specific messenger RNA, which in turn leads to the synthesis of a functional protein essential for active transport of calcium in the intestine. Other *in vitro* studies suggest a vitamin D-dependent, heat-labile factor in the serum which promotes the uptake of calcium by bone.

Two divergent views exist concerning the biochemical role of tocopherol (vitamin E) in animal metabolism. One is that tocopherol functions only as a physiological antioxidant and is necessary in the tissues to prevent the peroxidation of lipids in cellular and subcellular membranes (Tappel, 1965). Lack of sufficient tocopherol to prevent this uncontrolled oxidation causes the eventual rupture of membranes and release of enzymes peculiar to a subcellular particle. Release of these enzymes into unnatural environments leads to a chaotic condition within the tissues and finally to the gross deficiency consequences observed in various species of animals. The fact that other nutrients interrelated with tocopherol have antioxidant properties and that certain other antioxidants chemically unrelated to tocopherol can substitute for this vitamin in the prevention of some deficiency diseases in animals lends support to this theory of tocopherol function.

A second view of the biochemical role of tocopherol in animals is that this vitamin has at least two functions, one of which is not due to a nonspecific antioxidant property (Schwarz, 1965). Because of its reversible oxidation-reduction properties, investigators have attempted to find a role for tocopherol in electron transport, but there is little evidence to support such a function. The fact that certain vitamin E deficiencies in animals can also be prevented by selenium or sulfur-containing amino acids, while others, even within the same species, cannot, suggests that tocopherol may have more than one function. For example, in rabbits, nutritional muscular dystrophy is prevented by tocopherol, but not by selenium, while in turkeys either selenium or tocopherol is effective. If tocopherol and its interrelated nutrients (selenium and cystine) act only as nonspecific antioxidants in preventing peroxidation in the tissues, it is puzzling why selenium is ineffective in the rabbit.

Vitamin K plays an essential role in normal blood coagulation. Synthesis of three proteins—prothrombin, proconvertin, and thromboplastin—occurs in the liver and requires vitamin K. The exact biochemical mechanism by which vitamin K participates in this process is unknown. These proteins are transported in the blood plasma and all are involved in blood coagulation. Work with microorganisms suggests that vitamin K functions in a respiratory enzyme system by participating in coupled electron transport and phosphorylation.

The exact biochemical role of essential fatty acids in animal metabolism also remains to be determined. Essential fatty acids serve as structural components in various phospholipids. They may play a role in membrane permeability and in lipid transport in the body.

B. Water-soluble Vitamins

Many enzymes require a nonprotein prosthetic group or coenzyme associated with the apoprotein in order to function as biocatalysts. Some of these coenzymes such as heme can be synthesized within the tissues of higher animals. The coenzymes that cannot be completely synthesized in the tissues require one of the B-complex vitamins,

which must then be provided in the diet of the animal (Fig. 18–3). All of the eight B-complex vitamins are integral parts of various coenzymes which are necessary for catalyzing a broad spectrum of biochemical reactions (Table 18–4). The B-complex vitamins are components of enzymes that are required in one or more reactions involved in every major metabolic pathway in living organisms. Some vitamins such as thiamine are specifically involved in carbohydrate metabolism while others such as vitamin B_6 are more specifically involved in protein and amino acid metabolism.

A simplified representation of the various intermediate compounds formed during the catabolism of glucose in the glycolytic and tricarboxylic acid pathways and the steps in which various B-complex vitamins are required are shown in Figure 18–4. The only B-complex vitamin directly involved in any of the catalytic reactions in glycolysis is niacin, but the coenzymes of four of the B-complex vitamins participate in the conversion of pyruvate to acetyl-Coenzyme A. Thiamine pyrophosphate is one of the coenzymes essential for this extremely important reaction in metabolism; it is also essential in some of the reactions of a second major pathway for hexose catabolism—the hexose monophosphate shunt or pentose phosphate pathway. Thiamine thus plays a key role in carbohydrate metabolism, and a lesser role in metabolism of lipids and amino acids. It is not surprising, therefore, that the dietary requirement for this vitamin

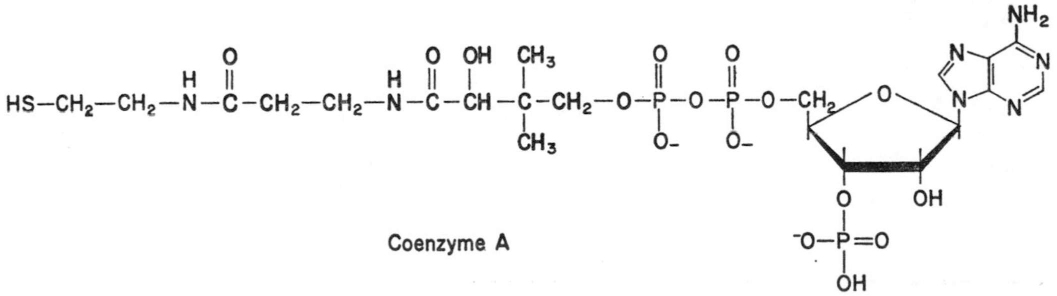

Coenzyme A

FIG. 18–3. An example of a coenzyme having one of the B-complex vitamins as an integral part of its structure. Higher animals are able to synthesize all parts of coenzyme A except for pantothenic acid, which is shown in **bold type.**

Table 18–4. Summary of the Coenzyme Forms of the B-complex Vitamins and the General
Reactions Catalyzed by Enzymes Containing Them

Vitamin	Coenzymes	Enzymatic Reactions	General Functions
Thiamine	Thiamine pyrophosphate (Cocarboxylase)	Nonoxidative decarboxylations of α-keto acids Oxidative decarboxylations of α-keto acids Formation of α-ketols	Carbohydrate metabolism
Riboflavin	Riboflavin mononucleotide (FMN) Flavin adenine dinucleotide (FAD)	Dehydrogenation of α-amino acids, aldehydes, α-hydroxy acids, pyridine nucleotides and other substrates	Electron transport system General energy metabolism
Niacin	Nicotinamide-adenine dinucleotide (NAD) Nicotinamide-adenine dinucleotide phosphate (NADP)	Oxidation-reduction reactions catalyzed by dehydrogenases	Electron transport system Lipid, carbohydrate and amino acid metabolism
Vitamin B$_6$	Pyridozal phosphate (codecarboxylase)	Nonoxidative amino acid transformations such as transamination, decarboxylation, racemization, β-elimination, and α-elimination	Amino acid metabolism
Pantothenic acid	Coenzyme A (CoA)	Two-carbon transfer Reactions involving nucleophilic attack at the acyl carbon atom, additions to unsaturated acyl derivations of acyl-coenzyme A, condensation at the α-carbon of acyl-coenzyme A and acyl group interchange	Carbohydrate and lipid metabolism
Cobalamin	Adeninylcobamide	One carbon metabolism Conversion of methylmalonyl CoA into succinyl-SCoA, biosynthesis of methionine from homocysteine, conversion of glutamate to B-methyl aspartate and conversion of 1,2-propanediol into propionaldehyde	Amino acid biosynthesis Protein and nucleic acid metabolism
Folic acid (pteroylmonoglutamic acid)	Tetrahydrofolic acid	Conversion of glycine to serine, methylation of ethanolamine to choline, methylation of homocystine to methionine, methylation of a pyrimidine intermediate to thymine, introduction of C-2 and C-8 in purine biosynthesis	Transfer of single carbon units at oxidation level of formate, formaldehyde and methanol Biosynthesis of amino acids, choline, purines, and pyrimidines
Biotin	Biotin coenzyme	ATP-dependent carboxylations such as formation of malonyl-CoA from acetyl-CoA and transcarboxylation such as between methylmalonyl-CoA and oxalacetate	Fatty acid synthesis Carbohydrate metabolism Utilization of propionate by ruminants

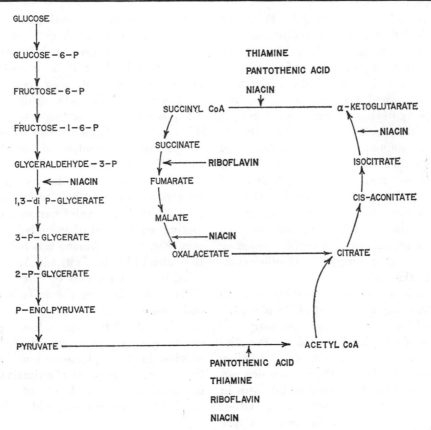

FIG. 18–4. A diagram of the glycolysis pathway and citric acid cycle illustrating points at which vitamins participate directly in chemical reactions.

is higher when animals are fed a high-carbohydrate diet than when fed a high-fat diet. Pyruvic acid and lactic acid accumulate in the blood of animals deficient in thiamine.

Details of the chemical reactions involving coenzymes containing B-complex vitamins are given in modern biochemistry textbooks. Most of these reactions also require certain metal ions as enzyme activators, thus illustrating the close integration of all of the classes of nutrients required by animals for maintenance and growth of new tissues. Although the general mechanism by which B-complex vitamins participate in animal metabolism has been determined, all of the specific reactions catalyzed by enzymes containing these coenzymes probably have not been identified. For example, nutritional observations indicate the involvement of

vitamin B_{12} and biotin in certain aspects of metabolism for which reactions catalyzed by coenzymes containing these vitamins are not now known.

Ascorbic acid has been known for a long time to prevent scurvy in animals, but the exact biochemical mechanism by which it functions in the tissues of animals to prevent this disease is still not elucidated. Mapson (1967) reviewed the status of this vitamin in biochemical systems. Ascorbic acid is readily oxidized and reversibly reduced. Most investigations attempting to establish its metabolic role have involved experiments around this chemical property. Coenzymes containing niacin and riboflavin are also reversibly oxidized and reduced, and participate in the electron transport system in animals. Attempts to establish a specific role for ascorbic acid in electron transfer in the cell have been unsuccessful. Feeding a guinea pig deficient in vitamin C abnormally high levels of tyrosine results in the appearance of abnormal metabolites of tyrosine in the urine. Administration of ascorbic acid maintains normal tyrosine oxidation, but this property of the vitamin does not appear to be an important factor in prevention of scurvy in animals fed normal levels of tyrosine. Ascorbic acid is necessary for formation and maintenance of collagen and thus connective tissues in animals. It is involved in the hydroxylation of proline to hydroxyproline, which is present in high quantities in collagen. Ascorbic acid may function in the hydroxylation of other compounds in the body.

Choline is a structural constituent of all cells of the body and is part of the phospholipid, lecithin. It is also necessary for formation of acetylcholine, which is essential for transmission of nerve impulses. Choline is a source of labile methyl groups which can be transmethylated to other metabolites in the body to form new compounds. Other nutrients such as methionine can also supply labile methyl groups and thereby reduce the dietary requirement for choline. Choline also has a specific role in the prevention of a leg deformity in birds called perosis.

III. DEFICIENCY AND IMBALANCE

A. Growth

Before they were obtained in crystalline form most of the vitamins were called "unidentified growth factors," a name which attests to their crucial participation in growth regulation. Even though all the raw materials such as amino acids for protein synthesis, minerals for bone growth, and carbohydrates and lipids for energy are available to an animal, growth cannot occur unless adequate levels of the vitamins are also provided. This fact is not surprising in view of the critical role of many of the vitamins as constituents of a host of enzymes catalyzing various reactions which are essential for biosynthetic processes.

Some vitamins are much more critical for growth than others. A deficiency of vitamin K does not greatly influence rate of growth in animals until the animal begins to hemorrhage internally as a result of inadequate synthesis of blood coagulation factors. Adequacy of a diet for this fat-soluble vitamin can be determined better by estimating the prothrombin content of the blood than by weighing the animal. On the other hand, vitamin A markedly influences rate of growth, and average body weight at various time intervals is a valid parameter for estimating requirement of animals for this vitamin (Fig. 18–5). Among rats fed diets with the different levels of vitamin A, growth rate (Fig. 18–5) for the first two weeks does not differ greatly because the animals are using body stores of vitamin A obtained prior to weaning. After two weeks, however, rate of growth greatly tapers off at the lower levels of vitamin A supplementation. This is a typical growth response of animals to administration of varying levels of a vitamin. Similar response curves could be illustrated

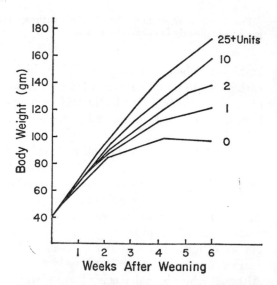

Fig. 18–5. Growth response of weanling rats to different levels of vitamin A. The numbers corresponding to the curve were the number of international units of vitamin A given each rat daily. The upper curve is a composite of those fed 25, 50, 100, and 1,000 international units daily. Approximately 20 rats were fed each level of the vitamin. (*Adapted from Lewis et al., 1942. J. Nutr. 23:351.*)

PLATE 24

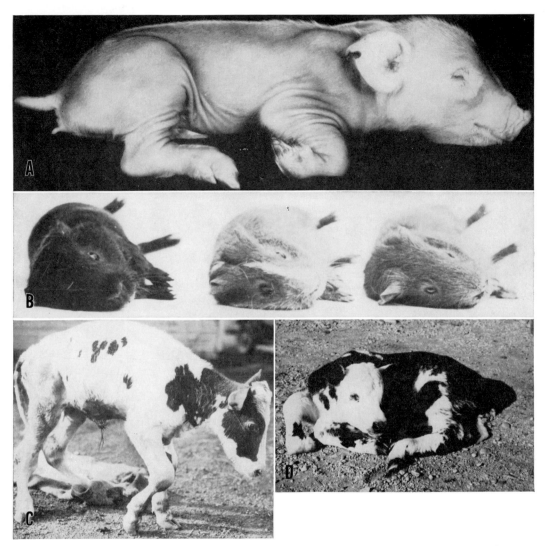

A, This pig was born blind and with other abnormalities as a result of maternal vitamin A deficiency. (*Courtesy, Professor Fred Hale, Texas Agri. Exp. Sta.*)

B, When guinea pigs are fed a diet deficient in tocopherol and supplemented with a source of highly unsaturated fat (cod liver oil), the skeletal muscles degenerate. Because of such a myopathy, these guinea pigs were unable to right themselves after being placed on their sides. (*Courtesy, Dr. L. L. Madsen, Washington State University.*)

C, This calf had rickets which was produced experimentally by feeding a ration low in vitamin D and keeping the animal away from direct sunlight. Note the crippled condition: bowed legs, enlarged joints, and humped back. (*Courtesy, C. F. Huffman, Michigan Agri. Exp. Sta.*)

D, This calf was born of a cow in good condition with no evidence of vitamin A deficiency other than low-blood carotene and vitamin A. The calf was normal in size but was unable to stand alone until 4 days of age and had frequent convulsions. Note the abnormal position of the front legs. It survived only 4 weeks. (*Courtesy, Dr. L. L. Madsen, Washington State University.*)

PLATE 25

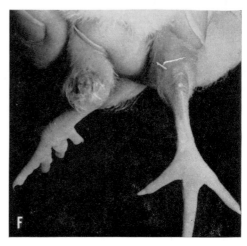

A, Chicks fed a diet deficient in pantothenic acid grow very slowly and develop a characteristic dermatitis. Note severe dermatitis at corner of beak and on eyelid of chick on the left and its smaller size in comparison with the chick of the same age (on the right) fed the same diet but supplemented with pantothenic acid.

B, A characteristic sign of riboflavin deficiency in the chick is "curled toe paralysis." This condition is depicted in the chick on the right. Note that this chick is resting on its hocks in contrast to the upright position of the control bird on the left.

C, The Broad Breasted Bronze turkey poult on the left was fed a diet deficient in folic acid. In addition to a slower rate of growth, the deficiency also causes poor feather development, feather depigmentation, and abnormal leg development. Note the curvature of the right back. The control poult on the right had normal feather and leg development.

D, A vitamin B_6 (pyridoxine) deficiency markedly reduces growth rate. A deficient chick is hyperexcitable and frequently enters a convulsive-like state. Note the uncoordinated appearance of the chick on right in comparison with the normal chick on left.

E, The chick on the right is severely deficient in niacin. Growth rate is markedly depressed and the chick has a severe "watery" dermatitis in the wing region. It also has a severe inflammation of the tongue and the mouth area, which is the most common sign seen in a group of deficient chicks.

F, Chicks hatched from eggs laid by hens fed diets deficient in vitamins frequently show characteristic deficiency signs. Here is a newly hatched chick from an egg with an inadequate level of vitamin B_{12}. Note the malformed left leg (perosis).

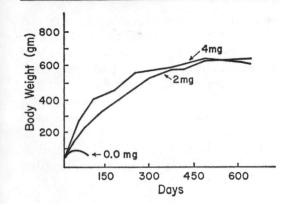

FIG. 18–6. Growth response of rats to pantothenic acid. The levels indicate concentration in mg/kg diet. Those given no supplement were all dead within a few weeks. Note that the growth differential between the two higher levels of vitamin supplementation was greatest during the early stages of growth and disappeared for the most part after one year. (*Adapted from Barboriak et al., 1957. J. Nutr. 61:13.*)

for most of the other vitamins and with many other animal species. The growth response of rats to different levels of pantothenic acid over a longer period of time is shown in Figure 18–6. A much greater spread in growth rate between adequate and marginal levels of pantothenic acid is evident during the stage of rapid growth than during the later stages of growth. In fact, by adulthood the difference in body weight has almost disappeared. These data illustrate a principle which holds for most of the vitamins, that the dietary requirement per unit of food is higher for growth than for maintenance.

A reduced overall rate of growth in vita-

min deficiencies does not imply that rate of growth is slowed in the same relative proportion in all tissues and organs in an individual. For example, in a vitamin B_{12} deficiency in rats, the size of kidneys, liver, and thyroid glands is significantly greater per unit of body weight than that of the same organs in vitamin B_{12} supplemented rats (Table 18–5). Differences are not due to inanition because they persist even when the two groups of animals are limited to the same food intake. Reasons for a disproportionate growth of organs in vitamin deficiencies undoubtedly are varied. In some cases, the disparity reflects the development of mechanisms through evolution which reserve the minimal vitamin supply for synthesis of organs that are most critical for survival. On the other hand, enlargement of the thyroid gland in a vitamin B_{12} deficiency may be a compensating hypertrophy caused by the essentiality of vitamin B_{12} for synthesis of thyroxine.

Just as with minerals, an imbalance of some of the vitamins in the diet can lead to a reduction in growth rate and death of the animal. Toxic levels of two of the fat-soluble vitamins lead to diseases known as hypervitaminosis A and D. Danger of overdosage with other vitamins is negligible. The water-soluble vitamins are not toxic at high levels because in most cases they can be readily excreted in the urine after the immediate needs of the body have been satisfied. An excess of vitamin A in the diet of animals causes an acceleration of bone resorption in

Table 18–5. Effect of Vitamin B_{12} Deficiency on Organ Weights of Male Rats
(*Adapted from Dryden & Hartman, 1966. J. Nutr. 90:377.*)

Organ	Ad Libitum Fed		Pair Fed	
	Without B_{12}	With B_{12}	Without B_{12}	With B_{12}
	gm/100gm Body Weight		gm/100gm Body Weight	
Testes	1.3	1.1	1.1	1.2
Thyroid	0.013	0.008	0.013	0.010
Kidneys	1.8	1.1	1.6	1.1
Liver	7.4	6.9	7.0	5.6
Heart	0.45	0.39	0.44	0.41

the remodeling process and results in bone fragility and spontaneous fractures. Exophthalmia, temporary thickening of the skin, mucous cell formation in keratinized membranes, and hemorrhaging are also consequences of excess vitamin A. Hemorrhaging is associated with a reduced synthesis of prothrombin and can be corrected by elevating the level of vitamin K in the diet. Hypervitaminosis D is characterized by resorption of bone salts and abnormal deposition of calcium in soft tissues of the body. Abnormal calcium deposits are commonly found in the smooth muscle, vascular system, and the urinary and respiratory tracts. Brittle, deformed bones and spontaneous fractures are also a consequence of excess vitamin D. Excess levels of vitamin A and D are more apt to be observed in man than in domestic animals.

B. Characteristic Symptoms

In addition to a generalized effect on rate of growth, deficiencies of vitamins also lead to specific disease conditions which are characteristic of a particular vitamin (Plates 24 & 25). These deficiency signs develop because the vitamin is needed more critically in a particular tissue or organ than in other tissues or organs in the body. A summary of some of the characteristic symptoms, signs, and diseases of vitamin deficiencies in several species of animals is given in Tables 18–6 and 18–7. A marked similarity in deficiency consequences for a vitamin is evident among the species. For example, a deficiency of vitamin B_6 leads to convulsions or epileptiform fits in all animals including man. An epidemic of convulsions in infants in the early 1950's was shown to be due to the use of a proprietary milk substitute which was low in vitamin B_6.

In contrast to the other vitamins, a deficiency of tocopherol takes various forms in different species of animals. This is caused in part by tocopherol's interrelationship with selenium and sulfur amino acids, but even in a combined deficiency of tocopherol and selenium, the deficiency consequences vary considerably among species. In rats and pigs, for example, a degeneration of the liver is observed while in chicks, turkeys, and monkeys, but in no other species, anemia is seen. Even muscular dystrophy is not uniformly observed in all species as a result of a combined deficiency of tocopherol and selenium.

Certain tissues are particularly affected by vitamin deficiencies. Changes in skin (dermatitis) are observed in deficiencies of vitamins A, B_6, riboflavin, niacin, biotin, B_{12}, pantothenic acids, and essential fatty acids. Defects of nervous tissue are seen in deficiencies of vitamins A, E, B_6, B_{12}, thiamine, niacin, biotin, and pantothenic acid. Deficiencies of folic acid, B_{12}, B_6, tocopherol, and niacin cause anemia in one or more species of animals. On the other hand, heart defects are usually only observed in tocopherol and thiamine deficiencies, hemorrhaging in vitamin E, K, and C deficiencies, and eye defects in vitamin A and riboflavin deficiencies.

Although a variety of consequences from vitamin A deficiency is noted among species, these can be accounted for generally by three basic changes. One is a metaplasia of the epithelia to a generalized, keratinized, stratified epithelia. Xerophthalmia, thickened skin, and urates in the kidneys are gross signs associated with this change. The second major change is in bone remodeling, which in turn causes nerve lesions and increased pressure in the cerebrospinal fluid. Thus blindness in calves and deafness in dogs are actually nerve lesions caused by faulty formation of the bones in the head. A third major change is a reduced synthesis of rhodopsin, which causes defective dark adaptation in all species.

Most vitamin deficiency symptoms in animals are reversible by vitamin administration. In some cases a dramatically rapid response to vitamin administration occurs, while in others a latent period of several weeks is required before a definite improvement is observed. In polyneuritis in birds, which is caused by a deficiency of thiamine, a chick in a coma and with retracted head

Table 18–6. Some Characteristic Consequences of Deficiency of the B-complex Vitamins in the Young of Several Species

Vitamin	Cat	Cattle	Chicken	Dog	Horse	Man	Pig	Rat
Thiamine	Anorexia Cardiac disorders	Head retraction Convulsions Incoordination	Polyneuritis	Anorexia Paralysis Convulsions Impaired gastric secretion	Anorexia Incoordination Enlarged heart	Beriberi Polyneuritis Edema Heart disorders	Anorexia Enlarged heart Slight staggering	Anorexia Bradycardia Head rotation "Walking in circles"
Riboflavin	Cataracts	Mouth and nose lesions Excessive lacrimation and salivation Hair loss	Curled toe paralysis	Watery eyes "Bloodshot eyeballs" Fatty liver	Periodic ophthalmia Catarrhal conjunctivitis	Cheilosis Seborrhea	Alopecia Dermatitis Lens opacities	Dermatitis over extremities Eye exudate Hyperkeratosis Alopecia
Niacin	Mouth lesions	Weakness Diarrhea Dehydration	Inflammation of mouth cavity Poor feathering	"Black tongue" Sore, purplish black tongue Nervous symptoms		Pellagra Dermatitis Diarrhea Dementia	Rough hair coat Normocytic anemia Alimentary tract lesions	Rough hair coat Alopecia
Vitamin B$_6$	Anemia Convulsions	Convulsions	Hyperexcitability Convulsions	Microcytic anemia Convulsions Improper heart function	Anemia	Skin lesions Spastic gait Convulsions	Microcytic anemia Spastic gait Convulsions Fatty liver	Acrodynia Microcytic anemia Convulsions
Pantothenic acid	Fatty liver	Inability to stand Diarrhea	Head dermatitis Poor feathering	Intestinal disorders Convulsions		Weakness Numbness in hands and feet	Excess lacrimation Alopecia Dermatitis "Goose stepping"	Achromotrichia Exfoliative dermatitis
Folic acid	Macrocytic anemia Leukopenia		Feather depigmentation Anemia Poor feathering	Anemia		Macrocytic anemia	Normocytic anemia	Leukopenia Anemia Rough hair coat
Vitamin B$_{12}$		Weakness Incoordination	Weakness			Pernicious anemia (adults)	Dermatitis Hyperirritability Posterior incoordination Voice failure	Weakness
Biotin		Paralysis in hind quarters	Food dermatitis Perosis Poor feathering			Exfoliative dermatitis Weakness	Alopecia Skin dermatosis Spasticity of hind legs	Exfoliative dermatitis "Spectacle eye" Achromotrichia Spastic gait

Table 18-7. Some Characteristic Consequences of Deficiency of the Non-B-complex Vitamins in the Young of Several Species

Vitamin	Cat	Cattle	Chicken	Dog	Horse	Man	Pig	Rat
Vitamin A	Weakness of hind legs	Night blindness Lacrimation Total blindness Incoordination	Ataxia Eye lesions Urates in kidneys	Xerophthalmia Deafness	Night blindness Lacrimation Incoordination	Xerophthalmia Night blindness Dry skin	Ataxia Paresis of hind legs Night blindness Spasms Total blindness	Ataxia Night blindness Skeletal defects Xerophthalmia
Vitamin D	Rickets	"Humped back" Bending of forelegs Swollen and stiff joints	Stiff-legged gait Rubbery beak	Bowed legs Irregular teeth		Rickets Bowed legs Bending of ribs	Posterior paralysis Bone deformities	Rickets Leg stiffness
Vitamin E High Se	Steatitis	Muscular dystrophy Heart failure	Encephalomalacia	Muscular dystrophy		Increased susceptibility of red blood cells to hemolysis		Late lactation paralysis Seminiferous epithelium degeneration
Vitamin E Low Se		Muscular dystrophy	Exudative diathesis Anemia		Muscular dystrophy		Hepatosis dietetica	Necrotic liver degeneration
Vitamin K			Hypoprothrombinemia Internal hemorrhages	Hypoprothrombinemia		Hypoprothrombinemia	Hypoprothrombinemia Anemia	Hypoprothrombinemia Internal hemorrhages
Essential fatty acids			Reduced resistance to diseases	Dry, scaly skin Dry hair Susceptibility to skin infections		Skin disorders	Dermatitis Alopecia Skin necrosis	Scaly skin Rough, thin hair root Tail necrosis
Vitamin C	Synthesized in tissues	Synthesized in tissues	Synthesized in tissues	Synthesized in tissues	Synthesized in tissues	Scurvy Internal hemorrhaging Swollen gums Failure of wound healing	Synthesized in tissues	Synthesized in tissues
Choline	Fatty liver	Inability to stand Mild fatty livers	Perosis	Fatty liver			Fatty liver Poor conformation Incoordination	Fatty liver
Inositol								Kidney degeneration Alopecia

can walk around normally within 60 minutes after an injection of thiamine. The drastic effect of thiamine deficiency on the chick results from an accumulation of pyruvic and lactic acids in the tissues (biochemical lesion). Within minutes after the injection of thiamine, cocarboxylase-containing enzymes are synthesized and begin to metabolize the acids. In a chronic deficiency, however, permanent changes in nerves occur which cannot readily be corrected by thiamine administration.

C. Reproduction

Inadequate vitamin nutrition can lead to sterility in both females and males. After fertilization, abnormal growth of the fetus, resorption, abortions, and birth of dead or malformed offspring are the consequences. A chronic vitamin A deficiency will lead to degeneration of the testes and a reduction in production of spermatozoa. In the rat, a deficiency of tocopherol also causes degeneration of the testes. Because of the early observations that this vitamin prevents sterility in this species, it became known as the "antisterility" vitamin. In domestic animals, however, sterility cannot be produced by feeding a diet deficient in tocopherol. An essential fatty acid deficiency also causes male rats to become sterile, but the testicular degeneration and failure of spermatogenesis can be reversed by administration of testosterone. Deficiencies of some of the B-complex vitamins also result in defective spermatogenesis.

Irregularities of estrus, suppression of ovulation, or failure of implantation may result from a vitamin A deficiency. Both rate of ovulation and oviposition are markedly affected in birds by a vitamin A deficiency. A deficiency of some of the B-complex vitamins results in anestrus and degeneration of ova. Although vitamin E deficiency will cause sterility in the male, there is little evidence of this in the female. A deficiency of all of the B-complex vitamins, and vitamins D and A causes a reduced rate of ovulation in birds.

After fertilization, the embryo must rely either on the mother for supply of vitamins or, in the case of birds, on the yolk and albumen of the egg. In mammals, a vitamin A deficiency produces fetal resorption and stillbirths. Young born alive are usually very weak. Experiments have demonstrated that deficiencies of most of the B-complex vitamins also lead to gestation resorption in mammals. In the rat, tocopherol deficiency is necessary for prevention of resorption, but in farm animals this vitamin does not seem to be necessary for normal growth of the fetus. In poultry, a deficiency of most of the vitamins markedly influences hatchability of fertile eggs. In marginal deficiencies of the B-complex vitamins, oviposition rate and fertilization are not necessarily affected, but embryonic mortality is greatly increased. Similarly, a deficiency of tocopherol does not reduce oviposition rate or fertility, but markedly increases embryonic mortality and reduces viability of chicks that do manage to hatch. (Table 18–8). Most of the embryonic mortality occurs during the latter stages of incubation.

Vitamin deficiencies during the early stages of gestation can lead to various defects in the embryo of a teratological nature. The classical studies of Hale (1935) revealed the great importance of vitamin nutrition in prevention of embryonic malformations. Hale fed female pigs vitamin A deficient diets for several months prior to breeding. All of the young from the pigs fed the deficient diets were abnormal in one way or another. Hairlip, cleft palate, accessory

Table 18–8. Effect of Vitamin E Deficiency on Reproduction in the Chicken

(Adapted from Jensen & McGinnis, 1960. J. Nutr. 72:23.)

	Percent with Vitamin E	Percent without Vitamin E
Oviposition rate	49	53
Fertility	68	82
Hatch of fertile eggs	83	55
Post-hatching mortality	5	49

Table 18–9. Teratological Effects Observed in Rats Fed Diets with Deficient or Excessive Levels of Vitamins Prior to and during Pregnancy

Vitamin	Defects
DEFICIENCY	
Riboflavin	Cleft palate, mandible shortening, syndactyly, general bone deformations
Pantothenic acid	Exencephaly, anophthalmia, thorax deformities
Folic acid	Anophthalmia, microphthalmia, cleft palate, spina bifida, hydrocephalus, cleft lip
Vitamin E	Hydrocephalus, retrolental fibroplasia, microphthalmia, cleft lip, cleft palate, syndactyly, ectocardia
Vitamin D	Skeletal defects
Vitamin A	Diaphragmatic hernia, ocular defects, hydrocephalus, cardiovascular anomalies, pseudohermaphroditism, kidney defects
Vitamin B_6	None
Choline	None
Biotin	None
Thiamine	Exencephaly
EXCESS	
Vitamin A	Exencephaly, spina bifida, hydrocephalus, anophthalmia, microphthalmia, exophthalmos cataracts, skeletal defects, cleft palate

ears, absence of eyes, abnormal smallness of the eyes, undescended testes, and ectopic ovaries were some of the abnormalities observed. These observations prompted considerable investigation on the influence of other vitamin deficiencies on production of embryonic malformations. A summary of the effects of deficiencies on rat embryos is given in Table 18–9. The teratogenic effect of vitamin deficiencies occurs during the period between implantation and completion of differentiation. A vitamin deficiency following differentiation may kill the embryo or cause the offspring to be born extremely weak, but will not usually result in malformations.

IV. FACTORS AFFECTING REQUIREMENT

Information on quantitative requirement of each vitamin for each species is necessary

in order to guide man in selecting appropriate rations of foods that supply adequate quantities of the vitamin for optimal performance and health. Rats and chicks are easy to work with in establishnig requirements and have served as pilot animals for basic research in nutrition and other fields of biology. Information concerning the quantitative requirements for some domestic animals such as the horse is very limited.

In order to estimate the minimum requirement of a vitamin, groups of animals are either fed rations containing different levels of the test vitamin or are administered daily dosages of the vitamin which vary in concentration. Certain criteria, depending upon species and age of the animal, are used to evaluate which level of the vitamin is adequate for optimal performance. Criteria includes rate of growth, efficiency of feed conversion, blood levels of the test vitamin, reproductive efficiency, enzyme levels in the blood, and prevention of characteristic deficiency signs such as night blindness, low percentage of bone ash, and short life span. Growth rate and feed efficiency have been major criteria used for estimating vitamin requirements. In man, in whom there is danger of permanent defects in the health of experimental subjects, nutritionists have relied on various biochemical measurements of the urine, the blood and the tissues which reflect changes in normal metabolism. For example, blood levels of transkelotase activity are used to estimate dietary adequacy of thiamine in human beings.

Quantitative requirement of animals for vitamins is expressed either as the amount of vitamin required per unit of food or per unit of body weight daily. Preference for method of expression relates to the manner in which the particular species is normally fed. In animals such as the rat or chick, that are fed *ad libitum*, requirements are expressed per unit of feed, but in animals such as dairy cattle that are fed a specified amount of feed each day, the daily requirement per animal is preferred.

A. Species

The qualitative requirement for vitamins is greatly influenced by the particular species involved. The presence or absence of specific enzymes in the tissues for synthesis of ascorbic acid from glucose determines whether this vitamin is qualitatively required by the species. All domestic animals are able to synthesize vitamin C in their tissues, and therefore do not require a dietary source of this vitamin. Vitamin D, choline, and perhaps inositol are also synthesized within the tissues under certain conditions. The quantities of these vitamins synthesized are not always adequate to meet the needs of the animal, however. All of the other vitamins are apparently not synthesized in the tissues of any species of domestic animal, and therefore must be provided from an exogenous source. In species with functioning reticulorumens, all of the vitamins other than vitamins A, D, E, choline, and the essential fatty acids can be obtained through synthesis of microorganisms in the alimentary tract so that no dietary source is required. Therefore, the quantitative dietary requirement of the B-complex vitamins for the ruminant is nil, but there is a qualitative requirement for the vitamins in the tissues of the animal.

The quantitative requirement of animals for vitamin A is influenced by the relative efficiency of conversion of carotene and other carotonoids to vitamin A among the species. An international unit of vitamin A is defined as the activity equivalent to 0.344 μg of vitamin A acetate. In studies with the rat at dosage levels near the minimum requirement, about twice as much β-carotene has been defined as an international unit of vitamin A. Unfortunately, with many species of domestic animals, efficiency of conversion of β-carotene to vitamin A is considerably less than with the rat. Furthermore, as the level of carotene is increased in the ration, efficiency of conversion also becomes less. In ruminants, 1 mg of β-carotene is considered to be equivalent to only 400 international units of vitamin A rather than 1,677, as defined by the international standard.

The rat and guinea pig do not require a dietary source of vitamin D even in the absence of sunlight if the diet contains adequate levels of calcium and phosphorus in an appropriate ratio. Other species such as the chick do require a dietary source of vitamin D under these same conditions.

B. Age and Physiological State

Within a species the age and physiological state of the animal influence the quantitative requirement for a vitamin. When the requirement is expressed as the daily quantity per animal, requirement of all the vitamins increases as the animal grows older. However, if the requirement is expressed per unit of feed, it is highest for all animals during the first stages of growth and gradually diminishes as the animal approaches maturity. The lowest requirement for vitamins per unit of feed is for maintenance. Requirements increase sharply during gestation and lactation, and, in general, are about the same per unit of feed as those required for early growth (Table 18–10).

Although requirement for vitamins for support of early rapid growth is very high, the level of vitamin A adequate for maxi-

Table 18–10. Effect of Age and Physiological State on Riboflavin Requirement of Swine
(*Adapted from NRC bulletin on Nutrient Requirements of Swine, 1964.*)

	Riboflavin Requirement	
Type of Pig	*mg/animal/day*	*mg/kg diet*
GROWING		
4–12 kg	11.8	3.3
12–24 kg	3.5	3.1
24–35 kg	4.4	2.6
FINISHING		
35–57 kg	5.2	2.2
57–80 kg	66.7	2.2
80–102 kg	7.8	2.2
SOW		
pregnant	9.8	3.3
lactating	18.8	3.3
BOAR, Adult	11.2	3.3

mum growth rate in rats is not adequate for longevity. As little as 2 international units per rat daily is adequate to restore growth in rats deprived of vitamin A, but a level several-fold larger than this is necessary for maximum longevity.

C. Diet Composition

Composition of the ration fed to animals has a marked influence on quantitative requirement of some vitamins. The function of certain vitamins is concerned more with metabolism of a single class of nutrients. When the level of these nutrients is changed, the requirement changes (for example, thiamine and carbohydrates).

The levels of calcium and phosphorus in the diet influence the requirement for vitamin D. With a marginal level of either calcium or phosphorus, increasing the level of the other will increase the requirement of vitamin D per unit of feed. The requirement for vitamin D is minimal, with a ratio of 1 to $1\frac{1}{2}$ part calcium to 1 part phosphorus. Rickets is not produced in rats even when no vitamin D is present in the diet if the calcium to phosphorus ratio ranges between 0.6 to 2.5 and the level of each mineral is adequate. Rickets occurs above and below these ratios, however, unless vitamin D is added to the diet. The requirement for vitamin D increases as the variation from the optimal calcium to phosphorus ratio becomes greater.

The requirement for tocopherol is influenced by the level of polyunsaturated fatty acids in the ration. As mentioned before, tocopherol serves as a physiological antioxidant which helps to prevent the uncontrolled oxidation of polyunsaturated fatty acids in the tissues. Increasing the level of these acids in the ration increases the level in the tissues in monogastric animals and raises the dietary requirement for tocopherol.

D. Nutrient Interrelationships

Several of the vitamins are interrelated with other vitamins and nutrients. Changes in the level of the interrelated nutrients will also affect the quantitative requirement. Need of young growing animals for choline is greater in diets marginal in vitamin B_{12} and folic acid. These latter vitamins are involved in methyl synthesis in the body and methyl groups are essential for choline synthesis. Level of methionine will also influence choline requirement. In baby pigs that were fed a diet containing 1.6 percent methionine, no evidence was obtained for a choline requirement (Nesheim & Johnson, 1950), but at normal levels of this sulfa-containing amino acid, a definite deficiency of choline was observed. Methionine also supplies a labile methyl group which accounts for its sparing action on the choline requirement.

The niacin-tryptophan interrelationship is another excellent example of the influence of diet composition on vitamin requirements. A diet high in tryptophan markedly lowers the requirement for niacin because this vitamin is synthesized in the tissues of animals from the amino acid. An interesting interrelationship exists between tocopherol and selenium. Selenium is completely effective in replacing tocopherol for certain functions such as prevention of muscular dystrophy in calves and lambs on natural diets, but is ineffective for other functions such as preventing gestation resorption in rats.

E. Natural Antivitamin Compounds

A number of factors produced by microorganisms, plants, and some animals are antagonistic to vitamins. Dicumarol, which was isolated from spoiled sweet clover, is a prime example. Dicumarol interferes with the normal functioning of vitamin K in the liver, and, therefore, reduces the level of prothrombin in the blood, resulting in hemorrhagic disease in cattle. Effect of this antivitamin K compound can be counteracted by increasing the level of vitamin K in the ration.

Several antithiamine compounds also are produced in nature. The first example of this was observed in foxes that were fed a ration containing uncooked fresh-water fish. The foxes developed a paralysis (Chastek paralysis) due to a thiamine deficiency. Fresh-water fish contain a thiaminase enzyme which splits the two-ring structures of the thiamine molecule, thus making the vitamin inactive for animals. A number of microorganisms that synthesize thiaminases have been isolated in Japan from the intestinal tracts of human beings with resistant beriberi. Horses that are fed a ration containing bracken fern also develop a thiamine deficiency, but this does not appear to be due to thiaminase since the factor is not destroyed by heat.

Linseeds contain a compound that is similar in structure to pyridoxine and acts as an antimetabolite for this vitamin. Including linseed meal in a ration for chicks markedly causes a reduced growth rate which can be partially counteracted by greatly elevating the vitamin B_6 level in the ration.

Including raw egg albumen in the ration is one way a deficiency of biotin can be developed in most species of animals not requiring a dietary source of the vitamin under normal conditions. Egg albumen contains avidin, a protein which binds biotin in the intestinal tract, and thus prevents absorption. Heat-treating the egg albumen inactivates avidin and allows normal absorption of biotin.

F. Chemotherapeutic Agents and Other Chemicals

Sulfonamides and drugs used to control some microbial diseases can influence the quantitative dietary requirement of animals for vitamins. The antibacterial activity of the sulfonamides is due to their competition with para-aminobenzoic acid for a site in an enzyme system essential for life of the bacterial cell. Para-aminobenzoic acid is part of the folic acid molecule. Sulfonamides influence vitamin requirements primarily by changing the microflora of the intestinal tract so that less synthesis of the B-complex

vitamins is obtained. The level of microbially synthesized vitamins absorbed directly from the alimentary tract or obtained through coprophagy is thus reduced. In growing pigs, for example, a biotin deficiency can be developed by inclusion of sulfathaladine in a purified ration, but no deficiency is observed without the sulfa drug.

Antibiotics such as penicillin and the tetracyclines also influence the requirement of animals for vitamins, but direction of effect cannot always be predicted. Use of antibiotics at nutritional levels to stimulate growth in animals has been shown to reduce the requirement for some B-complex vitamins. Since antibiotics used at these levels do not produce sterile alimentary tracts, a shift to a microbial population that permits a greater synthesis of a particular vitamin is entirely possible.

Other chemicals inadvertently included in the rations of animals can also influence vitamin requirement. A widespread occurrence of hyperkeratosis in cattle in the early 1950's, with symptoms similar to those of a vitamin A deficiency, was caused by highly chlorinated naphthalenes. These chemicals were included in lubricating oils and other materials used on farms which contaminate the animals and/or feed supply. A high nitrate content in certain forages also influences carotene and vitamin A metabolism in animals consuming these forages.

G. Temperature and Other Environmental Factors

Ambient temperature and humidity influence the requirement for vitamins. In pigs, for example, the riboflavin requirement was found to be in excess of 2 mg per kg of feed at an environmental temperature of 42° F while the requirement was about one half that per unit of feed at 85° F.

Animals exposed to sufficient ultraviolet irradiation either through sunlight or lamps producing ultraviolet rays do not require a dietary source of vitamin D. Some animals placed in cages which do not allow access to their feces may develop deficiencies of vitamins synthesized by microorganisms.

H. Other Factors

Genetic differences and presence of diseases and stresses are other factors affecting the quantitative requirement of animals for vitamins. Breeds of small laboratory animals differ in their quantitative requirement for a particular vitamin. Certain genetic mutants require high levels of certain vitamins. A mutant of White Leghorn chickens produces eggs with levels of riboflavin too low to allow normal development of the embryos. Chicks were hatched when riboflavin was injected into the eggs before incubation. Such a genetic defect is self-eliminating under normal conditions of animal breeding. The extent to which requirements for vitamins vary between breeds and between strains within the same breed of animals is not known.

Diseased animals may have higher requirements for the vitamins. A reduced efficiency of absorption of vitamins is evident in animals with intestinal diseases, and loss of vitamin reserves stored in the liver is accelerated in some diseased animals.

REFERENCES

Blondin, G. A., Kulkarni, B. D., & Nes, W. R. (1964). Concerning nonphotochemical biosynthesis of vitamin D_3 in fish. J. Amer. Chem. Soc. 86:2528–2529.

Hale, F. (1935). The relation of vitamin A deficiency to anophthalmia in pigs. Amer. J. Ophthal. 18: 1087–1093.

Mapson, L. W. (1967). Ascorbic Acid. IX. Biochemical Systems. In: *The Vitamins*. W. H. Sebrell, Jr., & R. S. Harris (Eds.), 2nd ed., Vol. I, pp. 386–398. New York, Academic Press.

Nesheim, R. O., & Johnson, B. C. (1950). Effect of a high level of methionine on the dietary choline requirement of the baby pig. J. Nutr. 41:149–152.

Schwarz, K. (1965). Role of vitamin E, selenium, and related factors in experimental nutritional liver disease. *Fed. Proc.* 24:58–67.

Silverman, M. L. (1958). *Magic in a Bottle.* 2nd ed., New York, Macmillan.

Tappel, A. L. (1965). Free-radical lipid peroxidation damage and its inhibition by vitamin E and selenium. Fed. Proc. 24:73–78.

Taylor, A. N., & Wasserman, R. H. (1967). Vitamin D_3-induced calcium-binding protein: partial purification, electrophoretic visualization, and tissue distribution. Arch. Biochem. Biophys. 119:536–540.

Wald, G. (1960). The visual functions of the vitamins A. Vitamins Hormones 13:417–430.

Chapter 19

Water Metabolism

By C. B. Roubicek

I. DISTRIBUTION OF BODY FLUIDS

A. Functions of Water

WATER is by far the largest single constituent of the body, making up approximately two thirds of the total mass of the mammalian organism. Furthermore, of the substances immediately essential for life, water stands second only to oxygen. From this it is evident that water plays a very important part in the existence and activity of the living being. This view is substantiated by the fact that while a fasting animal may survive a loss of practically all its fat and half of its protein, a loss of one fifth of its water content is fatal.

The amount of water in the body is very constant. Its concentration varies from one tissue to another, being least in the dentin of teeth (10 percent) and greatest in the gray matter of the brain (85 percent). The younger and more active the protoplasm, the greater is the amount of water it contains. Relatively small changes in body water cause profound changes in function, and in healthy individuals the fluctuations in water content are scarcely detectable in spite of the fact that the intake of fluid is

intermittent and quite variable. That we are dealing here with efficient regulatory mechanisms is apparent from such simple observations as the constancy of body weight from day to day, the need to void soon after imbibing a large quantity of fluid, and the increased water intake on hot summer days.

No other chemical compound has so many distinct and vital functions as water. This is largely due to water's great solvent power, the fact that water is chemically a neutral substance, and the fact that ionization of most materials takes place more freely in water than in any other medium.

Solvent. Living material, or protoplasm, is an intimate mixture of crystalloids and colloids. Water forms the solvent for the crystalloids and a medium for dispersion or suspension of the colloids.

Medium. Water furnishes a medium for digestion, absorption, metabolism, secretion, and excretion. All these processes, chemical or physical, can take place only in a water medium.

Transportation. Water furnishes a vehicle for the transportation of nutrients, wastes, hormones, gases, and other material.

Hydrolysis. Water takes part in hydrolytic cleavages, such as during digestion.

Temperature Regulation. Water plays a dominant part in equalizing the temperature throughout the body and in maintaining it at a fairly constant value. In this function three physical properties of water are concerned: (1) thermal conductivity, (2) specific heat, and (3) high latent heat of vaporization.

Sense Organs. Water plays an indispensable part in sense organs. Taste and smell are the result of stimulation by chemical compounds in solution. Sound is conducted through the inner ear by a liquid which is chiefly water. The function of the semicircular canals as sense organs of equilibrium depends upon the presence of water in these canals. The transparency of the media of the eye to light is maintained by water.

Lubricant and Cushion. Water serves as a lubricant for moving surfaces such as joints (synovial fluid), the heart, and intestine. Cerebrospinal fluid serves as a cushion for brain and spinal cord.

Respiratory Gas Exchange. Water moistens the surfaces of the alveoli in the lungs for gas diffusion.

B. Composition of Body Fluids

The body water is distributed in distinct compartments. The two major divisions are the *intracellular fluid* (inside the cells) and the *extracellular fluid* (outside the cells). The intracellular and extracellular fluids differ fundamentally in function and composition.

The extracellular fluid is divided into *blood plasma*, which is confined within the vascular system, and the *interstitial fluid*, which occupies the spaces between the cells.

The interstitial fluid surrounds the cells throughout the body, forming an internal environment in which the living cells are suspended. A fractional portion of the extracellular fluid also includes the *synovial, lymph, pleural, pericardial, peritoneal,* and *cerebrospinal fluids,* and the *aqueous humor.*

The rumen and the lumen of the gut normally contain a considerable quantity of fluid. The amount is quite variable, and although this is actually outside of body tissues it may be included in measurements of total body water or extracellular fluid volume. The turnover of digestive fluids per day is greater than the plasma volume.

The fluids in the different compartments are in a dynamic equilibrium with one another (Fig. 19–1). This results in a continual interchange of fluid among the various compartments. The division of water between intra- and extracellular compartments is primarily dependent on the quantity and distribution of the cations. Shifts of water into or out of the cells, without total volume change, can be as harmful to the animal as actual dehydration or overhydration.

The principal cation of the extracellular fluids is sodium. Within the cells, potassium, magnesium, and calcium are the principal cations. Loss of sodium chloride in interstitial fluid will necessitate an equivalent loss of water to keep the remaining fluid isotonic, and excess sodium in the tissue spaces will result in an increase of water volume for the same reason. An important aspect of sodium-ion concentration is in osmotic pressure regulation. The total osmotic pressure of a fluid such as blood plasma is due to the total osmotic effectiveness of all the ions present. Since sodium is the principal cation of plasma, changes in the osmotic pressure of plasma are largely due to, and may be caused by, changes in the concentration of sodium.

Translocations of water in the animal can be explained only to a certain extent by known physical and chemical forces. Osmotic and hydrostatic pressures are important forces in the movement of water. Cells

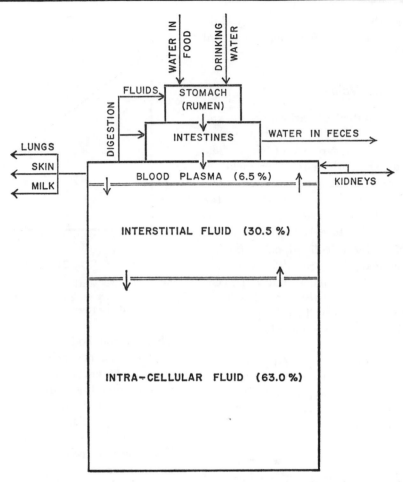

Fig. 19-1.   Fluid compartments and water exchange of a typical mammal.   The maintenance of proper water balance depends upon equal values of water intake and water loss.

lose water and diffusible solutes to a hypertonic medium and gain volume in a hypotonic medium.   The permeability of the cell membrane to the solutes must also be considered.   However, the living membrane may also utilize energy to participate actively in the fluid exchange, causing water and solutes to move in a direction opposite to the concentrate gradients.   The Donnan equilibrium has been associated with a variety of physiological phenomena including certain aspects of the difference in concentration of diffusible ions.   Chemical hydration and electrical potentials are also suspected of affecting the exchange of water within and among tissues, but these mechanisms are only poorly understood.

The inorganic compounds of the body are known to influence such phenomena as cell membrane permeability and hydrogen-ion equilibrium.   The dissolved salts also affect the secretion of digestive fluids, perspiration, and urine, and thereby directly influence water balance.

The stability of volume and composition of the body fluids appears to be based upon the total intracellular water content.   That is, when the fluid balance of the animal is subjected to sudden or extreme stress conditions, the volume and composition of the intracellular compartment are strongly defended aganst change at the expense of the extracellular compartment.   Also, the intracellular compartment is not altered with age.

The relationship of the various body fluid compartments is shown for a mature animal (Fig. 19–1).

C. Body Fluid Volume

The water content of various organs and tissues is very similar in all mature warm-blooded animals regardless of size. The muscles and skin, which together make up 50 to 60 percent of the body weight, account for nearly three quarters of the water in the entire body. Comparative studies of various species from mice to cattle show that the percent of water calculated on a fat-free basis is practically identical. The average is 73.2 percent, with a range of 70 to 75 percent (Chew, 1965). The percent of chemically combined nitrogen as well as the percent of water in the fat-free body mass is essentially constant.

The water content of the fat-free weight of the embryo or newborn animal is much higher than for the mature animal. The point of fat-free water content stability is reached at a constant 4.6 percent of the total life span for all species.

Since adipose tissues have a water content of only about 30 percent, the percent of total body water is greatly influenced by the amount of adipose tissue present. Fat acts as a diluent, taking up space and weight that completely confounds interpretation of data expressed as the water content of total weight. Thus, body water content should be considered on a fat-free basis (Brozek, 1961).

Because of the constancy of water content of the fat-free weight, the determination of the water content of total weight can be used to estimate fat content of the body (Fig. 19–2). That is, the relation between total body water and the percent of total body fat can be expressed as:

$$\text{Percent fat} = 100 - \frac{\text{percent water}}{0.732}$$

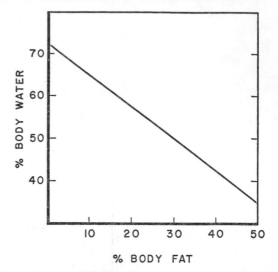

FIG. 19–2. The relationship of body water and body fat content. Water content of the body decreases proportionally with increased fat deposition.

Thus, if the body water content of an animal can be accurately determined, the body fat content can be established. It should be noted that individual animals with nutritional or glandular disturbances which alter the composition of the lean body mass will show deviation from this pattern.

The dilution technique is the standard procedure for estimating total body water content in living animals. A prescribed amount of water-soluble substance (*e.g.*, antipyrene) is injected into the blood stream, and after an equilibration period, the concentration is measured. This dilution is used to estimate the total body water (Kleiber, 1961). Isotype dilution, with deuterium or tritium, has also been used. Since potassium seems to be present almost exclusively in the intracellular fluid, the gamma radiation resulting from K^{40} has been successfully measured in a whole-body scintillation counter to determine intracellular fluid content and, thereby, total body water. The various techniques for estimating extracellular water are more difficult to perform in ruminants because of the large amount of water that may be in the rumen or other parts of the gut (Till & Downs, 1962). This may include 25 percent of total body water.

II. WATER INTAKE

It must be recognized that water requirements of farm animals cannot be specified for a species or even for an individual animal. The water of feces, the water of urine, the water vaporized, and the water of milk are all influenced by such factors as environmental temperature, age and size of animal, diet, activity, and many others. Although information on the water requirements of farm animals is difficult to determine, it is possible to obtain information on actual water consumption. Individual animal total water intake has been obtained in temperature and metabolism chambers. The average water consumption of a group of range animals may be obtained by calculating the water content of stock tanks between given periods, with adjustment for evaporative loss. Total free water intake will include the additional water obtained from the feed consumed. An example of the results obtained from these types of studies has been summarized for cattle (Winchester & Morris, 1956). The average water intake of cattle is a function of dry matter consumption and ambient temperature. At a temperature range of −18° to 5° C, the water intake of cattle is about 1 ml per calorie of heat produced, or, it is at a constant ratio of 3 liters of water/kg of dry matter ingested. Ambient temperature has no influence on water consumption within this temperature range. In sheep, the water to food ratio is much more variable, ranging from 0.114 ml/gm on a low nutritional plane to 0.053 on a high plane of nutrition. Water intake for horses at rest is about 54 ml of water/kg of body weight.

The quantity of water intake per unit of dry matter consumed by cattle increases at an accelerated rate with increasing temperatures above 5° C (Table 19–1). The liters of water consumed per kilogram of dry matter intake more than double between 5° and 30° C. The actual water consumption of nonlactating range cows on a Southwestern range (Table 19–2) shows the im-

Table 19–1. Total Water Intake (Liters) per Kilogram of Dry Matter Consumed for Cattle
(*Data from Winchester & Morris, 1956. J. Anim. Sci. 15:728.*)

Temperature (°C)	4.4	10.0	15.6	21.1	26.7	32.2
Liters of water per kg of feed	3.08	3.33	3.83	4.49	5.18	7.33

portance of adequate potable water during the warm summer months. During this period the water consumption of lactating cows would be expected to increase 4 liters for each kilogram of milk produced, resulting in an intake of over 45 liters per day. Merino sheep kept in unshaded yards drink twelve times more in summer than in winter. In this regard, it has been noted that for range cattle, a distance greater than 1.6 km from water results in a percent utilization of forage of 25 percent compared to 50 percent utilization at 0.3 to to 0.5 km. At 3 km from water, the percent utilization is 15 percent (Valentine, 1947).

There are enormous individual differences in water-consumption response to increasing ambient temperatures. This may in part be explained by differences in lactation status. However, even at moderate temperatures there is over 50 percent difference in water consumption among individual animals (Kelley, 1945; Thompson *et al.*, 1949).

Water consumption increases at a slower rate with rising temperatures for *Bos indicus*

Table 19–2. Water Consumed by Nonlactating Range Cows*

Months with Mean Temperature °C	Water Consumption Daily (liters)
7	14.5
13	17.6
15	26.0
20	33.5

* The importance of adequate water for cattle was recognized as early as 1557 when Tusser in his "A hundreth good pointes of husbandrie" stated: "In summer-time daily, in winter in frost, if cattle lack drink, they be utterly lost." (*Data from Stanley & McCall, 1945. Ariz. Agr. Expt. Sta., Tech. Bull. No. 79.*)

than for *Bos taurus* (Winchester & Morris, 1956). This difference is evident if based on weight or unit of body surface area. In all comparable conditions, the ratio of water drunk to hay eaten is significantly smaller for the Zebu as compared to the Hereford. The decrease in hay eaten resulting from restriction is less for the Zebus. The Zebus drink significantly less water than do Herefords in relation to body weight (Horrochs & Phillips, 1961).

Studies with sheep indicate interactions of water intake with type and quantity of forage and ambient temperature (Kelley, 1945). Goats show increasing water consumption with rising temperature up to 40° C, when failure of the heat regulating system occurs. Dogs, rabbits, rats, and mice all show different levels of maximum temperatures for water consumption increases. Desert mammals are probably not exceptionally well adapted to enduring high temperatures. Many are burrowing animals and remain under ground during the daytime. It does seem likely that desert rodents do not differ materially from their nondesert relatives in their water requirements although many obtain their water from succulent plants or metabolic sources.

At ambient temperatures of $-12°$ C the temperature of the drinking water does not influence the amount of water consumed by cattle and sheep. Warm drinking water in cold weather has no apparent beneficial effects, so that water should be maintained at a temperature no higher than is necessary to prevent freezing (Altman, 1955). Range sheep in Utah show no detrimental effects from using snow exclusively as their source of water. Humidity differences at moderate temperatures have no appreciable effect on water consumption. Increasing humidity and temperatures result in depressed water consumption but increased frequency of drinking.

A. Regulation of Drinking

Since the body water content of the animal is maintained within very narrow limits, it is reasonable to look for some mechanism that regulates water intake and adjusts it to body needs. Water enters the body as a fluid, as an integral part of food, and as water of oxidation. The water obtained from food and oxidation is highly variable in availability, so the immediate regulation of body water constancy is determined mainly by the fluid intake.

The stimuli to drinking may result from several conditions. An increased concentration of body fluids resulting from the loss of water can be quickly corrected by drinking. Transitory shifts of body water can also result in increased concentration in some fluid compartments which will in time be remedied by reserve shifts but may also be immediately corrected by drinking. This type of "fine adjustment" in water balance is most likely to occur when the animals are under confined conditions with water available *ad libitum*. Drinking may also be induced through the nervous system when there is no need to correct water balance. Factors such as emotional stress or taste stimulus may result in the wasteful consumption of inordinate quantities of liquid (Falk, 1961).

Detailed reviews concerned with the subject of thirst show that multiple factors relate stimuli to drinking response (Wayner, 1964; Wolf, 1958). The precise stimuli will vary under different conditions and in different species. However, there are "urge" and "satiety" mechanisms that appear to be generally applicable to mammals.

An association between thirst and reduced salivary flow has long been recognized. This resulted in the formulation of the Cannon "dry throat" theory of thirst. The theory as proposed by Cannon in 1918 stated that thirst is of local origin, a dryness of the mouth and throat, and caused by decreased salivary flow. Subsequent experimental work did show that a decrease in plasma volume caused by a depletion of water content of the body does result in decreased salivary flow and consequent dryness of the mouth. Thus, the dry throat theory has often been used to explain the drinking urge. However, the basic premise that thirst and dry mouth are colligated also has the corollary that thirst should not exist when the membranes are wet. Wolf cites many trials involving various drugs, esophageal or gastric fistulas, and motivational tests showing that salivary flow is not a critical factor in water balance, thirst, or satiety. The dry mouth, then, may be more accurately considered as a cardinal symptom of thirst rather than the actual seat of the thirst mechanism.

Evidence for an osmotic stimuli for thirst is described in detail by Wolf (1958). Early studies showed that osmotic pressure of blood serum rises during water deprivation. It was postulated that this was the result of alteration of osmotic pressure throughout the tissues and that general tissue receptors transmitted the thirst stimuli. Subsequent studies with sodium chloride and urea injections showed that the loss of cellular water is more important than an increase in plasma concentration. Sodium chloride solutions draw water from the cells while urea solutions actually penetrate the cells. Isosmotic solutions of sodium chloride and urea cause about the same rise in plasma pressure, but the sodium chloride solutions result in twice as much drinking. Chew (1965) cites research showing that sodium chloride induces drinking in dogs when it is injected while other solutions of sugars and electrolytes do not. During sodium chloride deficiency, persistent drinking is induced even though body fluids are hypotonic.

Under nonstress conditions the quantity of water ingested by the animal is accurately metered during drinking so that intake offsets deficit. Even after a period of dehydration the animal will ingest rapidly the proper quantity of water and then abruptly stop drinking.

The drinking stops long before the ingested water reaches any tissue in which water is deposited. In fact, drinking may stop before water even reaches the stomach. Dogs with esophageal fistulas drink in defi-

nite proportion to their water deficit and then stop. The water merely passes through the mouth and throat and never reaches the stomach. However, the dogs repeat the sham-drinking process again in a few minutes. Animals deprived of the senses of taste and feeling in the mouth and pharnyx by nerve resection still drink in a normal manner. Water placed directly into the stomach of a thirsty animal greatly reduces the drinking intake but leaves the animal restless and unsatisfied. Thus, the immediate satisfaction of thirst involves an accurate metering of water by the muscles of the mouth and pharynx and a stretch reflex initiated by distention of the stomach.

Since there are marked individual variations in caloric expenditure, it is obvious that the amount of water required to maintain normal hydration will vary in different individuals, and in the same individual from day to day. The actual amount of water required to maintain normal hydration can be ascertained best by noting the amount of water required to maintain a normal or nearly normal urine volume.

B. Free Water in Feed

The total water intake includes the water contained in feed. For "dry" rations such as hay and grain, cattle ordinarily obtain 2 liters or less of water in a day's feed consumption. Because this amount is small compared to the total water intake it is often ignored in determining total water consumption.

In contrast to the small amount of water included in "dry" rations, the water included in the feed when cattle or sheep are on pasture, silage, or other succulent feed may amount to a substantial part of the total water intake of the animal.

In order to estimate the drinking water required by animals while on succulent feeds, the moisture in the feed must be included in the estimate of total water intake. For example, if animals are being fed a ration containing silage with 25 percent dry matter, the animals will ingest 1 liter of

24

water for each 1.33 kg of silage consumed. Depending upon the other ration constituents and total dry matter intake, the water obtained from the silage will account for from 20 to over 60 percent of the total water intake (Table 19–1).

The high water content of immature pasture and range forages may actually be a limiting factor in animal performance. The feed capacity of the animal may not be adequate to permit the intake of sufficient dry matter for the energy requirements of maintenance and growth.

C. Oxidation Water

Although most of the water utilized by the animal is obtained by drinking or as a component of food, there is a further available source provided by the oxidation process. This is known as *oxidation* or *metabolic water*. When an organic molecule is oxidized, its hydrogen goes into the formation of water. The potential water of oxidation depends upon the number of hydrogen atoms and the completeness of the oxidation process of the substance. On the average, carbohydrates will yield 0.50 ml/gm; fat, 1.07; and protein, which is incompletely oxidized, 0.40.

Metabolic studies with cattle and horses show that about 15 percent of the total water intake is obtained from the water of oxidation (Chew, 1956; Tasker, 1967). This is more than is obtained from the feed in a normal "dry" ration.

Although the oxidation process provides water, it also results in additional water loss. Respiration, heat dissipation, and urine excretion all increase because of oxidation. The principal end product of protein catabolism, urea, requires large amounts of water for dilution and excretion. Thus the end result of protein catabolism is that oxidation water is not adequate to meet the additional metabolic needs so the animal must draw on body water stores. Under conditions of acute water shortage a high protein diet will only aggravate the deteriorating body water stores, resulting in a negative water balance. Under certain physiological conditions, met-

abolic water plays an important role in the animal's water economy. It is adequate to meet the needs of hibernating animals and some species of desert rodents. For most mammals, oxidation water production is not a net gain so that increasing metabolism for the purpose of increasing oxidation water is not profitable.

III. WATER LOSS

The constancy of body water content depends upon the fine balance of water intake and water loss. Several sources for water intake have been noted; correspondingly, there are several avenues of water loss.

A. Urine Water Loss

Urine water loss is a direct consequence of kidney function. Urine formation begins with the production of an ultrafiltrate of plasma at the glomerular membrane of the kidney, the energy for this being derived from the work of the heart. In its passage through the tubule of the nephron this filtrate is modified by having substances removed from it by a process of reabsorption or by having material added to it by a process of excretion.

Water is excreted in the urine as a solvent for catabolic products, such as urea, which leave the body through this channel. It also carries a quantity of excess minerals, such as salt, which are excreted by the kidneys. The higher the proportions of minerals and proteins in the diet, the larger is the quantity of water required for excretion in the urine. The type of nitrogenous end product excreted largely determines the species differences in urine water excretion. In mammals the principal end product of protein catabolism is urea which is soluble in water and toxic to the tissues in concentrated solution. Thus much water is required to dilute it to a harmless concentration, remove it from the tissues, and excrete it. Uric acid, the principal nitrogenous end product in chickens and other birds, is excreted in a nearly solid form with a minimum loss of water.

The transport of water and material by the plasma to the kidneys for processing requires a very abundant supply of blood to the kidneys. It is estimated that a quantity of blood equal to all of the blood in the body flows through the kidneys every four minutes. The blood in its passage through the kidneys must traverse two capillary beds in series. In the first set of capillaries, 20 percent of the fluid in the blood is filtered off and undergoes a "purification process" by the kidney cells. After this process the major portion of the fluid is resorbed into the blood stream by the second set of capillaries. Ullrich and Marsh (1963) present a detailed discussion of renal circulation, urine dilution and concentration, and electrolyte transport.

Urine water loss may be considered under the general categories of *obligatory* and *regulatory* (Chew, 1965). The concentrating ability of the kidney and the amount of solutes to be excreted largely determine the obligatory water loss. The concentrating ability can be conditioned by certain previous physiological conditions of diet and rate of metabolism. The neonatal calf appears to be unique in the ability to produce a distinctly hypertonic urine.

Since the quantity of water ingested is normally in excess of bodily needs, the kidneys also function as a spillway for water not dissipated through other channels. This is the regulatory urine volume essential in the balancing exchange of intake and loss. When drinking water is not available, urine flow is reduced to the minimum required for the obligatory functions. Since sweat glands are also a source of water loss, urine loss in herbivores is inversely related to the functional development of sweat glands. Horses which sweat more profusely than cattle or sheep have a correspondingly lower water loss through the urine.

B. Fecal Water Loss

Under conditions of adequate availability of water, fecal water loss varies with species and diet. On a hay ration, sheep, horses, and camels have a drier feces than do cattle. While cattle feces average 80 to 85 percent moisture, the other species range from 50 to 65 percent (Chew, 1965).

The high crude fiber content of the diet of herbivores results in a relatively large fecal water loss. With hay rations, the fecal water loss of cattle is about equal to the urine output. Increasing the fiber content of the diet results in a proportional increase in fecal water content. The use of very succulent or laxative feeds may result in fecal water loss exceeding urine output. All species show a reduction in fecal water content with even moderate water restriction. Under conditions of acute water shortage the fecal water content decreases to about 40 percent.

The mechanisms involved in regulating fecal water loss are not clearly defined. Excess water is removed from the feces while it is in the digestive tract. Osmotic pressure studies in the intestines of the cow show that in regard to blood, the content of abomasum is slightly hypotonic; the content of small intestine, strongly hypertonic; the content of cecum, isotonic; and the content of large intestine, hypotonic. There is a difference between the physiology of the large intestines of European and Zebu cattle. The large intestinal contents have a higher dry matter content in Zebu than in European cattle. This may be a partial explanation of the greater water economy of Zebu cattle.

C. Sweating

The apocrine glands found on many mammals are a source of cutaneous water loss. The principal function of the glands is to provide a source of water for evaporative cooling for body temperature control. The apocrine glands are generally distributed over the skin in association with hair follicles. In cattle and sheep, the total number of follicles is present at birth, and as the surface area of the animal increases during growth, gland density decreases. In mature cattle the number of sweat glands per square centimeter of surface area is about 800 for European breeds and 1500 for Zebu. There is also considerable variation in shape and size of the glands. In Sindhi and Sahiwal cattle the sweat glands are so large that at normal densities of 1500 glands per cm², the glands touch and form a continuous secretion layer.

The loss of heat by evaporation depends on the fact that a certain amount of heat is required to change water to water vapor at the same temperature. More heat is required for the vaporization of water than for an equivalent amount of any other liquid. In the human body, each gram of water evaporated from the skin at room temperature results in a loss of 500 calories. This amount of heat is 50 percent more than that required to evaporate 1 gm of water's closest competitor.

In man the heat loads handled by sweat glands and cutaneous blood vessels are impressive. Sweating pours out water at the rate of more than 8 liters a day. The resulting increased conductance is equivalent to more than 1000 kg of blood circulated through the skin each day.

Warming drinking water in the body and excreting it at body temperature dissipates only 3 or 4 percent as much body heat as do both warming and vaporizing the same amount of water. Thus the use of water by sweating for evaporative cooling conserves water and energy economically compared to respiratory cooling or water loss via urine and feces. This is an important part of adaptation and water conservation by heat-tolerant animals.

The evaporation of water as a means of heat loss becomes increasingly important as environmental temperature approaches or exceeds body temperature. The ability of animals to withstand high temperature is proportional to their capacity to vaporize water. The limiting factors of water evaporation are vapor pressure gradient (wind and relative humidity), vaporizing surface area, and the secretion rate of water.

In domestic animals, sweating has been observed in cattle, sheep, horses, donkeys, camels, and in some breeds of dogs. In

sheep, sweating is important to the regulation of the temperature of peripheral regions, especially the scrotum, but of less importance in total body heat loss. The camel has a relatively high sweat secretion rate, up to 280 ml/m²/hour. This wastes less energy and water than does the more expensive respiratory cooling of sheep. That is, in the camel 95 percent of the heat loss by vaporization is from the skin; 5 percent is from respiration. In the sheep, less than half of the evaporative heat loss is from the body surface. In cattle, skin evaporation accounts for 16 to 26 percent of the losses of body heat at air temperatures of 10° C, 40 to 60 percent at 27° C, and over 80 percent at air temperatures above 38° C.

Bos taurus and *Bos indicus* differ significantly in their sweating rate. At ambient temperatures of 45° C and higher, Zebu and Zebu-cross cattle evaporate over 600 ml/m² of skin per hour. Values for the European breeds range from 140 to 400 ml. However, at ambient temperatures of 20° to 35° C, the sweating rate of the Zebu is less than 50 percent of that of the temperate breeds. This ability of the Zebu to conserve body water by restricting cutaneous water loss at moderate temperatures but increase evaporative cooling at high temperatures is an important factor in establishing the heat tolerant superiority of the Zebu compared with the heat tolerance of temperate breeds.

Knowledge of the innervation and endocrinological relations of the sweat glands of domestic animals is not complete. Sweating is generally considered to be under humoral control with epinephrine acting as the most potent sudomotor drug, effecting control by an adrenergic mechanism. Temperature receptors in the skin, hypothalamus, and other body areas participate in body temperature regulation, including the activation of the sweat gland mechanism. Generally, sweating is initiated when the skin temperature reaches 34° or 35° C. However, transient and long-term thermal stress affects the receptors differently and results in different physiological responses.

D. Respiratory Water Loss

Water is lost through the normal process of breathing. This is due to the changes in water content of air as it passes to and fro along the respiratory tract. Incoming air is warmed and wetted as it passes over the turbinates and then becomes nearly saturated in the lungs. The outgoing air gives up some of its moisture, again to the turbinates, and is expired at about 90 percent saturation. The amount of water lost depends upon the absolute humidity (vapor pressure) of the incoming and expired air and on the respiratory volume of air.

Respiration also acts as a cooling mechanism. Evaporative cooling occurs from the surfaces of the respiratory passages; the cooled blood drains into the right side of the heart. The amount of cooling obtained by respiration depends upon temperature and vapor pressure of the inspired air, respiratory volume, body temperature, and vapor pressure of expired air. For animals maintained at high environmental temperatures, the beneficial effects of respiratory evaporative cooling are partly offset by increased heat production owing to greater muscular exertion required for the increased respiratory activity. Cooling by respiration requires four times more energy than does the secretion of sweat. A net increase in cooling effect is accomplished by decreases in thyroid activity (metabolic rate) and consumption of total digestible nutrients.

Respiratory water loss in beef cattle ranges from about 23 ml/m²/hr at 27° C to 47 ml at 40° C. The respiratory vaporization accounts for 5 percent of total heat dissipation at 25° C to 17 at 40° C. Respiratory water loss in sheep reaches a maximum of 95 gm/m²/hr, which is two to three times the sweat secretion rate. Factors such as high plane of nutrition, lactation, pregnancy, animal movement, and light intensity increase respiratory rate at the same environmental temperature.

E. Insensible Water Loss

Diffusion of water through the skin, apart from the sweat gland secretion, is referred to as *insensible water loss*. Since it is a diffusion process it is directly affected by the vapor pressure of the skin surface compared with the adjacent air. In domestic animals the air at skin surface is trapped by body hair so that it is saturated at skin temperature. Thus changes in ambient air temperature, humidity, and movement directly affect the vapor pressure gradient and thereby alter water loss by diffusion.

In cattle, approximately three times as much water is lost by skin diffusion as by respiration. The water loss by diffusion and respiration is about equal in sheep (Chew, 1965). At moderate temperatures, insensible water loss in cattle is about 120 ml/hr, 50 ml/hr in sheep, 40 ml/hr in pigs, and near 400 ml/hr in horses.

Since the diffusion process of insensible water loss depends upon physical factors, there is very limited physiological control. Thus, under conditions of water restriction, insensible water loss becomes critically important as the major loss of water for the dehydrating animal.

Licking of saliva on the fur is used as a method of evaporative heat loss for body temperature regulation by various desert mammals. Drooling of saliva is very common in some domestic animals. This serves no useful function and is another source of water loss. Cattle drool 10 to 15 liters of saliva per day. This loss includes not only water but also other saliva ingredients including salt.

IV. FACTORS AFFECTING WATER BALANCE

A. Body Temperature Control

In all normal mammals, the body temperature remains constant within relatively narrow limits despite widely changing external conditions. This ability to maintain a constant temperature (homoiothermy) has been of great importance in establishing the superiority of the higher forms of animal life. Homoiothermy is attained by maintaining an equilibrium between the heat produced by the animal, together with any heat it may absorb from its environment, and the heat which it emits.

It is virtually impossible to discuss the water metabolism of mammals without simultaneously considering problems of temperature regulation since these two are interrelated. Water metabolism is a function of water intake and water loss, and these are influenced by the amount of heat, both exogenous and endogenous, which must be dissipated in order to keep the body temperature within the limits typical of a homeotherm. Thermal stress complicates water balance since heat regulating mechanisms of sweating, panting, and salivation result in additional water losses.

Water alone possesses certain qualities which are essential to body temperature control. (1) The specific heat of water is considerably higher than that of any other liquid or solid. Because of this great heat storage power of water, sudden changes of body temperature are avoided. (2) More heat is required for the vaporization of water than for an equivalent amount of any other liquid. This is especially important for heat loss through evaporation from the lungs and skin. (3) The thermal conductivity of water is greater than of any other ordinary liquid. This is important for the dissipation of heat from deeply situated regions of the body.

These physical properties of water which make it ideal as a heat-regulating medium are enhanced by other purely physiological factors. The mobility of the blood and the rapidity with which it may be quickly redistributed in the body, together with the special physical properties of the fluid itself, result in a highly efficient body temperature regulator. At a moment's notice the blood may shift from deep body regions to superficial regions and spread out in fine blood vessels over a broad area just beneath the skin. This provides for maximum radiation surface for heat loss. At another instant, in order to conserve body heat, blood is drained from the surface areas and collected in the deep regions of the body.

There is a close relationship between temperature regulation and water economy, and there may be a quantitative shift with respect to the effectiveness of any particular mechanism in adapting an animal to its environment. This shift, when it occurs, is in the direction of increasing water conservation and reducing water requirements for the dissipation of metabolic heat. Thus, water is not used for evaporative cooling in small mammals. The high ratio of surface to mass does not provide adequate water storage for surface evaporative water loss. Sweating would rapidly result in dehydration or circulatory failure. Small animals avoid high heat stress by burrowing or seeking cool microhabitats (Schmidt-Nielsen, 1964). Thus, small mammals which have no great evaporative water loss during heat stress do not show the large increase in water intake noted for large mammals that do sweat profusely.

In larger mammals, sufficient "spare" water can be stored to permit evaporative use for thermoregulation. After a degree of water depletion, evaporative loss is curtailed and other adjustments in metabolism and behavior are utilized. Drinking allows the larger animals to maintain evaporative heat loss. Also, the body temperatures of large animals such as cattle or horses lag behind changing ambient temperature because of the animals' large thermal capacity (weight times specific heat). Therefore, cattle can withstand very high day temperatures provided the nights are cool. The animal can dispose of the stored heat into the cool atmosphere, especially if it is exposed to the cold sky.

In some animals exposed to heat stress there is an immediate hemodilution, which results from a shift of water from tissues into plasma. The greater blood volume increases cutaneous circulation, sensible heat loss, and sweating. With continuing stress the blood may become concentrated as water is lost by evaporation. Exposure to cold causes a moderate loss of water from the blood to the tissues, but the total water content of the body remains unaltered.

Since water is essential to thermoregulation in domestic animals, the mechanisms controlling body temperature and water balance are essentially the same. In acute dehydration there is a failure of temperature control and an explosive rise in body temperature. In acute heat stress, if body cooling mechanisms are not adequate to prevent accumulating heat stores, drinking may actually cease, possibly due to overheating of the hypothalamic centers. In either case, continuing exposure to the heat stress will be fatal to the animal.

B. Dehydration

When the output of water exceeds the intake, water content of the body obviously will suffer reduction. That is, the body is in negative water balance and the condition known as dehydration results. For all animals, fluid intake is intermittent while loss of water is continuous. Thus dehydration is a part of everyday life. The animal response to dehydration depends upon the rate at which the dehydration occurs, the ambient conditions, the previous physiological condition of the animal, and the physiological characteristics of the animal.

Under temperature conditions the main requirement for water is to meet metabolic demands. Thus, moderate restriction may have no appreciable effect on animal performance since voluntary water intake is usually in excess of body needs. Under tropical conditions the supply of drinking

water and water contained in the food may be severely restricted while at the same time the water demand for thermoregulation is at a maximum. Under such conditions, water becomes the limiting factor in animal performance.

In domestic animals the initial response to water restriction is to reduce all pathways of water loss. The greatest absolute means of conserving water is usually by the reduction of urine and fecal water loss. Also water may be conserved by changes in the water content of the body fluid compartments. When dehydration results from complete water restriction in the absence of heat stress, there is a gradual decline in both extracellular and intracellular water, with the greatest loss in the intracellular portion. Part of the extracellular body water is continually circulating as blood plasma. The maintenance of plasma volume is essential to body temperature control. In sheep exposed to the sun during complete water restriction, 12 percent of the total weight loss is plasma water, 35 percent interstitial, and over 50 percent intracellular and gut water (MacFarlane *et al.*, 1963). In the dehydrated camel, plasma loss is 5 percent and interstitial fluid 9 percent.

In cattle injected with tritium-labeled water, it was determined that the half-life of water in cattle weighing 150 to 750 kg is about 3.5 days. This is shorter than that found for other mammalian species. These studies indicate that water turnover is 14.2 l/24 hr for cattle watered once daily, and 12.4 l/24 hr for cattle watered each 4 days. The *Bos indicus* types use less water than the *Bos taurus* in the same environment. At high environmental temperatures, water turnover for *Bos taurus* is 600 ml/kg$^{0.82}$/hr and less than 400 for *Bos indicus* (Table 19–4).

In addition to the blood plasma portion of extracellular water, another portion can be considered to be in "storage." This includes the water in connective tissues, that cyclically accumulated in the reproductive tract, and water in the rumen and lumen of

Table 19–3. Blood Volume of Brahman and Hereford Female Cattle

(*Data from Howes et al., 1963.*
J. Anim. Sci. 22:184.)

Values as Percent Body Weight	Hereford		Brahman	
	2-Year	*4-Year*	*2-Year*	*4-Year*
Plasma	2.8	3.1	2.8	3.9
Erythrocyte	2.1	2.7	2.9	4.5
Blood	4.9	5.8	5.6	8.4

Table 19–4.   Water Turnover Rate of Animals at High Environmental Temperature (40°C)

Animal	Water Turnover $ml/kg^{0.82}/24\ hr$
Camel	190
Sheep	230
Cattle, Brahman . . .	350
Horse	400
Cattle, European . . .	600

the gut. An increase in total pool size associated with a greater water flux provides greater water reserves for hot weather conditions when losses due to respiration and body cooling increase. The larger pools with their greater capacities resist, for a longer time than smaller pools, the adverse physiological effects of depletion associated with high fluxes through these pools. The camel has a water turnover about half that of Merino sheep and one third that of Shorthorn cattle.

The camel can withstand considerable dehydration. In a hot environment it can tolerate a loss of at least 27 percent of body weight, twice the dehydration that is lethal for other mammals. When the camel becomes dehydrated the loss of water is not accompanied by a proportional loss in plasma volume. The maintenance of a high plasma volume facilitates circulation, which is one of the first functions to suffer during dehydration in other animals in hot environments. Also, the camel can tolerate a body temperature increase of over 6° C. This rise in body temperature as heat stored reduces the body-to-air temperature gradient, and thus reduces the gain of heat. The high plasma volume serves to transport stored heat to the body surface so that it can be disposed of by radiation during the cold nights. At sunrise on the following day there is a sudden warming of the skin and a drop in rectal temperature as a result of cutaneous vasodilation. This prepares the animal for another day of heat storage.

Water restriction in mammals results in a voluntary decrease in food consumption. The rate of dehydration determines the de-

crease in caloric intake for a given species. In temperate cattle breeds, water deprivation results in a decline of 50 percent from the food consumption of each preceding day during the period of water deprivation. Zebu and Zebu-cross cattle do not show a loss of appetite during the first 24 to 48 hours of water deprivation, but after two days without water they also have a considerably depresssed appetite. The decrease in food consumption also decreases fecal output. In cattle, the resulting decline in fecal water loss may actually exceed the decreased urine output (Weeth et al., 1967).

Dairy cows which have water available continually drink 18 percent more water and give 3.5 percent more milk than cows which are watered twice daily. Beef cattle that are permitted to drink twice a day take 13 percent more than when they are watered once a day. Cattle and sheep which are watered every second or third day drink only 89 percent and 68 percent, respectively, of what they take when watered every day (Altman, 1955).

Laboratory animals on chronic restricted water ration undergo a gradual weight loss that recovers, or nearly recovers, the original ratio of body water to body weight. The amount of weight loss becomes a measure of the degree of water depletion which they have undergone. These mammals show great ability to stabilize body weight on water intakes much below those they voluntarily take. Rats about 1 month old can be held at constant weight for several weeks by restricting the amount of liquid in a diet otherwise adequate for growth.

In domestic animals, thirst is a lagging guide to water requirements during hot weather. Animals allowed water ad libitum do not drink enough to maintain body weight until they become well acclimated (Adolph, 1952). The loss of weight in prolonged dehydration is due to the reduction in tissue water as well as to the actual breakdown of body substances which occurs in the effort to provide water for the maintenance of the normal fluid content. Fat

PLATE 26

A, In warm summer months, range cattle tend to stay close to water. Range grass distant from water is not, therefore, fully utilized. This can be disastrous when range feed is scarce during drought periods. (*Courtesy, C. R. Whitfield.*)

B. Careful water management improves both range utilization and animal performance. The proper distribution of potable water throughout the range area is a basic part of water management. (*Courtesy, B. N. Freeman.*)

C, An illustration of the genetic influence on water metabolism. An eightfold difference in 10-week water intake resulted with 4 generations of selection for this trait in the laboratory rat. With proper selection procedures, similar results may be expected in domestic animals. (*Courtesy, John Burnham.*)

and carbohydrate stores are used first, and eventually, protein.

An investigation was made of the effects of dehydration induced by withholding drinking water on various metabolic and thermoregulatory functions of steers both in a temperate (15° C) and in a hot (40° C) environment (Bianca *et al.*, 1965). During 4 days of dehydration at 15° C, decreases occurred in food intake, the excretion of feces and urine, evaporation, body weight, heat production, respiratory ventilation, plasma potassium, and urinary potassium output. At the same time, increases were found for hematocrit levels, plasma total solids, plasma sodium, plasma chloride, blood urea, and the output of urinary sodium. During 2 days of dehydration at 40° C, qualitatively similar changes occurred in most of these variables.

Dehydration is an important factor affecting heat tolerance. The rise in body temperature during heat exposure is proportional to the water deficit, although sweat secretion is active at rather high levels of dehydration. Nondesertic mammals such as man, dog, rat, and cattle are not able to adapt to the water shortages experienced under desert conditions by repeated or continued dehydration. The actual survival time for domestic animals deprived of water and exposed to solar radiation depends largely upon water reserves in the digestive tract and extracellular compartment. Digestive tract water reserves in sheep may be 6 or 7 liters and 80 to 100 liters for mature cattle and horses.

The repayment of water debt accumulated during dehydration occurs very rapidly. When offered water, dehydrated cattle will drink 15 percent of their weight within 30 minutes. The camel can consume 20 percent of its body weight in 10 minutes. The donkey can drink at the rate of 8 liters per minute and take up 20 percent of its body weight in about 2 minutes. Dehydrated sheep will consume 7 to 9 liters in one quick drink. Ruminants have a distinct advantage in making up a water deficit quickly. The

water is stored in the rumen and is then slowly released into the abomasum for absorption. The rapid water intake is an obvious advantage for grassland and wild desert animals that must obtain their water from a potentially dangerous water hole. Domestic cattle also follow the drinking behavior of game animals in that they rarely drink at night, even when grazing. Their daytime consumption on the range is maximal in early morning and late afternoon.

The restoration to normal of the various physiological values deranged by dehydration requires more than one day in most instances. There do not, however, appear to be permanent changes or any impairment of health due to relatively short periods of dehydration. Horses without water for 3 days replace over 90 percent of their water deficit the first day but only 60 percent of their total weight loss. The high salt loss associated with dehydration in horses is probably a factor in their slower recovery.

C. Saline Drinking Water and Salt Consumption

In many arid regions the available drinking water is often present in small water holes. The high evaporative rate concentrates the dissolved salts, resulting in saline drinking water. The utilization of a saline solution requires special effort by the kidney. To provide for minimum intake of saline drinking water the kidney must reabsorb all water but the obligatory urine loss. In addition, it must also concentrate and excrete the ingested sodium chloride.

Salt retention results in intracellular dehydration and electrolyte shifts. Salt excretion without additional urine concentration results in osmotic diuresis and extracellular dehydration. The result in either case can be just as harmful to the animal as dehydration resulting from water restriction. Wolf (1958) describes many experiences of shipwrecked people using sea water for drinking. Any prolonged use of sea water for this purpose has always proved to be fatal.

Some small desert animals do have the

ability to use saline water by excreting a very highly concentrated urine. Domestic animals have a limited ability to conserve water by excreting a high osmolal urine. Sheep and cattle can tolerate a 1.5 percent sodium chloride drinking solution during cool weather but are adversely affected by even 1 percent in the drinking water during the summer months. A solution of 2 percent sodium chloride is definitely toxic. It causes severe anorexia, weight loss, and anhydremia (Weeth & Lesperance, 1965; Wolf, 1958). A high content of sodium chloride and calcium chloride in drinking water renders a cow unable to suckle her calf before any other injury to the cow is apparent.

The effect of water restriction is much more detrimental when the available water is even slightly saline than when it contains no salt. When water is of good quality once a day watering is adequate. When the water is saline, drinking induces diuresis; the water consumed is apparently involved in renal osmoregulation rather than tissue rehydration. Watering only once in 48 hours is inadequate, even with water of high quality. With this watering frequency the effects of solute loading are accentuated. Recovery from salt dehydration is much slower than recovery from dehydration due to water restriction. When salt-dehydrated animals are allowed to drink an unlimited amount of pure water they often collapse and show tetany. Limiting the water intake for 2 or 3 days prevents this occurrence.

The sodium content of many of the range grasses is very low, thus the herbivore will travel many miles to a salt lick to replete body stores of sodium. Salt is generally provided *ad libitum* to domestic animals. Range cattle have a voluntary average salt consumption of 48 gm. They consume 85 gm per head daily in the winter and less than 28 gm daily in spring and summer. Increasing the salt intake to 770 gm per animal increases water consumption 40 percent, but no benefits can be ascribed to the salt. High-salt diets increase water consumption,

urine flow, and mineral balance in sheep. Water consumption is not adequate to compensate for urine volume for sheep on 11 percent salt diet.

The practice of feeding a mixture of salt and protein supplement to range cattle is used extensively in some areas. At a ratio of 1 part salt to 2 parts meal, weaner calves will consume 0.5 kg of meal per day. Range cows will eat 1.0 to 1.4 kg daily of a 30 to 70 salt-meal mixture. On a mixture of 25 percent salt and 75 percent cottonseed meal, the average daily intake of salt per ewe ranges from 26 to 53 gm.

D. Diet

Control centers for feeding and drinking are adjacently located in the hypothalamus. The integration of these centers results in an interdependence of feeding and drinking intakes so that there is a voluntary reduction in one whenever the other is restricted. Water restriction results in reduced feed intake, and, in most species, limiting food intake reduces drinking. Animals on restricted feed intake simply need less water. The survival of starving animals is not enhanced by forcing the intake of more water than is taken voluntarily.

When unlimited water is available, the ratio of water to feed consumed by temperate breeds of cattle ranges from 2.76 to 3.81, with an average value of 2.96 ml/gm (Winchester & Morris, 1956). Zebu cattle have a lower ratio of water to feed than do the European breeds, with a range of 2.41 to 3.01 and a mean value of 2.70.

Cattle drink more on a high-protein than on a low- or average-protein diet. With diets of equal protein content they drink the least with high carbohydrate and low fat. The higher protein level results in more nitrogenous end products which require a higher obligatory urine volume for excretion. Carbohydrates in the diet provide the most oxidation water per calorie.

Cattle are better able to reduce their nitrogen output on a low-protein diet when water consumption is restricted. Thus cattle on a low-protein roughage diet might be better able to maintain positive energy balance if subjected to some degree of water restriction. There is also evidence that although water restriction reduces feed intake, digestion and absorption are increased.

Excretion of urea nitrogen via urine was studied during water diuresis in cattle (Weeth & Lesperance, 1965). Water treatments consisted of voluntary consumption and intake of 150 percent of this amount. Water diuresis had no effect on digestibility of dry matter, nitrogen balance, urine urea nitrogen or total nitrogen excretion, or blood urea nitrogen concentration. Since there was no washing out of urea nitrogen by water diuresis, it is suggested that the bovine kidneys are capable of reabsorbing normal amounts of urea even when presented with extreme water loads.

The higher the proportions of minerals in the diet, the larger the urine excretion, and accordingly, the larger the water requirement. Cattle drink an additional 230 to 440 ml of water for each gm of salt ingested when 1 percent or 2 percent salt is added to a diet of chopped alfalfa hay. Water consumption and urine volume are increased significantly when cattle are fed high levels of potassium, but water balance is not affected. The addition of 500 ppm of manganese to the drinking water of cattle has no effect on feed consumption, water consumption, or growth. Sodium supplementation will increase water consumption and urine volume of cattle and sheep at high environmental temperatures.

There is a highly significant relationship between water intake and the level of inorganic phosphorus in the blood. Water starvation causes a steady increase in the level of inorganic phosphorus. When water is again available for the cattle to drink the inorganic phosphorus level falls rapidly to a subnormal level before it returns to normal. Cattle that are not watered daily show less efficient mineral absorption than do those receiving daily amounts. Some foods have diuretic properties that are responsible for

increased drinking. The higher water intake of sheep on lucerne hay compared with grass hay is attributed to diuretic principles in the lucerne. There is indirect evidence that fat in the diet may have some antidiuretic properties.

E. Pregnancy and Lactation

Heifers drink almost 50 percent less water on the day of estrus than on the other days of the estrous cycle. It is not known if this is due to estrogenic effects on the thirst mechanisms or to sexual excitement at estrus. The fetal tissues and associated embryonic fluids do increase the total body water content of the pregnant female, especially during the latter stages of pregnancy. There are also increased physiological demands on the pregnant animal which result in additional water intake. Many mammals including cows, ewes, and sows, increase water intake over 50 percent prior to parturition.

The production of milk obviously increases the water requirements of the lactating female. In addition to the water of milk, there is an increased need for water to meet the requirements of higher food consumption and heat production. In beef cows at moderate environmental temperatures, each liter of milk produced requires an additional water intake of over 2 liters. This requirement will be doubled at high environmental temperatures (Winchester & Morris, 1956). This does become a critical factor influencing milk production for range animals under hot arid conditions.

F. Genetic Influences

Artificial selection for various traits in domestic animals undoubtedly results in genetic changes in water metabolism as a correlated response. Indirect evidence of this is shown for various components of water balance in cattle and sheep. A good example is the superior heat tolerance of *Bos indicus* compared to *Bos taurus*. This is due to a combination of factors including more efficient evaporation, lower water turnover rate, and drier feces. These characteristics of the Zebu are transmitted to their progeny when they are crossbred with *Bos taurus*.

Some breeds of sheep can maintain water balance with only the water of feed and oxidation. The Barbary sheep, which were imported to New Mexico from Africa, went for a 6-week period of the hottest summer weather with only the water obtained by grazing. MacFarlane and his associates (1963) noted that, during the winter in Australia, Merino sheep, which originated in a hot, arid environment, do not drink if adequate grazing is available. Direct selection for high and low water consumption in the laboratory rat resulted in a tenfold difference in water intake between selected animals (Plate 26). Heritability estimates of water intake exceed the estimate for diuresis of 0.60 reported by Stewart and Spickett (1967) for mice. The high and low water consumption groups are very similar in growth, feed efficiency, and reproductive performance.

There are also genetically influenced metabolic disturbances which directly effect water balance. Diabetes insipidus and diabetes mellitus in man are two well-known examples but there are many others. Congenital dropsy or edema in cattle is inherited as a simple recessive trait. Any disease or condition that affects the kidneys or endocrine system will almost certainly influence the water metabolism of the animal. Likewise, inherent differences in temperament, phlegmatic or choleric, directly influence the animal's water balance processes.

Twin studies in a series of monozygotic cattle were conducted by Hancock (1953) and others to determine the genetic influences on behaviorial characteristics. Many items of behavior, including grazing time, the number of drinks, the distance an animal will walk in a day, and frequency of defecation or urination, are simply inherited. In turn, many of these traits are directly associated with animal performance and water

balance processes. This helps to explain the enormous differences in water intake that are found for any group of animals under apparently comparable conditions.

The high degree of genetic variability for components of efficient water metabolism in our domestic animals indicates that much progress should be possible in selecting for these traits. Until more exact methods for evaluating efficient water metabolism for individual animals are available, improvement in this characteristic may be affected by selecting for adaptability to tropical or desert conditions.

G. Neural and Hormonal Mechanisms

The regulation of water balance is primarily a matter of adjusting intake and output for deficits and excesses of the normal water level in the body. The physical loss of fluid through the skin and lungs goes on continuously and at approximately a constant rate. The loss of fluid in feces is more variable, depending largely upon the diet. The losses through the kidneys and sweat glands, however, fluctuate greatly, bearing a rough inverse relationship to each other. Conservation of water in time of need is affected principally through the reduction of excretion by the kidney. Under normal circumstances the intake of water exceeds the requirements, causing the kidneys to excrete more water than is essential for elimination of solutes.

A lowering of the water level, either because of too great loss or too limited intake, increases the osmotic pressure of the blood. As a result, water is drawn from the tissue spaces and from the tissues themselves into the blood. This stimulates the osmoreceptors (sense organs of thirst). From there the impulse is conveyed to the hypothalamus and the renal loss of water is adjusted. The hypothalamus has been identified as the site of both receptors for osmotic stimuli and centers for regulation of renal output and water intake. The hypothalamus is a basal part of the diencephalon that lies just below the thalamus and above the pituitary body.

It consists of several distinct masses of gray matter or nuclei. The hypothalamus receives nerve stimuli from the medulla oblongata, the spinal cord, the thalamus, and, indirectly, from the cerebral cortex. It is also efferently connected with the pituitary body. The introduction of specific chemicals into precise areas of the hypothalamus or bilateral lesions of these areas will stimulate or inhibit drinking with no specific effect on food intake. Cholinergic stimulation through implanted cannulas in the hypothalamus elicits a rapid and marked increase in water intake without a corresponding increase in food consumption.

It appears that a generalized nervous system network is specifically and functionally sensitive to specific drug action and that other primary drives depend closely on parallel neural circuits partitioned both structurally and biochemically (Fisher & Coury, 1962). Studies with bilateral lesions show that an important pathway runs from the anterior thermoregulatory region caudally past the ventromedial region to the lateral hypothalamus, which is involved in the regulation of food as well as water intake. These pathways may be essential to the effects of environment and body temperature on food and water intake (Stevenson *et al.*, 1964).

Water intake can be altered by stimulation of the taste buds. Taste seems to be different among the various mammals. Some species have separate taste buds that respond to salt or water, others respond only to water. Thus the drinking response to different saline solutions will vary greatly among the various species. Saline solutions increase water intake in rats but decrease water consumption in cattle. A dilute sucrose solution is much preferred by calves, with degrees of rejection for various acid and salt solutions subject to genetic control.

The water loss via the kidneys must be carefully controlled to maintain the body water content constant. The excretion of excess water or the production of hypertonic urine to conserve body water stores is accomplished by the reabsorption process in the kidneys. Water reabsorption is controlled by the antidiuretic hormone of the neurohypophysis, vasopressin. The antidiuretic hormone (ADH) functions to increase the water reabsorption rate in the kidneys, thereby conserving water. The supraoptic and paraventricular nuclei of the hypothalamus synthesize and release ADH. Unmyelinated nerve fibers run from these centers and terminate in the posterior lobe of the pituitary. These neural axons then serve to transport the newly synthesized ADH to their terminal endings in the hypophysis. The hypothalamus functions as the control unit for ADH release by the posterior pituitary. Antidiuretic material appears in the urine during dehydration, the quantity being directly related to the need for water conservation. When drinking water is not available, ADH functions to reduce urine output to the minimal obligatory level. In the absence of ADH the tubular cells of the kidneys lose, to a degree, their ability to reabsorb water from the glomerular filtrate.

There is another hormone, aldosterone, secreted by the adrenal cortex, which influences water excretion by the kidneys. Aldosterone is chiefly concerned with electrolyte homeostasis, which in turn is closely associated with water balance. Aldosterone production is stimulated by high-potassium, and low-sodium electrolyte levels. The function of aldosterone in the renal system is to retain sodium. It acts to resorb sodium from the urine, and to maintain proper electrolyte balance and osmotic pressure. An exchange occurs between sodium and intracellular potassium. Sodium is the principal cation of the extracellular fluids, while potassium is found chiefly within the cells. Chronic overproduction of aldosterone results in excess sodium retention and chronic potassium loss. This may damage the kidneys and impair their ability to produce concentrated urine.

Other hormones, including oxytocin, epinephrine, reserpine, and intermedin, act with ADH or aldosterone in controlling water balance under particular physiological conditions. Even such factors as length of daylight have been noted to play a large part in the causation and timing of circadian rhythms of urine and electrolyte metabolism.

V. RESEARCH TECHNIQUES

Techniques of experimental procedures and the determination of physiological responses in water metabolism of domestic animals are well illustrated in the reports by Bianca and associates (1965), Brozek (1961), Kleiber (1961), MacFarlane and coworkers (1963), Thompson and colleagues (1949), and Till and Downs (1962), among others. The physiological determinations include the measurement of cutaneous evaporation, respiratory vaporization, insensible water loss, changes in body fluid compartments, and body water turnover.

The general procedures involved in measurement of cutaneous evaporation usually include covering a small skin area with a cup containing a desiccant and determining the weight gain of the container. Care must be taken in selecting proper sampling sites. Polyethelene covers that enclose the entire animal have also been used. The trapped air is periodically withdrawn for moisture analysis and fresh air introduced under the cover. Infrared gas analysis has also been adapted to measure evaporative water loss. Insensible water loss can be determined as: change in body weight + (weight of: food + water intake − feces − urine). This does require very precise measurements, within 100 gm for a 500 kg animal for body weight (Bianca et al., 1965). Respiratory water loss is determined with a face mask. A device called a pneumotachometer will measure frequency and minute volume of respiration.

Volume and turnover of body water pools are determined by tracer dilution techniques. Water soluble tracers such as antipyrene or isotopes such as tritium are injected into the blood stream. Body water volume and turnover are determined from the time-

concentration and disappearance rates (Kleiber, 1961; MacFarlane et al., 1963).

Urine and feces can be quantitatively collected from grazing cattle (Border et al., 1963). These collection techniques are also useful in animal chambers or confined areas. Blood flow can be determined from catheters placed in portal or mesenteric blood vessels (Waldern et al., 1963; Kolin, 1959).

The role of the hypothalamus as a control center for many of the body functions has been emphasized. The techniques and procedures used by researchers in studying the functions of the hypothalamus provide an outstanding example of ingenuity, imagination, and skill in physiological research (Fisher & Coury, 1962; Stevenson et al., 1964; Wayner, 1964). Sophisticated techniques for studying renal function and electrolyte transport are cited by Ullrich and Marsh (1963). However, much of this type of research is conducted with laboratory animals, and suitable modifications for large animal research are lacking in many areas.

REFERENCES

Adolph, E. F. (1952). Tolerance to heat and dehydration in several species of animals. *Amer. J. Physiol. 151*:564–575.

Altman, L. B. (1955). Automatic livestock waterers. U.S. Dept. Agri. Leaflet No. 395.

Bianca, W., Findlay, J. D., & McLean, J. A. (1965). Responses of steers to water restriction. Res. Vet. Sci. *6*:38–55.

Border, J. R., Harris, L. E., & Butcher, J. E. (1963). Apparatus for obtaining sustained quantitative collections of urine from male cattle grazing pasture or range. J. Anim. Sci. *22*:521–525.

Brozek, J. (1961). Body composition. Science *134*: 920–930.

Chew, R. M. (1965). Water metabolism of mammals. Chap. 2. In: *Physiological Mammalogy*. Vol. II, W. V. Mayer & R. G. Van Gelder (Eds.), New York, Academic Press.

Falk, J. L. (1961). Production of polydipsia in normal rats by an intermittent food schedule. Science *133*: 195–196.

Fisher, A. E., & Coury, J. N. (1962). Cholinergic tracing of a central neural circuit underlying the thirst drive. Science *138*:691–692.

Hancock, J. (1953). Grazing behavior of cattle. Anim. Breed Abstr. *21*:1–13.

Horrochs, D., & Phillips, G. D. (1961). Factors affecting the water and food intake of European and Zebu-type cattle. J. Agri. Sci. *56*:379–381.

Kelley, J. T. (1945). Water requirements of farm animals. Victoria Dept. Agri. J. *43*:158–159.

Kleiber, M. (1961). *The Fire of Life*. New York, Wiley.

Kolin, A. (1959). Electromagnetic blood flow meters. *Science 130*:1088–1097.

MacFarlane, W. V., Morris, R. J. H., & Howard, B. (1963). Turn-over and distribution of water in desert camels, sheep, cattle and kangaroos. Nature (Lond.) *197*:270–271.

Schmidt-Nielsen, K. (1964). *Desert Animals. Physiological Problems of Heat and Water*. New York, Oxford University Press.

Stevenson, J. A. F., Box., B. M., & Montemurro, D. G. (1964). Evidence of possible association pathways for the regulation of food and water intake in the rat. Canad. J. Phys. Pharm. *42*:855–860.

Stewart, J., & Spickett, S. G. (1967). Genetic variation in diuretic responses: further and correlated responses to selection. Genet. Res. *10*:95–106.

Tasker, J. B. (1967). Fluid and electrolyte studies in the horse. III. Intake and output of water, sodium and potassium in normal horses. Cornell Vet. *57*:649–657.

Thompson, H. J., Worstell, D. M., & Brody, S. (1949). Environmental physiology with special reference to domestic animals. V. Influence of temperature 50° to 105° F. on water consumption in dairy cattle. Mo. Agri. Exp. Sta. Res. Bull. No. 436.

Till, A. R., & Downs, A. M. (1962). The measurement of total body water in the sheep. Aust. J. Agri. Res. *13*:335–342.

Ullrich, K. J., & Marsh, D. J. (1963). Kidney, water and electrolyte metabolism. Ann. Rev. Physiol. *25*:91–142.

Valentine, K. A. (1947). Distance from water as a factor in grazing capacity of rangeland. J. Forestry *45*:749–754.

Waldern, D. E., Frost, O. L., Harsch, J. A., & Blosser, T. H. (1963). Improved equipment for catheterization of blood vessels of the portal system of cattle. Amer. J. Vet. Res. *23*:212–214.

Wayner, M. J. (Ed.) (1964). *Thirst*. New York, Macmillan.

Weeth, H. J., & Lesperance, A. L. (1965). Renal function of cattle under various water and salt loads. J. Anim. Sci. *24*:441–447.

————, Sawhney, D. S., & Lesperance, A. L. (1967). Changes in body fluids, excreta and kidney function of cattle deprived of water. J. Anim. Sci. *24*:418–423.

Winchester, C. F., & Morris, M. J. (1956). Water intake rates of cattle. J. Anim. Sci. *15*:722–740.

Wolf. A. V. (1958). *Thirst*. Springfield, Ill., Charles C Thomas.

Chapter 20

Mathematical Models of Growth

By T. S. Russell

SEVERAL equations have been developed to analyze and predict growth patterns and relationships during different phases of animal development (Brody, 1945; Guttman & Guttman, 1965; Fabens, 1965; Laird *et al.*, 1965; Laird, 1966). In the use of any equations for analysis of growth, one must keep in mind that the growth pattern is not composed of point-to-point changes, but is a continual balance of gain and loss of total animal mass.

During the early phase of growth, increase in mass greatly outweighs the loss. The equilibrium between increase and loss of animal mass remains relatively constant during this phase and increase in weight with age is linear. However, as the animal attains maturity the rates of gain and loss show a shift in equilibrium and the growth curve becomes curvilinear. After the inflection period, change in weight with change in age again follows a linear pattern to maximum weight.

Even though changes in weight with increased age tend to follow the same pattern (Brody, 1945), there is variation between species (Fig. 20–1). The Rhesus monkey has a gradual increase through 20 percent of its

life span, while in the other animals shown in Figure 20–1 the initial linear phase of growth is completed before 10 to 15 percent of their life span is completed. The simi-

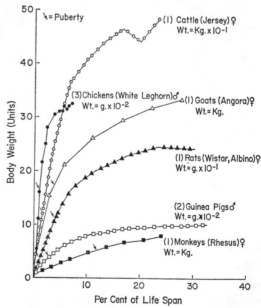

FIG. 20–1. Comparative growth in six species. Note comparative early sexual maturity in the chick and guinea pig. (*From Altman & Dittmer, 1964. Biology Data Book, Amer. Soc. Exp. Biol.; Spector, 1956. Handbook of Biological Data. Saunders, Philadelphia; and Tyler & Stearner, 1966. Radiat. Res. 29:257.*)

larity of the growth pattern of chickens and cattle may be the effect of domestication and selection for rapid growth and early maturity.

In most of the animals shown in Figure 20-1, inflection in the growth pattern occurs soon after puberty. However, in Jersey cattle and White Leghorn chickens, puberty occurs near the center of early linear growth phase. Both of these animals have been subjected to intense genetic selection for productive traits. This intense selection appears to have resulted in sexual precocity in these animals. This is especially understandable for the Leghorn, which has been subjected to intensive selection for reproductive capacity. The animals which have been domesticated but not selected for food production purposes (rats and guinea pigs) show intermediate growth patterns.

Growth of the total animal is due to continuous changes occurring in different constituent tissues of the animal. These changes are reflected in the pattern of growth for a particular animal. In general, the order of tissue contribution to the growth of an animal from embryo to mature individual is: nervous system, bone and tendon, muscle, and fat, respectively (Pálsson & Verges, 1952a, 1952b; McMeekan, 1959). An excellent review of the changes occurring in tissue constituents of domestic animals during growth and development has been provided by Callow (1948).

The desire to predict biological phenomena and the ability to process data electronically have increased man's use of mathematical formulas for estimating biological functions. The need for formulas to estimate growth is no exception. It is especially important in animals that are used for food production. This chapter provides a limited discourse in regression analysis (or least squares), a method by which mathematical formulas are obtained and may be objectively evaluated. The use of regression analysis is suited for examination of the linear phase of growth, which is of economic importance in meat-producing animals.

The definition of growth indicates that it is a quantitative characteristic even though no units of measure are given. One explanation for the lack of units is that growth can be measured by both quantitative and qualitative characteristics. Weight, height, and the number of cells are quantitative characteristics. Prepuberty and puberty are qualitative characteristics. Thus, one should be aware that many measurable characteristics represent growth.

It is also important to know that the units of time in growth are not necessarily equally spaced at intervals or mathematically related such as the relationship of minutes to hours to days. Time may be characterized as egg, larva, nymph, or adult stages, which are qualitative measurements. Although characteristics other than quantitative units such as weight, area, volume, number of cells, weeks, months, and years have been suggested, only the quantitative units will be used for illustration.

The statistical method for determining the equation for predicting a quantitative variable such as weight or height as a function of time is regression analysis. Thus, weight is called the dependent variable and time is the independent variable. The purpose of the method is to predict the weight at a specified time and/or to predict the rate of change in growth for a unit of time. Two assumptions in regression analysis are that (1) the independent variable has no measurable error, and (2) the errors about the regression line (deviations of the observed from the predicted) are equal. There are other assumptions, but only these two will be discussed.

The assumption that the independent variable has no measurable error is interpreted to mean that the error in the independent variable is very small relative to the dependent variable. Error is measured by the variation between two (or among three or more) values which supposedly are evaluating the same condition. If there were no error in the system, the two measurements, being under the same condition, would have

25

the same value. Any difference between the two values is error. In fitting a linear function (straight line) to a set of values, the deviation of the observed points from the fitted line is measured as error. If the relationship of the dependent variable to the independent variable is exactly linear, the observed points will be on the line. Thus the deviations will be zero, the symbol which corresponds to no error.

The assumption regarding the distribution of errors is that the error is the same at any value of the independent variable. That is, the error in estimating weight, for example, is the same at any age during the prediction time. Another way of stating this fact is that the deviation from the estimated line for weight when the animal is young is approximately the same as the deviation when the animal is older.

If the time interval (or any other independent variable) is short enough, a linear relationship can be assumed. This type of

relation is called linear regression and will be discussed first. Weight is the dependent variable and time is the independent variable in the example to be discussed.

The researcher desires to know if there is a *meaningful* relationship between weight and time. If so, what are the predicted weight at a given time, the rate of increase, and the error in both the predicted weight and rate of increase.

An objective criterion for determining if there is a meaningful relationship is to find out if the regression equation is statistically significant at some level, for example 10, 5, or 1 percent. The level of significance is the probability of falsely saying that the relationship between the dependent variable and independent variable is not due to chance.

If the regression is statistically significant at the 1 percent level, then the probability that the relationship is due to chance alone is 1 percent or less while if the regression is statistically significant at the 10 percent level, then the probability that the relationship is due to chance is 10 percent or less. In general, a regression equation that is significant at the 1 percent level is considered more likely to be a meaningful relationship between the two variables than one that is only significant at the 10 percent level.

The formulas that will be used in this test for linear regression of a dependent variable on an independent variable are:

$$Y = a + bX \qquad (1)$$

$$b = \frac{\Sigma Y_i Y_i - \dfrac{\Sigma X_i \Sigma Y_i}{n}}{\Sigma X_i^2 - \dfrac{(\Sigma X_i)^2}{n}} = \frac{\Sigma xy}{\Sigma x^2} \qquad (2)$$

$$s_{y.x} = \sqrt{\frac{\Sigma Y_i^2 - \dfrac{(\Sigma Y_i)^2}{n} - \dfrac{\left(\Sigma X_i Y_i - \dfrac{\Sigma X_i Y_i}{n}\right)^2}{\Sigma X_i^2 - \dfrac{(\Sigma X_i)^2}{n}}}{n-2}} \qquad (3)$$

$$= \sqrt{\left(\Sigma y^2 - \dfrac{(\Sigma xy)^2}{\Sigma x^2}\right) \Bigg/ n-2}$$

$$a = \bar{y} - b\bar{x} \qquad (4)$$

$$r = \frac{\Sigma xy}{\sqrt{\Sigma x^2 \Sigma y^2}} \qquad (5)$$

and

$$s_b = \frac{s_{y.x}}{\sqrt{\Sigma x^2}} \qquad (6)$$

in which:

X_i is i^{th} observation for the independent variable,

Y_i is i^{th} observation for the dependent variable,

a is the estimate of the dependent variable when the independent variable is zero,

b is the estimated slope of the line, the rate of change of the dependent variable for a unit change in the independent variable, and is called the regression coefficient,

$Y = a + bX$ is the estimated regression equation,

$\sum X_i, \sum Y_i, \sum (X_i Y_i), \sum X_i^2, \sum Y_i^2$ are the sums, (uncorrected) sum of cross products and (uncorrected) sum of squares for $i = 1, 2, 3, \ldots, n$ for the independent and dependent variables, and $\sum x^2, \sum y^2$, and $\sum xy$ are the sum of squares and cross products of the deviations from the mean of the variables (corrected for the mean).

$s_{y.x}$ is the standard error of estimate (also called the standard deviation of Y for a fixed X or the standard deviation of Y holding X constant),

r is the correlation coefficient, and

s_b is the standard error of b.

The calculated values from data in Table 20–1 are:

$\sum X_i = 260.791$,
$\sum Y_i = 890.46$,
$\sum (X_i Y_i) = 13{,}787.43272$,
$\sum X_i^2 = 4{,}515.480681$,
$\sum Y_i^2 = 44{,}380.5780$.

Table 20–1. Age and Corresponding Weight of Man

(Data from *Handbook of Biological Data.* 1956. *William S. Spector*, Ed., W. B. Saunders Co., Philadelphia, p. 132.)

Age	Body Weight kg.
Birth	3.54
0–3 mo.	5.92
3–6 mo.	8.4
6–12 mo.	10.8
1.5 yr.	12.2
2.5 yr.	14.4
3.5 yr.	16.3
4.5 yr.	18.5
5.5 yr.	20.6
6.5 yr.	22.7
7.5 yr.	25.1
8.5 yr.	28
9.5 yr.	31
10.5 yr.	33
11.5 yr.	37
12.5 yr.	41
13.5 yr.	46
14.5 yr.	52
15.5 yr.	58
16.5 yr.	62
17.5 yr.	65
18.5 yr.	67
19.5 yr.	69
25 yr.	71
35 yr.	72

A method to test whether the relationship is statistically significant is to compute the t test statistic in which

$$t = \frac{b}{s_b} \qquad (7)$$

$$b = \frac{13{,}787.43272 - \dfrac{(260.791)(890.46)}{25}}{4{,}515.480{,}681 - \dfrac{(260.791)^2}{25}} = \frac{4498.47457}{1795.002854} = 2.506 \text{ kg/year.}$$

$$s_{y.x} = \sqrt{\frac{44{,}380.5780 - \dfrac{(890.46)^2}{25} - \dfrac{\left[3{,}787.43272 - \dfrac{(260.791)(890.46)}{25}\right]^2}{4515.480681 - \dfrac{(260.791)^2}{25}}}{25-2}}$$

$$s_{y.x} = 7.77 \text{ kg.}$$

$$s_b = \frac{7.77}{\sqrt{4515.480681 - \dfrac{(260.791)^2}{25}}} = \frac{7.7743}{42.3674} = 0.1835 \text{ kg/year.}$$

The computed t value is compared with the tabled t value, with n-2 degrees of freedom at the level of significance desired. (The distribution of the t-test statistics and assumptions regarding the distribution necessary to use a t test are in many statistical texts.) The tabled t (two-sided) at 0.1 percent level of significance is 3.767. Thus if the calculated t is less than or equal to −3.767 or greater than or equal to +3.767, the regression coefficient is said to be statistically significant at the 0.1 percent level.

kg/year. The standard error of estimate is $s_{y.x} = 7.77$ kg and the standard error of the regression coefficient is $s_b = 0.1835$ kg/year. The smaller the standard errors, the better the fit. That is, as the distance parallel to the Y axis between the observed data and the estimated line decreases, the smaller the standard errors will be.

Another method used to determine whether the relationship is a meaningful one is that of the correlation coefficient, r. For this example,

$$r = \frac{12{,}787.4372 - \dfrac{(260.791)\,(890.46)}{25}}{\sqrt{\left(4515.480681 - \dfrac{(260.791)^2}{25}\right)\left(44{,}380.5780 - \dfrac{(890.46)^2}{25}\right)}} = 0.94$$

The calculated value for Equation 7 is $t = \dfrac{2.506}{0.1835} = 13.66$ with 23 degrees of freedom. The magnitude of 13.66 is greater than 3.767. Thus we conclude that the relationship between weight and age is statistically significant at the 0.1 percent level. If statistical significance at this level is the criterion to be used to determine the relationship between weight and age, it is concluded that the equation $Y = 9.48 + 2.506X$ is appropriate for estimating the weight, Y, in kg for the age, X, in years. The numerical value of 9.48 is obtained from Equation 4 as follows:

$$a = \frac{890.46}{25} - (2.506)\frac{(260.791)}{25} = 9.48 \text{ kg}$$

The numerical value of the Y intercept, a in this example, could be interpreted as an estimate of the birth weight of a male baby. However, the reader is cautioned that there are better ways to obtain estimates of birth weight than by this particular technique. The regression coefficient, b = 2.506 kg/year, is the slope of the line and the rate of growth per year as can be seen from the units

This calculated value may be compared with tabled values of r to determine the level of significance.

The absolute numerical value of the correlation coefficient measures how the data fit a linear relationship. The closer the numerical value is to either plus or minus 1, the better is the fit of the data to a linear relationship. The square of the correlation coefficient is a measure of the variation in the dependent variable (weight in this example) that is due to variation in the independent variable; r^2 equals 0.89, so that 89 percent of the variation in weight is explained by the variation in age.

The standard error of estimate for a predicted age is determined by

$$s_{y.x}\sqrt{1 + \frac{1}{n} + \frac{(X_0 - \bar{x})^2}{\Sigma x^2}}$$

for an individual estimate, and by

$$s_{y.x}\sqrt{\frac{1}{n} + \frac{(X_0 - \bar{x})}{\Sigma x^2}}$$

for the mean weight for a group. X_0 is the particular predicted age. The standard error of estimate for an individual at age 21 is

$$s_{y.x}\sqrt{1 + \frac{1}{n} + \frac{(X_0 - \bar{x})^2}{\Sigma x^2}} =$$

$$7.77\sqrt{1 + \frac{1}{25} + \frac{\left(21 - \frac{260.791}{25}\right)^2}{1795.002854}} = 8.15 \text{ kg}$$

and the standard error of estimate for the mean at age 21 is

$$7.77\sqrt{1 + \frac{\left(21 - \frac{260.791}{25}\right)^2}{1795.002854}} = 2.48 \text{ kg}$$

The weight at age 21 is estimated by

$$Y = a + bX_0 = 9.47 + 2.506(21) = 62.10 \text{ kg}$$

Although the fit is statistically significant, the standard error for an individual is 13 percent (8.15 kg/62.10 kg) of the predicted weight at age 21 years. This error may be larger than desired for practical use. The point is that a relationship may be statistically significant but the error may be such that little practical use can be made of the estimating equation.

The researcher should be aware of several principles. The type of fit (regression) proposed should be based upon the biological or physical model proposed, not on what is the best fit *per se*. The errors given are for the dependent variable, not for the independent variable. That is, the equation derived is not the same equation if age had been the dependent variable and weight the independent variable.

If the researcher desired to fit an exponential equation of

$$Y' = a'X'^b$$

it must first be expressed as a linear equation. A logarithmic transformation will accomplish this:

$$\log Y' = \log a' + b \log X' \quad \text{(Base 10)}$$
$$\text{or } \operatorname{Ln} Y' = \operatorname{Ln} a' + b \operatorname{Ln} X' \quad \text{(Base e)}$$

which is of the same form as

$$Y = a + bX$$
$$\text{in which } Y = \log Y' \text{ (or } \operatorname{Ln} Y')$$
$$a = \log a' \text{ (or } \operatorname{Ln} a')$$
$$\text{and} \quad X = \log X' \text{ (or } \operatorname{Ln} X').$$

The data in Table 20–1 could be considered to be of the form $\log Y' = a + bX$

$$\operatorname{Ln} Y' = a + bX$$

in which Y' is the weight and X is the age. The results (natural logarithm of weight) on age,

$$Y = 2.417 + 0.0846 \, X,$$
$$s_{y.x} = 0.4175 \operatorname{Ln} \text{kg},$$
$$s_b = 0.00985 \operatorname{Ln} \text{kg/year},$$
$$t = \frac{0.0946}{0.00985} = 8.59,$$
$$\text{and} \quad r = 0.87.$$

At age 21

$$\ln Y = 2.417 + (0.0846)(21) = 4.194 \operatorname{Ln} \text{kg}$$
$$Y = 66.3 \text{ kg}$$

The standard error of an individual 21 years of age is

$$0.4175\sqrt{1 + \frac{1}{25} + \frac{\left(21 - \frac{260.791}{25}\right)^2}{1795.0028}} = .4383 \text{ kg}$$

Table 20–2 summarizes the statistics that are often used in regression analysis for the two types of regression equations in this chapter. The anti-ln of 0.4175 equals 1.52 kg. If the standard errors of estimate of the two

Table 20–2. Calculated Statistics of Data in Table 20–1 for Linear and Logarithmic Regression of Weight of Man on Time

(kg) = 9.48 + 2.506 × (age)	Ln kg = 2.417 + 0.0846 × (age)
r = 0.94	r = 0.87
$s_{y.x}$ = 7.77 kg	$s_{y.x}$ = 0.4175 Ln kg
t = 13.66	t = 8.59

equations are compared, $s_{y.x}$ for the logarithmic transformed data is smaller and thus one might conclude that the equation, in weight = a + b (age), is the better fit. If the absolute values of the correlation coefficients or the t statistic are compared, the equation, weight = a + b (age), is the better fit. It could now be asked: Which of the criteria should be used to determine which line is the more appropriate for the data? An answer to this question is that statistics measures probabilities, not appropriateness. This answer, although correct, does not aid the researcher; in general, however, the comparison of the standard error of estimates is a more important criterion than the correlation coefficient. The comparison, *per se*, of numerical values of the standard error of estimates in the two types of regression here does not exactly present the whole truth even when the same units are used (that is, 1.52 kg for the logarithmic equation and 7.77 kg for nontransformation equation). Although the logarithmic equation has the smaller value, the difference in the predicted

value and the observed value is larger for the logarithmic equation than for the non-transformed curve. The point is that a single criterion, except in games, is at best a poor statistic.

The graph of the data and the respective equations are presented in Figure 20–2. Observe that between the ages 12.5 and 19.5 years for the nontransformed data and between 3.5 and 19.5 years for the transformed data, the regression line is below the data points; for the other years, the regression line is above the points. It is the author's opinion that neither of the equations is a good fit, and that the proposed mathematical model does not portray the biological phenomena even though the statistical tests are both significant.

More complex equations can be fitted to data by an extension of the principles set forth for linear regression of one dependent variable on one independent variable. For example,

$$\text{weight} = b_0 + b_1 \text{ (age)} + b_2 \text{ (age)}^2$$

In particular,

$$Y = 5.219 + 5.955X - 1.08489X^2 \\ + 0.11575X^3 - 0.00459X^4 + 0.00006X^5$$

is the fifth degree polynomial equation for the data in Table 20–1, with Y the weight in kg and X the age in years. The standard error of estimate for this equation, $s_{y.x}$, is 1.12 kg. There are statistical rules for determining what degree of the polynomial is significant; however, statistical significance should not be construed to imply the relationship is right, appropriate, or best, and nonsignificance to imply that it is incorrect, poor, or not the best. The use of complex equations becomes important when an estimate of the entire life span is needed. However, for analysis of growth over a short period of time for food-producing animals, a regression equation using a single independent variable is usually adequate.

Formulas 1 through 6 summarize the necessary statistics for estimating a linear rela-

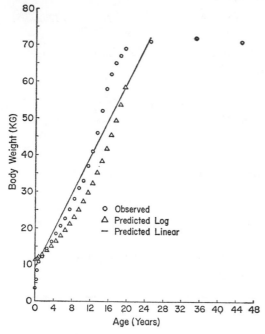

FIG. 20.2   Observed and predicted growth of man. (*Data from Spector, 1956.   Handbook of Biological Data. Saunders, Philadelphia.*)

tionship. The numerical data considered are those of man. The reader should understand that *if* a linear relationship exists between two variables, such as between age and weight of cattle, of pigs, of sheep, or some particular organ of an animal, the procedures should be the same. The procedures are as given when it is necessary to transform a curvilinear relationship to a linear relationship. Criteria for statistically determining the appropriateness of the models are the standard error of estimates and the correlation coefficient. A subjective criterion for judging the appropriateness of a model is a graph of the data.

REFERENCES

Brody, S. (1945). *Bioenergetics and Growth.* New York, Reinold Publishing Corp.

Callow, E. H. (1948). Comparative studies of meat. II. The changes in the carcass during growth and fattening, and their relation to the chemical composition of the fatty and muscular tissue. J. Agri. Sci. *38*:174–199.

Fabens, A. J. (1965). Properties and fitting of the Von Bertalanffy growth curve. Growth *29*:265–289.

Guttman, R., & Guttman, L. (1965). A new approach to the analysis of growth patterns: the simplex structure of intercorrelations of measurements. Growth *29*:219–232.

Laird, A. K. (1966). Postnatal growth of birds and mammals. Growth *30*:349–363.

Laird, A. K., Sylvanus, A. T., & Barton, A. D. (1965). Dynamics of normal growth. Growth *29*:233–248.

McMeekan, C. P. (1959). *Principles of Animal Production.* Whitcombe & Tombe, Ltd., London, Melbourne, Sydney, Perth.

Mitchell, H. H., Card, L. E., & Hamilton, T. S. (1931). A technical study of the growth of White Leghorn chickens. Ill. Agri. Exp. Sta. Bull. No. 367.

Pálsson, H., & Verges, J. B. (1952a). Effect of plane of nutrition on growth and development of carcass quality in lambs. Part I. The effect of high and low planes of nutrition at different ages. J. Agri. Sci. *42*:1–92.

Pálsson, H., & Verges, J. B. (1952b). Effect of plane of nutrition on growth and the development of carcass quality in lambs. Part II. Effects on lambs of 30 lb. carcass weight. J. Agri. Sci. *42*:93–149.

Appendix

I. Gross Energy of Nutrients and Feeds

(Data from Brody, 1945; Nehring, 1965; Hartman, 1968.)

	kcal/gm		kcal/gm
Carbohydrates		*Proteins*	
Glucose ($C_6H_{12}O_6$)	3.75	Gluten	6.0
Sucrose ($C_{12}H_{22}O_{11}$)	3.94	Casein	5.7
Starch ($C_6H_{10}O_5$)x	4.18	Egg albumin	5.7
Cellulose ($C_6H_{10}O_5$)x	4.18		
		Amino Acids and Nitrogenous Compounds	
Fats		Alanine ($C_3H_7NO_2$)	4.35
Coconut oil	8.9	Tyrosine ($C_9H_{11}O_3N$)	5.92
Butter fat	9.1	Creatinine ($C_4H_2N_3O$)	4.60
Safflower oil	9.4	Urea (CH_4ON_2)	2.53
Corn oil	9.4		
Olive oil	9.4	*Feeds*	
Peanut oil	9.5	Corn meal	4.4
		Soybean	5.5
Fatty Acids		Wheat bran	4.5
Acetic acid (CH_3COOH)	3.49	Linseed oil meal	5.1
Propionic acid (C_2H_5COOH)	4.96	Straw	4.4
Butyric acid (C_3H_7COOH)	5.95	Hay	4.4
Palmitic acid ($C_{16}H_{32}COOH$)	9.35		
Stearic acid ($C_{17}H_{35}COOH$)	9.53		
Oleic acid ($C_{18}H_{34}COOH$)	9.50		

II. The Amino Acid Content of Plants, Plant Proteins, or Animal Tissues

Protein (%)	Alfalfa 17	Corn Germ Meal 20	Zein 100	Potato Dry Matter 8	Rice— Polished 7.9	Soybean Meal 50	Wheat Flour 10.7
			Amino Acids as Percent Protein				
Arginine	4.8	7.5	1.7	5.0	8.2	7.3	3.4
Histidine	1.8	2.9	1.3	2.2	2.4	2.9	2.0
Isoleucine	4.8	3.8	7.3	3.7	4.4	6.0	3.9
Leucine		14.0	23.7	9.6	8.0	8.0	7.3
Lysine	5.0	6.0	0	8.3	3.3	6.8	2.1
Methionine	1.9	1.5	2.4	2.0	2.5	1.7	1.0
Cystine	1.8	1.6	0.8	1.3	1.4	1.9	2.0
Methionine + cystine	3.7	3.1	3.2	3.3	3.9	3.6	3.0
Phenylalanine	4.4	5.5	6.2	5.4	4.6	5.3	5.3
Tyrosine	4.6	6.0	5.3	2.5	5.6	4.0	3.3
Phenylalanine + tyrosine	9.0	11.5	11.5	7.9	10.2	9.3	8.6
Threonine	3.4	4.6	3.5	6.9	3.4	3.9	2.5
Tryptophan	1.5	1.3	0.1	2.1	1.6	1.4	0.8
Valine	4.8	6.0	3.5	5.3	6.3	5.3	3.9
Aspartic acid			6.6	11.5		3.7	3.8
Glutamic acid			26.9	7.4		18.4	33.0
Glycine			0.4	1.9	10.0	4.0	7.2
Proline			10.5	3.0		5.0	14.0
Serine			8.3	2.6		4.2	4.0

Protein (%)	Egg 13	Milk 3.5	Wool 100	Beef Muscle 13.8	Liver	Rumen Bacteria
			Amino Acids as Percent Protein			
Arginine	6.7	3.6	10.7	6.8	6.6	4.0
Histidine	2.5	2.3	1.1	3.4	2.5	1.5
Isoleucine	6.9	6.2	2.4	5.4	4.8	5.0
Leucine	9.0	10.0	4.9	8.1	8.4	6.0
Lysine	0.8	7.5	3.2	10.1	7.0	7.0
Methionine	3.4	2.5	0.6	2.6	3.6	1.9
Cystine	2.2	0.8	7.5	1.3	1.5	1.5
Methionine + cystine	5.6	4.3	8.1	3.9	5.1	3.4
Phenylalanine	5.8	5.0	4.0	4.4	6.0	4.0
Tyrosine	4.4	5.5	5.2	2.8	3.9	4.0
Phenylalanine + tyrosine	10.2	10.5	4.7	7.2	9.9	8.0
Threonine	5.3	4.4	5.1	5.2	4.9	5.5
Tryptophan	1.5	1.4	1.0	1.1	1.5	0.7
Valine	7.5	6.6	3.9	7.0	6.0	4.8
Aspartic acid	8.0	4.8	4.4	7.9	7.0	11.0
Glutamic acid	12.4	19.8	8.5	15.0	11.0	14.3
Glycine	3.6	0.3	5.5	4.8	7.9	4.7
Proline	4.4	7.3	5.5	6.1		9.4
Serine	7.7	4.6	8.6	6.0	7.3	4.6

III. Computed Weights, Energy Contents, and Composition of the Total Products of Conception for a Litter of Eight

(Mitchell et al., 1931. Ill. Agri. Exp. Sta. Bull. No. 375.)

Week of Gestation	Total Fresh Weight gm	Crude Protein gm	Gross Energy cal	Ash gm	Calcium gm	Phosphorus gm	Iron mgm
2	366	8.5	29	.6	.005	.028	4.2
4	1,354	47	209	5	.14	.36	22
6	2,909	130	662	18	.96	1.61	57
8	5,005	265	1,499	45	3.8	4.7	113
10	7,625	462	2,825	92	10.9	10.6	191
12	10,755	726	4,742	165	26	21	294
14	14,385	1,065	7,347	269	54	37	423
16	18,505	1,483	10,736	411	101	60	581

IV. A Comparison of the Relative Concentrations[1] of Essential Amino Acids in Swine to Their Estimated Requirement

	Requirement[2]	Muscle	Visceral	Liver	Kidney
Arginine	27	88	105	85	94
Histidine	27	36	47	39	38
Isoleucine . . .	80	73	87	75	70
Leucine	93	99	127	136	125
Lysine	100	100	100	100	100
Methionine + cystine . . .	75	55	74	31	31
Phenylalanine + tyrosine . . .	67	114	116	77	68
Threonine . . .	60	57	82	63	64
Tryptophan . .	21	16	26	20	21
Valine	60	74	87	94	87

[1] Lysine taken as 100 for each comparison.
[2] University of Missouri, unpublished.

V. Macroelement Content of Certain Feeds and Mineral Supplements (Percent)

(Adapted from N.R.C. Publication No. 449, 1956; No. 585, 1958.)

Feeds and Supplements	Calcium	Phosphorus	Magnesium	Sulfur	Potassium	Sodium	Chlorine
Feeds							
Alfalfa hay	1.64	0.26	0.32	0.36	1.77	0.16	0.28
Barley	0.09	0.47	0.14	0.19	0.63	0.02	0.13
Corn	0.03	0.32	0.17	0.12	0.35	0.01	0.05
Meat and bone meal. .	10.57	5.07	1.13	—	1.46	0.73	0.75
Molasses, beet . . .	0.16	0.03	0.23	0.48	4.77	1.17	1.27
cane . . .	0.89	0.08	0.35	0.34	2.38	0.17	2.75
wood . . .	1.41	0.05	0.07	0.03	0.04	0.03	0.13
Sorghum grain . . .	0.05	0.35	0.19	0.18	0.38	0.05	0.10
Soybean meal, solvent .	0.32	0.67	0.27	0.43	1.97	0.34	—
Wheat	0.06	0.41	0.18	0.19	0.58	0.10	0.08
Supplements							
Bone meal, steamed .	28.98	13.59	0.61	0.22	0.18	0.46	—
Dicalcium phosphate .	27.00	19.07	—	—	—	—	—
Rock phosphate . . .	35.94	14.87	0.80	0.28	—	0.03	0.03
Limestone.	33.84	0.02	2.97	0.04	0.12	0.06	0.02
Dolomitic limestone .	22.26	0.04	9.97	—	0.36	—	0.12
Oyster shell	38.50	0.07	0.30	—	0.10	0.21	0.01
Monosodium phosphate.	—	22.45	—	—	—	16.66	—
Sodium tripolyphosphate	—	25.28	—	—	—	31.25	—

VI. Microelement Content of Certain Feeds and Supplements

(Adapted from N.R.C. Publication No. 449, 1956; No. 585, 1958.)

Feeds and Supplements	Cobalt ppm	Copper ppm	Iodine ppm	Iron ppm	Manganese ppm	Selenium ppm	Zinc ppm
Feeds							
Alfalfa hay	0.13	13.64	0.12	240	51.70	—	16.94
Barley	0.12	8.60	—	60	18.25	—	17.16
Corn	0.02	2.86	0.33	30	5.94	—	19.58
Distillers, dried grain, corn	0.09	44.66	—	195	18.92	—	—
Meat and bone meal. .	0.18	1.54	—	407	12.92	—	—
Molasses, beet . . .	0.38	17.6	—	47	4.62	—	—
cane . .	0.90	59.62	—	187	42.24	—	—
wood . . .	—	—	—	—	13.42	—	—
Sorghum grain . . .	0.30	10.78	—	50	16.28	—	15.40
Soybean oil meal, solvent	0.09	14.30	—	131	27.50	—	—
Wheat	0.08	8.41	—	60	54.78	—	15.40
Bone meal, steamed . .	0.06	16.28	—	840	30.36	—	424.60
Limestone.	—	—	—	3297	279.62	—	—
Limestone, dolomitic .	—	—	—	767	—	—	—
Oyster shell	—	—	—	2862	133.32	—	—
Rock phosphate . . .	—	—	—	—	1247.40	—	—
				Percent			
Supplements							
Cobalt chloride . . .	24.08	—	—	—	—	—	—
Copper sulfate . . .	—	25.20	—	—	—	—	—
Calcium iodate . . .	—	—	32.56	—	—	—	—
Ferrous sulfate . . .	—	—	—	20.10	—	—	—
Manganese sulfate . .	—	—	—	—	25.50	—	—
Sodium selenite . . .	—	—	—	—	—	46.00	—
Zinc carbonate . . .	—	—	—	—	—	—	52.10

VII. Mineral Requirements for Growth (Including Maintenance)[1,2]

	Cattle						
Element	Beef	Dairy	Horse	Sheep	Swine	Rat	Dog
A. Macroelements (percent of air-dry ration)							
Calcium	0.25	0.21	0.30	0.20	0.50	0.06	1.00
Chloride	0.15	0.15*	0.09	0.15*	0.30	0.05	0.88
Magnesium	0.10*	0.15*	—	—	0.04	0.04	0.04
Phosphorus	0.21	0.19	0.30	0.18	0.40	0.50	0.80
Potassium	0.75	—	—	0.75	0.25	0.18	0.80
Sodium	0.10	0.10*	0.07	0.10*	0.20	0.05	0.52
Sulfur	0.15*	—	—	—	—	—	—
B. Microelements (ppm in air-dry feed)							
Cobalt	0.1	0.1	0.05	—	—	—	2.5
Copper	8	8	8	—	10	5	7
Fluorine	—	—	—	—	—	—	—
Iodine	0.3*	0.3	0.013*	—	0.20*	15	1.5
Iron	—	—	40	—	80	25	50
Manganese	25	20	—	—	40	50	5
Molybdenum	—	—	—	—	—	—	—
Selenium	0.1	0.1*	—	0.1	0.10	0.104	—
Zinc	—	—	—	—	50	12	5

1. Growth at 50 percent mature weight in horses (light breeds), dairy cattle, dogs, and rats; 50 percent market weight in swine, sheep, and beef cattle.
2. Values represent requirements, not recommended allowances.
* Estimated.
— Requirements unknown.

Index

Epinephrine, 141, 144
Epiphysis, 149
   cartilage of, 150, 155
   dysplasia of, 299
   plate of, 145, 148, 149
Epistasis, 70
Epithelial tissue, 4
Ergosterol, 334
Erythrocytes, 325
Erythromycin, growth promotion and, 153
Erythropoiesis, 7, 140
Erythropoietin, 140
Esophagus, 126, 134
   cellular growth of, 139
Estrogen, 144, 150, 151, 163, 185
Evaporation, 85, 93, 94
Evolution, 101, 102
Exercise, 173, 174
   fattening, 249
Exophthalmos, 161
Exostosis, 229
Extracellular fluid, 354
Eyelids, cellular growth of, 138

FALSE hellebore, 160
Fat, 78, 188, 209, 210, (*see* lipids and adipose tissue)
   absorption of, 241
     bile salts, 241
     lipases, 241
   gross energy of, 382
   in swine, 79, 99
   subcutaneous, 101
   synthesis of, in liver, 242
   transport of, to fetus, 46
Fattening (*see* growth and fattening)
   effects of age, 248
   effects of caloric intake, 249
   effects of exercise, 249
   effects of insulin, 249
   precursors for synthesis, 250
   synthesis, 245, 247, 250
   in types of lipids, 240
Fatty acid, 100, 122, 188
   in adipose tissue, 239, 240
   in different species, 239, 240
   essential, 334, 335
   free, 128
   gross energy of nutrients, 382
   species differences in synthesis, 249
   volatile (VFA), 128
Feces, 122
Feed (*see* food)
   additive, 151–154
   efficiency of, 111
   intake of, 126, 127
     glucagon, 127
     as a homeostatic mechanism, 121–130
     hypothalamus, 368
     in lactation, 129, 130
   preference, 130–135
   processing, 136, 137
Feedback system, 122, 128
Feeding center, 124
Fermentation, 137
Ferroprotoporphyrin, 325

Fetal growth, 21–39
   cleavage and differentiation, 21–24
   differential growth, 26–32
   factors affecting, 34–38
   fetal hormones and, 32–34
   measurement of, 25–26
   organogenesis, 24–25
Fetal nutrition, 40–59
   anabolism of pregnancy, 40–41
   dietary manipulations, 57
   endocrine adjustments to pregnancy, 41–42
   enzyme activity during gestation, 42–47
   factors affecting placental nutrient transfer, 47–57
   prenatal influence upon postnatal performance, 57, 58
Fetus, 144
   anencephaly in, 32
   growth in, 51–59
   nutrient accretion in, 51–59
   prenatal development of, 51–59
Fever, 181, 188
FFA (*see* free fatty acid)
Fiber (*see also* hair and wool)
   connective tissues, 261
   environmental influence, 263
   genetic influence, 263
   hormonal influence, 263
   nutrition, 262
Fish, epidermal choline in, 141
Fleece (*see* wool)
Flour beetle, 80
Fluoride, in bone, 226, 234
Fluorine, 162
   absorption, 313, 315
   excretion, 324
   function, 317, 324
   interrelationships, 313
Folic acid, 335
Follicle, 258, *et seq.*
Food intake, 103
   appetite for specific substances and, 133–135
   central nervous system and, 123–124
   chemostatic regulators, 126–128
   effect of, on body temperature, 87–88
   feed processing and, 136, 137
   gastric distension and, 126
   glucostatic regulators, 126, 127
   as a homeostatic mechanism, 121–130
   lipostatic regulation, 127, 128
   management and feeding practice, 135–137
   palatability and feed preference, 130–135
   palatability versus caloric intake, 132–133
   taste and odor, 130–132
   thermostatic signals, 128, 129
   voluntary, 121–137
Fowl, mitosis in, 4
Fox, maximum life span, 165
Freeze-branding, 119
Fruit fly, 80

GAMMA irradiation, 162
Gas, transport of, to fetus, 45
Gastroenteritis, 185
Gastrointestinal tract, 4
   infections of, 177
Gastrula, 23

Maintenance, of water requirement, 355
Malaria, 182
Malformation, 58 (*see* deformity)
Malnutrition, 38, 58, 182, 187 (*see* specific nutrients)
  during gestation, 50
Mammal, 55, 91, 92
  moulting of, 260
  tropical, 92
Man, 48, 52, 55, 57
  placenta in, 48
  plasma lipids in, 243
Management and nutrition, influence of fat utilization on, 253
Manganese, 315, 326
  absorption of, 314, 326
  in bone composition, 326
  enzyme activation, 326
  excretion of, 326
  function of, 317
  interrelationships, 313
  in placental transfer, 326
  requirement for, 386
  source of, 385
Marbling, 78, 238
Marrow, of bone, 170, 224
Mass spectroscopy, 330
Maternal size, influence on size of foal, 34, 35
Mating systems, 71-74
Matrix, 55
Maturity, 80
Maximum life span, 165
Mechanism, for defense, 296
  neural, 370
    water balance and, 370
Melanocyte, 119
Melengestrol acetate, 152
Mesoderm, 23
Metabolic adaptation, 101
Metabolism, 16, 63-65, 85, 87, 97, 101, 102, 122, 127, 129, 138
  of adipose tissue, 241
  of amino acids, 299, 300
  energy pathways of, 279-282, 285-289
  infections, 177, 178
  fecal nitrogen of, 294
  of lipids, 241, 248, 250, 252, 254, 255
  of muscle, 210, 211
  summit, 87
  vitamins and, 336-352
  water, 353-372
Metalloproteins, 292
Methionine, 116, 293
Methylglyoxal, 142
Microbial fermentation, 296, 297
Microclimate, 84
Microelements, 46, 56, 322-328
Microorganisms, 133
  drug-resistant, 178, 179
Microphthalmia, 161
Milk, 122, 133, 135
  in dietary control of fat, 250
  in fat production, 250
  species differences in fat, 250
  synthesis of, 280, 281
  synthesis of fat in, 250
  as a tissue, 249

Milking, 68
Minerals, 53, 102, 122, 186-188, 232, 312-331
  absorption of, 314, 315
  deficiency, hair development of, 263
    wool development of, 263
  deficiency signs, 329
  excess of, 329
  function of, 317
  requirements for, 386
  research techniques, 330
Mitosis, 1, 4, 64, 143
Mitotic inhibitors, 141
Moiety, of carbohydrate, 140
Molybdenum, 326, 327
  absorption of, 315
  body composition of, 326
  in enzyme activation, 317
  excretion of, 327
  function of, 317
  interrelationships, 313
  source of, 385
Monkey, life span of, 165
Monoglycerides, 241
Morphogenesis (*see* bone morphogenesis)
Morphology, adipose tissue, 236
  characteristics of, in germ-free condition, 180
  glands of skin, 258
  hair and wool, 257
  skin, 256
Mothering, 74
Mouse, 4, 11, 55, 57, 81
Movement, 104
Muscle, 1-16, 29-31, 58, 100, 193-215
  blood supply in, 197
  changes in, 213
  color of, 195
  composition of, 193, 202-206
  connective tissue in, 194
  contractions of, 205
  dietary control of lipid utilization, 248
  gracilus, 13
  in hindquarters, 13
  innervation, 198
  lipids in, 209, 210
  longissimus dorsi, 14, 16
  macrostructure of, 194
  metabolism of, 210, 211
  microstructure of, 196
  nomenclature of, 194
  percentage of carcass, 194
  semitendinosus, 16
  ultrastructure of, 199-202
Muscular dystrophy, 168
Muscular work and energy requirement, 284
Musculoskeletal deformities, 159
Mutation, 171, 172
  somatic theory, 169
Myofibrillar protein, 203, 206
Myofibrils, 201
Myoglobins, 208
Myosin, 204
Myristoleic acid, 14

NECROBIOLOGY, enzyme change in, 212-214
  extensibility, loss of, 213
  heat effects on lipids, 253